BASIC SOLID STATE ELECTRONIC CIRCUIT ANALYSIS
THROUGH EXPERIMENTATION

Lorne MacDonald

College of San Mateo

2nd Edition

The Technical Education Press
Seal Beach, California

BASIC SOLID STATE ELECTRONIC CIRCUIT ANALYSIS
Through Experimentation

Copyright © 1981, 1984 by The Technical Education Press. All rights reserved. Printed in the United States of America. No part of this publication may be reproduced, stored in retrieval systems, or transmitted, in any form or by any means, electronic, mechanical, photocopying, recording, or otherwise, without the prior written permission of the publisher. The Technical Education Press, P. O. Box 342, Seal Beach, California 90740.

ISBN: 0-911908-12-9

MANUFACTURED IN THE UNITED STATES OF AMERICA

The Day

Once upon a time it happened to me
The very sweetest thing that could ever be.
It was a fantasy, a dream come true,
It was the day I came to meet you.

Preface

BASIC SOLID STATE ELECTRONIC CIRCUIT ANALYSIS THROUGH EXPERIMENTATION begins with a review of selected direct current and alternating current circuits and proceeds through the analysis of basic solid state circuits that use diodes, zener diodes, bipolar transistors, field effect transistors, and silicon controlled rectifiers. The book, intended for college level students who have had DC and AC circuit theory and algebra, uses approximation methods in most of the analysis. However, exact techniques and equivalent circuits are also used to provide comparison with the approximation method.

The book combines both theory and laboratory experiments. The theory is complete for the introductory electronic course and it is reinforced by the experiments. All of the circuits in the laboratory portion of the text have been thoroughly tested and measured values will be very close to calculated values. If they are not, the reason will be miscalculations, poor measuring techniques, or an incorrectly constructed circuit.

The number of laboratory hours required to do all of the experiments is between 45 and 60 hours. Setup and breakdown time can be minimized if the components in the circuit do not have to be soldered. Proto, EL, and spring boards can be used to accommodate solderless connections.

A debt of gratitude is owed to the hundreds of students who, through feedback and constructive criticism, have helped develop this book to its present form. I am especially indebted to my colleague Albert Camps, who was my "sounding board" throughout the writing of this book. His inputs were invaluable. I also appreciate the suggestions of my colleagues Dick Raines of Shasta College, Leon Lucchesi of Yuba College, Jean Louis Garand of Ecole Louis Real, Ted Kirsch of San Francisco State University, and Jim Petromilli of The College of San Mateo. Lastly, I would like to acknowledge John A. Scullion, Donald Snell, and LeRoy Olson, who collectively provided inspiration and encouragement.

Lorne MacDonald
1984

ELECTRONIC CIRCUIT ANALYSIS SERIES
from
The Technical Education Press

DIRECT CURRENT CIRCUIT ANALYSIS
Through Experimentation - 4th Edition
    Kenneth A. Fiske and James H. Harter

ALTERNATING CURRENT CIRCUIT ANALYSIS
Through Experimentation - 3rd Edition
    Kenneth A. Fiske and James H. Harter

BASIC SOLID STATE ELECTRONIC CIRCUIT ANALYSIS
Through Experimentation - 2nd Edition
    Lorne MacDonald

PRACTICAL ANALYSIS OF AMPLIFIER CIRCUITS
Through Experimentation - 3rd Edition
    Lorne MacDonald

PRACTICAL ANALYSIS OF ELECTRONIC CIRCUITS
Through Experimentation - 2nd Edition
    Lorne MacDonald

DIGITAL CIRCUIT LOGIC AND DESIGN
Through Experimentation
    Darrell D. Rose

Table of Contents

CHAPTER 1. DIRECT AND ALTERNATING CURRENT CIRCUIT ANALYSIS:
A Review of Selected Topics   7

CHAPTER 2. SOLID STATE SEMICONDUCTOR DIODES   24

CHAPTER 3. ZENER DIODES   54

CHAPTER 4. BIPOLAR TRANSISTORS — AN INTRODUCTION   71

CHAPTER 5. TRANSISTOR MEASUREMENT AND TESTING TECHNIQUES   110

CHAPTER 6. SINGLE STAGE TRANSISTOR BIASING   127

    SECTION I: Base-Biased Transistor Circuits   128
    SECTION II: Two-Supply, Emitter-Current-Controlled Circuits   138
    SECTION III: Universal Circuit Biasing   147
    SECTION IV: The PNP Universal Biased Circuit   154
    SECTION V: The Collector Feedback Biased Circuit   161
    SECTION VI: Analyzing Circuits with Various Configurations   162
    SECTION VII: Exact Techniques Reviewed   164
    SECTION VIII: Stability Considerations and Circuit Choice   166
    SECTION IX: Reflected Resistance Circuit Analysis   170

CHAPTER 7. INTRODUCTION TO ALTERNATING CURRENT AMPLIFIER CIRCUIT ANALYSIS — USING BASE-BIASED AMPLIFIER CIRCUITS   173

    SECTION I: The Common-Emitter, NPN, Base-Biased Circuit   174
    SECTION II: The PNP, Common Emitter, Base-Biased Circuit   174
    SECTION III: The Common-Collector, Base-Biased Circuit   214
    SECTION IV: The Common-Base, Base-Biased Circuit   222
    SECTION V: Equivalent Circuits   231

CHAPTER 8. THE UNIVERSAL-BIASED AMPLIFIER CIRCUITS   240
    SECTION I: The Common Emitter Circuit   240
    SECTION II: The Common-Collector, Universal Circuit   264
    SECTION III: The Common-Base, Universal Circuit   275

CHAPTER 9. THE TWO POWER SUPPLY CIRCUITS   282
    SECTION I: The Two-Power-Supply, Common-Emitter Circuit   282
    SECTION II: The Common-Collector, Two-Power-Supply Circuit   292
    SECTION III: The Common-Base, Two-Power-Supply Circuit   297

CHAPTER 10.	**INTRODUCTION TO JUNCTION FIELD EFFECT TRANSISTORS**	**302**
CHAPTER 11.	**SINGLE STAGE JFET BIASING**	**323**
	SECTION I: Self-Biased Circuits	323
	SECTION II: Two-Power-Supply Circuits	324
	SECTION III: Universal-Biased Circuits	326
CHAPTER 12.	**THE COMMON SOURCE JFET CIRCUIT**	**335**
	SECTION I: Introduction to Common-Source Circuits	335
	SECTION II: The Self-Biased, JFET, Amplifier Circuit	344
	SECTION III: The Universal JFET Amplifier Circuit	350
	SECTION IV: The Common-Source, Amplifier Circuit	350
	SECTION V: The Two-Power-Supply, JFET Amplifier	351
CHAPTER 13.	**THE COMMON DRAIN JFET CIRCUITS**	**359**
	SECTION I: The Self-Biased, Common-Drain Amplifier	359
	SECTION II: The Universal Circuit	361
	SECTION III: The Two-Power-Supply Circuit	362
CHAPTER 14.	**THE COMMON GATE JFET CIRCUITS**	**369**
	SECTION I: The Self-Biased, Common Gate Amplifier	369
	SECTION II: The Universal Circuit	371
	SECTION III: The Two Power Supply Circuit	372
CHAPTER 15.	**DIRECT CURRENT POWER SUPPLIES — UNREGULATED AC TO DC CONVERTERS**	
	SECTION I: Halfwave Rectifier Circuits	378
	SECTION II: Full-Wave Rectifier Circuits	390
CHAPTER 16.	**INTRODUCTION TO REGULATED POWER SUPPLIES**	**405**
CHAPTER 17.	**SEMICONDUCTOR SWITCHES AND LATCHES**	**419**
CHAPTER 18.	**INTRODUCTION TO MULTISTAGE CIRCUITS**	**445**
APPENDIX A.	Components Required to Work all the Experiments	465
APPENDIX B.	Power Gain in dB Form	466
APPENDIX C.	Device Data Sheets	467
APPENDIX D.	Waveform Rise Time	476
APPENDIX E.	Power Dissipation Calculations	476
APPENDIX F.	Problem Answers [even]	477
Index		**479**

1 | DIRECT CURRENT AND ALTERNATING CURRENT CIRCUIT ANALYSIS: A Review

GENERAL DISCUSSION

Some of the analytical techniques learned during the study of direct current and alternating current circuits are very important in studying active devices in working circuits, and are used again and again. This chapter reviews these techniques in certain selected areas.

SECTION I: DIRECT CURRENT CIRCUIT ANALYSIS

Ohm's law and common sense are the basic tools used in analyzing the simplest to the most complex of electric circuits. Additionally, Kirchhoff's loop equations, the voltage divider equation, and Thevenin's theorem can provide mathematical short cuts, or exactness, when required.

SERIES RESISTORS AND VOLTAGE DIVIDER EQUATION CIRCUIT ANALYSIS

The standard approach to solving voltage drops across two resistors in series is to add the total resistance of the two resistors and, then, divide the applied voltage by the total resistance to solve the circuit current. The individual voltage drop across each series resistor is solved, at this point, by multiplying the circuit current by the individual resistor values, as shown in the following example.

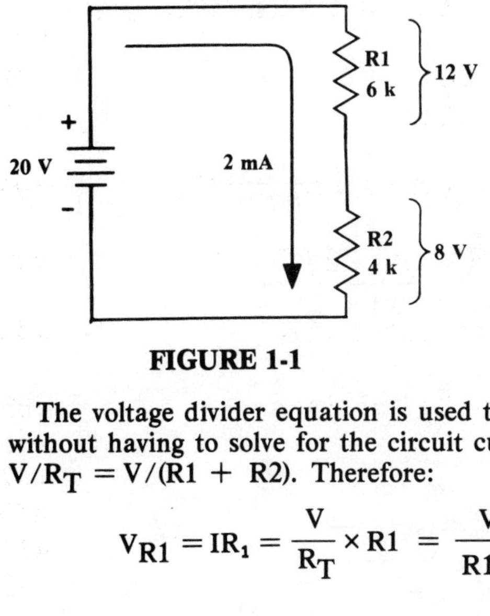

$$I = \frac{V}{R_T} = \frac{V}{R1 + R2} = \frac{20 \text{ V}}{6 \text{ k}\Omega + 4 \text{ k}\Omega} = 2 \text{ mA}$$

$$V_{R1} = IR_1 = 2 \text{ mA} \times 6 \text{ k}\Omega = 12 \text{ V}$$

$$V_{R2} = IR_2 = 2 \text{ mA} \times 4 \text{ k}\Omega = 8 \text{ V}$$

$$V = V_{R1} + V_{R2} = 12 \text{ V} + 8 \text{ V} = 20 \text{ V}$$

FIGURE 1-1

The voltage divider equation is used to solve for the voltage drops across the series resistors directly, without having to solve for the circuit current, because the current I is substituted for by V/R_T, where $V/R_T = V/(R1 + R2)$. Therefore:

$$V_{R1} = IR_1 = \frac{V}{R_T} \times R1 = \frac{V \times R1}{R1 + R2} = \frac{20 \text{ V} \times 6 \text{ k}\Omega}{6 \text{ k}\Omega + 4 \text{ k}\Omega} = \frac{20 \text{ V} \times 6 \text{ k}\Omega}{10 \text{ k}\Omega} = 12 \text{ V}$$

$$V_{R2} = \frac{V \times R2}{R1 + R2} = \frac{20 \text{ V} \times 4 \text{ k}\Omega}{6 \text{ k}\Omega + 4 \text{ k}\Omega} = \frac{20 \text{ V} \times 4 \text{ k}\Omega}{10 \text{ k}\Omega} = 8 \text{ V}$$

PARALLEL RESISTORS CIRCUIT ANALYSIS

The total resistance of a parallel combination of two resistors can be solved from:

$$R_T = R1 \parallel R2 = \frac{R1 \times R2}{R1 + R2} \quad \text{or from} \quad R_T = \frac{1}{1/R1 + 1/R2}$$

However, if more than two resistors are connected in parallel, the latter equation requires the fewest steps. Too, if the applied voltage is known, further simplification can be made by totaling the current flow in each resistive branch and dividing by the applied voltage.

1. The total resistance of two resistors in parallel can be solved from:

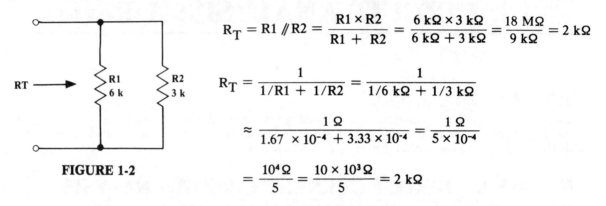

FIGURE 1-2

$$R_T = R1 \,/\!/\, R2 = \frac{R1 \times R2}{R1 + R2} = \frac{6\,k\Omega \times 3\,k\Omega}{6\,k\Omega + 3\,k\Omega} = \frac{18\,M\Omega}{9\,k\Omega} = 2\,k\Omega$$

$$R_T = \frac{1}{1/R1 + 1/R2} = \frac{1}{1/6\,k\Omega + 1/3\,k\Omega}$$

$$\approx \frac{1\,\Omega}{1.67 \times 10^{-4} + 3.33 \times 10^{-4}} = \frac{1\,\Omega}{5 \times 10^{-4}}$$

$$= \frac{10^4\,\Omega}{5} = \frac{10 \times 10^3\,\Omega}{5} = 2\,k\Omega$$

2. Total resistance of two resistors in parallel — with applied voltage.

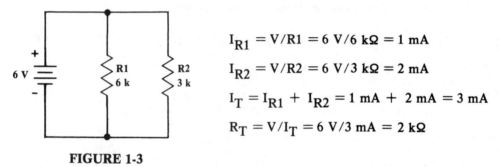

FIGURE 1-3

$$I_{R1} = V/R1 = 6\,V/6\,k\Omega = 1\,mA$$

$$I_{R2} = V/R2 = 6\,V/3\,k\Omega = 2\,mA$$

$$I_T = I_{R1} + I_{R2} = 1\,mA + 2\,mA = 3\,mA$$

$$R_T = V/I_T = 6\,V/3\,mA = 2\,k\Omega$$

3. Total resistance of three resistors in parallel.

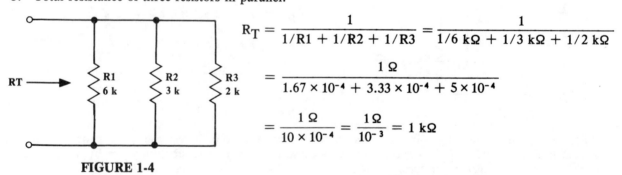

FIGURE 1-4

$$R_T = \frac{1}{1/R1 + 1/R2 + 1/R3} = \frac{1}{1/6\,k\Omega + 1/3\,k\Omega + 1/2\,k\Omega}$$

$$= \frac{1\,\Omega}{1.67 \times 10^{-4} + 3.33 \times 10^{-4} + 5 \times 10^{-4}}$$

$$= \frac{1\,\Omega}{10 \times 10^{-4}} = \frac{1\,\Omega}{10^{-3}} = 1\,k\Omega$$

4. Total resistance of three resistors in parallel — with applied voltage.

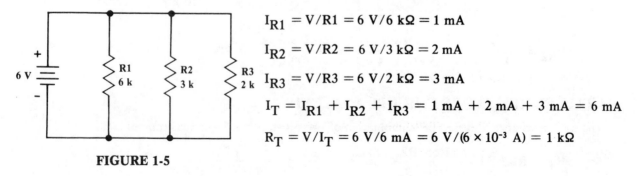

FIGURE 1-5

$$I_{R1} = V/R1 = 6\,V/6\,k\Omega = 1\,mA$$

$$I_{R2} = V/R2 = 6\,V/3\,k\Omega = 2\,mA$$

$$I_{R3} = V/R3 = 6\,V/2\,k\Omega = 3\,mA$$

$$I_T = I_{R1} + I_{R2} + I_{R3} = 1\,mA + 2\,mA + 3\,mA = 6\,mA$$

$$R_T = V/I_T = 6\,V/6\,mA = 6\,V/(6 \times 10^{-3}\,A) = 1\,k\Omega$$

SERIES-PARALLEL RESISTOR CIRCUIT ANALYSIS

Several methods can be used to solve for the voltage drops around a series-parallel circuit. If the circuit has only one voltage source, they can be solved by the methods used in the previous examples, or they can be solved by using loop equations or Thevenin's equivalent theorem. Also, lesser used techniques such as Norton's theorem or the superposition theorem can be used successfully.

In this section, a single voltage source circuit will be analyzed using Ohm's law, loop equations, and Thevenin's equivalent theorem. Each method of analysis will be used on the same circuit so that a comparison of each can be made.

OHM'S LAW ANALYSIS

Ohm's law analysis for the circuit of Figure 1-6 begins by solving for the parallel resistance of R2 and R3 and using the parallel resistance in series with R1 to solve for the current flow through R1, the voltage drop across R1, and the voltage drops across the parallel resistors R2 and R3. Also, the current flow through the R2 and R3 resistive branches can be noted. These steps are shown by the following examples.

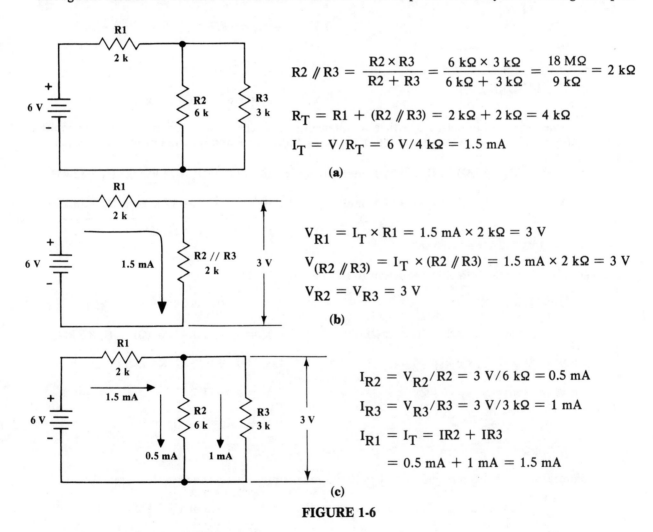

FIGURE 1-6

LOOP EQUATIONS ANALYSIS

Loop equations are the most mechanized of the circuit analysis techniques. They can be used on the simplest to the most complex circuit and the structure of the equation does not change when the number of voltage sources increases. Also, the current flow through each branch of the circuit can be solved by either simultaneous equations or determinants.

For the circuit of Figure 1-7, start the loop equation by writing the effective voltage drops around the first loop, where the effect of the single 6 volt voltage source causes I_1 and I_2 to oppose each other. Therefore:

$$V = I_1 R_1 + (I_1 - I_2)R_2 = I_1 R_1 + I_1 R_2 - I_2 R_2 = I_1(R_1 + R_2) - I_2 R_2$$

While there is no voltage source in the second loop, the I_2 current does cause a voltage drop across R2 and R3 and, again, since I_1 and I_2 are in opposition in R2, the equation reads:

$$0\,V = I_2 R_3 + (I_2 - I_1)R_2 = I_2 R_3 + I_2 R_2 - I_1 R_2 = I_2(R_2 + R_3) - I_1 R_2$$

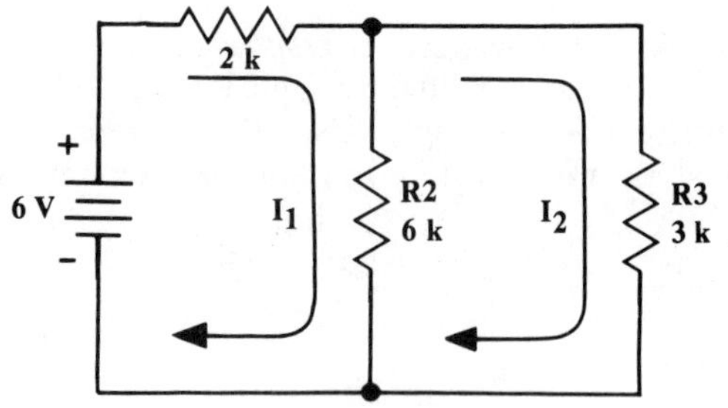

$$V = I_1(R_1 + R_2) - I_2 R_2$$

$$0\,V = -I_1 R_2 + I_2(R_2 + R_3)$$

$$6\,V = I_1(2\,k\Omega + 6\,k\Omega) - (I_2 \times 6\,k\Omega)$$

$$0\,V = -(I_1 \times 6\,k\Omega) + I_2(6\,k\Omega + 3\,k\Omega)$$

$$6\,V = (I_1 \times 8\,k\Omega) - (I_2 \times 6\,k\Omega)$$

$$0\,V = -(I_1 \times 6\,k\Omega) + (I_2 \times 9\,k\Omega)$$

FIGURE 1-7

The two unknowns can be solved by either simultaneous equations or determinants. In using simultaneous equations, I_1 is solved by using multiplication and addition to make I_2 drop out of the equation.

$6\,V = + (I_1 \times 8\,k\Omega) - (I_2 \times 6\,k\Omega)$ multiply by 3: $18\,V = + (I_1 \times 24\,k\Omega) - (I_2 \times 18\,k\Omega)$

$0\,V = - (I_1 \times 6\,k\Omega) + (I_2 \times 9\,k\Omega)$ multiply by 2: $0\,V = - (I_1 \times 12\,k\Omega) + (I_2 \times 18\,k\Omega)$

$$18\,V = \quad I_1 \times 12\,k\Omega$$

$$I_1 = 18\,V/12\,k\Omega = 1.5\,mA$$

Then, I_1 is substituted into either equation to solve for I_2.

1. $6\,V = (I_1 \times 8\,k\Omega) - (I_2 \times 6\,k\Omega)$

    $6\,V = (1.5\,mA \times 8\,k\Omega) - (I_2 \times 6\,k\Omega)$

    $6\,V = 12\,V - (I_2 \times 6\,k\Omega)$

    $6\,V - 12\,V = -(I_2 \times 6\,k\Omega)$

    $-6\,V = -(I_2 \times 6\,k\Omega)$

    $6\,V/6\,k\Omega = I_2 = 1\,mA$

2. $0\,V = -(I_1 \times 6\,k\Omega) + (I_2 \times 9\,k\Omega)$

    $0\,V = -(1.5\,mA \times 6\,k\Omega) + (I_2 \times 9\,k\Omega)$

    $0\,V = -9\,V + (I_2 \times 9\,k\Omega)$

    $9\,V = I_2 \times 9\,k\Omega$

    $9\,V/9\,k\Omega = I_2 = 1\,mA$

Therefore, since $I_1 = 1.5\,mA$ and $I_2 = 1\,mA$, the voltage drops across the resistors are:

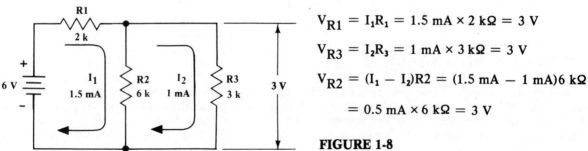

$V_{R1} = I_1 R_1 = 1.5\,mA \times 2\,k\Omega = 3\,V$

$V_{R3} = I_2 R_3 = 1\,mA \times 3\,k\Omega = 3\,V$

$V_{R2} = (I_1 - I_2)R_2 = (1.5\,mA - 1\,mA)6\,k\Omega$

$\quad\quad = 0.5\,mA \times 6\,k\Omega = 3\,V$

FIGURE 1-8

THEVENIN'S THEOREM ANALYSIS

Thevenin's theorem, or Thevenin's equivalent as it is sometimes called, is a method of analysis that can be used to solve for the load current which, with reference to Figure 1-8, is I_2. Then, the remaining circuit currents and the voltage drops around the circuit can be calculated. First, as shown in Figure 1-9, the load resistor R3 is removed and the voltage is calculated or measured across the output terminals. The output voltage is labelled V_{TH}. Next, the input resistance, from the load looking in as shown in Figure 1-10, is solved. Since the DC resistance of the DC voltage source is (theoretically) zero ohms, the Rin = R1 // R2. This input resistance is called R_{TH}. Once the V_{TH} and R_{TH} are known, they along with R_L can be used to solve the load current $I_L = I_{R3}$ from $I_L = V_{TH}/(R_{TH} + R_L)$. Then, the remaining voltage drops around the circuit can be solved, as shown with the calculations included with Figure 1-12.

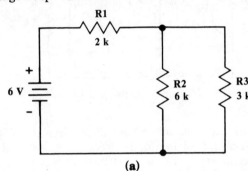
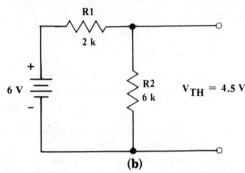

FIGURE 1-9

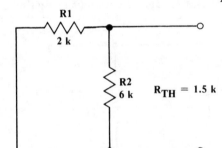

$$V_{TH} = \frac{V \times R2}{R1 + R2} = \frac{6\,V \times 6\,k\Omega}{2\,k\Omega + 6\,k\Omega} = 4.5\,V$$

$$R_{TH} = \frac{R1 \times R2}{R1 + R2} = \frac{2\,k\Omega \times 6\,k\Omega}{2\,k\Omega + 6\,k\Omega} = 1.5\,k\Omega$$

FIGURE 1-10

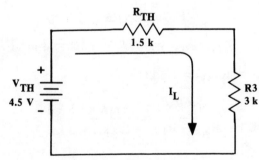

$$I_L = \frac{V_{TH}}{R_{TH} + R_L} = \frac{4.5\,V}{1.5\,k\Omega + 3\,k\Omega} = 1\,mA$$

where: $R3 = R_L$

FIGURE 1-11

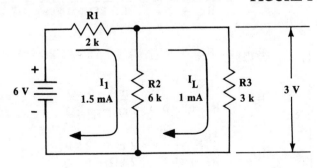

$I_L = I_{R3} = 1\,mA$

$V_{R3} = I_2 R_3 = 1\,mA \times 3\,k\Omega = 3\,V$

$V_{R2} = V_{R3} = 3\,V$

$V_{R1} = V - V_{R2} = 6\,V - 3\,V = 3\,V$

$I_{R2} = V_{R2}/R2 = 3\,V/6\,k\Omega = 0.5\,mA$

$I_{R1} = V_{R1}/R1 = 3\,V/2\,k\Omega = 1.5\,mA$

FIGURE 1-12

NORTON'S THEOREM ANALYSIS

Norton's Theorem, like Thevenin's theorem, is a circuit analysis technique that can reduce a complex circuit to a simplified equivalent circuit. For instance, Thevenin's equivalent circuit reduces the basic circuit to a Thevenized voltage and a Thevenized resistance in series with the load resistor, while Norton's equivalent circuit reduces the basic circuit to a current source in parallel with the Nortonized resistance in parallel with the load, as shown in Figure 1-13, the equivalent circuits for Figure 1-12.

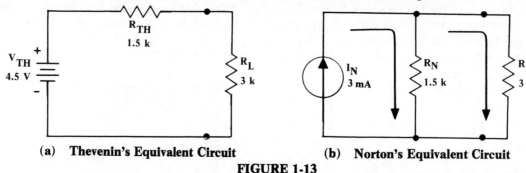

(a)   Thevenin's Equivalent Circuit   (b)   Norton's Equivalent Circuit

FIGURE 1-13

Norton's equivalent circuit can be obtained by converting the Thevenin's equivalent circuit of Figure 1-13(a), by shorting the load resistor as shown in Figure 1-14, and solving for I_N from $I_N = V_{TH}/R_{TH}$. Then, I_N and R_N, where $R_N = R_{TH}$ and is solved in the same way (see Figure 1-10), can be connected as shown in Figure 1-15, and the load is added as shown in Figure 1-16. Also, if needed, Norton's equivalent can be converted back to the basic circuit with all the current and voltage calculations using the techniques shown in Figure 1-17.

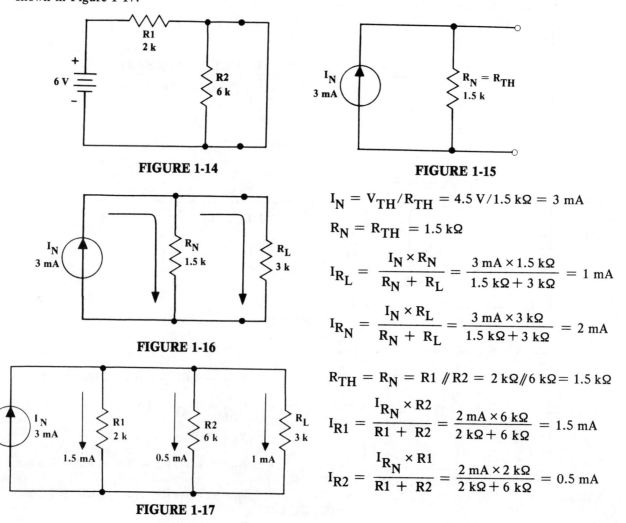

FIGURE 1-14   **FIGURE 1-15**

$$I_N = V_{TH}/R_{TH} = 4.5\,V/1.5\,k\Omega = 3\,mA$$

$$R_N = R_{TH} = 1.5\,k\Omega$$

$$I_{R_L} = \frac{I_N \times R_N}{R_N + R_L} = \frac{3\,mA \times 1.5\,k\Omega}{1.5\,k\Omega + 3\,k\Omega} = 1\,mA$$

$$I_{R_N} = \frac{I_N \times R_L}{R_N + R_L} = \frac{3\,mA \times 3\,k\Omega}{1.5\,k\Omega + 3\,k\Omega} = 2\,mA$$

FIGURE 1-16

$$R_{TH} = R_N = R1 \,/\!/\, R2 = 2\,k\Omega /\!/ 6\,k\Omega = 1.5\,k\Omega$$

$$I_{R1} = \frac{I_{R_N} \times R2}{R1 + R2} = \frac{2\,mA \times 6\,k\Omega}{2\,k\Omega + 6\,k\Omega} = 1.5\,mA$$

$$I_{R2} = \frac{I_{R_N} \times R1}{R1 + R2} = \frac{2\,mA \times 2\,k\Omega}{2\,k\Omega + 6\,k\Omega} = 0.5\,mA$$

FIGURE 1-17

POWER SOURCES

The ideal battery or power supply has an internal resistance of zero ohms. Therefore, whether the voltage source is loaded or unloaded, the voltage remains relatively constant. For instance, if the internal resistance of a power supply is 10 ohms and the open circuit voltage is 6 volts, then, connecting a 10 ohm load resistance will cause the voltage at the output to drop to 3 volts. However, if the internal resistance is 0.1 ohms, then the loaded voltage drops to only 5.94 volts. Obviously, a zero volt internal resistance would maintain the loaded voltage at 6 volts. Figure 1-18 shows the effect of a 10 ohm load on power supplies with 10 ohm and 0.1 ohm internal resistances.

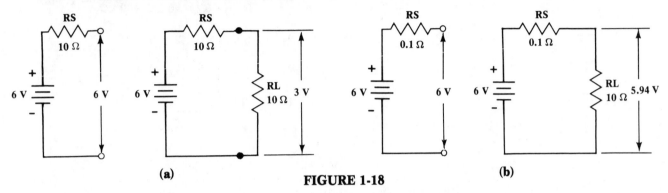

FIGURE 1-18

CIRCUITS WITH TWO POWER SUPPLIES

Connecting two power supplies in a series aiding connection will produce the sum of the two supply voltages, regardless of where the ground connection is made. For instance, referring to Figure 1-19(a), the total voltage across the 6 volt supplies is 12 volts. Since the ground connection is made at the negative terminal of the bottom supply (point C), the voltage at the junction of the two supplies (point B), with respect to ground, is 6 volts. The voltage at the positive terminal of the top supply (point A), with respect to ground, is 12 volts.

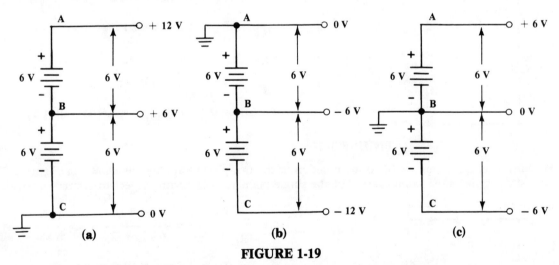

FIGURE 1-19

However, when the ground is connected to the positive terminal of the top supply (point A), as shown in Figure 1-19(b), the voltage at point B is − 6 volts, with respect to ground, and the voltage at point C is − 12 volts, with respect to ground.

If the ground is connected to the junction of the two supplies (point B), as shown in Figure 1-19(c), the voltage at point A, with respect to ground, is + 6 volts and the voltage at point C, with respect to ground, is − 6 volts.

NOTE: In all three connections shown in Figure 1-19, the voltage across the two series supplies remains at 12 volts and does not change as the ground is moved from point A to point B to point C. What does change, however, is the voltage at the respective A, B, and C terminals, with respect to ground.

CIRCUITS WITH TWO POWER SUPPLIES AND LOAD RESISTORS

When two series resistors are connected across two power supplies in a series aiding connection, the voltage across the two series load resistors is the same, regardless of where the ground connection is made. However, as shown in Figure 1-20, the single point voltages, with respect to ground, change.

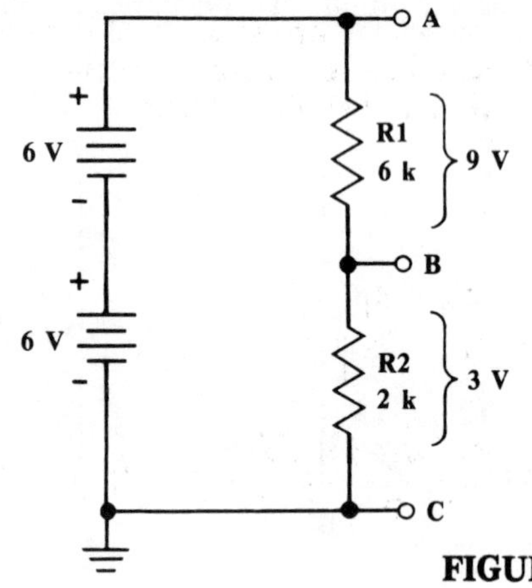

$$I = V_T/R_T = \frac{V_T}{R1 + R2} = \frac{12\text{ V}}{6\text{ k}\Omega + 2\text{ k}\Omega} = 1.5\text{ mA}$$

$$V_{R1} = IR_1 = 1.5\text{ mA} \times 6\text{ k}\Omega = 9\text{ V}$$

$$V_{R2} = IR_2 = 1.5\text{ mA} \times 2\text{ k}\Omega = 3\text{ V}$$

$$V_A = V_{R1} + V_{R2} = 9\text{ V} + 3\text{ V} = 12\text{ V}$$

$$V_B = V_{R2} = 3\text{ V}$$

$$V_C = 0\text{ V}$$

FIGURE 1-20

When the ground is connected to point A, as shown in Figure 1-21, the voltage drops across R1 and R2 remain the same, but the voltages at points A, B, and C, with respect to ground, change.

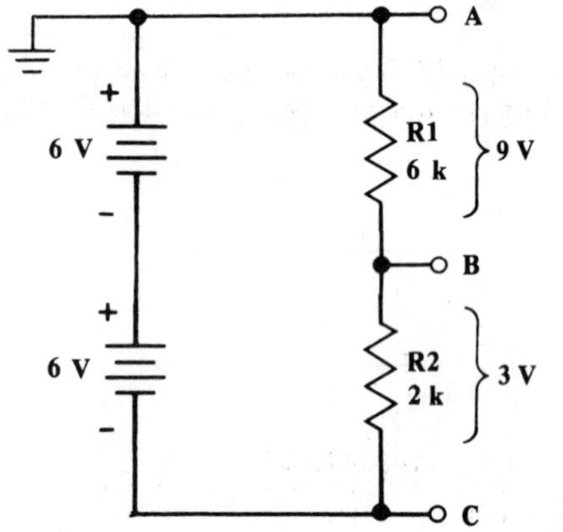

$$I = \frac{V_T}{R1 + R2} = \frac{12\text{ V}}{6\text{ k}\Omega + 2\text{ k}\Omega} = \frac{12\text{ V}}{8\text{ k}\Omega} = 1.5\text{ mA}$$

$$V_{R1} = IR_1 = 1.5\text{ mA} \times 6\text{ k}\Omega = 9\text{ V}$$

$$V_{R2} = IR_2 = 1.5\text{ mA} \times 2\text{ k}\Omega = 3\text{ V}$$

$$V_A = 0\text{ V}$$

$$V_B = -(V_{R1}) = -9\text{ V}$$

$$V_C = -(V_{R1} + V_{R2}) = -(9\text{ V} + 3\text{ V}) = -12\text{ V}$$

FIGURE 1-21

When the ground is connected to the junction of the two power supplies, as shown in Figure 1-22, the voltage drop remains the same again, but the single point voltages, with respect to ground, change.

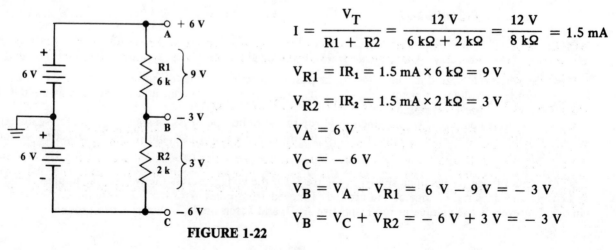

$$I = \frac{V_T}{R1 + R2} = \frac{12\text{ V}}{6\text{ k}\Omega + 2\text{ k}\Omega} = \frac{12\text{ V}}{8\text{ k}\Omega} = 1.5\text{ mA}$$

$$V_{R1} = IR_1 = 1.5\text{ mA} \times 6\text{ k}\Omega = 9\text{ V}$$

$$V_{R2} = IR_2 = 1.5\text{ mA} \times 2\text{ k}\Omega = 3\text{ V}$$

$$V_A = 6\text{ V}$$

$$V_C = -6\text{ V}$$

$$V_B = V_A - V_{R1} = 6\text{ V} - 9\text{ V} = -3\text{ V}$$

$$V_B = V_C + V_{R2} = -6\text{ V} + 3\text{ V} = -3\text{ V}$$

FIGURE 1-22

SECTION II: ALTERNATING CURRENT CIRCUIT ANALYSIS

INTERNAL RESISTANCE OF SIGNAL GENERATORS

Most laboratory signal generators have internal resistance of either 600 ohms or 50 ohms. This internal resistance, depending on the load the signal generator is driving, causes a certain amount of the signal voltage to be lost.

For instance, Figure 1-23 shows a circuit with a 1 kΩ load resistor connected across a signal generator with an internal resistance of 600 ohms. If the signal across the 1 kΩ load is to be 1 Vp-p, the open circuit voltage will have to be 1.6 Vp-p because, obviously, the internal resistance of 600 ohms drops 600 mVp-p.

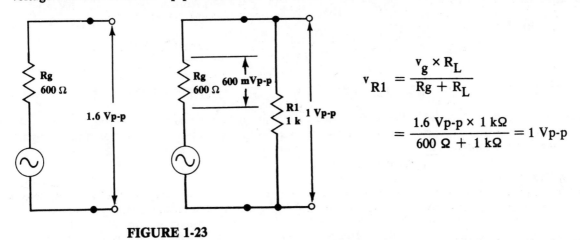

$$v_{R1} = \frac{v_g \times R_L}{R_g + R_L}$$

$$= \frac{1.6 \text{ Vp-p} \times 1 \text{ k}\Omega}{600 \text{ }\Omega + 1 \text{ k}\Omega} = 1 \text{ Vp-p}$$

FIGURE 1-23

If the load resistor is changed to 10 kΩ, as shown in Figure 1-24, then 1.06 Vp-p of open circuit voltage is required to deliver 1 Vp-p to the load, because 60 mVp-p is lost to the internal resistance of 600 ohms.

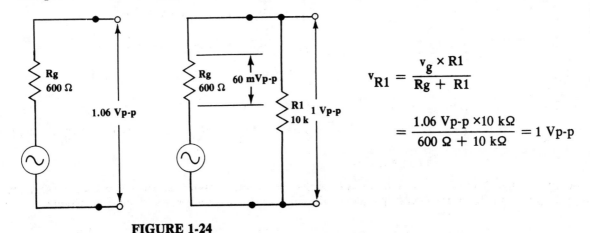

$$v_{R1} = \frac{v_g \times R1}{R_g + R1}$$

$$= \frac{1.06 \text{ Vp-p} \times 10 \text{ k}\Omega}{600 \text{ }\Omega + 10 \text{ k}\Omega} = 1 \text{ Vp-p}$$

FIGURE 1-24

NOTE: A good laboratory procedure is to connect the signal generator to the circuit before adjusting the signal level of the generator. Then, the internal resistance of the generator does not have to be included in the circuit calculations.

ALTERNATING CURRENT SIGNAL DISTRIBUTION

The AC signal is distributed in a manner similar to the DC voltage. If a signal is applied to two series resistors, the applied voltage is shared by the two resistors. For instance, as shown in Figure 1-25, if 6 Vp-p is applied to R1 and R2, 4 Vp-p is developed across the 6 kΩ resistor R1 and 2 Vp-p is developed across the 3 kΩ resistor R2. The amount of signal voltage absorbed by the internal resistance of the signal generator is 0.4 Vp-p and the open circuit voltage of the generator is calculated to be approximately 6.4 Vp-p.

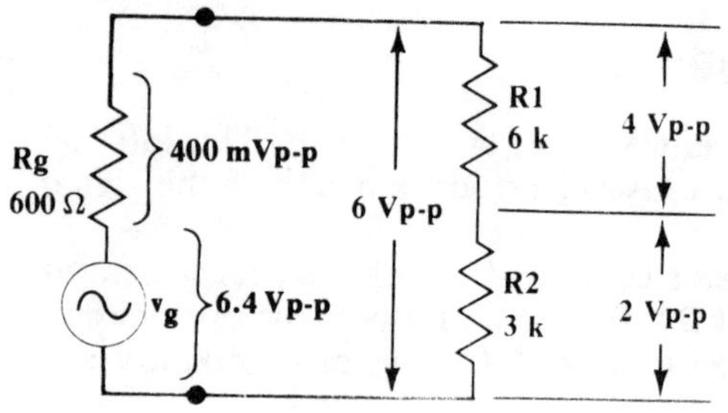

$$v_{R1} = \frac{v_{in} \times R1}{R1 + R2} = \frac{6\ \text{Vp-p} \times 6\ k\Omega}{6\ k\Omega + 3\ k\Omega} = 4\ \text{Vp-p}$$

$$v_{R2} = \frac{v_{in} \times R2}{R1 + R2} = \frac{6\ \text{Vp-p} \times 3\ k\Omega}{6\ k\Omega + 3\ k\Omega} = 2\ \text{Vp-p}$$

$$i_{R_T} = \frac{v_{in}}{R1 + R2} = \frac{6\ \text{Vp-p}}{6\ k\Omega + 3\ k\Omega} \approx 667\ \mu\text{Ap-p}$$

$$v_{Rg} = i_{R_T} \times Rg \approx 667\ \mu\text{Ap-p} \times 600\ \Omega = 400\ \text{mVp-p}$$

$$v_g = v_{Rg} + v_{R1} + v_{R2}$$
$$= 0.4\ \text{Vp-p} + 4\ \text{Vp-p} + 2\ \text{Vp-p} = 6.4\ \text{Vp-p}$$

FIGURE 1-25

SUPERIMPOSING ALTERNATING CURRENT ON DIRECT CURRENT

Much like oil on water, the AC voltage rides on the DC voltage as long as the DC voltage is not at AC ground. Two methods of superimposing an AC signal on a DC voltage are by superposition and by capacitor coupling. An example of an AC ground is the junction of resistors R2 and R3 of Figure 1-27.

SUPERPOSITION

Superposition is when the DC power supply, the AC signal generator, and the load resistors are connected in series, as shown in Figure 1-26. Then, since the DC and AC voltages do not mix, the respective voltages are calculated separately. With reference to Figure 1-26, if the DC voltage is 12 volts, the DC voltage developed across R1 is 8 volts and across R2 is 4 volts. Since the AC signal voltage is 9 Vp-p, 6 Vp-p is developed across R1 and 3 Vp-p is developed across R2.

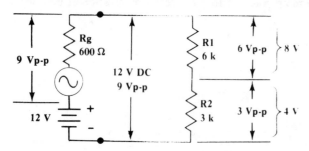

FIGURE 1-26

DIRECT CURRENT VOLTAGE DISTRIBUTION

$$V_{R1} = \frac{V\ DC \times R1}{R1 + R2} = \frac{12\ V \times 6\ k\Omega}{6\ k\Omega + 3\ k\Omega} = 8\ V$$

$$V_{R2} = \frac{V\ DC \times R2}{R1 + R2} = \frac{12\ V \times 3\ k\Omega}{6\ k\Omega + 3\ k\Omega} = 4\ V$$

ALTERNATING CURRENT SIGNAL VOLTAGE DISTRIBUTION

$$v_{R1} = \frac{v_{in} \times R1}{R1 + R2} = \frac{9\ \text{Vp-p} \times 6\ k\Omega}{6\ k\Omega + 3\ k\Omega} = 6\ \text{Vp-p}$$

$$v_{R2} = \frac{v_{in} \times R2}{R1 + R2} = \frac{9\ \text{Vp-p} \times 3\ k\Omega}{6\ k\Omega + 3\ k\Omega} = 3\ \text{Vp-p}$$

Then, if another series resistor is added to the circuit and a bypass capacitor is connected across that resistor, as shown in Figure 1-27, the DC voltage will be distributed across the three series resistors. However, the capacitor C1 is an effective short to the AC signal because the signal voltage frequency is 1 kHz and the 100 μF capacitor offers only 1.59 ohms of reactance to the signal voltage. Therefore, the AC signal voltage is distributed only across R1 and R2, where C1 bypasses R3 to ground.

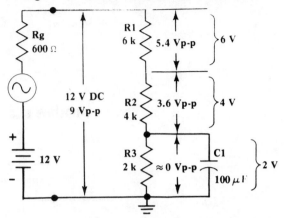

FIGURE 1-27

DIRECT CURRENT VOLTAGE DISTRIBUTION

$$V_{R1} = \frac{V\ DC \times R1}{R1 + R2 + R3} = \frac{12\ V \times 6\ k\Omega}{6\ k\Omega + 4\ k\Omega + 2\ k\Omega} = 6\ V$$

$$V_{R2} = \frac{V\ DC \times R2}{R1 + R2 + R3} = \frac{12\ V \times 4\ k\Omega}{6\ k\Omega + 4\ k\Omega + 2\ k\Omega} = 4\ V$$

$$V_{R3} = \frac{V\ DC \times R3}{R1 + R2 + R3} = \frac{12\ V \times 2\ k\Omega}{6\ k\Omega + 4\ k\Omega + 2\ k\Omega} = 2\ V$$

ALTERNATING CURRENT SIGNAL VOLTAGE DISTRIBUTION

$$X_{C1} = \frac{1}{2\pi fC} \approx \frac{0.159}{1\ kHz \times 100\ \mu F} = \frac{0.159}{10^3 \times 10^{-4}} = \frac{0.159}{10^{-1}} = 1.59\ \Omega$$

$$v_{R1} = \frac{v_{in} \times R1}{R1 + R2} = \frac{9\ Vp\text{-}p \times 6\ k\Omega}{6\ k\Omega + 4\ k\Omega} = 5.4\ Vp\text{-}p$$

$$v_{R2} \approx \frac{v_{in} \times R2}{R1 + R2} = \frac{9\ Vp\text{-}p \times 4\ k\Omega}{6\ k\Omega + 4\ k\Omega} = 3.6\ Vp\text{-}p$$

$$v_{R3} \approx 0\ Vp\text{-}p$$

CAPACITOR COUPLING

The second method of superimposing an AC signal on a DC voltage is by capacitor coupling, as shown in Figure 1-28. In this circuit, the DC voltage is developed across the R1 and R2 resistors, but it is isolated from the signal generator by the coupling capacitor C1. Then, because the DC resistance of an ideal power supply is zero ohms, the applied AC signal is delivered to the parallel combination of R1 and R2. As shown in Figure 1-29, the AC signal is developed equally across R1 and R2 at 6 Vp-p and the open circuit signal voltage is 7.8 Vp-p.

$$V_{R1} = \frac{V\ DC \times R1}{R1 + R2} = \frac{12\ V \times 6\ k\Omega}{6\ k\Omega + 3\ k\Omega} = 8\ V$$

$$V_{R2} = \frac{V\,DC \times R2}{R1 + R2} = \frac{12\,V \times 3\,k\Omega}{6\,k\Omega + 3\,k\Omega} = 4\,V$$

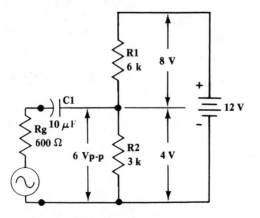

FIGURE 1-28

$v_{R1} = v_{R2} = 6\,V\text{p-p}$

$R(\text{equiv}) = R1 \,//\, R2 = 6\,k\Omega \,//\, 3\,k\Omega = 2\,k\Omega$

$$v_{OC} = \frac{v_{in}(R_g + R[\text{equiv}])}{R(\text{equiv})} = \frac{6\,V\text{p-p}(600\,\Omega + 2\,k\Omega)}{2\,k\Omega} = 7.8\,V\text{p-p}$$

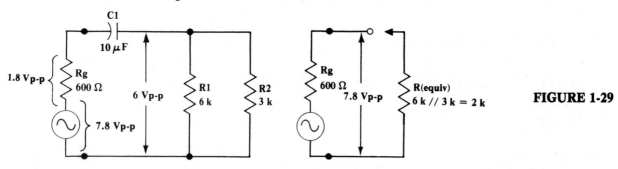

FIGURE 1-29

Connecting a resistor in series with the coupling capacitor, as shown in Figure 1-30, will not effect the DC voltage because the coupling capacitor prevents DC current flow through resistor RS. However, the AC signal voltage will be affected and only 4 Vp-p is developed across R1 and R2, as shown in Figure 1-31 and the calculations following that illustration.

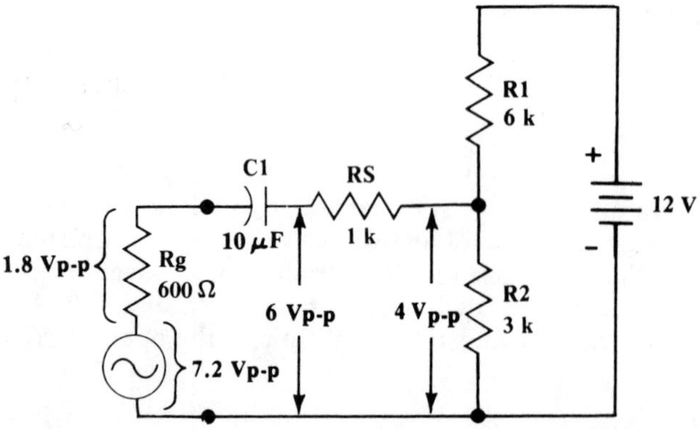

$$V_{R1} = \frac{V \times R1}{R1 + R2} = \frac{12\,V \times 6\,k\Omega}{6\,k\Omega + 3\,k\Omega} = 8\,V$$

$$V_{R2} = \frac{V \times R2}{R1 + R2} = \frac{12\,V \times 3\,k\Omega}{6\,k\Omega + 3\,k\Omega} = 4\,V$$

$V_{RS} = 0\,V$

FIGURE 1-30

NOTE: Removing R1 and R2 and measuring the generator signal voltage indicates a larger signal voltage than when loaded. Essentially, the 6 Vp-p jumps to 7.2 Vp-p because, with no signal current, no signal voltage is dropped across the 600 ohm resistance Rg.

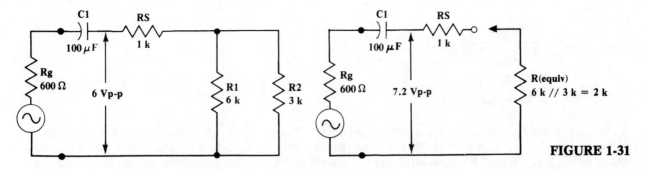

FIGURE 1-31

NOTE: The reason the AC equivalent circuit has R1 in parallel with R2 is because, to the AC signal, the battery is an effective AC short. Therefore, if the battery lead is ground, the positive lead is at AC ground.

$$v_{R1} = v_{R2} = \frac{v_{in} \times R(equiv)}{RS + R(equiv)} = \frac{6 \text{ Vp-p} \times 2 \text{ k}\Omega}{1 \text{ k}\Omega + 2 \text{ k}\Omega} = 4 \text{ Vp-p}$$

$$v_{OC} = \frac{v_{in}(Rg + RS + R[equiv])}{RS + R(equiv)} = \frac{6 \text{ Vp-p} \times (600 \text{ }\Omega + 1 \text{ k}\Omega + 2 \text{ k}\Omega)}{1 \text{ k}\Omega + 2 \text{ k}\Omega} = 7.2 \text{ Vp-p}$$

LABORATORY EXERCISE

OBJECTIVES

To review basic concepts and measuring techniques by investigating the DC and AC parameters of the voltage divider circuit.

LIST OF MATERIALS AND EQUIPMENT

1. Resistors (one each):   1 kΩ 2.2 kΩ 3.3 kΩ 4.7 kΩ
2. Capacitors (25 volt — one each):   10 μF   100 μF
3. Power supply: 24 volt
4. Voltmeter
5. Signal generator
6. Oscilloscope

PROCEDURE

1. Connect the circuit shown in Figure 1-32.

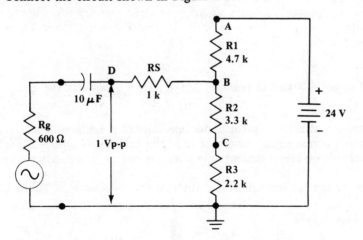

FIGURE 1-32

2. Calculate the DC voltage drops around the circuit from:

   a. $V_{R1} = \dfrac{V \times R1}{R1 + R2 + R3}$ where: V = 24 V

19

b. $V_{R2} = \dfrac{V \times R2}{R1 + R2 + R3}$

c. $V_{R3} = \dfrac{V \times R3}{R1 + R2 + R3}$

3. Measure the DC voltage drops around the circuit. However, instead of measuring the DC voltage across each resistor, measure the single point voltages at points A, B, C, and D, and then solve the voltage drops of R1, R2, R3, and RS by subtracting the known voltages. This procedure will avoid ground loop problems that can occur if the voltmeter used is grounded. Therefore, connect the negative lead of the voltmeter to circuit ground and measure point A (the positive supply lead), point B (the junction of R1 and R2), point C (the junction of R2 and R3), and point D (the junction of RS and C1). Use the measured single point voltage values to calculate the "measured" values of V_{R1}, V_{R2}, V_{R3}, and V_{RS}.

$V_{R1} = V_A - V_B$

$V_{R2} = V_B - V_C$

$V_{R3} = V_C$

$V_{RS} = V_B - V_D$ where: $V_B \approx V_D$

4. Insert the calculated and "measured" values, as indicated, into Table 1-1.

TABLE 1-1	V_{R1}	V_{R2}	V_{R3}	V_{RS}	V_A	V_B	V_C	V_D
CALCULATED				≈ 0 V	/////	/////	/////	/////
MEASURED								

5. SIGNAL VOLTAGE DISTRIBUTION:
   a. Calculate the signal voltage magnitude at point B and point C, where the signal at point B is developed across the parallel resistor R1 and the total of resistors R2 in series with R3. Therefore, resistors R2 and R3 share the signal voltage at point B, with respect to ground.

   1. $v_B = \dfrac{v_{in} \times R1 \,/\!/\, (R2 + R3)}{RS + (R1 \,/\!/\, [R2 + R3])}$ where: $v_{in} = 1$ Vp-p

   2. $v_C = \dfrac{v_B \times R3}{R2 + R3}$

   b. Measure the signal voltage level at point B and at point C. Make sure v_{in} equals 1 Vp-p at a frequency of 1 kHz.

NOTE: Do not make the 1 Vp-p input signal measurement under open circuit conditions because some signal will be lost to the internal resistance of the signal generator once the external circuit is connected. Therefore, make the 1 Vp-p input signal voltage measurement only after the circuit is properly connected.

6. Insert the calculated and measured values, as indicated, into the first three columns of Table 1-2.

TABLE 1-2	Figure 1-32			Figure 1-33	
	v_{in}	v_B	v_C	v_{in}	v_B
CALCULATED	1 Vp-p			1 Vp-p	
MEASURED					

7. Connect a 100 µF capacitor across resistor R3, as shown in Figure 1-33.

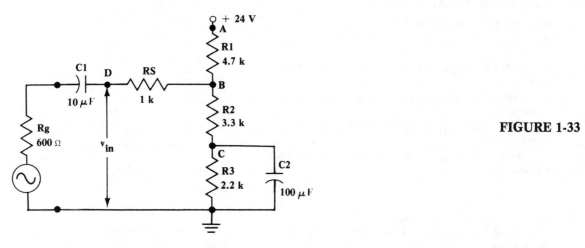

FIGURE 1-33

8. Monitor the DC voltages around the circuit to make sure the DC voltage distribution remains the same.

NOTE: A leaky C2 capacitor or a C2 capacitor connected backwards can act resistive and redistribute the DC voltages. Remember, no DC voltage variations should occur by connecting the C2 capacitor across resistor R3. Its effect should be on the AC only.

9. SIGNAL VOLTAGE DISTRIBUTION:

   a. Calculate the signal voltage level at point B from:

   $$v_B = \frac{v_{in} \times (R1 \; // \; R2)}{RS + (R1 \; // \; R2)}$$

   b. Measure the signal voltage at point B, where v_{in} equals 1 Vp-p at a frequency of 1 kHz.

NOTE: Point A and point C are, theoretically, at AC ground conditions. Monitor these points for a trace of AC signal voltage to make sure they are indeed at AC ground.

10. Insert the calculated and measured values, as indicated, into the last two columns of Table 1-2.

LABORATORY QUESTIONS (Refer to Figure 1-33)

1. Does the addition of capacitor C2 across resistor R3 cause a DC voltage change in the circuit? Why or why not?

2. Why was the value of the C2 capacitor selected at 100 µF instead of 0.1 µF, for instance?

3. Does the effect of connecting the C2 capacitor cause a change in the distributed AC signal. If so, why?

4. How much DC voltage is dropped across the RS resistor? What component, in series with RS, insures this condition, and how does it provide the condition?

5. Why is it advisable for the circuit to be fully connected before Vin is measured?

6. Why is the positive 24 volt connection of the circuit considered as an AC ground? Does it measure as an AC ground?

7. Why is it a good idea to make all DC voltage and AC signal measurements with respect to ground, instead of measuring directly across the components?

CHAPTER PROBLEMS

1. With reference to Figure 1-34:
   a. Solve for I_{R1}, the DC current flow through resistor R1.
   b. Solve for V_{R1}, the DC voltage drop across resistor R1.
   c. Solve for V_{R2}, the DC voltage drop across resistor R2.
   d. Solve for V_{R3}, the DC voltage drop across resistor R3.
   e. Solve for V_{R4}, the DC voltage drop across resistor R4.

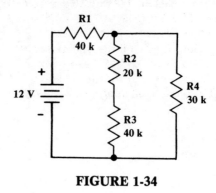

FIGURE 1-34

2. Thevenize the circuit of Figure 1-35.
   a. Solve for V_{TH}, the Thevenized voltage for the circuit.
   b. Solve for R_{TH}.
   c. Solve for $I_L = I_{R3}$.
   d. Draw the Thevenin's equivalent circuit.

3. Nortonize the circuit of Figure 1-35. Use the Thevenin's equivalent circuit, and then solve for:
   a. I_N, the Norton equivalent current.
   b. R_N, which equals R_{TH}.
   c. Draw the Norton equivalent circuit, complete with load resistor.

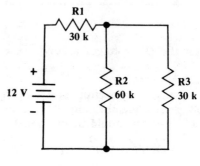

FIGURE 1-35

4. With regard to Figure 1-36:
   a. Solve for V_{R1}, the DC voltage dropped across R1.
   b. Solve for V_{R2}, the DC voltage dropped across R2.
   c. Calculate the DC voltage at point B, with respect to ground.
   d. Solve the DC voltage at point C, with respect to ground.

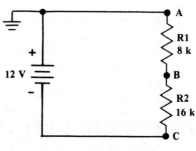

FIGURE 1-36

5. With reference to Figure 1-37:
   a. Solve for the DC voltages dropped across R1, R2, and R3.
   b. Solve for the voltages at points A, B, C, and D, with respect to ground.

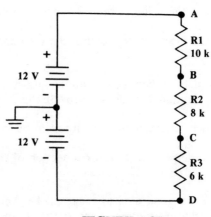

FIGURE 1-37

6. The unloaded 600 Ω signal generator of Figure 1-38 puts out 2 Vp-p. A 1.8 kΩ load resistor is connected across the output terminals.
   a. How much of the signal voltage is developed across the 1.8 kΩ load resistor?
   b. How much of the signal voltage is dropped across the 600 Ω internal resistance of the generator.

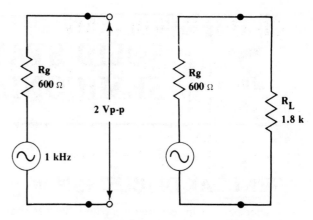

FIGURE 1-38

7. With reference to the circuit of Figure 1-39, the input 1 kHz signal at point D, with respect to ground, is 5 Vp-p.
   a. Calculate the DC voltage drops across R1, R2, R3, and RS.
   b. Draw the equivalent AC circuit and solve for the peak-to-peak signals developed across RS, R2, and R3.
   c. Calculate the 10 μF X_{C1} at 1 kHz.

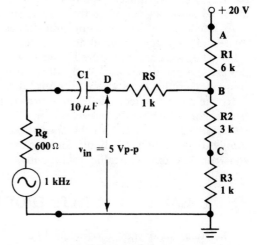

FIGURE 1-39

8. With reference to Figure 1-40, the input 1 kHz AC signal at point D, with respect to ground is 5 Vp-p.
   a. Calculate the DC voltage drops across R1, R2, R3, and RS.
   b. Calculate the 100 μF X_{C2} at 1 kHz.
   c. Draw the equivalent AC circuit, and solve the peak-to-peak signals developed across R1, RS, and R3.

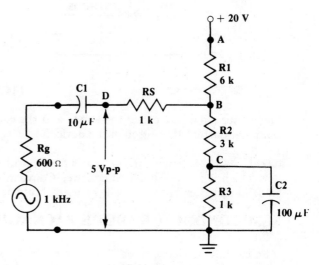

FIGURE 1-40

23

2 | SOLID STATE SEMICONDUCTOR DIODES

GENERAL DISCUSSION

Diodes are two terminal electronic devices that allow the free flow of electrical current in one direction only. Therefore, they are used as rectifiers in power supply applications, as detectors in communication receiver circuits, and as switches in computer applications. Most general purpose diodes in use today are solid state devices made from silicon materials.

As late as the 1950's, vacuum tube diodes were still in use but were being replaced rapidly by solid state diodes. The general acceptance of solid state diodes was inevitable because, in comparison to vacuum tubes, they were smaller, were lighter in weight, were more rugged, cost less to manufacture, required less power, and could perform nearly all the functions of the vacuum tube except in some high power and high frequency applications where vacuum tubes remain superior.

By the 1960's, silicon diodes emerged and began to replace germanium diodes as the electronic industries general purpose diode because, in all but a few parameters, silicon is superior to germanium. Silicon materials are also used in such specialized devices as the zener diode, silicon controlled rectifier (SCR), tunnel diode, four-layer diode, and varicap. The light emitting diode (LED), a non-semiconductor diode constructed from bimetallic compounds, finds a wide variety of applications as readouts and in light sensing.

THE IDEAL SOLID STATE DIODE

The ideal solid state diode acts like a closed switch in one direction and like an open switch when connected in the opposite direction. See Figure 2-1. As shown, the symbol for the diode is an arrow and

FIGURE 2-1

the two terminals are called the anode and the cathode. The arrow symbol points in the direction of conventional current flow through the device.

NOTE: Conventional current flow is from positive to negative and is directly opposite to electron current flow which is from negative to positive. Conventional current flow is the standard used in the analysis of all solid state devices and it will be used throughout this book.

SEMICONDUCTOR DIODE PACKAGING AND MARKING

Solid state semiconductor diodes come in a variety of packages, but those most widely used are the cylindrical, the top hat, and the stud-mounted packages shown in Figure 2-2. The cylindrical package, Figure 2-2(a), can be marked with an arrow indicating the conventional current flow through the device or it can have a band at one end that indicates the cathode side. The top hat package also can have an arrow on the package (Figure 2-2(b), or, in all cases, the flange end indicates the cathode side. The stud-mounted package, Figure 2-2(c), is used in higher current applications and it, too, can be marked by an arrow or, in all cases, the body of the device, electrically connected to the thread, is the cathode side. Both the anode of the top hat and the anode of the stud-mounted packages are electrically insulated from the body of the device.

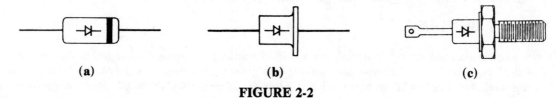

FIGURE 2-2

DIODE NUMBERING SYSTEMS

Most diodes are identified by 1N preceding a number, such as 1N914, 1N1244, or 1N4001. These are standardized diode types registered by the Electronic Industries Association (EIA) and the Joint Electron Device Engineering Council (JEDEC). The 1N indicates a semiconductor with one junction and the numbers that follow indicates the specific device. Other diodes have proprietary company numbers such as A14F, a General Electric number, or MR2271, a Motorola number. In most cases, these diodes can be cross-referenced in data books for purpose of substitution.

DIODE PARAMETERS

Diodes are generally classified under three basic categories: general purpose diodes, rectifier diodes, and switching diodes. The two main parameters used in selecting a general purpose diode are the maximum forward operating current I_F and the peak inverse voltage PIV. Exceeding either of these parameters can destroy the device. Obviously, for other applications, different parameters become important. For instance, diodes are used as mixers at high frequency, operate in high voltage circuits in excess of 20 kV, and process current greater than 100 amps. So, complete information on device parameters for specific applications must be found in the device data sheet of the manufacturer or in diode data books.

SEMICONDUCTOR MATERIALS

Solid state diodes are constructed from silicon and germanium semiconductor materials. As the name implies, the materials are rated between good electrical conductors, like copper or aluminum, and good electrical insulators, like glass or rubber.

Silicon and germanium, along with carbon and tin, belong to the carbon family which has four electrons in the outer shell of their respective atoms. However, only silicon and germanium can be used in the construction of diodes because, at room temperature, tin is too conductive and carbon is not conductive enough.

SILICON AND GERMANIUM ATOMS

In the Bohr models of the silicon and germanium atoms shown in Figure 2-3, the outer rings of orbiting

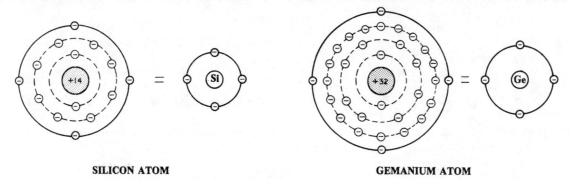

SILICON ATOM **GEMANIUM ATOM**

FIGURE 2-3

electrons have only four electrons occupying the eight available positions in their respective rings. Since these four electrons, called valence electrons, are farthest removed from the nucleus of their respective atoms, they are the most loosely bound. Hence, they can be liberated by thermal energy while the electrons in the complete inner rings remain intact. These four valence electrons are important because they are the current carriers in silicon and germanium crystal material.

COVALENT BONDING

Silicon and germanium crystal materials are made of millions of atoms held together by covalent bonding. Covalent bonding occurs because each silicon and germanium atom has only four electrons in their outer rings and they need eight to accomplish a balance. Hence, they share the four available positions with electrons from neighboring atoms. In turn, the neighboring atoms share valence electrons with four other atoms, and so on. This produces a lattice type pattern of millions of atoms bonded together to form crystal material. Figure 2-4 shows the covalent bonding of a single silicon atom with four of its neighbors.

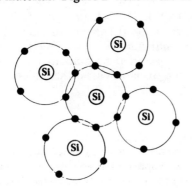

FIGURE 2-4

CURRENT FLOW IN A PURE CRYSTAL

A pure silicon or germanium crystal is a poor conductor because it has few free electrons to become current carriers. However, a voltage can be applied to the crystal, and valence electrons will be forced from their bond and enter into conduction. And in breaking away, the electrons cause holes to exist and the affected atoms assume a positive charge. (The holes are considered positively charged carriers.) Therefore, if a power supply is connected to the crystal with ohmic contacts on each side of the crystal, electrons will flow toward the positive terminal and holes will flow toward the negative terminal of the supply.

Nevertheless, the number of current carriers in a semiconductor crystal is still too low, even considering germanium which is 1000 times more conductive than silicon, to be an effective conductor. Therefore, by use of a process called doping, impurities are added to the crystal material to increase the conductivity.

DOPING THE SEMICONDUCTOR CRYSTAL

The current carriers in pure crystal material are free electrons and holes that have been liberated, mainly by thermal energy. However, if an impurity atom with five valence electrons in its outer ring is bonded with four silicon atoms, a free electron is produced, as shown in Figure 2-5(a). Further, if the added

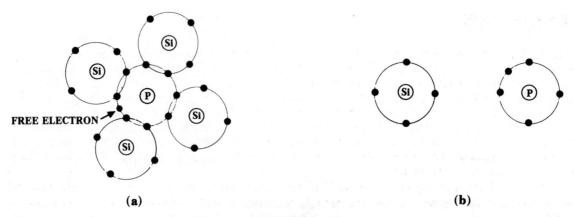

FIGURE 2-5

impurity atom has only three electrons in its outer ring, then covalent bonding of the impurity atom produces an absence of an electron, which is a free hole, as shown in Figure 2-6(a).

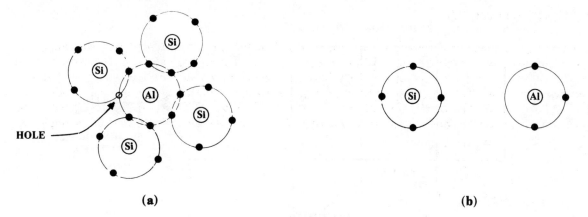

FIGURE 2-6

The atoms with five valence electrons in their outer ring (phosphorus, nitrogen, arsenic, and antimony) are called pentavalent atoms. In Figure 2-5, phosphorous was the impurity used to provide a free electron which carries a negative charge. Hence, by introducing more impurity atoms, more free current carrying electrons are made available. Silicon material that has been doped to have excess free electrons is called, appropriately, N type material.

The atoms with three valence electrons in their outer rings (aluminum, galium, indium, and boron) are called trivalent atoms. For instance, aluminum was the impurity atom used in Figure 2-6 to provide a free hole, which carries a postive charge. Silicon material that has been doped to have excess holes is labeled P type material.

Doping the pure crystal, by the controlled addition of impurity atoms at a ratio of one impurity atom to about 10 million semiconductor atoms greatly increases the number of current carriers and the conductivity of the material. Then, if the doped N type material is connected to a power supply, through ohmic contacts on each side of the crystal, the original current by the electron-hole flow is increased by the electron flow created by the free electrons of the impurity atoms. This increase in free electrons is rarely less than 10^6 times greater than the free electrons in the pure semiconductor crystal. Also, because both the silicon and phosphorus atoms are electrically neutral, the N type crystal is also electrically neutral, and the current flow through the crystal can flow in either direction. Similarly, if the P type material is connected to the power supply, the electron-hole flow of the crystal is established along with the great number of positively charged holes which, like the electrons in the N type material, are at least 10^6 times greater than the original pure crystal hole flow.

In the constructed diode, the majority carriers are effectively the electrons and holes created by the impurity atoms. The minority carriers are the electrons and holes created in the pure crystal by the covalent bonds being broken by thermal and electrical energy.

THE PN JUNCTION

Doping the P and N type materials on the same crystal, so that the crystal lattice is continuous across the junction of the P and N type materials, creates a PN junction diode. As previously stated, the majority carriers in the P type material are holes and the majority carriers in the N type material are electrons. Therefore, once the PN junction is formed, free electrons in the N region are attracted by the positive charges of the P region and cross the junction to combine with the holes in the P region. However, as the electrons move across the junction, the loss of negatively charged electrons on the N side of the junction causes a wall of positive charged ions to build on the edge of the N junction. Similarly, the combining of the electrons with holes on the P side of the junction causes a wall of negative charged ions to build along the edge of the P junction. See Figure 2-7(a).

Eventually, the polarized energy of the ion wall buildup causes further flow of current carriers across the junction to cease and the region between the walls to be depleted of all majority carriers. This region is called, rather obviously, the depletion region. Therefore, with the ionized walls and the depletion region, the diode is in equilibrium. However, the ion walls create a potential difference across the junction that is about 0.25 volts for germanium material and about 0.6 volts for silicon material. The schematic diagram of Figure 2-7(b) shows the junction potential simulated by a battery.

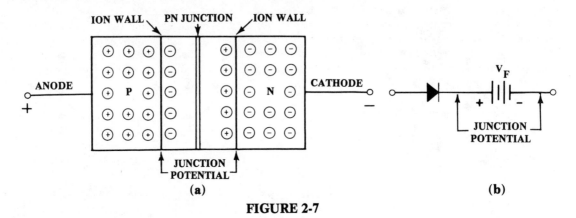

FIGURE 2-7

THE PN JUNCTION DIODE

Solid state semiconductor diodes are PN constructed and, depending on the semiconductor material used, can come close to approximating the ideal diode. For instance, in the forward biased connection, germanium material has a junction potential of about 0.25 volts and silicon has a junction potential of about 0.6 volts with 1 mA of current flow. However, in the reverse biased direction the silicon diode can exceed 1000 megohms while the germanium diodes are usually less than 10 megohms.

FORWARD BIASING THE PN JUNCTION DIODE

Forward biasing the PN junction diode decreases the junction potential and makes it easier for current carriers to cross the junction. It is achieved by connecting the positive voltage of the external power supply to the anode of the device and the negative voltage of the external supply to the cathode of the device. A series resistor RS also is used to limit the current flow through the device. The forward-biased PN junction diodes and the equivalent circuits for silicon and germanium diodes are shown in Figure 2-8.

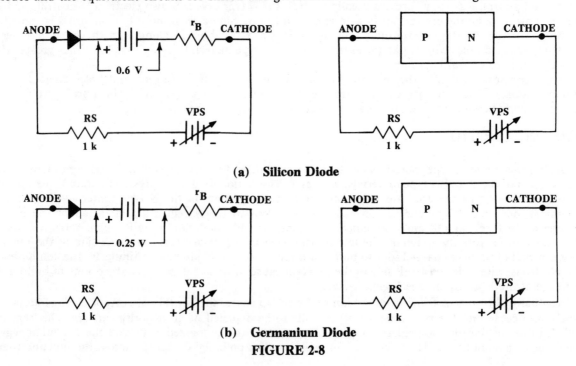

(a) Silicon Diode

(b) Germanium Diode

FIGURE 2-8

28

NOTE: The forward biased bulk resistance r_B, shown in Figure 2-8, will be discussed later.

The flow of current in a forward biased PN junction diode is primarily by electron and hole flow of majority carriers. For instance, electron flow is from the negative terminal to the positive terminal, and hole flow is from the positive terminal to the negative terminal. However, hole flow in one direction and electron flow in the other direction constitutes a current flow in one direction only. And since conventional positive to negative current flow is used, then the forward flow of current through the solid state diode is from positive to negative in the direction of the arrow and subscripted I_F, as shown in Figure 2-9.

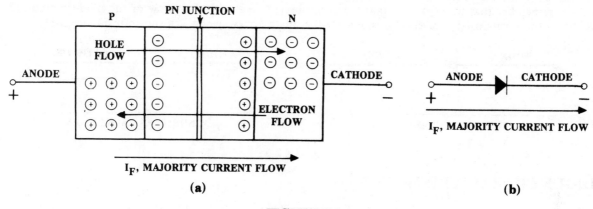

FIGURE 2-9

REVERSE BIASING THE PN JUNCTION DIODE

Reverse biasing the PN junction diodes is achieved by connecting the positive terminal of the power supply to the cathode and the negative terminal to the anode of the diode. Therefore, the reverse biased condition increases the width of the junction and the majority carriers do not have sufficient energy to get across the junction. However, in the reverse biased condition, minority carriers can cross the junction because the positive voltage on the anode attracts all the minority carrier electrons in the P type material across the junction, and the negative voltage on the cathode attracts all the minority carrier holes in the N region across the junction. Hence, the flow of current across the junction in the reverse biased connection is by minority carrier current and, since it includes all possible minority current, the subscript used to denote the reverse saturation current is I_S.

Since I_S is primarily the current flow of free electrons from covalent bonds being broken by thermal energy, its value doubles with every 10° C increase in temperature beyond 25° C(77° F). Also, I_S does not increase appreciably with large reverse biased voltage, but it does have a reverse voltage limitation, where the diode can break down and be destroyed. The reverse biased conditions of the PN junction diode are shown in Figure 2-10, where I_S also assumes a conventional current flow condition of from positive to negative against the diode arrow symbol.

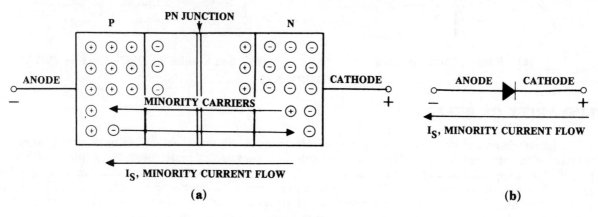

FIGURE 2-10

NOTE: The current through a PN junction is considered to flow in one direction only, because majority current carriers, which are at least 10^6 times more numerous than the minority carriers, flow in one direction only. Additionally, the voltage drop across the PN junction, because of the ionized walls, is about 0.25 volts for germanium and 0.6 volts for silicon materials.

In the reverse biased condition, the flow of current is by minority carriers which, as previously stated, cause the reverse saturation current I_S to double for every 10° C increase in temperature beyond 25° C. Also, the I_S in germanium materials is about 1000 times greater than the I_S in silicon materials, because the valence electrons of germanium are further removed from the atom nucleus, are consequently more loosely bound, and thus can be removed more easily by thermal energy.

Also, remember that the flow of majority carrier current I_F is with the arrow of the diode symbol, and the flow of minority carrier current I_S is against the diode arrow symbol, as shown in Figure 2-11.

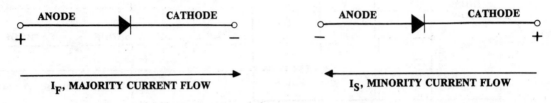

FIGURE 2-11

DIODE CHARACTERISTIC CURVES

The characteristics of electrical components can be graphically illustrated by plotting the voltage across versus the current through the component. For instance, the characteristics of a resistor is linear, as shown in Figure 2-12, because plotting the voltage-current relationship provides a straight line. However, the characteristics of diodes is non-linear in the forward biased directions and relatively linear in the reverse biased direction, prior to voltage breakdown conditions, as shown in Figure 2-12(b).

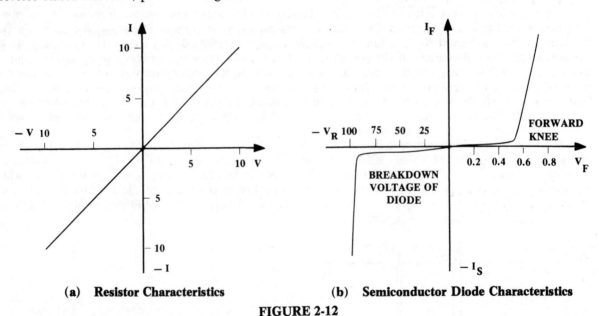

(a) Resistor Characteristics   (b) Semiconductor Diode Characteristics

FIGURE 2-12

LINEARITY OF RESISTORS

The linear characteristic of resistors is predictable because the resistance of resistors usually remains constant with increased voltage. Therefore, current increases proportionately with the applied voltage so that, if 0 V is applied to a series 1 kΩ resistor, 0 mA of current flows; if 1 V is applied to a series 1 kΩ resistor, 1 mA of current flows; and if 5 V is applied to a series 1 kΩ resistor, 5 mA of current flows. Likewise, if -1 V and -5 V are used, then a resultant -1 mA and -5 mA of current flows. The circuit connections and graphs for these connections are shown in Figure 2-13 and Figure 2-14.

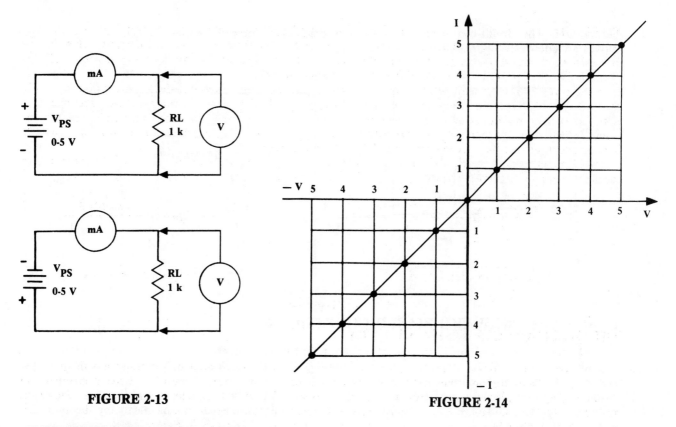

FIGURE 2-13 **FIGURE 2-14**

NONLINEARITY OF DIODES

In the forward-biased connection, the flow of current through the diode is not appreciable until the junction potential is overcome. Therefore, in silicon diodes current flow is extremely small, even with 0.4 volts dropped across the device. However, at about 0.5 V the current begins to increase rapidly and at about 0.6 V to 0.7 V the junction voltage is overcome. For germanium diodes with their lower junction potential, the rapid increase in current flow begins at about 0.25 to 0.3 V. Representative forward biased diode characteristic curves for silicon and germanium diodes are shown in Figure 2-15.

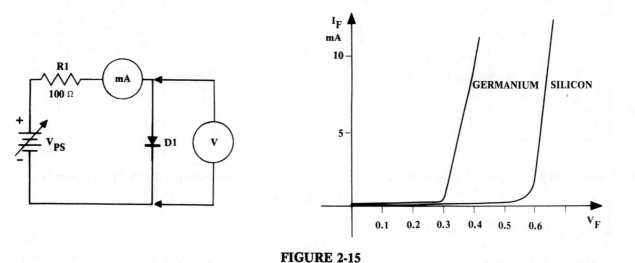

FIGURE 2-15

In the reverse biased connection of the diodes, large increases of voltage do not appreciably increase the saturation current until zener, or avalanche, breakdown in the device occurs. However, reverse biased saturation current I_S for silicon diodes is in the nano-ampere region (10^{-9} A) and for germanium diodes it is in the micro-ampere region (10^{-6} A). Also, silicon can withstand larger breakdown voltages than germanium. The circuit connections and characteristic curves for reverse biased silicon and germanium diodes are shown in Figure 2-16.

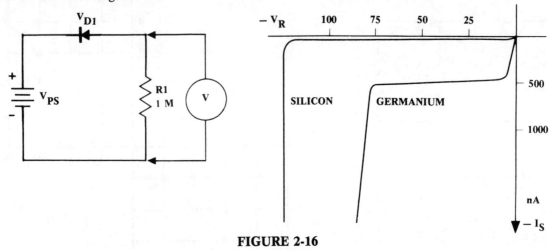

FIGURE 2-16

PRACTICAL FORWARD BIASED CONDITIONS

For silicon diodes, the knee of the curve in the forward-biased direction occurs at a nominal 0.6 V DC, and for germanium diodes it occurs at a nominal 0.25 V DC. However, the current flow through the diodes at the knee of the curve condition is relatively low, usually around 1 mA. As the current increases beyond the knee of the curve, the forward voltage drop across the diode increases slightly. And this slight voltage increase, tenths of volts, with increased current conditions, is caused primarily by the bulk DC resistance of the diode.

If the current through the diodes is kept relatively low, in the 1 mA region as illustrated in Figure 2-17, then the voltage drops of approximately 0.6 V for silicon diodes and 0.25 V for germanium diodes will remain relatively constant regardless of the applied voltage. Note in Figure 2-17, however, that the series resistor values were changed to accommodate the approximate 1 mA current condition. Regardless, even if the series resistors remained the same, the voltage drops across the diodes would not change appreciably. Also, notice that interchanging the position of the diode does not effect the voltage drops across series components.

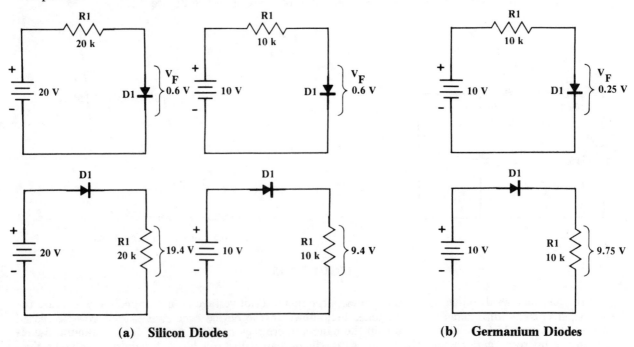

(a) Silicon Diodes (b) Germanium Diodes

FIGURE 2-17

PRACTICAL REVERSE BIASED CONDITIONS

Reverse biasing the diode is accomplished by reversing the polarity of the power supply or, more simply, by turning the diode around. For the circuit of Figure 2-18, if an I_S of 10 nano-amps is assumed for the silicon diode, then the voltage drop across the load resistor of 1 MΩ is 0.01 V. However, if the germanium diode has an I_S of 5 micro-amps, then the voltage drop across the load resistor of 1 MΩ is 5 V.

NOTE: Because the reverse biased diode has a high impedance, voltmeter connections are made across the series resistor. Additionally, a nominal resistor value of 1 MΩ is used because even high impedance voltmeters have impedances of 10 MΩ and increasing the value of the resistor would cause the voltmeter to absorb some of the load current and cause inaccurate voltage readings.

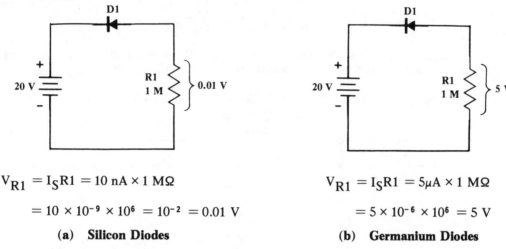

$$V_{R1} = I_S R1 = 10 \text{ nA} \times 1 \text{ M}\Omega$$
$$= 10 \times 10^{-9} \times 10^6 = 10^{-2} = 0.01 \text{ V}$$

(a) Silicon Diodes

$$V_{R1} = I_S R1 = 5\mu\text{A} \times 1 \text{ M}\Omega$$
$$= 5 \times 10^{-6} \times 10^6 = 5 \text{ V}$$

(b) Germanium Diodes

FIGURE 2-18

Therefore, for the example of Figure 2-18, the static resistance of the silicon diode is about 1999 MΩ from $R_D = V_D/I_S = 19.99 \text{ V}/(10 \times 10^{-9}\text{A})$, while the static DC resistance of the germanium diode is 3 MΩ from $R_D = V_D/I_S = 15 \text{ V}/(5 \times 10^{-6}\text{A}) = 3 \times 10^6$. Obviously, silicon diodes have much lower saturation current than germanium diodes and, therefore, a much higher reverse biased resistance.

NOTE: Further analysis of diodes will be restricted to silicon diodes. In practical circuit applications germanium diodes are used only sparingly. Their use is restricted because of their high reverse-saturation, leakage current and they cannot withstand high operating voltages and current conditions like the silicon diodes. Therefore, only in the smaller forward voltage drops of about 0.25 V, as compared to 0.6 V, does the germanium diode excell. For this latter reason, it is used in diode detection applications where the smaller DC voltage drop is necessary for the detection of the higher-percent, modulated carriers.

CIRCUIT ANALYSIS OF SILICON DIODES

Silicon diodes, in the reverse biased connection, have high resistances that, in most instances, can be solved using static (DC) techniques. However, silicon diodes in the forward biased connection require static techniques to solve the DC voltage drops across the diodes and dynamic techniques to solve the AC resistance of the diode. Static techniques are also used to determine the bulk resistance of the diode, which is present in both DC and AC analysis, but which only has importance at higher operating DC current conditions.

STATIC DIRECT CURRENT CONDITIONS

A typical silicon diode has a forward voltage drop of about 0.6 volts with about 1 mA of current through it, but as the current increases, so does the DC voltage forward voltage drop. For instance, a 1N4001 general purpose diode can have a 0.6 volt forward voltage drop at 1 mA, 0.8 volt forward voltage drop at 100 mA, and >1 volt forward voltage drop at 1 ampere. The increase in the forward voltage drop at the lower current conditions is caused primarily by the barrier voltage of the diode, while at the higher current conditions above 100 mA, it is caused by the bulk resistance of the diode. Bulk resistance is calculated from the slope of the forward characteristic curve by dividing the change in voltage ΔV by the corresponding change in current ΔI. However, bulk resistance, to be accurately solved, must have as large a current change as possible so that the barrier voltage at the junction does not weigh too heavily in the calculations. For instance, if 0.7 V at 1 mA and 1.1 V at 1 A are used, the bulk resistance is about 0.4 Ω from:

$$r_B = \Delta V / \Delta I = (1.1\ V - 0.7\ V)/(1\ A - 10\ mA) \approx 0.4\ V/1\ A = 0.4\ \Omega$$

CALCULATING V_F USING PIECEWISE LINEAR ANALYSIS

If the r_B of 0.4 ohms and the junction voltage of 0.7 V are used, the forward voltage drop can be solved from:

$$V_F = V_j + I_F r_B = 0.7\ V + (1\ A \times 0.4\ \Omega) = 0.7\ V + 0.4\ V = 1.1\ V$$

However, at lower than 1 ampere, at 100 mA for instance, the V_F, which is 0.8 V on the characteristic curve of Figure 2-19(a), is calculated at 0.74 V, which shows up as a discrepancy.

$$V_F = V_j + I_F r_B = 0.7\ V + (100\ mA \times 0.4\ \Omega) = 0.7\ V + 0.04\ V = 0.74\ V$$

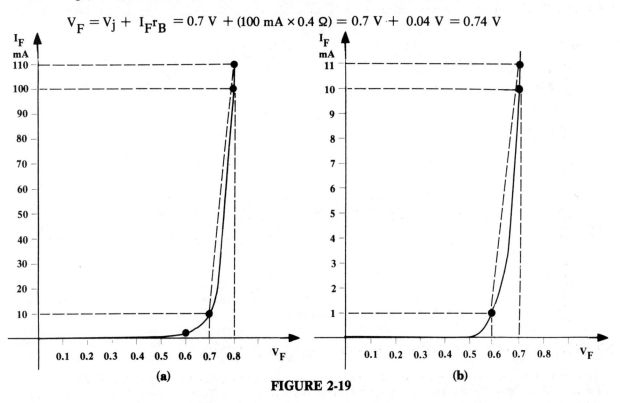

FIGURE 2-19

The discrepancy caused by using large linear samples on non-linear curves can be made practical by using a technique called piecewise linear analysis. For example, drawing a straight line between 1 mA at 0.6 V and 10 mA at 0.7 V on the characteristic curve of Figure 2-19(b) does not differ too greatly from the non-linear curve. Then, if an effective "bulk resistance" can be found for the 1-10 mA region of the curve, the forward voltage drop between 0.6 V and 0.7 V can be calculated. Therefore, a new term r_B' is introduced and will be used to determine, through calculations, the approximate V_F voltage drop across the diode.

For the 1 mA through 11 mA region, r_B' is about 10 ohms from:

$$r_B' = \Delta V / \Delta I \approx (0.7\ V - 0.6\ V)/(11\ mA - 1\ mA) = 0.1\ V/10\ mA = 10\ \Omega$$

Then, using the (lowest voltage) junction voltage of $V_j = 0.6$ V and sample currents of 3 mA, 6 mA, and 9 mA, the V_F at each current is:

$$V_F = V_j + I_F r_B' = 0.6\ V + (3\ mA \times 10\ \Omega) = 0.63\ V$$

$$V_F = V_j + I_F r_B' = 0.6\ V + (6\ mA \times 10\ \Omega) = 0.66\ V$$

$$V_F = V_j + I_F r_B' = 0.6\ V + (9\ mA \times 10\ \Omega) = 0.69\ V$$

Next, if a straight line is drawn between 10 mA and 110 mA, the ΔI is 100 mA and the ΔV is approximately 0.1 V. Therefore, the $r_B' = 1$ ohms from: $r_B' = \Delta V/\Delta I = 0.1\ V/100\ mA = 1\ \Omega$. Then, any current increase between 0.7 V and 0.8 V can be calculated beginning at the "new" junction voltage of 0.7 V. Therefore, using the (lowest voltage) junction voltage of 0.7 V and sample current of 20 mA, 50 mA, and 80 mA, the V_F at each current is:

$$V_F = V_j + I_F r_B' = 0.7\ V + (20\ mA \times 1\ \Omega) = 0.72\ V$$

$$V_F = V_j + I_F r_B' = 0.7\ V + (50\ mA \times 1\ \Omega) = 0.75\ V$$

$$V_F = V_j + I_F r_B' = 0.7\ V + (80\ mA \times 1\ \Omega) = 0.78\ V$$

In Figure 2-20, representative circuits are illustrated showing the voltage drop of 0.65 V at 5 mA and 0.8 V at 100 mA. Again, piecewise linear analysis is used in solving for the individual voltages.

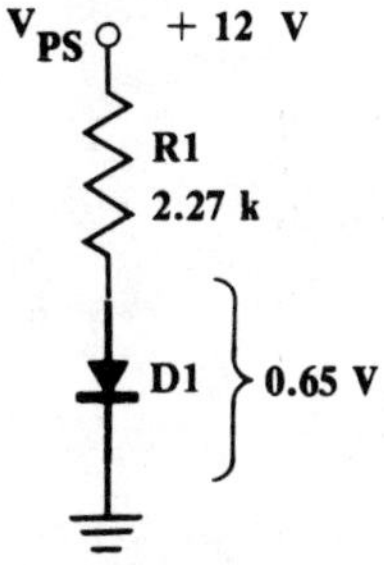

$$I_F = \frac{V_{PS} - V_F}{R1} = \frac{V_{R1}}{R1} = \frac{12\ V - 0.65\ V}{2.27\ k\Omega} = \frac{11.35\ V}{2.27\ k\Omega} = 5\ mA$$

where: $V_F = V_j + V_{r_B'} = V_j + I_F r_B' = 0.6\ V + (5\ mA \times 10\ \Omega)$

$$= 0.6\ V + 0.05\ V \approx 0.65\ V$$

(a)

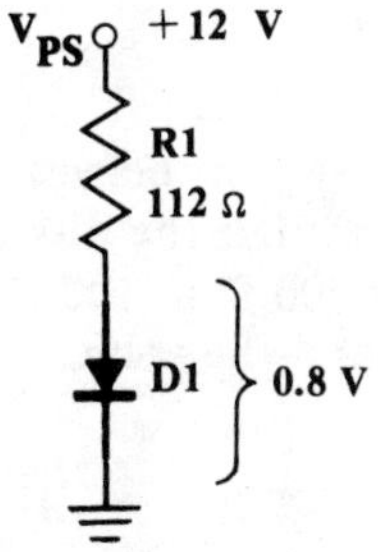

$$I_F = \frac{V_{PS} - V_F}{R1} = \frac{V_{R1}}{R1} = \frac{12\ V - 0.8\ V}{112\ \Omega} = \frac{11.2\ V}{112\ \Omega} = 100\ mA$$

where: $V_F = V_j + V_{r_B'} = V_j + I_F r_B' = 0.7\ V + (100\ mA \times 1\ \Omega)$

$$= 0.7\ V + 0.1\ V = 0.8\ V$$

(b)

FIGURE 2-20

STATIC RESISTANCE OF FORWARD BIASED DIODES

The static forward biased resistance of diodes is a parameter that has little importance other than an exercise in Ohm's law. For instance, with reference to the circuits of Figure 2-20, the forward-biased static resistance at 5 mA is 130 ohms; at 50 mA, it is 15 ohms, and at 100 mA, it is 8 ohms; and each describes a single point on the characteristic curve of the diode. The static resistance at each of the operating points (quiescent points) is solved from:

$$R_D = V_F/I_F = 0.65\ V/5\ mA = 130\ \Omega$$

$$R_D = V_F/I_F = 0.75\ V/50\ mA = 15\ \Omega$$

$$R_D = V_F/I_F = 0.8\ V/100\ mA = 8\ \Omega$$

OHMMETER TEST TECHNIQUES

The simplest way to test a diode is with an ohmmeter. A diode that is working will show a low forward biased resistance and a high reverse biased resistance. Low resistance in both directions usually means the diode is shorted, and high resistance in both directions means that the diode is probably open. Each R × setting on the ohmmeter will provide different resistor readings because at each setting a different current will flow through the diode. At best, ohmmeter testing of diodes can only provide a go or no-go check.

DIRECT CURRENT LOAD LINES

Direct curent load lines provide graphical solutions to series diodes and resistor voltage drops, where both Kirchhoff's and Ohm's laws are satisfied. For instance, if the diode characteristic curve is plotted, and a straight line is drawn between the extreme DC voltage and current conditions of the circuit, then the straight line intersects the characteristic curve at the DC operating point of the diode. Therefore, the DC operating, or quiescent, point (Q point) describes the voltage across and the current through the diode for the power supply voltage and series resistor used. The straight line that intersects the diode characteristic curve is called the DC load line.

The extreme load line conditions of a series resistor-diode circuit exist when $I_F = 0$ mA and when $V_F = 0$ V. Therefore, if R1 = 100 Ω and VPS = 3 V, then I(MAX) occurs when $V_F = 0$ V and I(MAX) = VPS/R1 = 3 V/100 Ω = 30 mA, a condition that occurs when D1 is temporarily shorted. $I_F = 0$ mA occurs if D1 is temporarily disconnected, and then $V_{R1} = 0$ V. These two extreme conditions of $I_F = 0$ mA and $V_F = 0$ V for the circuit of Figure 2-21 are shown calculated.

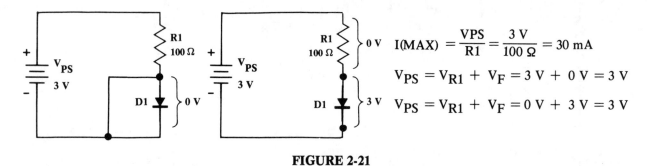

FIGURE 2-21

Once the load line is drawn between the two extreme conditions, the point of intersection between the load line and the diode characteristic curve will show the voltage drop and current through the diode for the load resistor used. Therefore, with reference to Figure 2-21, if R1 is changed from 200 Ω to 100 Ω and to 60 Ω, the I(MAX) will be 15 mA, 30 mA, and 50 mA with each respective change, and the voltage drop

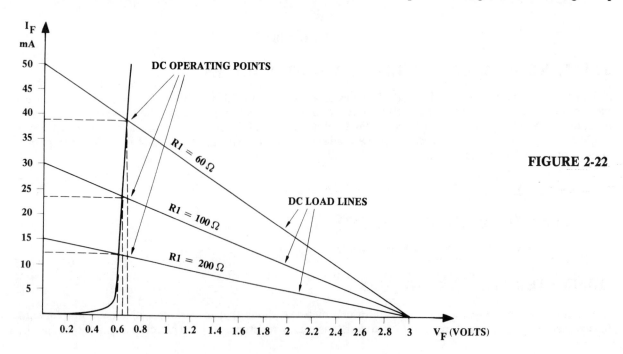

FIGURE 2-22

across the diode will change from 0.715 V to 0.73 V to 0.75 V. The characteristic curve and the respective load lines for these circuit conditons are shown in Figure 2-22 and the circuits and calculated values are given in conjunction with Figure 2-23.

LOAD LINE — CIRCUIT CALCULATIONS AND VERIFICATION

Load lines provide a graphical analysis of voltage drops around circuits. The calculated values for the circuits described in the load line analysis follow to provide a verification of the graphical solutions.

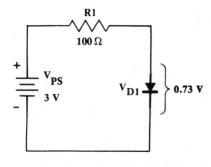

$I(MAX) = V_{PS}/R1 = 3\,V/200\,\Omega = 15\,mA$

$V_{R1} = V_{PS} - V_F = 3\,V - 0.715\,V = 2.285\,V$

$I_F = I_{R1} = (V_{PS} - V_F)/R1 = V_{R1}/R1 = 2.285\,V/200\,\Omega = 11.425\,mA$

$V_{PS} = V_F + I_F R1 = 0.715\,V + (11.425\,mA \times 200\,\Omega) = 3\,V$

$V_{PS} = V_F + V_{R1} = 0.715\,V + 2.285\,V = 3\,V$

(a)   R1 = 200 ohms

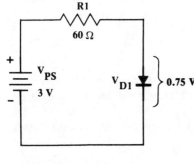

$I(MAX) = V_{PS}/R1 = 3\,V/100\,\Omega = 30\,mA$

$V_{R1} = V_{PS} - V_F = 3\,V - 0.73\,V = 2.27\,V$

$I_F = I_{R1} = V_{PS} - V_F/R1 = V_{R1}/R1 = 2.27\,V/100\,\Omega = 22.7\,mA$

$V_{PS} = V_F + I_F R1 = 0.73\,V + (22.7\,mA \times 100\,\Omega) = 3\,V$

$V_{PS} = V_F + V_{R1} = 0.73\,V + 2.27\,V = 3\,V$

(b)   R1 = 100 ohms

$I(MAX) = V_{PS}/R1 = 3\,V/60\,\Omega = 50\,mA$

$V_{R1} = V_{PS} - V_F = 3\,V - 0.75\,V = 2.25\,V$

$I_F = I_{R1} = V_{PS} - V_F/R1 = V_{R1}/R1 = 2.25\,V/60\,\Omega = 37.5\,mA$

$V_{PS} = V_F + I_F R1 = 0.75\,V + (37.5\,mA \times 60\,\Omega) = 3\,V$

$V_{PS} = V_F + V_{R1} = 0.75\,V + 2.25\,V = 3\,V$

(c)   R1 = 60 ohms

FIGURE 2-23

DYNAMIC RESISTANCE OF DIODES

The dynamic resistance of diodes is the AC resistance of the diode at the operating point determined by the DC current conditions. Therefore, an AC signal developed across a diode at a particular DC operating point will deviate about the operating point and, therefore, is determined by the change in voltage ΔV, divided by the change in current ΔI. Therefore, $r_{AC} = \Delta V/\Delta I$, where the dynamic resistance of diodes includes both bulk and junction resistance, $r_{AC} = r_j + r_B$. However, bulk resistance does not change appreciably with wide changes in the DC operating current, but the junction resistance does and it changes inversely proportional to changes in the DC operating current.

The dynamic resistance can be solved in two ways: from the $\Delta V/\Delta I$ data obtained off the forward-biased characteristic curve, or from $r_{AC} = r_j + r_B$, where r_B is considered as a constant and r_j is solved from Shockley's junction resistance relationship where $r_j = 26\ mV/I_F$, and where I_F is in milliamperes.

GRAPHICAL SOLUTION OF THE AC DYNAMIC RESISTANCE OF DIODES

The AC dynamic resistance of diodes is obtained through graphical analysis techniques, when a straight line is drawn tangent to the characteristic curve at the DC operating current of the diode. Then, if a relatively small current change is used, that is equally above and below the DC current condition, and the corresponding voltage conditions obtained, the r_{AC} can be solved from $r_{AC} = \Delta V/\Delta I$.

For example, at the 40 mA DC operating current, the characteristic curve is fairly linear. If the change in current is taken at ± 5 mA of the operating 40 mA condition from 35 mA to 45 mA, then the corresponding 45 and 35 mA voltage conditions are obtained from $r_{AC} = \Delta V/\Delta I = (0.765\ V - 0.75\ V)/(45\ mA - 35\ mA) = 1.5\ \Omega$. This example is illustrated in Figure 2-24(a).

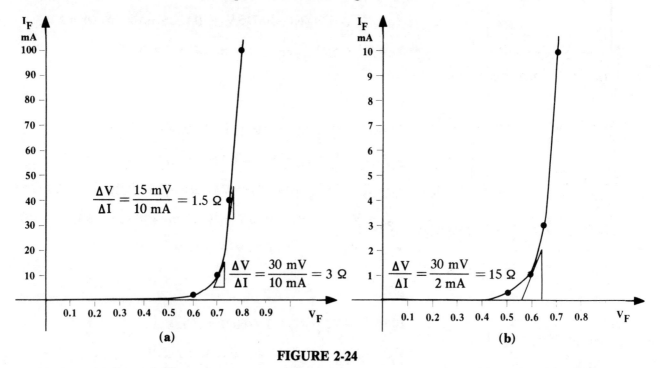

FIGURE 2-24

Likewise, if the DC operating current is changed to 10 mA and the change in current is maintained at ± 5 mA, then the $r_{AC} = \Delta V/\Delta I = (0.72 - 0.69\ V)/(15\ mA - 5\ mA) = 3\ \Omega$. And if the DC operating current is changed to 1 mA and the change in current is now ± 1 mA, the $r_{AC} = \Delta V/\Delta I \approx (0.62 - 0.58)/(2\ mA - 0.0\ mA) = 15\ \Omega$. The 10 mA operating condition is illustrated in Figure 2-24(a), and the 1 mA operating condition is illustrated in Figure 2-24(b).

A. 40 mA DC Condition:

$$r_{AC} = \frac{\Delta V}{\Delta I} = \frac{0.765\ V - 0.75\ V}{45\ mA - 35\ mA} = \frac{0.015\ V}{10\ mA} = \frac{15\ mV}{10\ mA} = 1.5\ \Omega$$

B. 10 mA DC Condition:

$$r_{AC} = \frac{\Delta V}{\Delta I} = \frac{0.72\ V - 0.69\ V}{15\ mA - 5\ mA} = \frac{0.03\ V}{10\ mA} = \frac{30\ mV}{10\ mA} = 3\ \Omega$$

C. 1 mA DC Condition:

$$r_{AC} = \frac{\Delta V}{\Delta I} \approx \frac{0.62\ V - 0.59\ V}{2\ mA - 0\ mA} = \frac{0.03\ V}{2\ mA} = \frac{30\ mV}{2\ mA} = 15\ \Omega$$

MATHEMATICAL SOLUTION OF THE AC DYNAMIC RESISTANCE OF DIODES

The graphical technique for finding the dynamic resistance of the diode was to draw a line tangent to the DC operating point on the curve and solve r_{AC} from the inverse slope of the line, where $r_{AC} = \Delta V/\Delta I$. However, mathematically, the slope of the line tangent to the DC operating point on the curve can be found by taking the derivative of the theoretical diode equation which produces:

$$\frac{dI}{dV} = \frac{q(I_F + I_S)}{\eta kT}, \text{ and since } r_{AC} = \Delta V/\Delta I, \text{ and } r_{AC} \approx r_j \text{ at 1 mA, then}$$

$$r_j = \frac{\Delta V}{\Delta I} \cong \frac{\eta kT/q}{I_F} \qquad \text{where: } I_F \gg I_S \text{ and } \eta = 1$$

$$r_j = \frac{\eta kT/q}{I_F} = \frac{1.38 \times 10^{-23} j/°k \times 298\ C/1.6 \times 10^{-19°}\ C}{I_F}$$

$$r_j = \frac{25.7 \times 10^{-3}\ V}{I_F} \approx \frac{26\ mV}{I_F}$$

NOTE: $0°k = 273°C$ and room temperature is considered $77°F$, which equals $25°C$ and, therefore, $T = 273°C + 25°C = 298°C$. Also, $k = 1.38 \times 10^{-23} j/°k$ is called the Bolzmann's constant and $q = 1.6 \times 10^{-19} C$ is the electron charge of the theoretical diode equation. In addition, $\eta = 1$ is a condition for silicon and germanium diodes beyond the knee of the curve where r_j is being calculated. However, while $r_j = 26\ mV/I_F$ in theory, the practical r_j resistance can vary over a 2 : 1 range, from $r_j = 26\ mV/I_F$ to $r_j = 52\ mV/I_F$.

APPLYING THE r_j FORMULA

In the following, the $r_j = 26\ mV/I_F$ formula is applied to four circuits. Each has a different operating current and r_B, necessary for solving the total AC resistance of the diode, is given at $0.4\ \Omega$. Although of no significant consequence at low current conditions, the $0.4\ \Omega$ bulk resistance is important at higher current conditions.

$$I_F = \frac{V_{PS} - V_F}{R1} = \frac{12\ V - 0.6\ V}{11.4\ k\Omega} = \frac{11.4\ V}{11.4\ k\Omega} = 1\ mA$$

$$r_j = 26\ mV/I_F = 26\ mV/1\ mA = 26\ \Omega$$

$$r_{AC} = r_j + r_B = 26\ \Omega + 0.4\ \Omega = 26.4\ \Omega$$

(a)

$$I_F = \frac{V_{PS} - V_F}{R1} = \frac{12\ V - 0.7\ V}{1130\ \Omega} = \frac{11.3\ V}{1130\ \Omega} = 10\ mA$$

$$r_j = 26\ mV/I_F = 26\ mV/10\ mA = 2.6\ \Omega$$

$$r_{AC} = r_j + r_B = 2.6\ \Omega + 0.4\ \Omega = 3\ \Omega$$

(b)

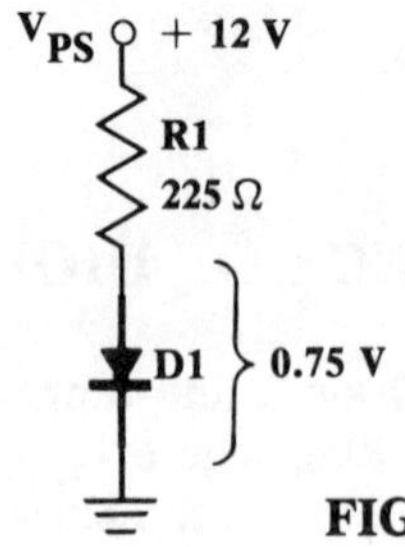

$$I_F = \frac{V_{PS} - V_F}{R1} = \frac{12\text{ V} - 0.75\text{ V}}{225\ \Omega} = \frac{11.25\text{ V}}{225\ \Omega} = 50\text{ mA}$$

$$r_j = 26\text{ mV}/I_F = 26\text{ mV}/50\text{ mA} = 0.52\ \Omega$$

$$r_{AC} = r_j + r_B = 0.52\ \Omega + 0.4\ \Omega = 0.92\ \Omega$$

FIGURE 2-25(c)

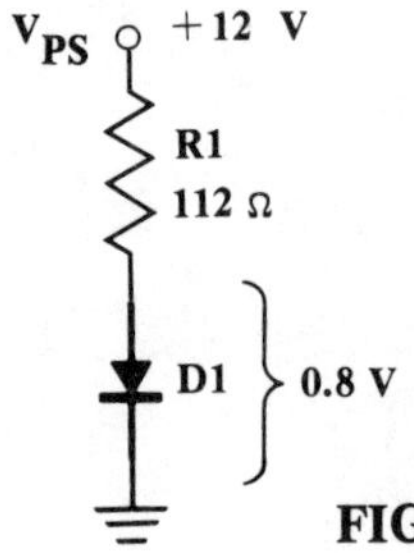

$$I_F = \frac{V_{PS} - V_F}{R1} = \frac{12\text{ V} - 0.8\text{ V}}{112\ \Omega} = \frac{11.2\text{ V}}{112\ \Omega} = 100\text{ mA}$$

$$r_j = 26\text{ mV}/I_F = 26\text{ mV}/100\text{ mA} = 0.26\ \Omega$$

$$r_{AC} = r_j + r_B = 0.26\ \Omega + 0.4\ \Omega = 0.66\ \Omega$$

FIGURE 2-25(d)

NOTE: Graphical methods are tedious and time-consuming and not practical to use. Consequently, in the remaining portion of this book, only the mathematical method will be used to solve for r_{AC}.

SMALL SIGNAL DIODE APPLICATIONS

The DC and AC resistance of forward biased diodes are, essentially, independent of each other. The DC or static resistance is dictated by Ohm's law, and the AC or dynamic resistance is dictated by the slope of the curve about the DC operating point of the forward biased characteristic curve.

The DC resistance is found by dividing the DC voltage drop across the diode by the DC current flow through it, and its only importance is that it helps explain the Ohm's law condition. However, AC resistance is important with reference to solving the amount of small signal that is developed across the diode with regard to series resistor RS. Hence, the $r_j = 26\text{ mV}/I_F$ formula will be used for conditions of 1 mA, 10 mA, 50 mA, and 100 mA to illustrate the point. Also, since the mathematical method of solution is being used, the bulk resistance of approximately 0.4 ohms must be included in the calculations.

A. 1 mA Quiescent Current Conditions: At 1 mA of DC current through the diode, $r_{AC} = 26.4$ ohms, and if 100 mVp-p of 1 kHz input signal is used, 21.83 mVp-p is developed across the diode and the remaining portion of the signal is dropped across the 100 ohm RS resistor. The circuit and equivalent circuit and calculations are given in Figure 2-26.

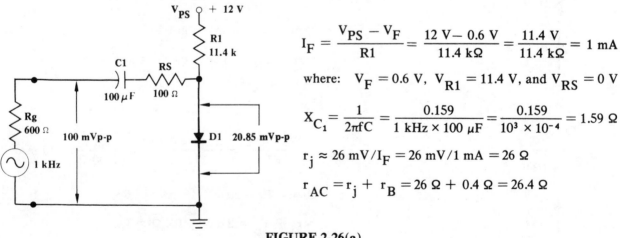

$$I_F = \frac{V_{PS} - V_F}{R1} = \frac{12\text{ V} - 0.6\text{ V}}{11.4\text{ k}\Omega} = \frac{11.4\text{ V}}{11.4\text{ k}\Omega} = 1\text{ mA}$$

where: $V_F = 0.6\text{ V}$, $V_{R1} = 11.4\text{ V}$, and $V_{RS} = 0\text{ V}$

$$X_{C_1} = \frac{1}{2\pi fC} = \frac{0.159}{1\text{ kHz} \times 100\ \mu F} = \frac{0.159}{10^3 \times 10^{-4}} = 1.59\ \Omega$$

$$r_j \approx 26\text{ mV}/I_F = 26\text{ mV}/1\text{ mA} = 26\ \Omega$$

$$r_{AC} = r_j + r_B = 26\ \Omega + 0.4\ \Omega = 26.4\ \Omega$$

FIGURE 2-26(a)

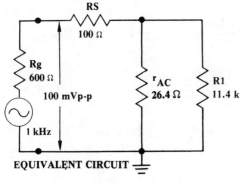

EQUIVALENT CIRCUIT

$r_{eq} = r_{AC} \parallel R1 = 26.4\ \Omega \parallel 11.4\ k\Omega \approx 26.34\ \Omega$

NOTE: In most instances $r_{eq} = r_{AC}$. Therefore,

$r_{eq} \approx r_{AC} \approx 26.4\ \Omega$

FIGURE 2-26(b)

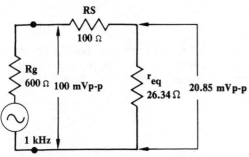

EQUIVALENT CIRCUIT INCLUDING R1

$v_{r_{eq}} = \dfrac{v_{in} \times r_{eq}}{r_{eq} + RS} = \dfrac{100\ mVp\text{-}p \times 26.34\ \Omega}{26.34\ \Omega + 100\ \Omega} = 20.85\ mVp\text{-}p$

$v_{RS} = v_{in} - v_{r_{eq}}\ p\text{-}p = 100\ mVp\text{-}p - 20.85\ mVp\text{-}p$

$= 79.15\ mVp\text{-}p$

$v_{RS} = \dfrac{v_{in} \times RS}{r_{eq} + RS} = \dfrac{100\ mVp\text{-}p \times 100\ \Omega}{26.34\ \Omega + 100\ \Omega} = 79.15\ mVp\text{-}p$

FIGURE 2-26(c)

NOTE: X_{C_1} at 1 kHz is 1.59 Ω and considered as a short circuit in the equivalent circuit. Also, R1 provides little AC loading and $r_{eq} \approx r_{AC}$ of the diode.

B. 10 mA Quiescent Current Conditions: At 10 mA of DC current flow thorough the diodes, $r_{AC} = 3$ ohms and the bulk resistance contributes significantly to the total resistance. The circuit, equivalent circuits, and calculations are given showing the AC signal distribution for the 100 mVp-p, 1 kHz, input signal. Note that the decreased value of R1 has only a slight effect in lowering r_{eq} of the circuit.

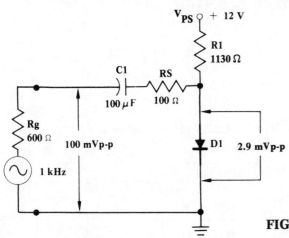

$I_F = \dfrac{V_{PS} - V_F}{R1} = \dfrac{12\ V - 0.7\ V}{1130\ \Omega} = \dfrac{11.3\ V}{1130\ \Omega} = 10\ mA$

where: $V_F = 0.7\ V,\quad V_{R1} = 11.3\ V,\quad V_{RS} = 0\ V,$

and $X_{C_1} = 1.59\ \Omega$

$r_j = 26\ mV / I_F = 26\ mV/10\ mA = 2.6\ \Omega$

$r_{AC} = r_j + r_B = 2.6\ \Omega + 0.4\ \Omega = 3\ \Omega$

FIGURE 2-27(a)

EQUIVALENT CIRCUIT

$r_{eq} = r_{AC} \parallel R1 = 3\ \Omega \parallel 1130\ \Omega = 2.99\ \Omega$

Therefore, $r_{eq} \approx r_{AC} \approx 3\ \Omega$

FIGURE 2-27(b)

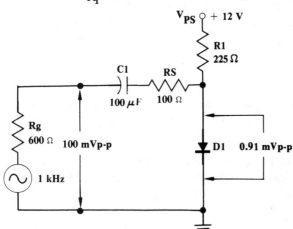

$$v_o = \frac{v_{in} \times r_{eq}}{r_{eq} + RS} = \frac{100 \text{ mVp-p} \times 2.99 \text{ }\Omega}{2.99 \text{ }\Omega + 100 \text{ }\Omega} \cong 2.9 \text{ mVp-p}$$

$$v_{RS} = \frac{v_{in} \times RS}{r_{eq} + RS} = \frac{100 \text{ mVp-p} \times 100 \text{ }\Omega}{2.99 \text{ }\Omega + 100 \text{ }\Omega} \cong 97.1 \text{ mVp-p}$$

$$v_{RS} = v_{in} - v_{r_{eq}} = 100 \text{ mVp-p} - 2.9 \text{ mVp-p}$$
$$= 97.1 \text{ mVp-p}$$

FIGURE 2-27(c)

C. 50 mA Quiescent Current Conditions: At 50 mA of DC current flow through the diode, $r_{AC} = 0.92 \text{ }\Omega$ and bulk resistance contributes a major portion of the total resistance. The circuit, equivalent circuit, and calculations are given in Figure 2-28. Note that the further reduced value of R1 does little to reduce the r_{eq} of the circuit.

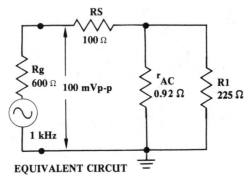

$$I_F = \frac{V_{PS} - V_F}{R1} = \frac{12 \text{ V} - 0.75 \text{ V}}{225 \text{ }\Omega} = \frac{11.25 \text{ V}}{225 \text{ }\Omega} = 50 \text{ mA}$$

where: $V_F = 0.75 \text{ V}$, $V_{R1} = 11.25 \text{ V}$, $V_{RS} = 0 \text{ V}$, and

$$X_{C_1} = 1.59 \text{ }\Omega$$

$$r_j \approx 26 \text{ mV}/I_F = 26 \text{ mV}/50 \text{ mA} = 0.52 \text{ }\Omega$$

$$r_{AC} = r_j + r_B = 0.52 \text{ }\Omega + 0.4 \text{ }\Omega = 0.92 \text{ }\Omega$$

FIGURE 2-28(a)

$$r_{eq} = r_{AC} \parallel R1 = 0.92 \text{ }\Omega \parallel 225 \text{ }\Omega = 0.916 \text{ }\Omega$$

Therefore: $r_{eq} \approx r_{AC} \approx 0.92 \text{ }\Omega$

FIGURE 2-28(b)

$$v_o = \frac{v_{in} \times r_{eq}}{r_{eq} + RS} = \frac{100 \text{ mV} \times 0.92 \text{ }\Omega}{0.92 \text{ }\Omega + 100 \text{ }\Omega} \approx 0.91 \text{ mVp-p}$$

$$v_{RS} = \frac{v_{in} \times RS}{r_{eq} + RS} = \frac{100 \text{ mV} \times 100 \text{ }\Omega}{0.92 \text{ }\Omega + 100 \text{ }\Omega} \cong 99.09 \text{ mVp-p}$$

$$v_{RS} = v_{in} - v_o = 100 \text{ mVp-p} - 0.91 \text{ mVp-p} = 99.09 \text{ mVp-p}$$

FIGURE 2-28(c)

D. 100 mA Quiescent Current Conditions: At 100 mA of DC current flow through the diode, $r_{AC} = 0.66$ ohms with bulk resistance being most of the total resistance. The circuit, equivalent circuit, and calculations are given in Figure 2-29. Again, note that further reduction of R1 only has minor effect on the r_{eq} of the circuit.

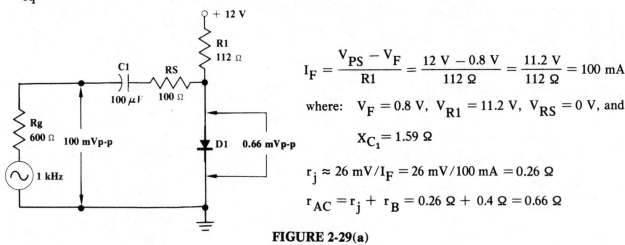

$$I_F = \frac{V_{PS} - V_F}{R1} = \frac{12\text{ V} - 0.8\text{ V}}{112\ \Omega} = \frac{11.2\text{ V}}{112\ \Omega} = 100\text{ mA}$$

where: $V_F = 0.8$ V, $V_{R1} = 11.2$ V, $V_{RS} = 0$ V, and

$$X_{C_1} = 1.59\ \Omega$$

$$r_j \approx 26\text{ mV}/I_F = 26\text{ mV}/100\text{ mA} = 0.26\ \Omega$$

$$r_{AC} = r_j + r_B = 0.26\ \Omega + 0.4\ \Omega = 0.66\ \Omega$$

FIGURE 2-29(a)

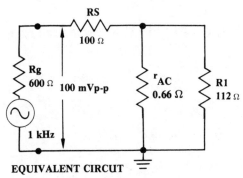

EQUIVALENT CIRCUIT

$$r_{eq} = r_{AC} \,/\!/\, R1 = 0.66\ \Omega \,/\!/\, 112\ \Omega \approx 0.656\ \Omega$$

Therefore: $r_{eq} \approx r_{AC} \approx 0.66\ \Omega$

FIGURE 2-29)b)

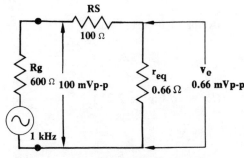

EQUIVALENT CIRCUIT INCLUDING R1

$$v_o = \frac{v_{in} \times r_{eq}}{r_{eq} + RS} = \frac{100\text{ mVp-p} \times 0.66\ \Omega}{0.66\ \Omega + 100\ \Omega} = 0.656\text{ mVp-p}$$

$$v_{RS} = \frac{v_{in} \times RS}{r_{eq} + RS} = \frac{100\text{ mVp-p} \times 100\ \Omega}{0.66\ \Omega + 100\ \Omega} \cong 99.34\text{ mVp-p}$$

$$v_{RS} = v_{in} - v_o = 100\text{ mVp-p} - 0.66\text{ mVp-p} = 99.34\text{ mVp-p}$$

FIGURE 2-29(c)

CALCULATING $r_{eq} \approx r_{AC}$ FROM MEASURED SIGNAL VOLTAGES

In a laboratory situation, if the series resistance RS is known, the input signal voltage can be adjusted and the signal across the diode measured to find the approximate AC resistance of the diode. In other words, the ratio of the two signal voltages equals the ratio of the two resistances. For instance, assuming that the signal voltage developed across the equivalent resistance is 10 mVp-p, the series resistance is 100 Ω, and the input voltage is 100 mVp-p, as shown in Figure 2-30, then 90 mVp-p is dropped across the 100 Ω series resistor. Since 10 mVp-p is dropped across the equivalent resistor, r_{eq} must be 11.1 Ω, because the same signal current flows through the series components. R1 at 4.4 kΩ is safely ignored in the AC calculations, but it provides an r_j of about 10 ohms, where $I_F = 2.6$ mA.

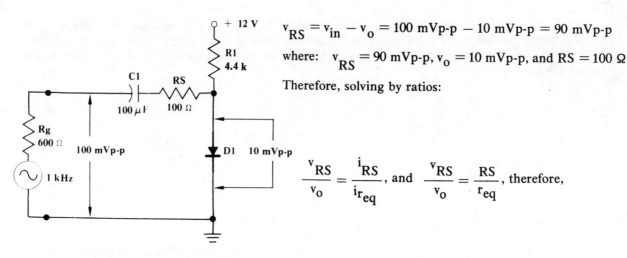

$$v_{RS} = v_{in} - v_o = 100 \text{ mVp-p} - 10 \text{ mVp-p} = 90 \text{ mVp-p}$$

where: $v_{RS} = 90$ mVp-p, $v_o = 10$ mVp-p, and RS = 100 Ω

Therefore, solving by ratios:

$$\frac{v_{RS}}{v_o} = \frac{i_{RS}}{i_{r_{eq}}}, \text{ and } \frac{v_{RS}}{v_o} = \frac{RS}{r_{eq}}, \text{ therefore,}$$

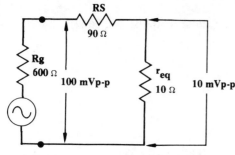

EQUIVALENT CIRCUIT INCLUDING R1

$$r_{eq} = \frac{v_o \times RS}{v_{RS}} = \frac{10 \text{ mVp-p} \times 100 \text{ Ω}}{90 \text{ mVp-p}} \approx 11.11 \text{ Ω}$$

Therefore: $r_{eq} \approx r_{AC} // R1 = 11.11 \text{ Ω} // 4.4 \text{ kΩ} \approx 11.1 \text{ Ω}$

Hence, $r_{eq} \approx r_{AC} \approx 11.1$ Ω

FIGURE 2-30

FURTHER EXAMPLES OF THE r_{eq} FORMULA

A. For 1 mA current conditions:

$$r_{eq} = \frac{v_o \times RS}{v_{RS}} = \frac{20.85 \text{ mVp-p} \times 100 \text{ Ω}}{79.15 \text{ mVp-p}} \cong 26.34 \text{ Ω} \qquad \text{(Refer to Figure 2-26.)}$$

B. For 10 mA DC current conditions:

$$r_{eq} = \frac{v_o \times RS}{v_{RS}} = \frac{2.9 \text{ mVp-p} \times 100 \text{ Ω}}{97.1 \text{ mVp-p}} \cong 2.99 \text{ Ω} \qquad \text{(Refer to Figure 2-27.)}$$

C. For 50 mA DC current conditions:

$$r_{eq} = \frac{v_o \times RS}{v_{RS}} = \frac{0.91 \text{ Vp-p} \times 100 \text{ Ω}}{99.09 \text{ mVp-p}} \cong 0.92 \text{ Ω} \qquad \text{(Refer to Figure 2-28.)}$$

D. For 100 mA DC current conditions:

$$r_{eq} = \frac{v_o \times RS}{v_{RS}} = \frac{0.66 \text{ mVp-p} \times 100 \text{ Ω}}{99.34 \text{ mVp-p}} \cong 0.66 \text{ Ω} \qquad \text{(Refer to Figure 2-29.)}$$

SUMMARY

The DC voltage drop across silicon diodes is about 0.6 V with 1 mA of current through them and about 0.8 V with 100 mA through them. The increase of voltage dropped across the device with increased current through it is caused mainly by bulk resistance. However, the amount of signal dropped across the diode, with reference to the applied signal, is caused by the dynamic resistance of the diode, which is both bulk and junction resistance. The bulk resistance is relatively constant and is one reason why the DC voltage across the diode changes with current flow through the diode, while the junction resistance can be varied by changing the amount of DC current flow through the diode. The dynamic impedance of the diode can be found by graphical or mathematical methods. Using the mathematical formula of $r_j \approx 26$ mV/I_F is the more practical of the two methods. Also, load lines intersect the characteristic curve of the diode at its operating, or DC quiescent point, and that point describes the ohmic value of the device, the foward-biased voltage V_F, and the current flow though to device I_F.

LABORATORY EXERCISE

OBJECTIVES
Investigate the DC characteristics of silicon diodes.

LIST OF MATERIALS AND EQUIPMENT
1. Diode: General purpose silicon diode such as the 1N4001-1N4004
2. Resistors (one each): 100 Ω   330 Ω   470 Ω   1 kΩ   4.7 kΩ   10 MΩ   1 kΩ pot
3. Capacitor: 10 µF (one)
4. Power Supply: 24 V DC Variable
5. Voltmeter
6. Oscilloscope

PROCEDURE
Section I: Fixed-Voltage, Forward-Biased Condition

1. Connect the circuit of Figure 2-31(a). The diode is forward biased and the power supply is set at 10 V DC.

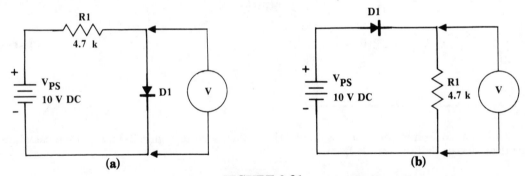

FIGURE 2-31

2. Measure the DC voltage drop across the forward biased diode D1, as shown in Figure 2-31(a), and then calculate the DC voltage drop across R1 from:

$$V_{R1} = V_{PS} - V_F, \quad \text{where:} \quad V_{PS} = 10 \text{ V DC and } V_{D1} = V_F$$

NOTE: The DC voltage drops across diodes are given, theoretically, at 0.7 volts, but in applications they will vary from 0.5 V to 0.7 V, where the higher DC voltage drop across the diode occurs with higher operating current. Also, all measurements are made with respect to ground to avoid ground loops and the resultant incorrect measurements. Therefore, when the power supply and diode voltages of Figure 2-31(a) are measured, the voltage drop across the R1 resistor is calculated by subtracting the diode voltage from the power supply voltage.

3. Calculate the current flow through the series R1 resistor and forward biased D1 diode from:

$$I_F = I_{R1} = I_{D1} = V_{R1}/R1, \quad \text{where:} \quad R1 = 4.7 \text{ k}\Omega$$

NOTE: A current meter could be used to measure the current directly, but the practice is to measure the voltage drop across a known resistor and then calculate the current using Ohm's law. This technique is used to eliminate having to break the circuit and insert an ammeter, which is a difficult approach to obtaining circuit currents in all but breadboarded circuit conditions.

4. Insert the calculated and measured values, as indicated, into Table 2-1(a).

5. Exchange the positions of R1 and D1, as shown in Figure 2-31(b), so that the voltage drop across R1 can be measured and the ground loop possibilities avoided. This exchange will verify the calculated result obtained for V_{R1}.

6. Measure the DC voltage drop across R1 and calculate the volage drop across diode D1 from:

$$V_F = V_{PS} - V_{R1} \quad \text{where:} \quad V_F = V_{D1}$$

7. Insert the calculated and measured values, as indicated, into Table 2-1(b).

TABLE 2-1	V_{PS}	(a)			(b)		(c)		
		V_F	V_{R1}	I_F	V_{R1}	V_F	V_F	V_{R1}	I_F
CALCULATED	10 V		9.37	2mA		.63		9.24	.77
MEASURED	10V	.63			9.54		10.02		

8. Connect the circuit of Figure 2-32, where the value of the R1 resistor is reduced to 330 Ω.

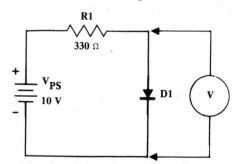

FIGURE 2-32

9. Measure the DC voltage drop across diode D1, as illustrated in Figure 2-32, and then calculate the DC voltage drop across the R1 resistor from:

$$V_{R1} = V_{PS} - V_F \quad \text{where:} \quad V_{PS} = 10 \text{ V DC} \quad \text{and} \quad V_F = V_{D1}$$

10. Calculate the forward current flow through R1 and D1 from:

$$I_F = I_{R1} = I_{D1} = V_{R1}/R1$$

11. Insert the calculated and measured values, as indicated, into Table 2-1(c).

Section II: Fixed-Voltage, Reverse-Biased Condition

1. Connect the circuit as shown in Figure 2-33. The diode is reverse biased and the power supply is set at 20 volts.

2. Measure the voltage drop across the 10 MΩ resistor R1 and calculate the approximate reverse current flow through the diode from:

$$I_S = I_{D1} = I_{R1} = V_{R1}/R1 \quad \text{where:} \quad R1 = 10 \text{ M}\Omega$$

NOTE: Use a voltmeter capable of measuring in the millivolt region and, if possible, having an input impedance of greater than 100 MΩ

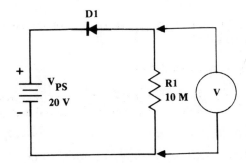

NOTE: Since the reverse saturation current is so low, in the nano-ampere region, a 10 MΩ resistor is used so the voltage drop can be measured. However, unless the input impedance of the meter is 100 MΩ or higher, the impedance will cause loading effect to the 10 MΩ R1 resistor, and this should be taken into account in the calculations.

FIGURE 2-33

3. Calculate the voltage drop across the reverse biased diode from:

$$V_{D1} = V_{PS} - V_{R1} \quad \text{where:} \quad V_{PS} = 20 \text{ V}$$

4. Calculate the DC static resistance of the reverse biased diode from:

   a. $R_{D1} \cong V_{D1}/I_{R1}$ where: $I_S = I_{R1} = I_{D1}$ or from;

   b. $R_{D1} = (V_{D1} \times R1)/V_{R1}$

NOTE: The voltage divider formula of step 4(b) enables R_{D1} to be solved without having to solve for $I_{R1} = I_{D1}$.

5. Insert the calculated and measured values, as indicated, into Table 2-2.

TABLE 2-2	VPS	V_{R1}	V_{D1}	I_S	R_{D1}
CALCULATED	20 V	/////			
MEASURED			/////	/////	/////

Section III: Characteristic Curve Using Bench Techniques

A. FORWARD BIASED CONDITIONS:

1. Connect the diode in a forward biased condition, as shown in Figue 2-34. Monitor the voltage drop across the diode at 0.4 V, 0.5 V, 0.55 V, 0.6 V, 0.65 V, and 0.7 V. Increase and decrease the Vin voltage by varying the 1 kΩ resistor RS. An ammeter is not used in the circuit of Figure 2-34 so that voltage measuring techniques used in industry can be practiced.

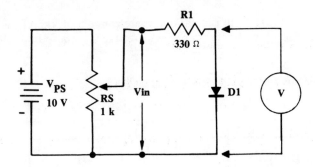

FIGURE 2-34

2. Increase the input voltage Vin until 0.7 V is dropped across the diode D1.
   (a) Measure the Vin and calculate the voltage drop across resistor R1 from:

   $$V_{R1} = V_{in} - V_F \quad \text{where:} \quad V_F = V_{D1}$$

47

(b) Calculate the forward current flow through series resistor R1 and diode D1 from:

$$I_F = I_{D1} = I_{R1} = V_{R1}/R1$$

3. Decrease the Vin voltage to provide conditions of 0.7 V, 0.65 V, 0.6 V, 0.55 V, 0.5 V, and 0.4 V. Measure Vin for each V_F condition and calculate V_{R1} and I_{R1}, as indicated, in Table 2-3.

NOTE: The current flow of silicon diodes will be extremely low for a V_F of 0.4 V and the voltage drop across R1 could be in the millivolt region.

4. Insert the calculated and measured values, as indicated, into Table 2-3.

MEASURED	V_F	0.7 V	0.65 V	0.6 V	0.55 V	0.5 V	0.4 V
MEASURED	Vin	6.09	2.31	1.07	.76	.54	.4
CALCULATED	V_{R1}	5.39	1.66	.47	.21	.04	0 V
CALCULATED	I_{R1}	16.33	5.03	4.424	636.36	121.2	0 V

TABLE 2-3

B. REVERSE BIASED CONDITIONS:

1. Connect the diode in a reverse biased (direction) condition, as shown in Figure 2-35.

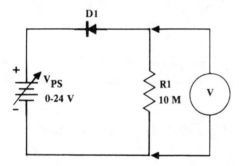

FIGURE 2-35

2. Increase the power supply voltage to 8 V, 16 V, and then 24 V.
3. Monitor the voltage drop across R1 for an applied 8 V of V_{PS} and calculate the reverse current through the diode D1 from:

$$I_S = I_{D1} = I_{R1} = V_{R1}/R1 \quad \text{where:} \quad R1 = 10\ M\Omega$$

NOTE: The reverse current flow of silicon diodes should be in the nano-ampere range and, therefore, the voltage drop across V_{R1} will be in the millivolt range. A voltmeter capable of measuring these low voltage readings must be used.

4. Repeat the calculated reverse current conditions for voltages of 16 V and 24 V.
5. Insert the calculated and measured values, as indicated, into Table 2-4.

TABLE 2-4	$V_{PS} = 8\ V$		$V_{PS} = 16\ V$		$V_{PS} = 24\ V$			
	V_{R1}	I_S	V_{R1}	I_S	V_{R1}	I_S	V_{D1}	R_{D1}
CALCULATED		1NA		1NA		8NA	11.83	1un
MEASURED	.01		.01V		.02			

NOTE: A reverse voltage of 24 volts should be enough to display the initial reverse current conditions and characteristics of the diode. However, a reverse voltage in excess of 100 volts is needed to reach the voltage breakdown for most general purpose diodes.

6. Use the data of Table 2-3 to plot the forward characteristics of the diode and Table 2-4 to plot the initial reverse characteristics of the diode on the graph of Figure 2-36.

NOTE: The graph given in Figure 2-36 is representative and further clarification or improvement may be needed, depending on the diode used.

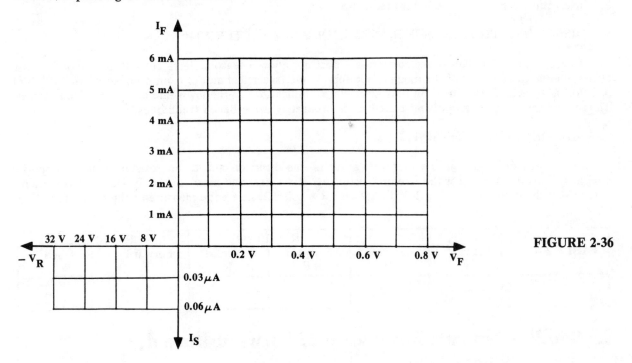

FIGURE 2-36

Section IV: Curve Tracer Techniques

A: DISPLAYING THE FORWARD BIASED CHARACTERISTIC CURVE OF THE DIODE:

1. The forward characteristics of the diode are displayed on the curve tracer by connecting the diode across the (marked) collector and emitter terminal and using the 0.1 V/division for the horizontal scale and 1 mA/division for the vertical scale. Use the polarity switch on the tracer to obtain the foward biased condition and then display the forward biased characteristic curve, by increasing the voltage.
2. Display the forward characteristic curve at the 1 mA/division conditions so the 1-10 mA can be monitored. Visually measure the associated voltages for 1 mA through 10 mA, as indicated in Table 2-5, and insert the values into the table.
3. Switch the vertical scale to 50 mA/division so that the displayed characteristic curve can monitor 50 mA and 500 mA. Insert the associated forward voltages for the visually measured current conditions, as indicated, into Table 2-5.

I_F	1 mA	2 mA	3 mA	4 mA	5 mA	10 mA	50 mA	500 mA	r_B
V_F									

TABLE 2-5

NOTE: 10 mA and its associated V_F voltage can be obtained off either the 0-10 mA or the 0-100 mA forward characteristic curve. If the curve tracer is properly calibrated, either measurement will be valid.

4. Use the measured 500 mA and associated voltage and the measured 50 mA and associated voltage to solve (approximately) the bulk resistance. Calculate from:

$$r_B \cong \frac{\Delta V_F}{\Delta I_F} = \frac{V_F(500\ mA) - V_F(50\ mA)}{500\ mA - 50\ mA}$$

5. Insert the calculated r_B value into table 2-5.

B. DISPLAYING THE REVERSE BIASED DIODE CHARACTERISTICS

The 0.01 mA division scale of the curve tracer does not allow a reasonable I_S current indication because of the nano-ampere region of the diode. Therefore, curve tracer techniques cannot do an adequate job of monitoring I_S accurately. They can only give relative indications of the low value of I_S for silicon devices. The reverse saturation current is measured more accurately using bench techniques.

C. OHMMETER TEST TECHNIQUES

A working diode should have a high resistance in one direction and a low resistance in the opposite direction. Use an ohmmeter with the ohms setting at $R \times 1\ k\Omega$ and measure the forward and reverse resistances of the diode. Repeat for $R \times 10\ k\Omega$ and $R \times 100\ k\Omega$. Insert the measured values into Table 2-6.

TABLE 2-6	$R \times 1\ k\Omega$		$R \times 10\ k\Omega$		$R \times 100\ k\Omega$	
	Forward R	Reversed R	Forward R	Reversed R	Forward R	Reversed R
MEASURED						

Section V: Dynamic Resistance of Forward-Biased, Silicon Diodes

1. Connect the diode in a forward biased (direction) condition and connect a signal generator to the circuit as shown in Figure 2-37. Use a Vin of 100 mVp-p at a frequency of 1 kHz.

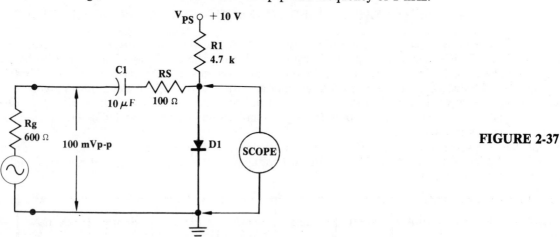

FIGURE 2-37

NOTE: Diodes are nonlinear devices and a Vin of much larger than 100 mVp-p will cause a larger signal to be developed across the diode and the possibility of distortion to occur.

2. DC CONSIDERATIONS:
   A. Measure the DC voltage drop across the diode D1 and the power supply voltage V_{PS}.

B. Calculate the DC current flow though diode D1 and series resistor R1 from:

$$I_F = I_{D1} = I_{R1} = (VPS - V_{D1})/R1 \quad \text{where:} \quad V_{PS} = 10 \text{ V}$$

3. ALTERNATING CURRENT RESISTANCE CONSIDERATIONS
   A. Calculate the junction resistance of the diode from:

   $$r_j = 26 \text{ mV}/I_F$$

   B. Calculate the dynamic resistance of the diode from:

   $$r_{AC} = r_j + r_B \quad \text{where:} \quad r_B \text{ was solved and entered into Table 2-5.}$$

NOTE: The actual measured r_B can be less than 1 ohm and, therefore, $r_{AC} \approx r_j$. Also, r_j can vary over a 2 : 1 range, with temperature, from $r_j = 26 \text{ mV}/I_F$ to $r_j = 52 \text{ mV}/I_F$.

   C. Measure v_o, where v_{in} is 100 mVp-p and calculate the approximate r_{AC} from:

   $$r_{AC} \approx v_o \times RS/V_{RS} \quad \text{where:} \quad V_{RS} = v_{in} - v_o$$

4. Insert the calculated and measured values, as indicated, into Table 2-7.

TABLE 2-7	V_{PS}	V_F	V_{R1}	I_F	r_j	r_{AC}	v_o
CALCULATED	////////	////////					////////
MEASURED			////////	////////	////////		

5. Decrease the value of series resistor R1 to 470 ohms, as shown in Figure 2-38. Calculate the DC and AC signal conditions of the circuit by repeating the procedures given in steps 2 and 3.

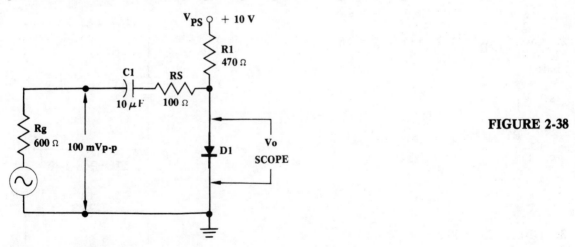

FIGURE 2-38

7. Insert the calculated and measured values, as indicated, into Table 2-8.

TABLE 2-8	V_{PS}	V_F	V_{R1}	I_F	r_j	r_{AC}	v_o
CALCULATED	////////	////////					////////
MEASURED			////////	////////	////////		

LABORATORY QUESTIONS

1. With reference to Figure 2-31, did interchanging the forward biased diode and resistor R1 effect the voltage drops across the components? Why or why not?
2. With reference to Figure 2-32, when R1 was decreased from 4.7 kΩ to 330 Ω, what happened to the voltage drop across the diode? Did it increase or decrease and why?
3. In the reverse biased diode circuit of Figure 2-33, why is it necessary to use a high impedance voltmeter and measure the voltage drop across the resistor instead of across the diode? Why, then, was 10 MΩ used instead of 470 kΩ?
4. Calculate the bulk resistance of the characteristic curve between 500 mA and 10 mA. Is the bulk resistance calculated higher than that over the 500 mA to 50 mA region? If so, is the difference enough to effect r_{AC} dramatically?
5. Draw the equivalent AC circuit for the circuit of Figure 2-38, where R1 equals 470 Ω. Does the value of R1 at 470 Ω effect the approximation of $r_{AC} \approx r_{eq}$, with reference to R1 at 7.7 kΩ?
6. Why is v_{in} kept at a low 100 mVp-p? If it is increased to 1 Vp-p, what would happen to the signal developed across the diode and why? (Refer to Figure 2-37.)
7. Why does the forward resistance of diodes measure differently at different R × settings of the ohmmeter? Explain.

NOTE: All of the questions can be answered from the laborabory data, from information contained in the text material, or through further investigation of the diode in the laboratory.

CHAPTER PROBLEMS

1. Given: $r_B' = 10\ \Omega$ and $V_F = 0.7$ V

   Find:

   a. V_j

   b. V_{R1}

   c. I_F

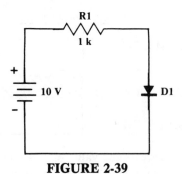

FIGURE 2-39

2. Given: $r_B' = 1\ \Omega$ and $V_F = 0.8$ V

   Find:

   a. V_j

   b. V_{R1}

   c. I_F

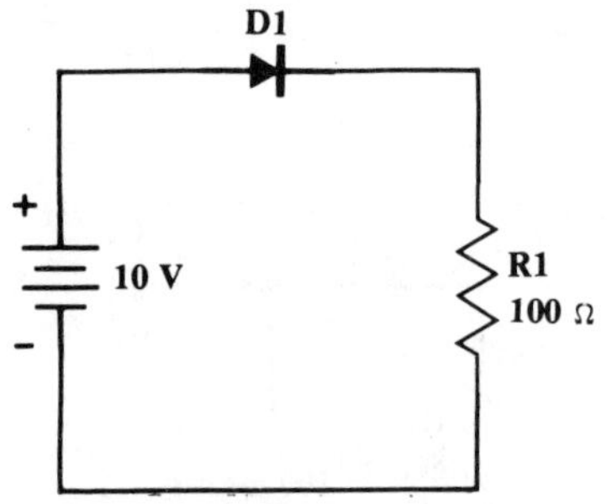

FIGURE 2-40

3. Given: $I_S = 50$ nA $= (50 \times 10^{-9})$

   Find:

   a. V_{D1} (reverse biased)

   b. V_{R1}

   c. R_{D1} (reverse biased)

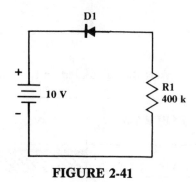

FIGURE 2-41

4. Given: $r_B' = 0.5\ \Omega$, Vin = 100 mVp-p, and $V_F = 0.7$ V

Find:

   a. I_F
   b. r_{AC}
   c. r_{eq}
   d. Vo

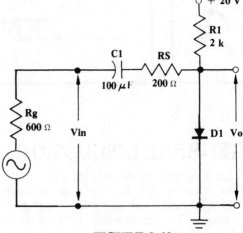

FIGURE 2-42

5. Given: Vin = 100 mVp-p and $I_S = 50$ nA $= (50 \times 10^{-9})$

Find:

   a. V_{D1} (reverse biased)
   b. Vo

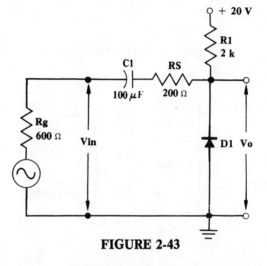

FIGURE 2-43

NOTE: Draw the equivalent circuit first in solving for Vo. At 1 kHz, only RS and R1 should be factors in solving for Vo.

3 ZENER DIODES

GENERAL DISCUSSION

All solid state diodes exhibit a voltage breakdown when the device is subjected to reverse biased conditions. General purpose diodes, for instance, have breakdown voltages that usually exceed 100 volts. Zener diodes, however, are purposely doped to break down at lower voltages. The zener diode symbol and a representative characteristic curve are shown in Figure 3-1. Note that once the zener is in the breakdown voltage condition, the voltage remains constant over a wide range of current conditions.

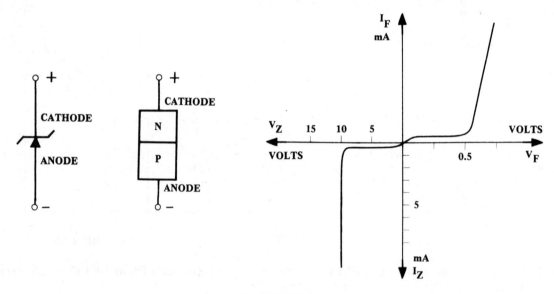

FIGURE 3-1

Zener diode reference voltages vary from below 4 volts to greater than 50 volts and, like general purpose diodes, zeners are primarily constructed from silicon material. However, zener diode breakdown can be caused by either of two breakdown phenomena: zener breakdown or avalanche breakdown.

Essentially all zener diode breakdown, below 5 to 6 volts, is subjected to the soft knee characteristic of the zener breakdown effect, and all zener diode breakdown, above 5 to 6 volts, is subjected to the sharp knee characteristic of the avalanche breakdown effect. Both characteristics are shown in Figure 3-2(a).

Zener diodes that have avalanche breakdown conditions above 6 volts are subject to positive temperature coefficients, while zener diodes that have zener breakdown below 5 volts are primarily subject to negative temperature coefficients, similar to the forward-biased general purpose diode. For instance, if the temperature about a 10 volts zener diode increases, the voltage of the 10 volt zener will also increase. However, a similar temperature increase about a 4 volt zener diode will cause the zener voltage to decrease. Therefore, to create a closer to zero temperature coefficient, so that the voltage reference remains relatively stable with temperature change, a negative coefficient foward-biased diode can be connected in series with a positive coefficient zener diode, as shown in Figure 3-2(b). Also, in some instances, two diodes, instead of one, are used to accomplish the zero temperature coefficient effect.

The series zener diode-diode combination is manufactured in a single package, and manufacturer specifications and suggestions can produce optimum zero temperature coefficient conditions. Therefore, using zener diodes above 6 volts is advantageous because the sharper knee provides a constant voltage over a wide range of operating current conditions and a general purpose diode, in series with the positive coefficient zener, can minimize voltage drift caused by temperature change.

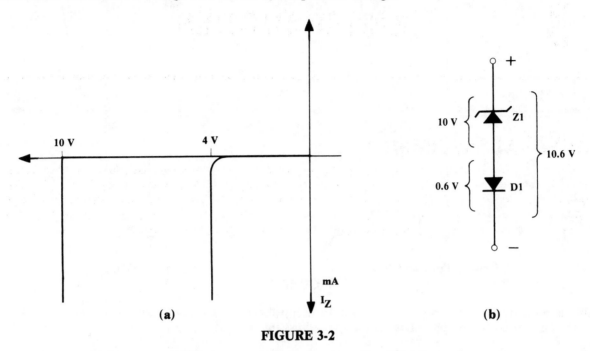

FIGURE 3-2

ZENER DIODE PACKAGING, NUMBERING, AND PARAMETERS

Zener diodes, like general purpose diodes, come in a variety of packages with power ratings from below 200 mW to above 50 watts. Also, like general purpose diodes, zener diodes are identified by 1N preceeding a series of numbers, such as 1N5563, or by a manufacturer's designated number such as CH4740 or MZ10B10.

The best method to select the proper zener diode is to consult a data book for the reference voltage, power dissipation at 25° C, percent tolerance, and type of packaging. Also, depending on the application, other important parameters can be the test current, the maximum dynamic impedance, the maximum operating temperature, and the temperature coefficient.

In almost all instances, the general purpose diode and zener diode have similar packages and markings on the package to designate the anode and cathode side. Therefore, unless the zener number is known, telling whether a given device is a zener of a general purpose diode can be difficult.

NOTE: Remember that a 15 volt zener with a 20 percent tolerance can be within plus-or-minus 3 volts of 15 volts or between 12 and 18 volts, while a 10 percent 20 volt zener will be within plus-or-minus 2 volts of 20 volts or between 18 and 22 volts.

ZENER DIODE AND GENERAL PURPOSE DIODE COMPARISON

Forward biasing a zener diode, or a general purpose silicon diode, will produce similar 0.6 volt voltage drops across the respective devices when about 1 mA of current flows through them. However, when the respective devices are reverse-biased, the zener diode will break down at its reference voltage, but the general purpose diode will not, because its breakdown region is, generally, considerably higher than the applied supply voltage which, for the chapter examples, is lower than 30 volts.

For example, if the zener diode reference voltage is 10 volts and the peak inverse voltage (PIV) of the general purpose diode is 100 volts, then the characteristic curves of the respective devices should have characteristics similar to those shown in Figure 3-3. Also, notice that the reverse biased breakdown characteristic of the general purpose diode has a softer knee characteristic than does the zener diode. This is caused by the doping technique used to obtain the high PIV breakdown voltage required for the general purpose diode.

Therefore, if the general purpose diode and zener diode are connected in the forward-biased direction, as

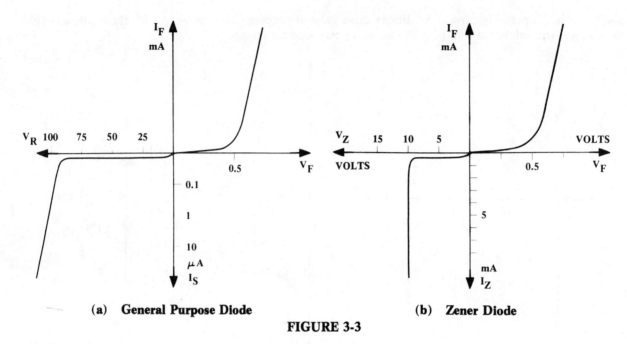

(a) General Purpose Diode (b) Zener Diode

FIGURE 3-3

shown in Figure 3-4, then about 0.6 volts will be dropped across each device and the remaining 19.4 volts of the applied 20 volt supply voltage is dropped across the respective series resistors.

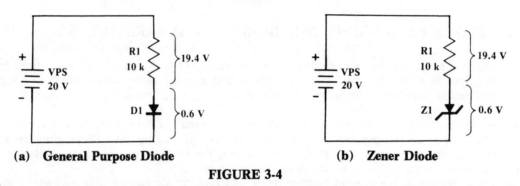

(a) General Purpose Diode (b) Zener Diode

FIGURE 3-4

However, if the devices are turned around, as shown in Figure 3-5, the zener diode will drop 10 volts and the R1 resistor will drop the remaining 10 volts of the applied 20 volt supply voltage. The general purpose diode, with a PIV of 100 volts or more, will not break down and the current flow through the series R1 and reverse-biased general purpose diode will be less than 100 nanoamps. Hence, almost no voltage will be dropped across the series R1 resistor. In fact, if a nominal 100 nanoamps are used, as shown in Figure 3-5(b), 0.001 volt is dropped across R1 and the remaining 19.999 volts are dropped across the reverse biased diode. In Figure 3-5(a), $I_{Z1} = V_{R1}/R1 = 10\text{ V}/10\text{ k}\Omega = 1\text{ mA}$.

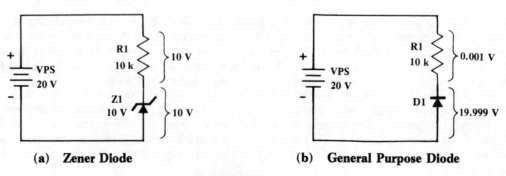

(a) Zener Diode (b) General Purpose Diode

FIGURE 3-5

56

THE CONDITIONS OF ZENER REGULATION

The zener diode limitations are determined by maximum and minimum current conditions of the device. The maximum current is limited by the power dissipating capabilities of the zener, and the minimum current is limited to a nominal low of 1 mA to make sure the zener will not come out of regulation. Therefore, selecting a series resistor is important, because it not only limits the current that can flow through the zener when no load is used, but it also limits the amount of current flow to the load resistor.

For instance, if a zener diode has a rating of 400 mW and 10 volts, then the series resistor should be selected to safely limit the power dissipation of the zener to about one-half of 400 mW or 200 mW. As shown in Figure 3-6, if the applied input voltage is 30 volts, then the current limiting resistor R1 is selected at 1 kohm.

$$V_{Z1} = 10 \text{ V}$$
$$V_{R1} = V_{PS} - V_{Z1} = 30 \text{ V} - 10 \text{ V} = 20 \text{ V}$$
$$I_{R1} = V_{R1}/R1 = 20 \text{ V}/1 \text{ k}\Omega = 20 \text{ mA}$$
$$I_{Z1} = I_{R1} = 20 \text{ mA}$$
$$P_{Z1} = I_{Z1}V_{Z1} = 20 \text{ mA} \times 10 \text{ V} = 200 \text{ mW}$$

FIGURE 3-6

As shown, R1 at 1 kΩ limits I_{R1} to 20 mA and P_{Z1} to 200 mW, but if R1 is decreased to 500 Ω, for instance, then I_{R1} would equal 40 mA and P_{Z1} would equal 400 mW and the zener diode, being at its power dissipation limit, would probably be destroyed. Likewise, if R1 remained at 1 kΩ and V_{PS} is increased to 50 volts, the zener would again have to dissipate a theoretical 400 mW limitation and would probably be destroyed. Therefore, for a 50 volt V_{PS}, an R1 of about 2 kΩ would have to be selected to make sure that the 10 volt zener would have about 20 mA of current flow and 200 mW of safe power dissipation.

VARYING THE POWER SUPPLY VOLTAGE

Decreasing the power supply voltage will not change the voltage drop across the zener as long as there is sufficient current flow through it. For instance, when V_{PS} is decreased to 20 volts, $I_{R1} = I_{Z1} = 10$ mA, and the zener diode remains in regulation. Likewise, if V_{PS} is decreased to 15 volts, $I_{R1} = I_{Z1} = 5$ mA and the zener diode continues to remain in regulation. However, as the V_{PS} approaches 10 volts, the voltage drop across R1 approaches zero volts, the current flow through the series R1 and Z1 approaches zero milliamps, the zener diode comes out of regulation, and it acts like any reverse biased diode. Therefore, decreasing the supply voltage further simply increases the out-of-regulation condition further and most of the supply voltage will be dropped across the reverse biased diode.

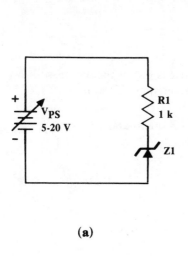

(a)

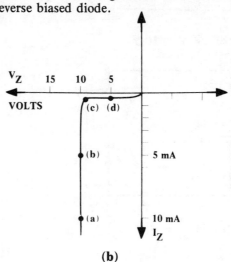

(b)

FIGURE 3-7

NOTE: An out of regulation zener, like a reverse biased diode, has a resistance that usually exceeds 10 MΩ and the current through $I_{R1} = I_{Z1}$ is extremely low, so most of the voltage drop is across the zener.

The current flow conditions from changing the supply voltage of Figure 3-7(a) from 20 volts, to 15 volts, to 10 volts, and to 5 volts are shown on the zener diode characteristic curve of Figure 3-7(b). Each change is individually shown and solved in Figure 3-8.

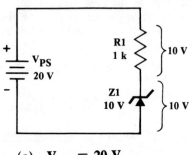

(a) $V_{PS} = 20$ V

$V_{Z1} = 10$ V

$V_{R1} = V_{PS} - V_{Z1} = 20$ V $- 10$ V $= 10$ V

$I_{R1} = V_{R1}/R1 = 10$ V$/1$ kΩ $= 10$ mA

$I_{Z1} = I_{R1} = 10$ mA

$P_{Z1} = I_{Z1}V_{Z1} = 10$ mA $\times 10$ V $= 100$ mW

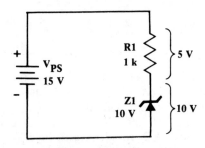

(b) $V_{PS} = 15$ V

$V_{Z1} = 10$ V

$V_{R1} = V_{PS} - V_{Z1} = 15$ V $- 10$ V $= 5$ V

$I_{R1} = V_{R1}/R1 = 5$ V$/1$ kΩ $= 5$ mA

$I_{Z1} = I_{R1} = 5$ mA

$P_{Z1} = I_{Z1}V_{Z1} = 5$ mA $\times 10$ V $= 50$ mW

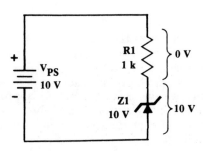

(c) $V_{PS} = 10$ V

$V_{Z1} = 10$ V

$V_{R1} = V_{PS} - V_{Z1} = 10$ V $- 10$ V $= 0$ V

$I_{R1} = V_{R1}/R1 = 0$ V$/1$ kΩ $= 0$ mA

$I_{Z1} = I_{R1} = 0$ mA

$P_{Z1} = I_{Z1}V_{Z1} = 0$ mA $\times 10$ V $= 0$ mW

NOTE: V_{Z1} is just out of regulation.

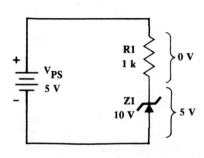

(d) $V_{PS} = 5$ V

$V_{Z1} = 5$ V

NOTE: Zener is out of regulation with reverse biased characteristics. Therefore:

$I_{Z1} = I_S \approx 100$ nano-amps

$V_{Z1} \approx 5$ V

$V_{R1} \approx 0$ V

FIGURE 3-8

LOADING THE ZENER DIODE

Placing a load resistor across the zener diode of a circuit configuration, such as that of Figure 3-9(a), causes the current flow through the series resistor to be distributed between the load resistor and the zener diode. However, if the I_{R1} current is 20 mA, then a load resistor connected across the zener diode must not draw more than 19 mA, because 1 mA must flow through the zener diode to keep it in regulation. Hence, as shown in the circuit sequence of Figure 3-9, if the load resistor is dropped below 526 Ω, the zener diode comes out of regulation.

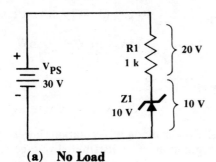

(a) No Load

$V_{Z1} = 10$ V

$V_{R1} = V_{PS} - V_{Z1} = 30$ V $- 10$ V $= 20$ V

$I_{R1} = V_{R1}/R1 = 20$ V$/1$ kΩ $= 20$ mA

$I_{Z1} = I_{R1} = 20$ mA

$P_{Z1} = I_{Z1}V_{Z1} = 20$ mA $\times 10$ V $= 200$ mW

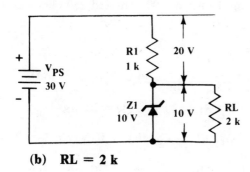

(b) RL = 2 k

$V_{Z1} = 10$ V

$V_{R1} = V_{PS} - V_{Z1} = 30$ V $- 10$ V $= 20$ V

$I_{R1} = V_{R1}/R1 = 20$ V$/1$ kΩ $= 20$ mA

$I_{RL} = V_{RL}/RL = V_{Z1}/RL = 10V/2$ kΩ $= 5$ mA

$I_{Z1} = I_{R1} - I_{RL} = 20$ mA $- 5$ mA $= 15$ mA

$P_{Z1} = I_{Z1}V_{Z1} = 15$ mA $\times 10$ V $= 150$ mW

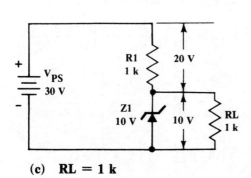

(c) RL = 1 k

$V_{Z1} = 10$ V

$V_{R1} = V_{PS} - V_{Z1} = 30$ V $- 10$ V $= 20$ V

$I_{R1} = V_{R1}/R1 = 20$ V$/1$ kΩ $= 20$ mA

$I_{RL} = V_{RL}/RL = 10$ V$/1$ kΩ $= 10$ mA

$I_{Z1} = I_{R1} - I_{RL} = 20$ mA $- 10$ mA $= 10$ mA

$P_{Z1} = I_{Z1}V_{Z1} = 10$ mA $\times 10$ V $= 100$ mW

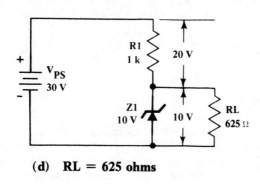

(d) RL = 625 ohms

$V_{Z1} = 10$ V

$V_{R1} = V_{PS} - V_{Z1} = 30$ V $- 10$ V $= 20$ V

$I_{R1} = V_{R1}/R1 = 20$ V$/1$ kΩ $= 20$ mA

$I_{RL} = V_{RL}/RL = 10$ V$/625$ Ω $= 16$ mA

$I_{Z1} = I_{R1} - I_{RL} = 20$ mA $- 16$ mA $= 4$ mA

$P_{Z1} = I_{Z1}V_{Z1} = 4$ mA $\times 10$ V $= 40$ mW

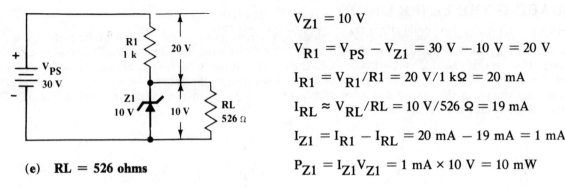

$V_{Z1} = 10\ V$

$V_{R1} = V_{PS} - V_{Z1} = 30\ V - 10\ V = 20\ V$

$I_{R1} = V_{R1}/R1 = 20\ V/1\ k\Omega = 20\ mA$

$I_{RL} \approx V_{RL}/RL = 10\ V/526\ \Omega = 19\ mA$

$I_{Z1} = I_{R1} - I_{RL} = 20\ mA - 19\ mA = 1\ mA$

$P_{Z1} = I_{Z1}V_{Z1} = 1\ mA \times 10\ V = 10\ mW$

(e) RL = 526 ohms

FIGURE 3-9

OUT OF REGULATION CONDITIONS

Since 1 mA is the lowest practical current flow that will keep the zener diode in regulation, lowering the load resistor below 526 ohms will cause the zener to come out of regulation. The distributed voltage then is determined by the series resistor and the load resistor and the zener is no more than a reverse-biased, high-impedance diode. Out of regulation conditions resulting from the use of 500 ohm and 250 ohm load resistors are shown and solved in Figure 3-10.

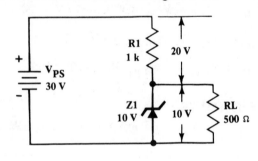

$V_{Z1} = 10\ V$

$V_{R1} = V_{PS} - V_{Z1} = 30\ V - 10\ V = 20\ V$

$I_{R1} = V_{R1}/R1 = 20\ V/1\ k\Omega = 20\ mA$

$I_{RL} = V_{RL}/RL = V_{Z1}/RL = 10\ V/500\ \Omega = 20\ mA$

$I_{Z1} = I_{R1} - I_{RL} = 20\ mA - 20\ mA = 0\ mA$

$P_{Z1} = I_{Z1}V_{Z1} = 0\ mA \times 10\ V = 0\ mW$

$V_{RL} = V_{Z1} = \dfrac{V_{PS} \times RL}{R1 + RL} = \dfrac{30\ V \times 500\ \Omega}{1\ k\Omega + 500\ \Omega} = 10\ V$

(a) RL = 500 ohms

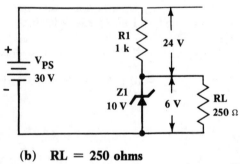

$V_{RL} = V_{Z1} = \dfrac{V_{PS} \times RL}{R1 + RL} = \dfrac{30\ V \times 250\ \Omega}{1\ k\Omega + 250\ \Omega} = 6\ V$

$V_{R1} = \dfrac{V_{PS} \times R1}{R1 + RL} = \dfrac{30\ V \times 1\ k\Omega}{1\ k\Omega + 250\ \Omega} = 24\ V$

$V_{R1} = V_{in} - V_{RL} = 30\ V - 6\ V = 24\ V$

(b) RL = 250 ohms

FIGURE 3-10

DYNAMIC RESISTANCE OF ZENER DIODES

The direct current static resistance and the alternating current dynamic resistance of the zener diode, like those of the general purpose diode, are essentially independent of each other. The static resistance is determined by Ohm's law at the DC operating point, while the dynamic resistance is determined by the slope of the reverse biased characteristic curve about the DC operating point. Therefore, while the static resistance helps explain the DC voltage across, and the current flow through, the zener diode, it is the dynamic impedance that determines the AC resistance and the portion of the applied AC signal actually dropped across the zener diode. The dynamic impedance, or the resistance associated with the zener diode equivalent circuit, is shown in Figure 3-11.

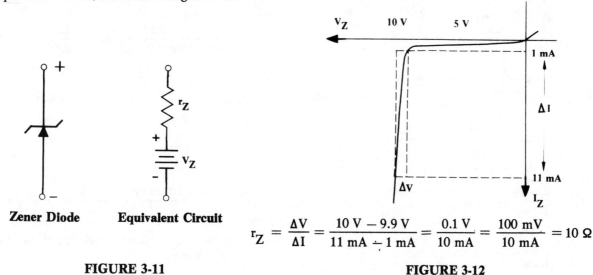

FIGURE 3-11   **FIGURE 3-12**

The dynamic resistance of zener diodes can be found in manufacturer data books or it can be calculated from the actual characteristic curve if ΔV and ΔI values are known. For example, as shown in Figure 3-12, if the reference voltage is 9.9 volts at 1 mA and 10 volts at 11 mA, then dynamic resistance R_Z is 10 Ω.

SMALL SIGNAL APPLICATIONS TO ZENER DIODES

Applying 100 mVp-p of input signal voltage to the circuit of Figure 3-13, where a 100 ohm resistor RS is in series with the parallel resistance of R1 and Z1, produces 9 mVp-p to be developed across Z1, and R1, if the dynamic resistance of the zener diode is 10 ohms.

NOTE: Remember that R1 and Z1 determine the DC voltage drops and, since the plus 30 volt supply connection is an AC ground, R1 and Z1 are seen in parallel, as shown in the equivalent circuit of Figure 3-13(b).

Therefore, the signal voltage of 100 mVp-p distributes 9 mVp-p across the parallel zener diode and resistor R1 and 91 mVp-p across the series RS resistor. Since capacitor C1 blocks DC current flow through resistor RS, there is no DC voltage drop across RS, and to the 1 kHz AC signal, X_{C1} is approximately 1.59 ohms and essentially a short circuit.

$V_{Z1} = 10 \text{ V}$ and $r_{Z1} = 10 \text{ }\Omega$

$$I_{Z1} = I_{R1} = \frac{V_{R1}}{R1} = \frac{V_{PS} - V_{Z1}}{R1} = \frac{30 \text{ V} - 10 \text{ V}}{1 \text{ k}\Omega} = 20 \text{ mA}$$

where: $V_{R1} = V_{PS} - V_{Z1} = 30 \text{ V} - 10 \text{ V} = 20 \text{ V}$

(a)

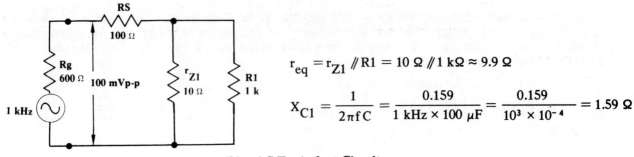

$$r_{eq} = r_{Z1} \mathbin{/\mkern-5mu/} R1 = 10\ \Omega \mathbin{/\mkern-5mu/} 1\ k\Omega \approx 9.9\ \Omega$$

$$X_{C1} = \frac{1}{2\pi f C} = \frac{0.159}{1\ kHz \times 100\ \mu F} = \frac{0.159}{10^3 \times 10^{-4}} = 1.59\ \Omega$$

(b) AC Equivalent Circuit

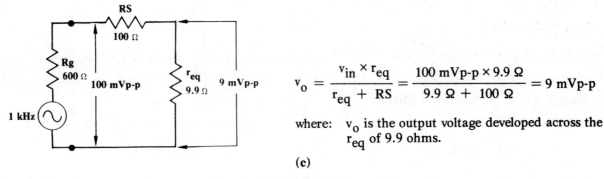

$$v_o = \frac{v_{in} \times r_{eq}}{r_{eq} + RS} = \frac{100\ mVp\text{-}p \times 9.9\ \Omega}{9.9\ \Omega + 100\ \Omega} = 9\ mVp\text{-}p$$

where: v_o is the output voltage developed across the r_{eq} of 9.9 ohms.

(c)

FIGURE 3-13

NOTE: If $r_{eq} = r_{Z1}$ is unknown but the input signal voltage and the signal voltage across the $r_{eq} = r_{Z1} \mathbin{/\mkern-5mu/} R1 \approx r_{Z1}$ can be measured, then the ratio of v_{RS}/v_o equals the ratio of RS/r_{eq}. For instance, if $v_{in} = 100\ mVp\text{-}p$ and $v_o = 9\ mVp\text{-}p$, then r_{eq} equals:

$$\frac{v_{RS}}{v_o} = \frac{RS}{r_{eq}} \quad \text{and} \quad r_{eq} = \frac{RS \times v_o}{v_{RS}} = \frac{100\ \Omega \times 9\ mVp\text{-}p}{91\ mVp\text{-}p} = 9.9\ \Omega \quad \text{or}$$

$$r_{eq} = \frac{RS \times v_o}{v_{in} - v_o} = \frac{100\ \Omega \times 9\ mVp\text{-}p}{100\ mVp\text{-}p - 9\ mVp\text{-}p} = \frac{100\ \Omega \times 9\ mVp\text{-}p}{91\ mVp\text{-}p} = 9.9\ \Omega$$

And since $r_{eq} \approx r_{Z1}$, then $r_{Z1} = \dfrac{RS \times v_{r_{Z1}}}{v_{in} - v_{r_{Z1}}} \approx \dfrac{100\ mVp\text{-}p \times 9\ mVp\text{-}p}{91\ mVp\text{-}p} = 9.9\ \Omega$

NOTE: Zener diodes are noisy devices and the lower the current through the zener, the closer the operating point gets to the knee of the curve, and the noisier the zener becomes. Therefore, not only is it important to keep sufficient current flow in the device to make sure it does not come out of regulation, but it is also important to minimize zener noise.

LABORATORY EXERCISE

OBJECTIVES
Investigate the DC and dynamic characteristics of zener diodes.

LIST OF MATERIALS AND EQUIPMENT
1. Zener Diode: 12 V ± 2 V, such as the 1N4742
2. Resistors (one each): 100 Ω   2.2 kΩ   4.7 kΩ   10 kΩ (potentiometer)
3. Capacitor: 10 μF (one)
4. Signal Generator
5. Oscilloscope

PROCEDURE
Section I: Forward and Reverse-Biasing the Zener Diode

1. Connect the circuit of Figure 3-14(a), where the zener diode is forward biased and the power supply is set at 20 V DC.

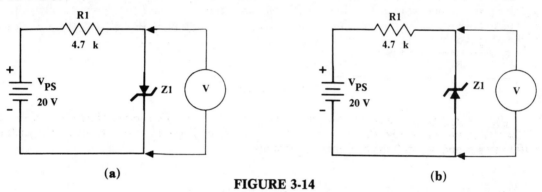

FIGURE 3-14

2. Measure the DC voltage across the forward-biased zener diode Z1, as shown in Figure 3-14(a). Then, calculate the DC voltage drop across resistor R1 from:

$$V_{R1} = V_{PS} - V_{DZ1} \quad \text{where:} \quad V_{PS} = 20 \text{ V}$$

NOTE: Forward biased zener diodes have approximately the same 0.5 V to 0.7 V voltage range as general purpose diodes.

3. Calculate the current flow through the series resistor R1 and the forward-biased zener diode R_{Z1} from:

$$I_{R1} = I_{D_{Z1}} = V_{R1}/R1$$

4. Insert the calculated and measured values, as indicated, into table 3-1(a).

5. Connect the zener diode in a reverse-biased condition (turn the zener around), as illustrated in Figure 3-14(b).

6. Measure the DC voltage across the reverse-biased zener diode Z1, as shown in Figure 3-14(b), and calculate the DC voltage drop across resistor R1 from:

$$V_{R1} = V_{PS} - V_{Z1} \quad \text{where:} \quad V_{PS} = 20 \text{ V}$$

7. Calculate the current flow through the series resistor R1 and the reverse-biased zener diode R_{Z1} from:

$$I_{Z1} = V_{R1}/R1$$

NOTE: The forward-biased zener diode behaves like a general purpose diode and is labeled V_{DZ1}.

However, in the reverse biased configuration it behaves like a zener diode and it is labeled V_{Z1}. Also, like the general purpose diode, the current flow through R1 establishes the current flow through the unloaded zener diode.

8. Insert the calculated and measured values, as indicated, into Table 3-1(b).

	(a)				(b)			
TABLE 3-1	V_{PS}	V_{Z1}	V_{R1}	I_{R1}	V_{PS}	V_{Z1}	V_{R1}	I_{R1}
CALCULATED	20 V	////			20 V	////		
MEASURED			////	////			////	////

Section II: Varying the Power Supply

1. Connect the circuit shown in Figure 3-15, and monitor the DC voltage across the zener diode. First, increase the power supply voltage to a high of 20 volts and, then, decrease the power supply voltage to a low of 2 volts above the zener voltage. For example, if the zener voltage is 10 volts, set the V_{PS} at 12 volts and, if the zener voltage is 12 volts, set the V_{PS} at 14 volts.

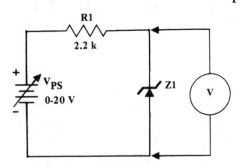

FIGURE 3-15

2. With the V_{PS} adjusted to 20 volts:
   A. Monitor V_{Z1} and calculate V_{R1} from:

   $$V_{R1} = V_{PS} - V_{Z1} \quad \text{where:} \quad V_{PS} = 20\text{ V}$$

   B. Calculate I_{R1} from: $I_{R1} = V_{R1}/R1$ where: $R1 = 2.2\text{ k}\Omega$

3. Adjust the V_{PS} to 2 volts above the zener voltage ($V_{Z1} + 2$ V DC):
   A. Monitor V_{Z1} and calculate V_{R1} from:

   $$V_{R1} = V_{PS} - V_{Z1} \quad \text{where:} \quad V_{PS} = V_{Z1} + 2\text{ V}$$

   B. Calculate I_{R1} from: $I_{R1} = V_{R1}/R1$ where: $R1 = 2.2\text{ k}\Omega$ and $V_{R1} \approx 2\text{ V}$

4. Insert the calculated and measured values, as indicated, into Table 3-2.

	$V_{PS} = 20$ V				$V_{PS} = V_{Z1} + 2$ V			
TABLE 3-2	V_{PS}	V_{Z1}	V_{R1}	I_{R1}	V_{PS}	V_{Z1}	V_{R1}	I_{R1}
CALCULATED	20 V	////			////			
MEASURED			////	////			////	////

Section III: Loading the Zener Diode

1. Connect the circuit of Figure 3-16. Reset V_{PS} to 20 volts and add a 4.7 kΩ load resistor across the zener diode.

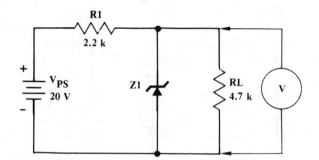

FIGURE 3-16

2. Measure the voltage drop across the parallel 4.7 kΩ load resistor and the zener diode, as shown in Figure 3-16. Then, calculate the voltage drop across resistor R1 from:

$$V_{R1} = V_{PS} - V_{Z1} \quad \text{where;} \quad V_{PS} = 20 \text{ V}$$

3. Calculate the current flow through resistor R1 from: $I_{R1} = V_{R1}/R1$ where: $I_{R1} = I_{Z1} + I_{RL}$

NOTE: The current flow through resistor R1 is shared by the parallel combination of Z1 and RL.

4. Calculate the current flow through the load resistor from: $I_{RL} = V_{RL}/RL$ where: $V_{RL} = V_{Z1}$

5. Calculate the current flow through the zener diode from: $I_{Z1} = I_{R1} - I_{RL}$

6. (Optional Step) If the $I_{R1} = I_{Z1} + I_{RL}$ concept is not fully understood, then, a current meter can be used as shown in Figure 3-17, where a single milliammeter can be moved around the circuit. While the use of the milliammeter is not a particularly good measurement technique, it does illustrate, clearly, how the current of I_{R1} is shared by Z1 and R_L.

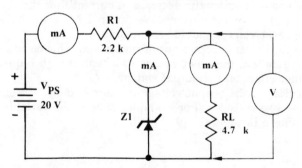

FIGURE 3-17

NOTE: If Step 6 is done, make sure that all the connections are made as the meter is moved to the various branches of the circuit.

7. Insert the calcuated and measured values, as indicated, into Table 3-3.

TABLE 3-3	V_{PS}	V_{Z1}	V_{R1}	I_{R1}	I_{Z1}	I_{RL}
CALCULATED	20 V	/////				
MEASURED			/////	Measurements Optional Step 6		

Section IV: Increased Loading Conditions

1. Connect the circuit of Figure 3-18(a), where a milliammeter is connected in series with the zener, as shown. Monitor the voltage across the zener diode and measure the $I_{R1} = I_{Z1}$ current flow on the milliammeter.

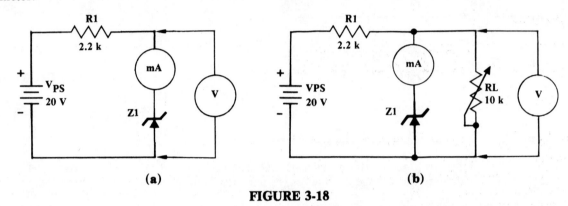

FIGURE 3-18

2. Verify the current flow by measuring the power supply voltage VPS and calculating $I_{R1} = I_{Z1}$ from:

$$I_{R1} = I_{Z1} = (V_{PS} - V_{Z1})/R1 \quad \text{where:} \quad V_{R1} = V_{PS} - V_{Z1}$$

NOTE: Exact measured-calculated current values can be achieved only if R1, V_{PS}, and V_{Z1} are known exactly.

3. Insert the calculated and measured values, as indicated, into Table 3-4.

4. Connect the 10 kΩ variable load resistor to the circuit, as shown in Figure 3-18(b). Initially, set the variable RL to about 10 kΩ. Monitor the voltage across the parallel zener diode and load resistor. (The effect of the current meter, in series with the zener diode, should have little effect on the monitored reference voltage.)
   A. Begin to vary (decrease) the load resistance and to monitor the decrease in current flow through the diode as the load current increases. Remember, as long as the current flow through Z1 exists, the zener remains in zener condition (regulation) and the zener voltage should remain stable.
   B. Continue to decrease the load resistance until the zener comes out of regulation. This occurs when I_Z approaches 0 mA. Therefore, when out of regulation, the zener reverts to a typical high impedance reverse biased diode condition and has no further effect on the output voltage.
   C. After the zener diode has gone out of regulation, the output voltage is controlled by resistors R1 and RL and is the voltage divider ratio of these two resistors. For example, remove the variable load resistor and replace it with a 1 kΩ resistor, as shown in Figure 3-19.

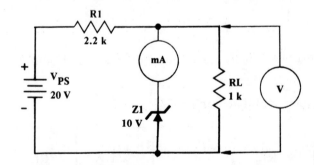

FIGURE 3-19

   D. Calculate the output voltage for the circuit of Figure 3-19 from:

$$V_O = V_{RL} = \frac{V_{PS} \times RL}{R1 + RL} \quad \text{where:} \quad RL = 1 \text{ k}\Omega$$

E. Measure the output voltage for the overloaded circuit of Figure 3-19, as shown, and monitor the zener current. Momentarily disconnect Z1 to prove it has no effect on Vo.
5. Insert the calculated and measured values, as indicated, into Table 3-4.

TABLE 3-4	FIGURE 3-18(a)			FIGURE 3-19	
	V_{Z1}	V_{PS}	$I_{R1} = I_{Z1}$	Vo	I_{Z1}
CALCULATED	/////////	/////////			
MEASURED					

Section V: Curve Tracer Techniques

1. FORWARD BIASED CONDITIONS
   A. Forward biased characteristics are obtained by connecting the diode across the (marked) collector and emitter terminals. Use the 0.1 V/division for the horizontal scale, and 1 mA/division for the vertical scale. Use the polarity switch on the curve tracer to obtain the forward biased condition and then display the characteristic curve.
   B. Measure 0.3 V, 0.4 V, 0.5 V, 0.55 V, 0.6 V, 0.65 V, and 0.7 V across the diode and monitor the associated currents. Insert the current values into Table 3-5.

TABLE 3-5							
V_{DZ}	0.3 V	0.4 V	0.5 V	0.55 V	0.6 V	0.65 V	0.7 V
I_{FZ}							

2. REVERSE BIASED CONDITIONS
   A. Reverse biased characteristics are obtained by using the polarity switch or by turning the device around. Use 2 V/division for the horizontal scale and 0.5 mA/division for the vertical scale. Measure the incremental 2 volt steps and the associated currents until zener breakdown occurs. Insert the measured current values, and then the zener breakdown voltage at 10 mA, as indicated, into Table 3-6.
   B. At the breakdown voltage condition, switch the vertical scale to 2 mA/division and reference the voltage at 2 mA and 20 mA. Insert the measured voltage values, as indicated, into Table 3-6.

TABLE 3-6						V_{Z1}	V1	V2
V_{DZ}	2 V	4 V	6 V	8 V	10 V			
I_Z						10 mA	2 mA	20 mA

NOTE: Depending on the zener voltage, some slots of the table may have to be deleted. Also, if a curve tracer is not available, bench measurement techniques can be used.

Section VI: Dynamic Impedance of Reverse-Biased, Zener Diodes

1. Connect the circuit of Figure 3-20 with the zener diode in the reverse biased condition.

2. DIRECT CURRENT CONSIDERATIONS
   A. Measure both the power supply voltage V_{PS} and the DC voltage drop across the zener diode V_{Z1}.
   B. Calculate the DC current flow through the zener diode Z1 and series resistor R1 from:

$$I_{D1} = I_{R1} = (V_{PS} - V_{Z1})/R1 \quad \text{where:} \quad V_{PS} = 20 \text{ V}$$

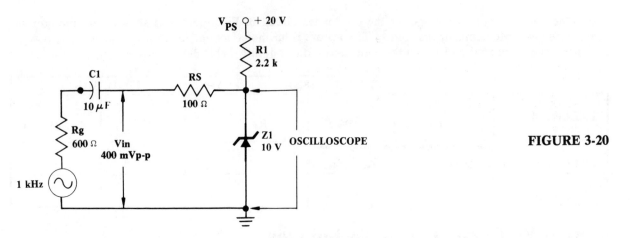

FIGURE 3-20

3. ALTERNATING CURRENT CONSIDERATIONS
A. Calculate the dynamic resistance of the zener diode from the reverse biased characteristic curve measurements of Table 3-6, where the associated voltages at 2 mA and 20 mA can be used to solve r_{Z1} from:

$$r_{Z1} = \frac{\Delta V}{\Delta I} = \frac{V_2 - V_1}{I_2 - I_1} = \frac{V_2 - V_1}{20 \text{ mA} - 2 \text{ mA}}$$

B. Use the circuit of Figure 3-20 and measure the signal voltage developed across the zener diode for an input signal voltage of 400 mVp-p at 1 kHz. Calculate r_{Z1} from:

$$r_{Z1} \approx r_{eq} = \frac{v_{Z1} \times RS}{v_{in} - v_o} \quad \text{where:} \quad v_o \approx v_{Z1}$$

NOTE: The R1 resistor at 1 kΩ is in parallel with r_{Z1}, but it is too high in value with comparison to r_{Z1} so it can be ignored. Also, zener noise can be apparent even with a v_{in} of 400 mVp-p.

4. Insert the calculated and measured values, as indicated, into Table 3-7.

TABLE 3-7	V_{PS}	V_{Z1}	V_{R1}	I_{R1}	3(a) r_{Z1}	3(b) v_{in}	v_o	v_{RS}	r_{eq}
CALCULATED							////		
MEASURED					////		////	////	

LABORATORY QUESTIONS

1. Once the zener diode is in breakdown condition, does varying the current flow through the zener diode, by increasing or decreasing the input voltage or series resistor R1, effect the reference voltage? Comment.
2. Can a 20 percent tolerance, 15 volt, zener diode at 12.5 volts still be within the rated tolerance of the device? Calculate the lower and upper voltage limits of a 20 percent, 15 volt, zener diode. Calculate the lower and upper voltage limits of a 10 percent, 15 volts, zener diode.
3. In the unloaded circuit conditions of Figure 3-15, the power supply voltage was dropped to about 2 volts above the zener diode voltage, so that slightly less than 1 mA of current would flow through the zener diode. Did the zener diode remain in regulation?
4. In the circuit of Figure 3-20, why is it necessary to increase Vin to about 400 mVp-p? Also, for the same reason, why is R1 = 2.2 kΩ and not, for instance, 10 kΩ? If a variable load is connected across the zener, will zener noise increase as the load resistance decreases? If uncertain, test a zener diode in the

laboratory.

5. Does the dynamic resistance of the zener diode change with an increase or decrease of current flow through the device? Test the dynamic resistance of the zener diode by increasing and then decreasing the zener current by varying the power supply voltage or resistor R1. Comment.

CHAPTER PROBLEMS

1. Given: $V_{Z1} = 10$ V

   Find:

   a. V_F

   b. V_{R1}

   c. I_F

   NOTE: In this circuit, Z1 is forward biased deliberately.

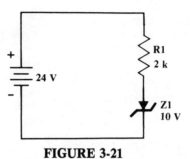

FIGURE 3-21

2. Given: $V_{Z1} = 10$ V

   Find:

   a. V_{Z1}

   b. V_{R1}

   c. I_{Z1}

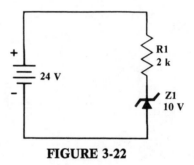

FIGURE 3-22

3. Given: $V_{Z1} = 10$ V

   Find:

   a. V_{Z1} d. I_{RL}

   b. V_{R1} e. I_{Z1}

   c. I_{R1}

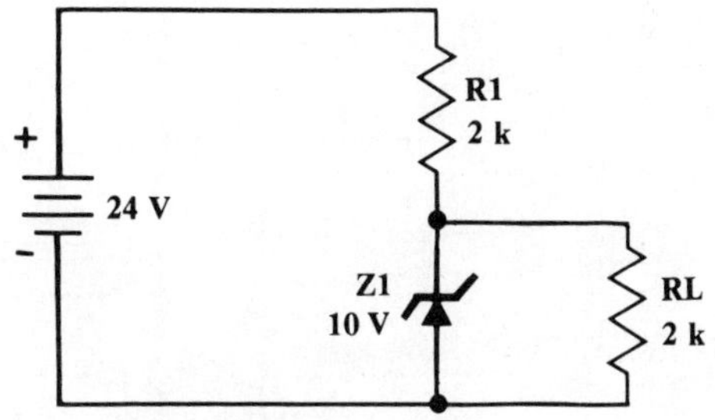

FIGURE 3-23

4. Given: $V_{Z1} = 10$ V and $P_{Z1}(\max) = 500$ mW

   Find:

   a. R1 minimum, to make sure the power dissipation limit of Z1 at 500 mW is not exceeded.

   b. I_{R1} maximum.

   NOTE: Z1 is 1 watt, so a 500 mW power dissipation limit is used, taking into account a 50% derating factor, to protect the zener against burnout.

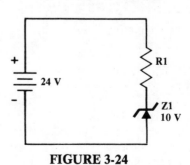

FIGURE 3-24

5. Given: $I_{Z1}(\text{min}) = 1$ mA and $V_{Z1} = 10$ V

Find:

a. RL minimum, while still maintaining Z1 in regulation with 1 mA of current flow through it.

b. I_{RL} maximum.

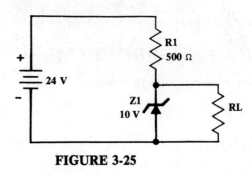

FIGURE 3-25

6. Given: $V_{Z1} = 10$ V, $r_{Z1} = 10\ \Omega$, and $v_{in} = 100$ mVp-p

Find:

a. V_{R1}

b. I_{R1}

c. r_{eq}

d. v_o

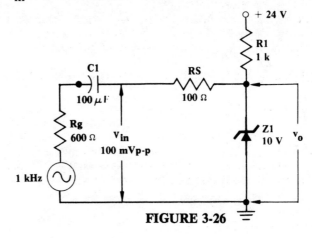

FIGURE 3-26

4 BIPOLAR TRANSISTORS: An Introduction

GENERAL DISCUSSION

In less than three decades, transistors have replaced vacuum tubes in all but a few high power and high frequency electronic circuit applications. The transistor, like the solid state diode, has contributed towards making products which are smaller, more rugged, consume less power, and cost less than products constucted from vacuum tubes. Additionally, semiconductor transistors and diodes have paved the way for miniaturization, where entire circuits such as the operational amplifier and the decade counter are fabricated on a single semiconductor chip.

The transistor, like the diode, can be fabricated from both silicon or germanium materials, but silicon is the material mostly used. Therefore, in this text, reference will be made to germanium devices, but all the analysis will be done with devices constructed from silicon.

TRANSISTOR STRUCTURE AND SCHEMATIC SYMBOLS

The bipolar transistor is a three terminal device that is constructed by forming three alternating layers of P, N, and P materials (or N, P, and N materials), as shown in Figure 4-1. The three alternating regions are called the emitter, base, and collector regions, and they create two PN junctions: one between the emitter and base regions and one between the base and collector regions. It is the interaction between these two junctions that creates transistor action. If interaction does not occur, the net effect is, effectively, two back-to-back diodes completely independent of each other.

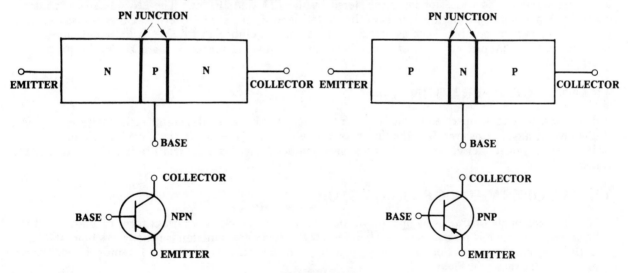

FIGURE 4-1

The schematic symbols for the NPN and PNP transistor devices also are shown in Figure 4-1. The arrow in the emitter load distinguishes the NPN from the PNP device, and the direction of the arrow shows the direction of current flow through the devices.

FABRICATION TECHNIQUES

Interaction of the PN junctions is physically achieved by making the base region very thin and lightly doped, the emitter region heavily doped, and the collector region (normally) moderately doped. Physically, the collector region is the largest, and it is usually electrically connected to the case in the higher powered devices, to facilitate the transfer of heat away from the collector region. Representative doping levels of the emitter, base, and collector regions are shown in Figure 4-2(a), where the collector region is 10 times more conductive than the base region and the emitter region is 10 times more conductive than the collector region.

Metal pads are bonded to each of the emitter, base, and collector regions, as shown in Figure 4-2(b), and leads are then connected to the pads. Also shown, but not to scale, are the physical areas of the emitter, base and collector regions, because the base region is usually less than one-hundredth of the total thickness of the combined emitter-base-collector "sandwich".

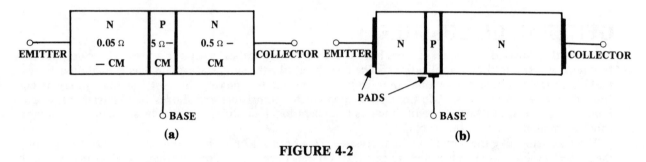

FIGURE 4-2

TRANSISTOR PACKAGING AND LEAD INDICATIONS

The bipolar transistor comes in a variety of metal and plastic packages. Some of the most widely used TO (transistor outline) packages are shown in Figure 4-3, and the examples show dimensions and designate the emitter, base, and collector leads. Data sheets can be found in Appendix C.

TRANSISTOR NUMBERING SYSTEMS

Most transistors are identified by 2N preceding a number, such as 2N3904, 2N3906, 2N4900, and 2N835. These are standardized transistor types registered by the EIA and JEDEC. The 2N indicates a semiconductor with two junctions and the numbers that follow identify the specific device. Other transistors have proprietary company numbers such as MJE5192, a Motorola number, or FA3333, a Fairchild number. In most cases, these proprietary transistor numbers can be cross-referenced in data books for purpose of substitution.

TRANSISTOR OPERATION

The transistor can be operated in four different modes: the linear mode, cutoff mode, saturation mode, or the inverse mode. However, it is the linear, or active region, where the device functions as an amplifier or impedance transfer device, and it is this region which best describes the "trans-resistance" action of the transistor.

LINEAR OPERATION OF TRANSISTOR

In the linear mode of transistor operation, the base-emitter junction is always forward biased and the collector-base junction is reverse biased. Therefore, the base-emitter junction is usually less than 100 ohms and the collector-base junction is usually greater than 100 kΩ. The plus and minus power supply connections are shown in Figure 4-4.

FIGURE 4-4

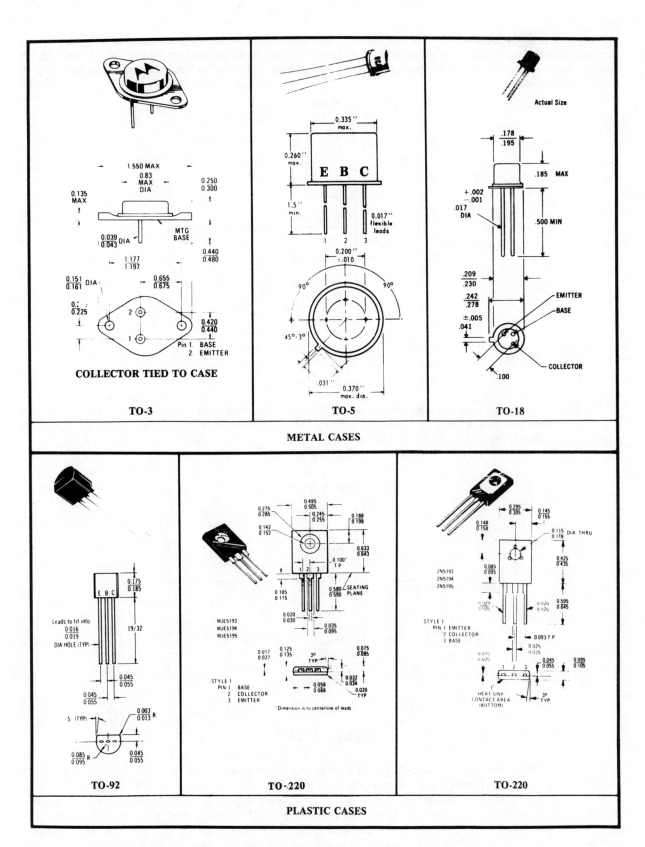

FIGURE 4-3

The concept of the forward biased base-emitter junction and reverse biased collector-base junction can be most clearly seen if the "thin" base region is shown divided, as illustrated in Figure 4-5. Then, the PN base-emitter junction of the NPN transistor is seen as one diode, and the NP collector-base junction is seen as the other diode.

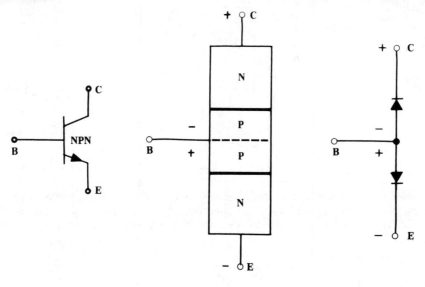

FIGURE 4-5

The PNP device can be illustrated in a similar manner. However, the power supply connections are reversed so that the base-emitter junction is forward biased and the collector-base junction is reverse biased. The PNP transistor is shown in Figure 4-6.

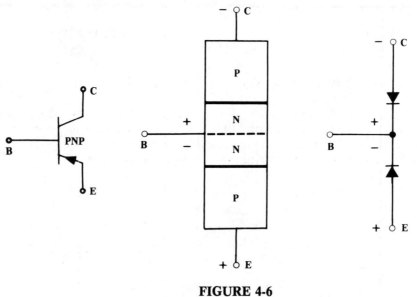

FIGURE 4-6

The base-emitter and collector-base junctions of transistors are diodes, and their resistance can be calculated the same way as forward and reverse-biased, general-purpose diodes. Recall that for the reverse-biased diode, the saturation current I_S was in the nano-ampere region and the resistance was extremely high but, when the diode was forward biased, the resistance could be calculated, theoretically, from $r_j = 26\ mV/I_F$. Therefore, the forward-biased, base-emitter, diode junction of the transistor, in a similar manner is solved from $r_e = 26\ mV/I_E$, where I_E is the current flow through the emitter of the device and r_e is the AC resistance of the base-emitter, diode junction. The forward and reverse-biased conditions of the NPN transistor are shown in Figure 4-7. For the linear-biased PNP transistors, the same conditions exist, but the polarities are reversed.

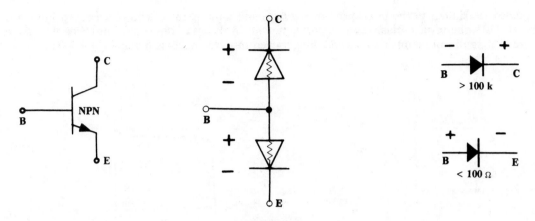

FIGURE 4-7

The DC voltage drop across the forward-biased, base-emitter, diode junction is approximately 0.6 volts for both the NPN and PNP silicon transistors and approximately 0.25 volts for germanium tranistors, as shown in Figure 4-8. This voltage is similar to the forward voltage dropped across general purpose silicon and germanium diodes.

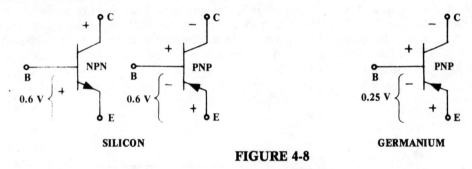

FIGURE 4-8

REVERSE BIASED JUNCTION VOLTAGE BREAKDOWN

Like reverse-biasing a general purpose diode, reverse biasing the collector-base junction of most general purpose silicon transistors produces a breakdown voltage that usually occurs at greater than 100 volts. This is true when the collector-base junction, alone, is under test and the emitter lead is left open. An illustrative circuit and the characteristic breakdown curve are shown in Figure 4-9.

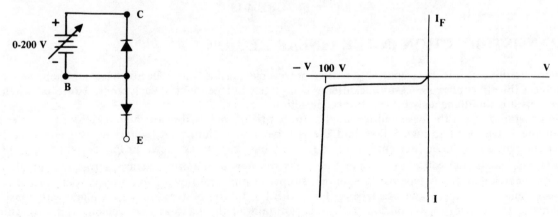

FIGURE 4-9

Like reverse-biasing a general purpose zener diode, reverse-biasing the base-emitter junction of most silicon general purpose transistors produces a breakdown voltage that usually occurs at much lower than 100 volts. This is true because the doping levels required to produce a low voltage, or avalanche effect, in zener diodes are similar to the levels used in the base and emitter regions to produce working transistors. For the 2N3904, the reverse-biased, base-emitter junction will break down at about 8 or 9 volts, as shown in Figure 4-10, and minimum breakdown voltage given in data sheets is 6 volts.

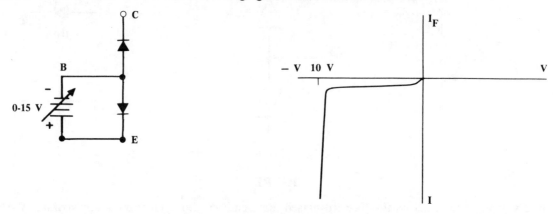

FIGURE 4-10

However, the breakdown voltage of the collector-emitter junction is almost always lower than the collector-base breakdown voltage and, in fact, can be as low as one-half. Therefore, regardless of whether the base-emitter leads are forward biased or the base lead is left unconnected, the collector-emitter, breakdown voltages are about equal. Further explanation of the breakdown characteristic curve is given later in the chapter. For the 2N3904, the minimum breakdown voltage of the collector-emitter junction given in the data sheets is 40 volts, as shown in Figure 4-11.

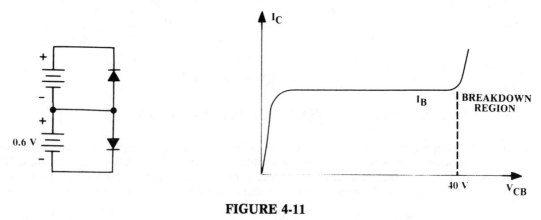

FIGURE 4-11

TRANSISTOR ACTION IN THE LINEAR REGION

Forward biasing the base-emitter junction or reverse biasing the collector-base junction, separately, produces the effect of two back-to-back diodes completely independent of each other. However, when both junctions are simultaneously biased, transistor actions occurs.

For instance, when the base-emitter junction of a PNP device is forward biased, majority carrier holes from the P type emitter region flow into the N type base region. And, when the collector-base is left unconnected, all of the emitter current flows into the base region, as shown in Figure 4-12.

Reverse biasing the collector-base junction stops the flow of majority carriers across the junction, but minority carriers in the N type base region are swept across. The flow of minority carriers in the N type base region are by holes and are labeled I_{co}. Both I_{co} for transistors and I_S for diodes are considered leakage current, and for silicon material, it is normally in the nano-ampere region. This condition is shown in Figure 4-13.

Applying both potentials simultaneously, as shown in Figure 4-14, causes the majority carrier holes from the P type emitter region to flow into the N type base region as minority carriers. However, as minority carriers, the holes are swept across the reverse-biased, collector-base junction and into the collector region. Essentially, few recombinations of the emitted emitter current take place in the thinly doped base region.

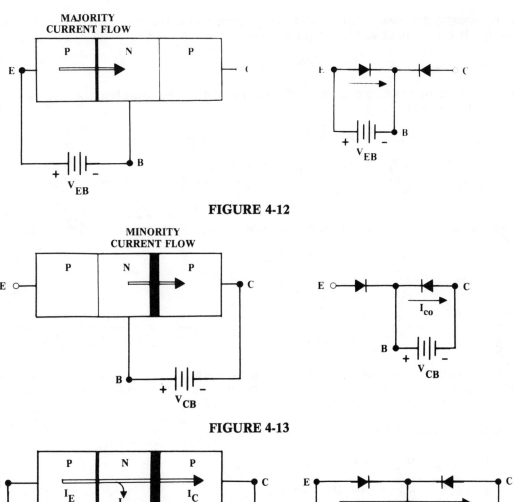

FIGURE 4-12

FIGURE 4-13

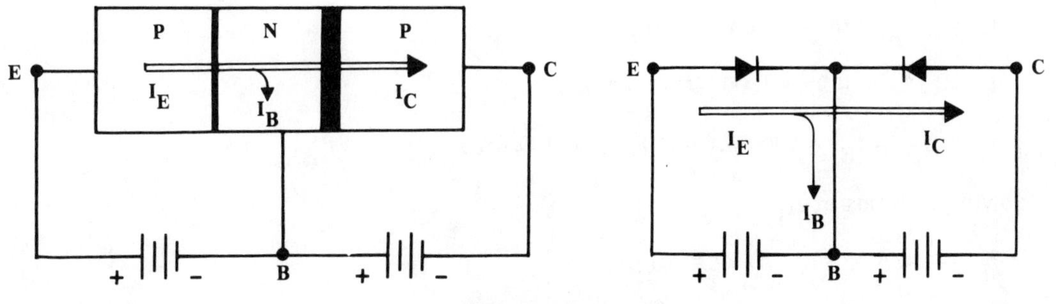

FIGURE 4-14

Therefore, almost all of the current emitted from the emitter region is collected in the collector region and only a small percentage of the emitter current flows in the base region. Further along in the text, it will be shown how the base current is used to control both the collector and emitter currents.

TRANSISTOR CURRENT RELATIONSHIPS

Standard notation for the emitter current is I_E, for the base current is I_B, and for the collector current is I_C. The emitter current equals the sum of the collector and the base currents or $I_E = I_C + I_B$, but only a very small percentage of the total is base current. Typically, if $I_E = 1000$ μA, then I_C would equal about 990 μA and I_B would equal about 10 μA. Hence, $I_E = I_C + I_B = 990$ μA $+ 10$ μA $= 1000$ μA. In the approximation techniques used for circuit analysis in this book, the collector current is considered equal to the emitter current, or $I_C = I_E$, because 990 μA is approximately equal to 1000 μA, and this small current difference does not significantly effect circuit calculations.

The relationship between the amount of current emitted from the emitter region to the amount reaching the collector region is called alpha (a), and $a = I_C/I_E$. If I_E is given as 1000 μA and I_C as 990 μA, then:

$$a = I_C/I_E = 990 \text{ μA}/1000 \text{ μA} = 0.99$$

The relationship between the amount of base current to the collector current is called beta (β) and $\beta = I_C/I_B$. If $I_C = 990$ μA and $I_B = 10$ μA, then:

$$\beta = I_C/I_B = 990 \text{ μA}/10 \text{ μA} = 99$$

The relationship of the emitter current to the base current can be solved from I_E/I_B. If $I_E = 1000$ μA and $I_B = 10$ μA, then:

$$I_E/I_B = 1000 \text{ μA}/10 \text{ μA} = 100, \text{ or}$$

$$I_E/I_B = \frac{I_C + I_B}{I_B} = \frac{\beta I_B + I_B}{I_B} = \frac{(\beta + 1)I_B}{I_B} = \beta + 1 = 99 + 1 = 100$$

Alpha and beta relationships can be derived from simple algebra if either alpha or beta are known. Hence:

A. $\quad a = \dfrac{I_C}{I_E} = \dfrac{\beta I_B}{(\beta + 1)I_B} = \dfrac{\beta}{\beta + 1} = \dfrac{99}{99 + 1} = 0.99$

B. $\quad \beta = \dfrac{I_C}{I_B} = \dfrac{a I_E}{I_E - I_C} = \dfrac{a I_E}{I_E - a I_E} = \dfrac{a I_E}{(1 - a)I_E} = \dfrac{a}{1 - a} = \dfrac{0.99}{1 - 0.99} = \dfrac{0.99}{0.01} = 99$

Further current relationships also can be solved by algebra once a few device parameters are known. Therefore, using $\beta = 99$ and $I_B = 10$ μA, then $I_C = 990$ μA and $I_E = 1000$ μA, as shown in the following equations.

A. Solving in terms of I_C:

$$I_C = \beta I_B = 99 \times 10 \text{ μA} = 990 \text{ μA}$$

$$I_C = a I_E = 0.99 \times 1000 \text{ μA} = 990 \text{ μA}$$

$$I_C = I_E - I_B = 1000 \text{ μA} - 10 \text{ μA} = 990 \text{ μA}$$

B. Solving in terms of I_B:

$$I_B = I_C/\beta = 990 \text{ μA}/99 = 10 \text{ μA}$$

$$I_B = I_E/(\beta + 1) = 1000 \text{ μA}/(99 + 1) = 10 \text{ μA}$$

$$I_B = I_E - I_C = 1000 \text{ μA} - 990 \text{ μA} = 10 \text{ μA}$$

C. Solving in terms of I_E:

$$I_E = (\beta + 1)I_B = (99 + 1)10 \text{ μA} = 1000 \text{ μA}$$

$$I_E = I_C/a = 990 \text{ μA}/0.99 = 1000 \text{ μA}$$

$$I_E = I_C + I_B = 990 \text{ μA} + 10 \text{ μA} = 1000 \text{ μA}$$

NOTE: $I_E = I_C + I_B = \beta I_B + I_B = I_B(\beta + 1)$

FIXED BASE CURRENT BIASING TECHNIQUES

Connecting an NPN transistor into a circuit like that of Figure 4-15, and slowly increasing the V_{BB} causes the base current to increase and, because $I_C = \beta I_B$ and $I_E = (\beta + 1)I_B$, the collector and emitter currents will also increase. The base resistor RB provides base current control.

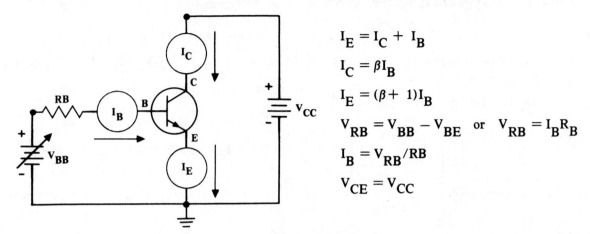

$$I_E = I_C + I_B$$
$$I_C = \beta I_B$$
$$I_E = (\beta + 1)I_B$$
$$V_{RB} = V_{BB} - V_{BE} \quad \text{or} \quad V_{RB} = I_B R_B$$
$$I_B = V_{RB}/RB$$
$$V_{CE} = V_{CC}$$

FIGURE 4-15

Therefore, if V_{BB} is fixed at 4 volts and RB is selected at 340 kΩ, then $I_B = 10$ μA. Also, if $\beta = 99$, then $I_C = 990$ μA and $I_E = 1000$ μA. Too, if $V_{CC} = 12$ volts, then $V_{CE} = 12$ volts. These various values are shown and solved in Figure 4-16.

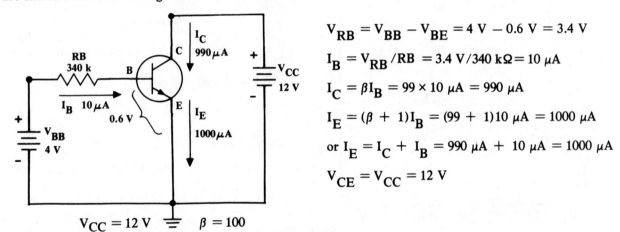

$$V_{RB} = V_{BB} - V_{BE} = 4\text{ V} - 0.6\text{ V} = 3.4\text{ V}$$
$$I_B = V_{RB}/RB = 3.4\text{ V}/340\text{ k}\Omega = 10\text{ μA}$$
$$I_C = \beta I_B = 99 \times 10\text{ μA} = 990\text{ μA}$$
$$I_E = (\beta + 1)I_B = (99 + 1)10\text{ μA} = 1000\text{ μA}$$
$$\text{or } I_E = I_C + I_B = 990\text{ μA} + 10\text{ μA} = 1000\text{ μA}$$
$$V_{CE} = V_{CC} = 12\text{ V}$$

FIGURE 4-16

Once the base current I_B is fixed, increasing or decreasing V_{CC} does not effect the I_C and I_E current flow. Therefore, if $\beta = 99$ remains constant, then I_C will remain at 990 μA and I_E at 1000 μA. This is shown in Figure 4-17, where $V_{CC} = V_{CE}$ is 6 volts and then is increased to 15 volts.

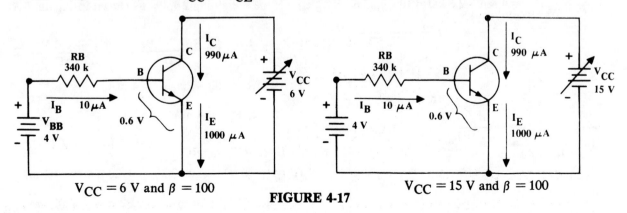

FIGURE 4-17

However, if a device with a different beta is used, I_B will continue to remain fixed if V_{BE} remains constant at 0.6 volts, but both I_C and I_E will change. For instance, if the beta of the device increases to 200, then I_C will be 2 mA and I_E will be 2.01 mA, as shown in Figure 4-18

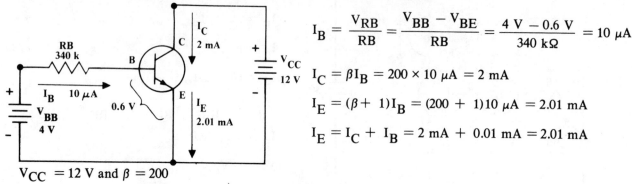

$$I_B = \frac{V_{RB}}{RB} = \frac{V_{BB} - V_{BE}}{RB} = \frac{4\text{ V} - 0.6\text{ V}}{340\text{ k}\Omega} = 10\ \mu A$$

$$I_C = \beta I_B = 200 \times 10\ \mu A = 2\text{ mA}$$

$$I_E = (\beta + 1)I_B = (200 + 1)10\ \mu A = 2.01\text{ mA}$$

$$I_E = I_C + I_B = 2\text{ mA} + 0.01\text{ mA} = 2.01\text{ mA}$$

$V_{CC} = 12$ V and $\beta = 200$

FIGURE 4-18

Therefore, a fixed base circuit is highly dependent on beta but not on V_{CC}, as long as the device remains in the linear region and does not exceed its power limitation. Therefore, if the base current is fixed by V_{BB} and RB, and I_C and I_E are fixed by constant beta conditons, then inserting a collector resistor will not change I_C or I_E, but it will change V_{CE}. The V_{CE} changes because the voltage drop across V_{RC} subtracts from the V_{CC} voltage, as shown in Figure 4-19.

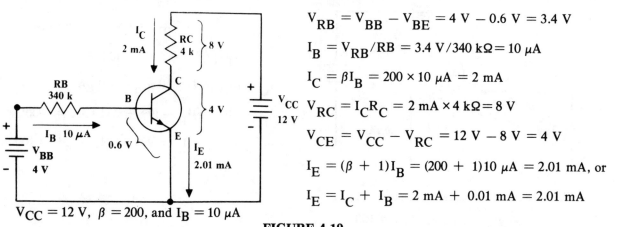

$$V_{RB} = V_{BB} - V_{BE} = 4\text{ V} - 0.6\text{ V} = 3.4\text{ V}$$

$$I_B = V_{RB}/RB = 3.4\text{ V}/340\text{ k}\Omega = 10\ \mu A$$

$$I_C = \beta I_B = 200 \times 10\ \mu A = 2\text{ mA}$$

$$V_{RC} = I_C R_C = 2\text{ mA} \times 4\text{ k}\Omega = 8\text{ V}$$

$$V_{CE} = V_{CC} - V_{RC} = 12\text{ V} - 8\text{ V} = 4\text{ V}$$

$$I_E = (\beta + 1)I_B = (200 + 1)10\ \mu A = 2.01\text{ mA, or}$$

$$I_E = I_C + I_B = 2\text{ mA} + 0.01\text{ mA} = 2.01\text{ mA}$$

$V_{CC} = 12$ V, $\beta = 200$, and $I_B = 10\ \mu A$

FIGURE 4-19

If the beta remains at 200 but RB is decreased to 170 kΩ, then I_B will increase to 20 μA and I_C to 4 mA. However, the RC resistor at 4 kΩ will then absorb all of the available V_{CC} voltage causing the collector-emitter voltage to approach a zero volt saturation condition. Therefore, in order to keep the circuit operating in the linear region, RC must be reduced in turn. For example, if RC is reduced to 2.5 kΩ, then, $V_{RC} = I_C R_C = 4\text{ mA} \times 2.5\text{ k}\Omega = 10$ volts and V_{CE} equals the remaining 2 volts, as shown in Figure 4-20.

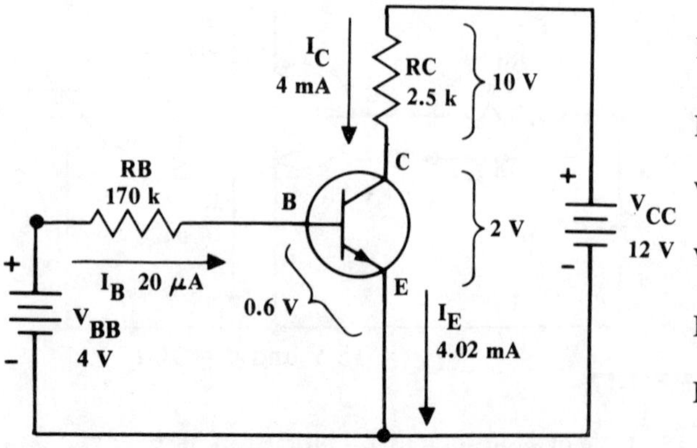

$$I_B = \frac{V_{RB}}{RB} = \frac{V_{BB} - V_{BE}}{RB} = \frac{4\text{ V} - 0.6\text{ V}}{170\text{ k}\Omega} = 20\ \mu A$$

$$I_C = \beta I_B = 200 \times 20\ \mu A = 4\text{ mA}$$

$$V_{RC} = I_C R_C = 4\text{ mA} \times 2.5\text{ k}\Omega = 10\text{ V}$$

$$V_{CE} = V_{CC} - V_{RC} = 12\text{ V} - 10\text{ V} = 2\text{ V}$$

$$I_E = (\beta + 1)I_B = (200 + 1)20\ \mu A = 4.02\text{ mA}$$

$$I_E = I_C + I_B = 4\text{ mA} + 0.02\text{ mA} = 4.02\text{ mA}$$

$V_{CC} = 12$ V, $\beta = 200$, and $I_B = 20\ \mu A$

FIGURE 4-20

OUTPUT CHARACTERISTIC CURVES — USING FIXED BASE CURRENT CONDITIONS

Characteristic curves of transistors, like those of diodes, provide graphical solutions to device currents and corresponding voltages. Two output characteristic curves of I_C versus I_B and I_C versus I_E can be plotted from the I_B, I_C, and I_E device currents obtained from just one circuit. Therefore, in the fixed biased circuits of Figure 4-21 through Figure 4-24, I_B will be changed to 10 μA by adjusting V_{BB} to 1.6 V, to 20 μA by adjusting V_{BB} to 2.6 V, to 30 μA by adjusting V_{BB} to 3.6 V, and to 40 μA by adjusting V_{BB} to 4.6 V.

Therefore, for the output characteristics, I_B will be fixed which, in turn, fixes I_C and I_E. Then, V_{CE} is varied from 0 volts to the collector-emitter breakdown voltage region for the device. Also, because the I_C versus I_B currents are primary to the fixed base current circuit, these characteristic curves are shown first.

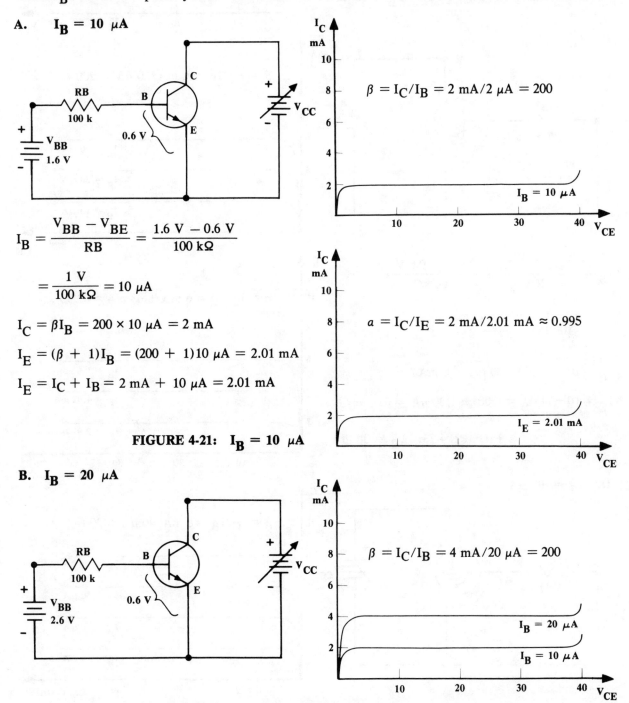

A. $I_B = 10$ μA

$$I_B = \frac{V_{BB} - V_{BE}}{R_B} = \frac{1.6 \text{ V} - 0.6 \text{ V}}{100 \text{ k}\Omega}$$

$$= \frac{1 \text{ V}}{100 \text{ k}\Omega} = 10 \text{ μA}$$

$I_C = \beta I_B = 200 \times 10$ μA $= 2$ mA

$I_E = (\beta + 1)I_B = (200 + 1)10$ μA $= 2.01$ mA

$I_E = I_C + I_B = 2$ mA $+ 10$ μA $= 2.01$ mA

$\beta = I_C/I_B = 2$ mA$/2$ μA $= 200$

$a = I_C/I_E = 2$ mA$/2.01$ mA ≈ 0.995

FIGURE 4-21: $I_B = 10$ μA

B. $I_B = 20$ μA

$\beta = I_C/I_B = 4$ mA$/20$ μA $= 200$

$$I_B = \frac{V_{BB} - V_{BE}}{RB} = \frac{2.6\text{ V} - 0.6\text{ V}}{100\text{ k}\Omega}$$

$$= \frac{2\text{ V}}{100\text{ k}\Omega} = 20\text{ }\mu\text{A}$$

$$I_C = \beta I_B = 200 \times 20\text{ }\mu\text{A} = 4\text{ mA}$$

$$I_E = (\beta + 1)I_B = (200 + 1)20\text{ }\mu\text{A} = 4.02\text{ mA}$$

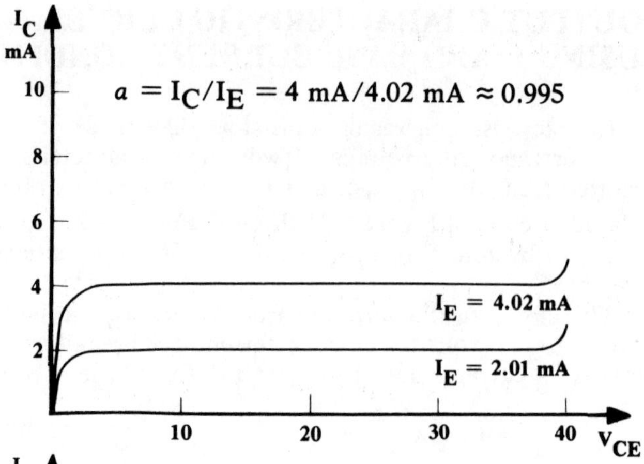

FIGURE 4-22: $I_B = 20\text{ }\mu\text{A}$

$a = I_C/I_E = 4\text{ mA}/4.02\text{ mA} \approx 0.995$

$I_E = 4.02\text{ mA}$
$I_E = 2.01\text{ mA}$

C. $I_B = 30\text{ }\mu\text{A}$

$\beta = I_C/I_B = 6\text{ mA}/30\text{ }\mu\text{A} = 200$

$I_B = 30\text{ }\mu\text{A}$
$I_B = 20\text{ }\mu\text{A}$
$I_B = 10\text{ }\mu\text{A}$

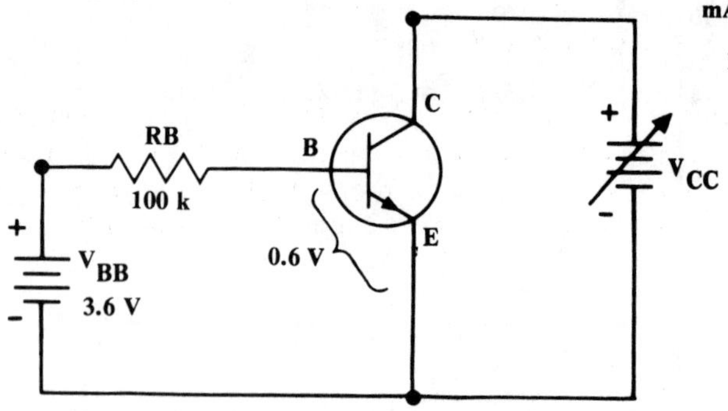

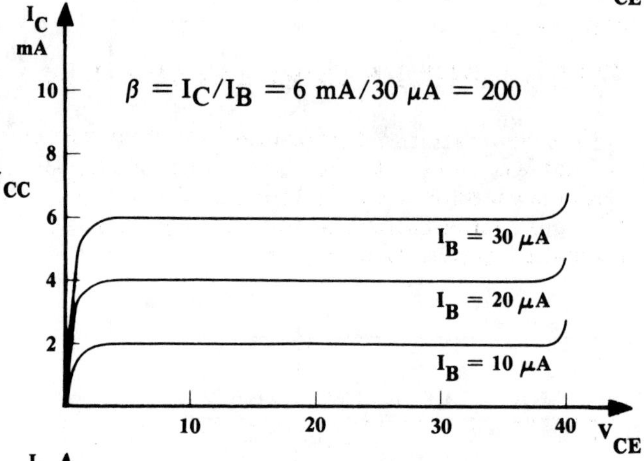

$$I_B = \frac{V_{BB} - V_{BE}}{RB} = \frac{3.6\text{ V} - 0.6\text{ V}}{100\text{ k}\Omega}$$

$$\frac{3\text{ V}}{100\text{ k}\Omega} = 30\text{ }\mu\text{A}$$

$$I_C = \beta I_B = 200 \times 30\text{ }\mu\text{A} = 6\text{ mA}$$

$$I_E = (\beta + 1)I_B = (200 + 1)30\text{ }\mu\text{A} = 6.03\text{ mA}$$

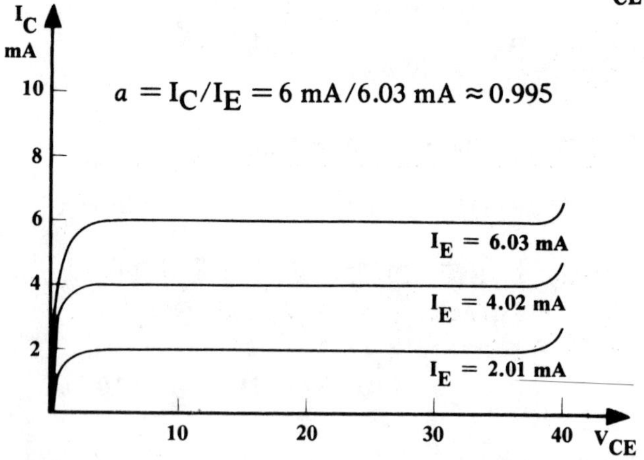

FIGURE 4-23: $I_B = 30\text{ }\mu\text{A}$

$a = I_C/I_E = 6\text{ mA}/6.03\text{ mA} \approx 0.995$

$I_E = 6.03\text{ mA}$
$I_E = 4.02\text{ mA}$
$I_E = 2.01\text{ mA}$

D. $I_B = 40\text{ }\mu\text{A}$

$\beta = I_C/I_B = 8\text{ mA}/40\text{ }\mu\text{A} = 200$

$I_B = 40\text{ }\mu\text{A}$
$I_B = 30\text{ }\mu\text{A}$
$I_B = 20\text{ }\mu\text{A}$
$I_B = 10\text{ }\mu\text{A}$

B

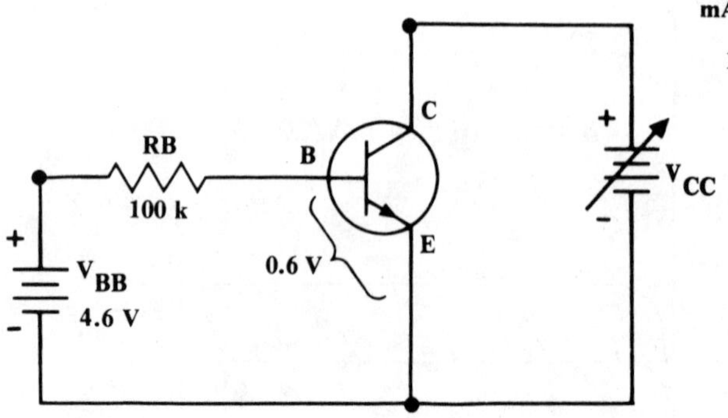

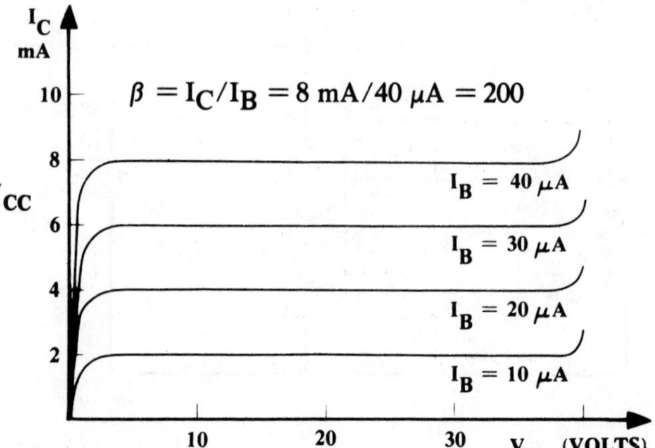

$$I_B = \frac{V_{BB} - V_{BE}}{RB} = \frac{4.6\text{ V} - 0.6\text{ V}}{100\text{ k}\Omega}$$

$$= \frac{4\text{ V}}{100\text{ k}\Omega} = 40\text{ }\mu\text{A}$$

$$I_C = \beta I_B = 200 \times 40\text{ }\mu\text{A} = 8\text{ mA}$$

$$I_E = (\beta + 1)I_B = (200 + 1)40\text{ }\mu\text{A} = 8.04\text{ mA}$$

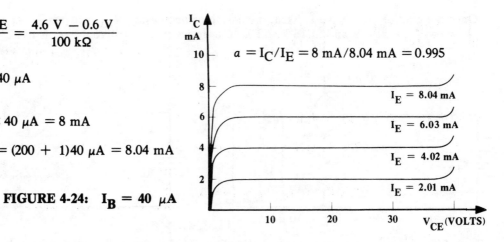

FIGURE 4-24: $I_B = 40\text{ }\mu\text{A}$

FIXED EMITTER CURRENT BIASING TECHNIQUES

Connecting an NPN transistor into a circuit configuration, such as that of Figure 4-25, allows the emitter resistor RE, along with the emitter voltage dropped across RE, to control the emitter current and, therefore, the collector and base currents of the device. Also, because $V_{RE} = V_{BB} - V_{BE}$ and $V_{BE} \approx 0.6$ V, then the emitter current is determined primarily by V_{BB} and RE, and only slightly by the parameters of the device, such as beta.

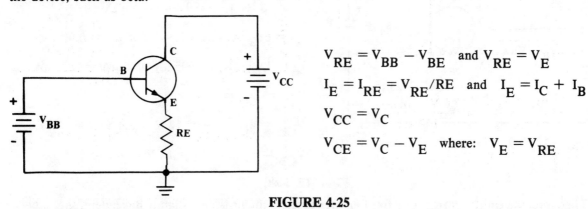

$$V_{RE} = V_{BB} - V_{BE} \quad \text{and} \quad V_{RE} = V_E$$

$$I_E = I_{RE} = V_{RE}/RE \quad \text{and} \quad I_E = I_C + I_B$$

$$V_{CC} = V_C$$

$$V_{CE} = V_C - V_E \quad \text{where: } V_E = V_{RE}$$

FIGURE 4-25

Therefore, if V_{BB} is fixed at 4 volts and RE is selected at 3.4 kΩ, then $I_E = 1$ mA. Again, if $\beta = 99$, then $I_C = 990\text{ }\mu\text{A}$ and $I_B = 10\text{ }\mu\text{A}$, and if $V_{CC} = 12$ V, then $V_{CE} = 8.6$ V. These values are shown and calculated in Figure 4-26.

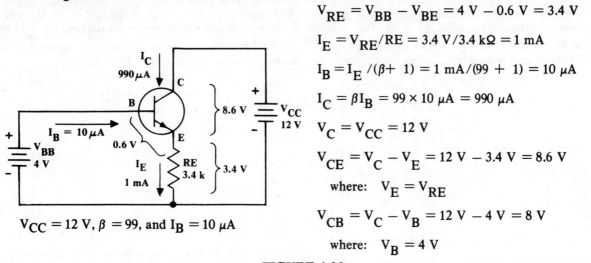

$V_{CC} = 12$ V, $\beta = 99$, and $I_B = 10\text{ }\mu\text{A}$

$$V_{RE} = V_{BB} - V_{BE} = 4\text{ V} - 0.6\text{ V} = 3.4\text{ V}$$

$$I_E = V_{RE}/RE = 3.4\text{ V}/3.4\text{ k}\Omega = 1\text{ mA}$$

$$I_B = I_E/(\beta + 1) = 1\text{ mA}/(99 + 1) = 10\text{ }\mu\text{A}$$

$$I_C = \beta I_B = 99 \times 10\text{ }\mu\text{A} = 990\text{ }\mu\text{A}$$

$$V_C = V_{CC} = 12\text{ V}$$

$$V_{CE} = V_C - V_E = 12\text{ V} - 3.4\text{ V} = 8.6\text{ V}$$

where: $V_E = V_{RE}$

$$V_{CB} = V_C - V_B = 12\text{ V} - 4\text{ V} = 8\text{ V}$$

where: $V_B = 4$ V

FIGURE 4-26

Once the emitter current I_E is fixed, increasing or decreasing V_{CC} will not effect I_E, I_C, or I_B. Figure 4-27 shows V_{CC} at 6 V and at 15 V. Too, if a device with a different beta is used, then I_E will remain

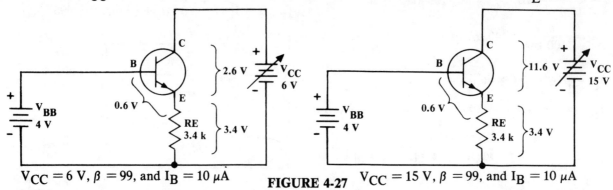

$V_{CC} = 6$ V, $\beta = 99$, and $I_B = 10$ μA **FIGURE 4-27** $V_{CC} = 15$ V, $\beta = 99$, and $I_B = 10$ μA

fixed if V_{BE} remains constant at 0.6 V. However, I_B will change because of beta, but I_C will be effected only slightly since $I_C = I_E - I_B$. For example, as shown in Figure 4-28, if the beta is 199, then $I_E = 1$ mA, $I_B = 5$ μA, and $I_C = 0.995$ mA.

$$I_E = \frac{V_{RE}}{RE} = \frac{V_{BB} - V_{BE}}{RE} = \frac{4\text{ V} - 0.6\text{ V}}{3.4\text{ k}\Omega} = 1\text{ mA}$$

$$I_B = \frac{I_E}{\beta + 1} = \frac{1\text{ mA}}{199 + 1} = \frac{1000\text{ μA}}{200} = 5\text{ μA}$$

$$I_C = \beta I_B = 5\text{ μA} \times 199 = 995\text{ μA} = 0.995\text{ mA}$$

$$I_E = I_C + I_B = 995\text{ μA} + 5\text{ μA} = 1000\text{ μA}$$

$$V_C = V_{CC} = 12\text{ V}$$

$$V_{CE} = V_C - V_E = 12\text{ V} - 3.4\text{ V} = 8.6\text{ V}$$

$$V_{CB} = V_C - V_B = 12\text{ V} - 4\text{ V} = 8\text{ V}$$

$V_{CC} = 12$ V, $\beta = 199$, and $I_B = 5$ μA

FIGURE 4-28

Therefore, the emitter current of a fixed emitter current circuit is only slightly dependent on the beta of the device and it is primarily determined by the voltage drop across the emitter resistor RE. Obviously, any form of biasing that minimizes the beta of the device provides DC stability to the circuit. Therefore, emitter current control is far superior to base current control and emitter current control is the method normally used in circuit design. (In Figures 4-26 to 4-28, beta change does not effect V DC distribution.)

Adding a collector resistor to the circuit, as shown in Figure 4-29, does not change I_E, I_C, or I_B; but it does effect the voltage dropped across the collector-emitter V_{CE} and collector-base V_{CB} junctions of the device, as shown in Figure 4-29 calculations. A beta of 99 is again assumed. Also, the reverse-biased, base-collector junction establishes a close to ideal constant current source, where $I_C \approx I_E$.

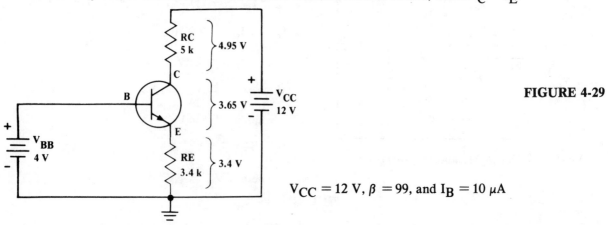

FIGURE 4-29

$V_{CC} = 12$ V, $\beta = 99$, and $I_B = 10$ μA

$V_{RE} = V_{BB} - V_{BE} = 4\text{ V} - 0.6\text{ V} = 3.4\text{ V}$

$I_E = V_{RE}/RE = 3.4\text{ V}/3.4\text{ k}\Omega = 1\text{ mA}$

$I_B = I_E/(\beta + 1) = 1\text{ mA}/(99 + 1) = 10\text{ μA}$

$I_C = \beta I_B = 99 \times 10\text{ μA} = 0.99\text{ mA}$

$V_{RC} = I_C R_C = 0.99\text{ mA} \times 5\text{ k}\Omega = 4.95\text{ V}$

$V_C = V_{CC} - V_{RC} = 12\text{ V} - 4.95\text{ V} = 7.05\text{ V}$

$V_{CE} = V_C - V_E = 7.05\text{ V} - 3.4\text{ V} = 3.65\text{ V}$

$V_{CB} = V_C - V_B = 7.05\text{ V} - 4\text{ V} = 3.05\text{ V}$

$V_B = 4\text{ V}$ and

$V_E = V_{RE} = 3.4\text{ V}$

$V_{CC} = V_{RC} + V_{CE} + V_{RE}$

$\quad\quad = 4.95\text{ V} + 3.65\text{ V} + 3.4\text{ V} = 12\text{ V}$

CALCULATIONS FOR FIGURE 4-29

OUTPUT CHARACTERISTIC CURVES — USING FIXED EMITTER CURRENT CONDITIONS

The output characteristic curves can also be obtained using the fixed emitter current circuit. Again, both sets of I_C versus I_E and I_C versus I_B characteristics are given to show that only one circuit is necessary for obtaining both sets, because of the relationship of I_B, I_C, and I_E. However, because I_C versus I_E currents are primary to this circuit, these characteristic curves are shown first. Therefore, in the circuits of Figure 4-31 through Figure 4-34, RE is 1 kΩ and V_{BB} is adjusted to provide emitter currents of 2 mA, 4 mA, 6 mA, 8 mA, and 10 mA. The V_{CC} is set at a constant 12 V and beta equals 99.

A. $I_E = 2\text{ mA}$

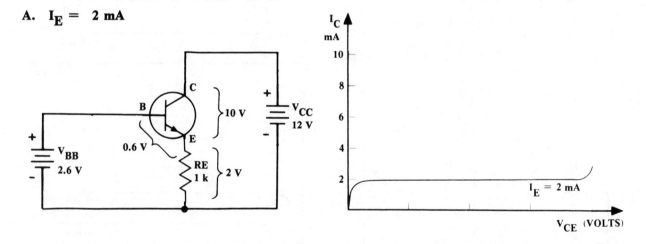

$I_E = \dfrac{V_{BB} - V_{BE}}{RE} = \dfrac{2.6\text{ V} - 0.6\text{ V}}{1\text{ k}\Omega}$

$\quad = \dfrac{2\text{ V}}{1\text{ k}\Omega} = 2\text{ mA}$

$I_B = I_E/(\beta + 1) = 2\text{ mA}/(99 + 1) = 20\text{ μA}$

$I_C = \beta I_B = 99 \times 20\text{ μA} = 1.98\text{ mA}$

$I_C = \alpha I_E = 0.99 \times 2\text{ mA} = 1.98\text{ mA}$

FIGURE 4-30: $I_E = 2\text{ mA}$

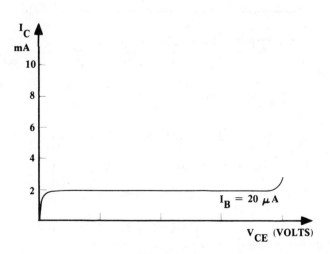

B. $I_E = 4$ mA

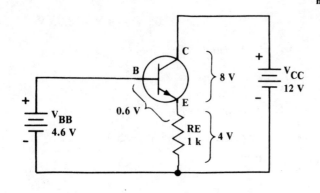

$$I_E = \frac{V_{BB} - V_{BE}}{RE} = \frac{4.6 \text{ V} - 0.6 \text{ V}}{1 \text{ k}\Omega}$$

$$= \frac{4 \text{ V}}{1 \text{ k}\Omega} = 4 \text{ mA}$$

$I_B = I_E/(\beta + 1) = 4 \text{ mA}/(99 + 1) = 40 \text{ }\mu\text{A}$

$I_C = \beta I_B = 99 \times 40 \text{ }\mu\text{A} = 3.96 \text{ mA}$

$I_C = \alpha I_E = 0.99 \times 4 \text{ mA} = 3.96 \text{ mA}$

FIGURE 4-31: $I_E = 4$ mA

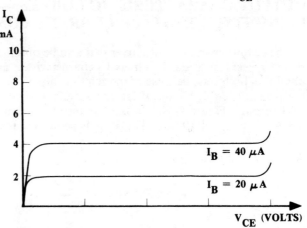

C. $I_E = 6$ mA

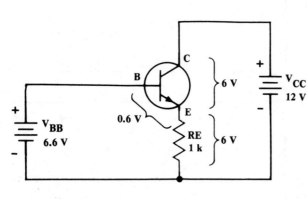

$$I_E = \frac{V_{BB} - V_{BE}}{RE} = \frac{6.6 \text{ V} - 0.6 \text{ V}}{1 \text{ k}\Omega}$$

$$= \frac{6 \text{ V}}{1 \text{ k}\Omega} = 6 \text{ mA}$$

$I_B = I_E/(\beta + 1) = 6 \text{ mA}/(99 + 1) = 60 \text{ }\mu\text{A}$

$I_C = \beta I_B = 99 \times 60 \text{ }\mu\text{A} = 5.94 \text{ mA}$

$I_C = \alpha I_E = 0.99 \times 6 \text{ mA} = 5.94 \text{ mA}$

FIGURE 4-32: $I_E = 6$ mA

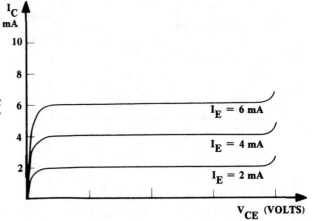

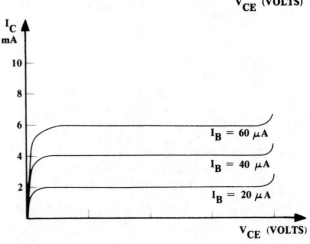

D. $I_E = 8$ mA

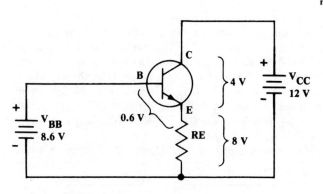

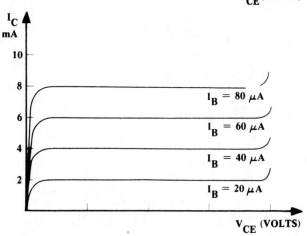

$$I_E = \frac{V_{BB} - V_{BE}}{R_E} = \frac{8.6 \text{ V} - 0.6 \text{ V}}{1 \text{ k}\Omega}$$

$$= \frac{8 \text{ V}}{1 \text{ k}\Omega} = 8 \text{ mA}$$

$I_B = I_E/\beta + 1) = 8 \text{ mA} / (99 + 1) = 80 \text{ }\mu\text{A}$

$I_C = \beta I_B = 99 \times 80 \text{ }\mu\text{A} = 7.92 \text{ mA}$

$I_C = \alpha I_E = 0.99 \times 8 \text{ mA} = 7.92 \text{ mA}$

FIGURE 4-33: $I_E = 8$ mA

E. $I_E = 10$ mA

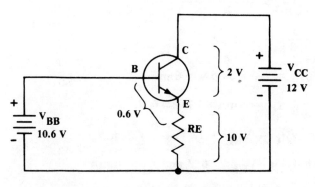

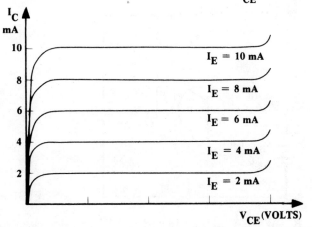

$$I_E = \frac{V_{BB} - V_{BE}}{R_E} = \frac{10.6 \text{ V} - 0.6 \text{ V}}{1 \text{ k}\Omega}$$

$$= \frac{10 \text{ V}}{1 \text{ k}\Omega} = 10 \text{ mA}$$

$I_B = I_E/(\beta + 1) = 10 \text{ mA}/(99 + 1) = 100 \text{ }\mu\text{A}$

$I_C = \beta I_B = 99 \times 100 \text{ }\mu\text{A} = 9.9 \text{ mA}$

$I_C = \alpha I_E = 0.99 \times 10 \text{ mA} = 9.9 \text{ mA}$

FIGURE 4-34: $I_E = 10$ mA

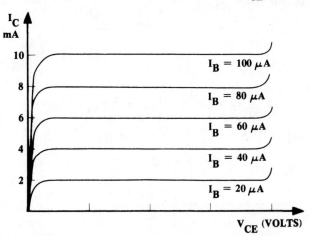

DIRECT CURRENT LOAD LINES

Graphical analysis, using load line techniques, can be used as a tool for solving the voltage drops and corresponding collector current of both the device and collector resistor RC of the circuit shown in Figure 4-35. In this example, I_C is varied from a condition of 0 mA through 12 mA, and the voltage drop across both the device and the collector resistor RC are plotted. As the load line of Figure 4-35 shows, when $I_B = 0$ μA, then $I_C = 0$ mA and the device is in the cutoff condition. When I_B is increased to 60 μA, then $I_C \approx 12$ mA, and almost all of the V_{CC} voltage is dropped across the collector resistor RC, forcing the device into saturation condition. However, in saturation V_{CE} rarely goes below 0.1 V and I_C will not reach the maximum 12 mA. The graphical analysis for 0 mA, 3 mA, 6 mA, 9 mA, and 12 mA of I_C are shown in the load line of Figure 4-35 and the individual circuit solutions are shown in Figures 4-36 through 4-40.

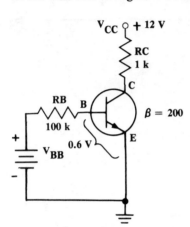

FIGURE 4-35

A. When $V_{BB} = 3.6$ V, then $I_B = 30$ μA, $I_C = 6$ mA, $V_{RC} = 6$ V, and $V_{CE} = 6$ V.

$$I_B = \frac{V_{RB}}{RB} = \frac{V_{BB} - V_{BE}}{RB} = \frac{3.6\text{ V} - 0.6\text{ V}}{100\text{ k}\Omega}$$

$$= \frac{3\text{ V}}{100\text{ k}\Omega} = 30\text{ μA}$$

$$I_C = \beta I_B = 200 \times 30\text{ μA} = 6\text{ mA}$$

$$V_{RC} = I_C R_C = 6\text{ mA} \times 1\text{ k}\Omega = 6\text{ V}$$

$$V_{CE} = V_{CC} - V_{RC} = 12\text{ V} - 6\text{ V} = 6\text{ V}$$

FIGURE 4-36: $V_{CE} = 6$ V and $I_C = 6$ mA

B. When V_{BB} is decreased to 2.1 V, I_B decreases to 15 μA, I_C to 3 mA, V_{RC} to 3 V, and V_{CE} to 9 V.

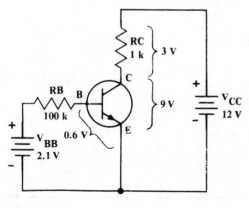

$$I_B = \frac{V_{RB}}{RB} = \frac{V_{BB} - V_{BE}}{RB} = \frac{2.1\text{ V} - 0.6\text{ V}}{100\text{ k}\Omega}$$

$$= \frac{1.5\text{ V}}{100\text{ k}\Omega} = 15\text{ μA}$$

$$I_C = \beta I_B = 200 \times 15\text{ μA} = 3\text{ mA}$$

$$V_{RC} = I_C R_C = 3\text{ mA} \times 1\text{ k}\Omega = 3\text{ V}$$

$$V_{CE} = V_{CC} - V_{RC} = 12\text{ V} - 3\text{ V} = 9\text{ V}$$

FIGURE 4-37: $V_{CE} = 9$ V and $I_C = 3$ mA

C. When V_{CC} is increased to 5.1 V, I_B increases to 45 μA, I_C to 9 mA, V_{RC} to 9 V, and V_{CE} to 3 V.

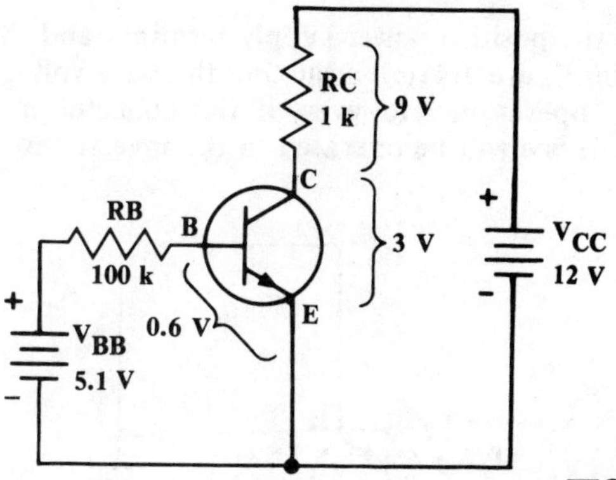

$$I_B = \frac{V_{RB}}{RB} = \frac{V_{BB} - V_{BE}}{RB} = \frac{5.1\ V - 0.6\ V}{100\ k\Omega}$$

$$= \frac{4.5\ V}{100\ k\Omega} = 45\ \mu A$$

$$I_C = \beta I_B = 200 \times 45\ \mu A = 9\ mA$$

$$V_{RC} = I_C R_C = 9\ mA \times 1\ k\Omega = 9\ V$$

$$V_{CE} = V_{CC} - V_{RC} = 12\ V - 9\ V = 3\ V$$

FIGURE 4-38: $V_{CE} = 3\ V$ and $I_C = 9\ mA$

D. When V_{BB} is increased to 6.6 V, I_B increases to 60 μA, I_C to 12 mA (ideally), V_{RC} to 12 V, and V_{CE} to 0 V.

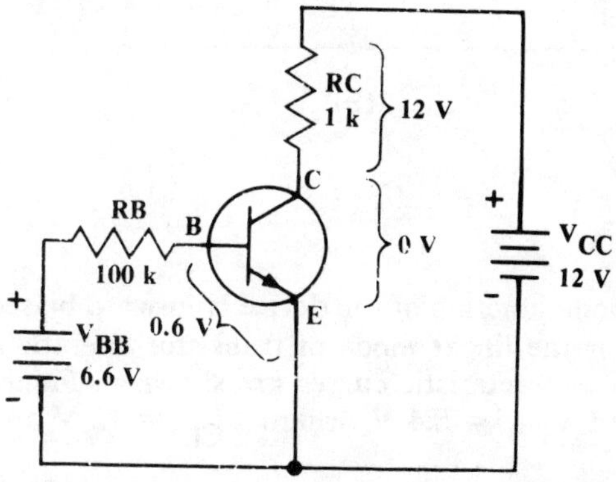

$$I_B = \frac{V_{RB}}{RB} = \frac{V_{BB} - V_{BE}}{RB} = \frac{6.6\ V - 0.6\ V}{100\ k\Omega}$$

$$= \frac{6\ V}{100\ k\Omega} = 60\ \mu A$$

$$I_C = \beta I_B = 200 \times 60\ \mu A \approx 12\ mA$$

$$V_{RC} = I_C R_C = 12\ mA \times 1\ k\Omega = 12\ V$$

$$V_{CE} = V_{CC} - V_{RC} = 12\ V - 12\ V = 0.0\ V \quad \text{(ideally)}$$

FIGURE 4-39: (SATURATION) $V_{CE} = 0.0\ V$ and $I_C = 12\ mA$

NOTE: V_{CE} (SAT) is usually between 0.0 V and 0.5 V. Therefore, if V_{CE} (SAT) = 0.2 V, then I_C = 11.8 mA, since V_{RC} = 11.8 V.

E. When V_{BB} is decreased to about 0.0 V, the base current decreases to 0.0 mA (ideally), I_C to 0.0 mA, V_{RC} to 0.0 V, and V_{CE} to 12 V.

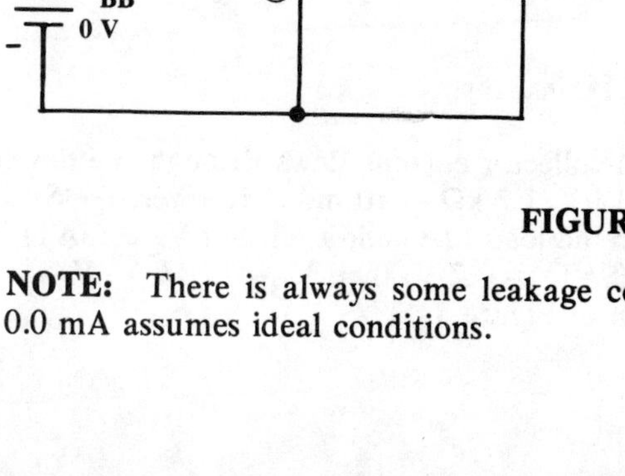

$$I_B = \frac{V_{RB}}{RB} = \frac{V_{BB} - V_{BE}}{RB} = \frac{0.6\ V - 0.6\ V}{100\ k\Omega}$$

$$= \frac{0\ V}{100\ k\Omega} = 0\ \mu A$$

$$I_C = \beta I_B = 200 \times 0\ \mu A = 0\ mA$$

$$V_{RC} = I_C R_C = 0\ mA \times 1\ k\Omega = 0\ V$$

$$V_{CE} = V_{CC} - V_{RC} = 12\ V - 0\ V = 12\ V$$

$$V_C = V_{CE} = 12\ V$$

FIGURE 4-40: (CUTOFF) $V_{CE} = 12\ V$ and $I_C = 0.0\ mA$

NOTE: There is always some leakage collector current, although extremely small, during cutoff. The 0.0 mA assumes ideal conditions.

MODES OF OPERATION — FURTHER INVESTIGATION

When the collector load of an NPN transistor is connected to the positive power supply terminal and the emitter load to the negative power supply terminal, as shown in Figure 4-41(a), adjusting the base voltage can place the device in saturation, cutoff, or linear modes of operation. However, if the collector and emitter loads are interchanged, as shown in Figure 4-41(b), the device will be operated in the inverse mode with resultant low beta and breakdown voltage conditions.

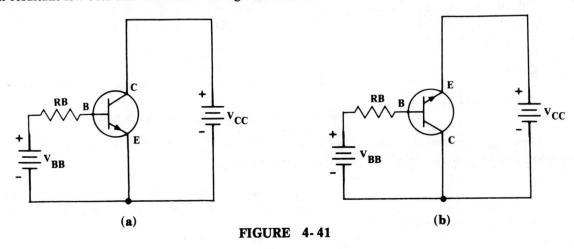

FIGURE 4-41

LINEAR MODE OF TRANSISTOR OPERATION

In the linear mode of transistor operation, the base-emitter diode junction of the device is forward biased and the collector-base junction is reverse biased. The circuit for the linear mode of transistor operation, the simulated diode configuration, and a respresentative set of characteristic curves are shown in Figure 4-42. Notice that $V_{BE} = 0.6$ V, $V_{RC} = 6$ V, $V_{CE} = 6$ V, and $V_{CB} = 5.4$ V, where $V_{CC} = 12$ V and $\beta = 100$.

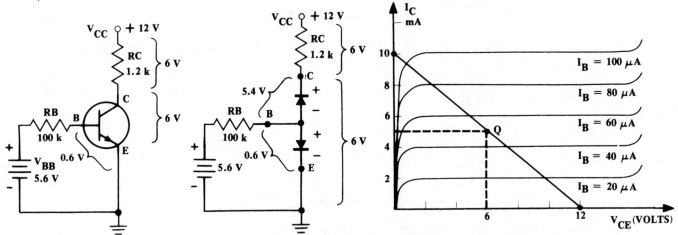

FIGURE 4-42

SATURATED MODE OF TRANSISTOR OPERATION

In the saturated mode of transistor operation, the maximum collector current flows through the device and is calculated, theoretically, from I_C (SAT) $= V_{CC}/RC = 12$ V$/1.2$ k$\Omega = 10$ mA. However, I_C (SAT) at 10 mA assumes that V_{CE} (SAT) $= 0$ V, to establish an extreme load line condition, but V_{CE} (SAT) is actually in the 0.2 V to 0.5 V range. For instance, when V_{CE} (SAT) $= 0.2$ V, then $V_{BE} = 0.6$ V, $V_{CB} = 0.4$ V, and both diodes are effectively forward biased, as shown in Figure 4-43.

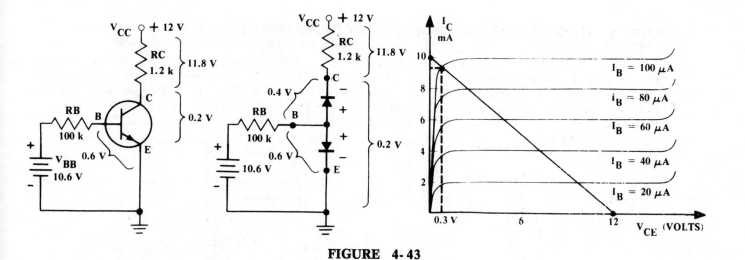

FIGURE 4-43

CUTOFF MODE OF TRANSISTOR OPERATION

In the cutoff mode of transistor operation almost no collector current flows through the device. Therefore, approximately zero volts are dropped across the collector-emitter of the device. However, actual cutoff conditions require that the base-emitter junction be reverse biased, and simply bringing the V_{BB} voltage down to zero volts allows the base to "float". Then, when the base load is open or floating, the I_{Co} collector current is increased by a factor of beta. Therefore, if $I_{Co} = 100$ nanoamps, $I_{CEo} \approx 100\, I_{Co} = 100 \times 100$ nA $= 10\,\mu$A. For 10 μA of collector current and an RC resistor of 1.2 kΩ, about 0.01 V is dropped across RC and about 11.99 V of the total 12 V across V_{CE}, as shown in Figure 4-44. I_{CEo} and I_{Co} are the reverse-biased, leakage currents of the device.

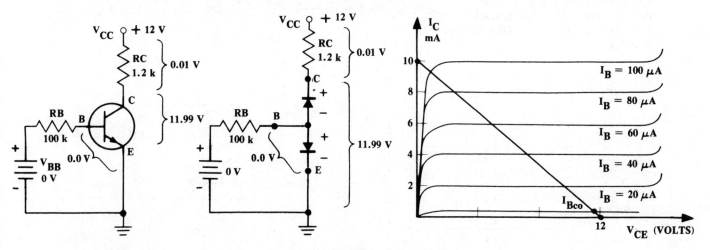

FIGURE 4-44

INVERSE MODE OF TRANSISTOR OPERATION

In the inverse mode of transistor operation, the collector and emitter terminals of the device are interchanged. Therefore, the base-collector junction is forward biased and the emitter-base junction is reverse biased and, at low V_{CE} voltages, the transistor will operate like a properly-biased, low-voltage device. However, as the V_{CE} voltage is increased much beyond 8 to 10 volts the device begins to break down, as shown in Figure 4-45.

Therefore, in order to provide an operating current of 5 mA for this circuit, 500 μA of base current is required for a beta of 10, and 5.6 V of V_{BB} with a base resistor of 10 kΩ must be used.

NOTE: Inverse connection usually requires correcting because, in most instances, it is the result of a wiring error. Obviously, the doping techniques used for the base-emitter and collector-base junctions are different enough so that the collector and emitter terminals are not successfully interchangeable.

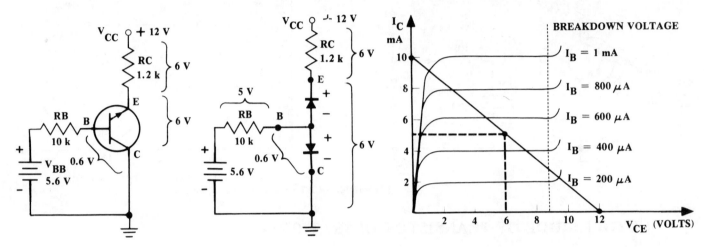

FIGURE 4-45

SINGLE POWER SUPPLY BENCH MEASUREMENT TECHNIQUES

Using two power supplies and current meters in the base and collector branches, as shown in Figure 4-46, provides a convenient method of directly monitoring the base and collector currents. And once I_C and I_B are known, beta is solved from: $\beta = I_C/I_B$.

NOTE: $\beta = I_C/I_B$, where I_C is in milliamperes and I_B is in microamperes.

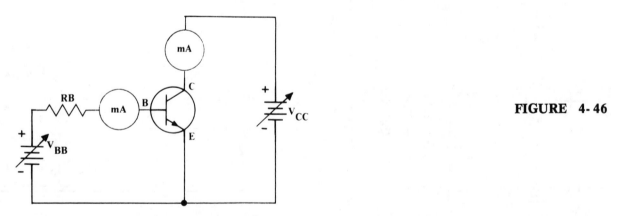

FIGURE 4-46

However, current meters are not normally used in practice. Instead, voltage drop measurements across known resistance values are used to find the current values. For example, if a collector resistor is added to the circuit, as shown in Figure 4-47, and V_{BB} is varied, then the voltage drop across V_{RB} and V_{RC} will vary and I_B is solved from $I_B = V_{RB}/RB$ and I_C is solved from $I_C = V_{RC}/RC$. Also, since all measurements are made with respect to ground, V_{BB} and V_{BE} can be measured directly but V_{RC} is solved from $V_{RC} = V_{CC} - V_{CE}$.

$$I_B = \frac{V_{BB} - V_{BE}}{RB} = \frac{V_{RB}}{RB} \quad \text{where:} \quad V_{BE} = 0.6 \text{ V}$$

$$I_C = \frac{V_{CC} - V_{CE}}{RC} = \frac{V_{RC}}{RC} \quad \text{where:} \quad V_C = V_{CE}$$

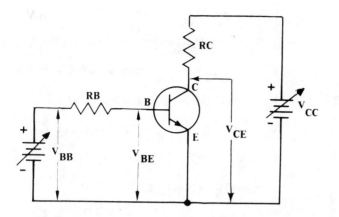

FIGURE 4-47

Additionally, two power supplies can be replaced by a single supply and a variable 10 kΩ resistor, as shown in Figure 4-48. In this circuit configuration, the V_{BB} voltage can be varied from zero volts to a high of V_{CC}.

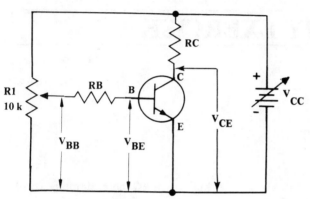

FIGURE 4-48

For example, if RB = 100 kΩ and RC = 2 kΩ, then setting V_{CC} to 12 V and adjusting V_{BB} to 3.6 V will provide an I_B of 30 μA and and I_C of 3 mA for a beta of 100. Therefore, V_{RC} will equal 6 V and V_{CE} will equal 6 V, as shown and solved in Figure 4-49.

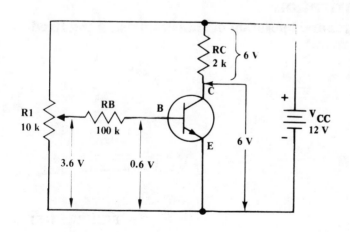

Given: $V_{CC} = 12$ V $\quad \beta = 100$

$V_{BB} = 3.6$ V $\quad$ RB = 100 kΩ

$V_{BE} = 0.6$ V $\quad$ RC = 2 kΩ

$V_{RB} = V_{BB} - V_{BE} = 3.6$ V $- 0.6$ V $= 3$ V

$I_B = V_{RB}/RB = 3$ V$/100$ kΩ $= 30$ μA

$I_C = \beta I_B = 100 \times 30$ μA $= 3$ mA

$V_{RC} = I_C R_C = 3$ mA $\times 2$ kΩ $= 6$ V

$V_{CE} = V_{CC} - V_{RC} = 12$ V $- 6$ V $= 6$ V

FIGURE 4-49

However, if the beta of the device is not known and V_{BB}, V_{BE}, and V_{CE} can be measured, then the remaining voltage drops and the currents of the circuit can be solved. Again, if $V_{BB} = 3.6$ V, $V_{BE} = 0.6$ V, and $V_{CE} = 6$ V, then:

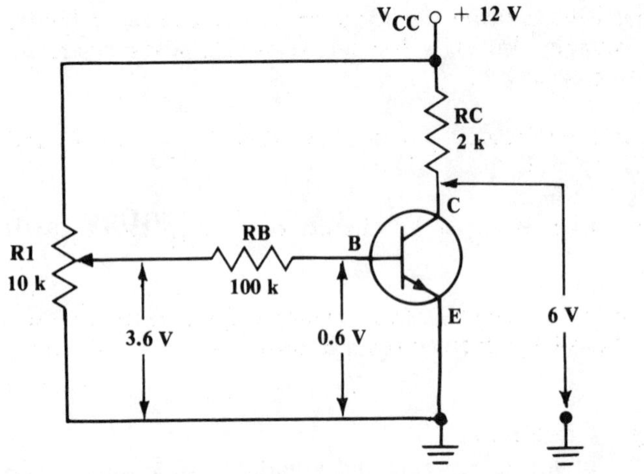

Measured: $V_{CC} = 12$ V $V_{BB} = 3.6$ V
$V_{CE} = 6$ V $V_{BE} = 0.6$ V

$V_{RB} = V_{BB} - V_{BE} = 3.6$ V $- 0.6$ V $= 3$ V

$I_B = V_{RB}/R_B = 3$ V$/100$ k$\Omega = 30$ μA

$V_{RC} = V_{CC} - V_{CE} = 12$ V $- 6$ V $= 6$ V

$I_C = V_{RC}/R_C = 6$ V$/2$ k$\Omega = 3$ mA

$\beta = I_C/I_B = 3$ mA$/30$ μA $= 100$

FIGURE 4-50

LABORATORY EXERCISE

OBJECTIVES

To investigate the basic DC characteristics of the transistor.

LIST OF MATERIALS AND EQUIPMENT

1. Transistor: 2N3904 or equivalent
2. Resistors (one each except where indicated):
   1 kΩ (two)   1.5 kΩ 2.2 kΩ 4.7 kΩ 10 kΩ 100 kΩ 10 kΩ potentiometer (two)
3. Power Supply
4. Voltmeter

PROCEDURE
Section I: Transistor Familiarization
(To become familiar with the general operation of transistors using the fixed-base-current, circuit configuration.)

Part 1:   Linear Mode of Operation

A. $V_{CE} = 6$ V, $V_{RC} = 6$ V, and $I_C = 4$ mA

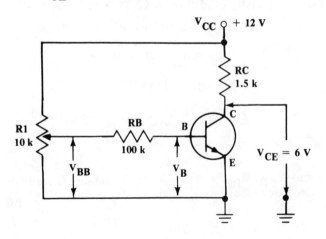

FIGURE 4-51

94

1. Connect the NPN 2N3904, or equivalent, transistor into the single power supply circuit of Figure 4-51, where resistor R1 is used to vary the V_{BB} voltage. Use the voltmeter, as shown in the figure, to monitor the voltage at the collector with respect to ground.

2. Adjust V_{BB} until V_{CE} is set to 6 volts. Therefore, since $V_{CC} = 12$ V, then $V_{RC} = 6$ V, and $I_C \approx 4$ mA, from $I_C \approx V_{RC}/RC = 6\text{ V}/1.5\text{ k}\Omega = 4$ mA

3. Monitor the V_{BB} and V_B voltages, with respect to ground. In this circuit, $V_B = V_{BE}$ and $V_{BE} \approx 0.6$ V for the silicon device.

NOTE: Current meters can be inserted in the base and collector leads to measure the current directly. However, the proper measuring technique is to use the voltage drop across a known resistance and calculate the current flow conditions.

4. Use the measured V_C, V_B, and V_{BB} voltages to calculate:

   a. V_{RC} from $V_{RC} = V_{CC} - V_C$ where: $V_C = V_{CE}$

   b. V_{RB} from $V_{RB} = V_{BB} - V_B$

   c. I_C from $I_C = V_{RC}/RC$

   d. I_B from $I_B = V_{RB}/RB$

   e. Beta from $\beta = I_C/I_B$

   f. V_{CB} from $V_{CB} = V_C - V_B$

   g. V_{BE} from $V_{BE} = V_B - V_E$ where: $V_E = 0$ V

5. Insert the calculated and measured values, as indicated, into Table 4-1.

TABLE 4-1	V_{CE}	V_{BB}	V_B	V_{RC}	V_{RB}	I_C	I_B	Beta	V_{CB}	V_{BE}
CALCULATED	6 V	/////	/////							
MEASURED				/////	/////	/////	/////	/////	/////	/////

B. $V_{CE} = 9$ V, $V_{RC} = 3$ V, and $I_C \approx 2$ mA

1. Monitor the collector voltage with respect to ground, as shown in Figure 4-52, and set the $V_C = V_{CE}$ voltage at 9 volts by adjusting the V_{BB} voltage. Therefore, since $V_{CC} = 12$ V, then $V_{RC} = 3$ V

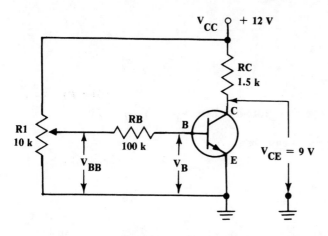

FIGURE 4-52

and $I_C \approx 2$ mA from $I_C \approx V_{RC}/RC = 3$ V$/1.5$ k$\Omega = 2$ mA.

2. Monitor the V_{BB} and V_B voltages, with respect to ground, where $V_B = V_{BE}$.

3. Use the measured V_C, V_B, and V_{BB} voltages to calculate:

   a. V_{RC} from $V_{RC} = V_{CC} - V_C$
   b. V_{RB} from $V_{RB} = V_{BB} - V_B$
   c. I_C from $I_C = V_{RC}/RC$
   d. I_B from $I_B = V_{RB}/RB$
   e. Beta from $\beta = I_C/I_B$
   f. V_{CB} from $V_{CB} = V_C - V_B$
   g. V_{BE} from $V_{BE} = V_B - V_E$ where: $V_E = 0$ V

4. Insert the calculated and measured values, as indicated, into Table 4-2.

TABLE 4-2	V_{CE}	V_{BB}	V_B	V_{RC}	V_{RB}	I_C	I_B	Beta	V_{CB}	V_{BE}
CALCULATED	9 V	/////	/////							
MEASURED				/////	/////	/////	/////	/////	/////	/////

C. $V_{CE} = 3$ V, $V_{RC} = 9$ V, and $I_C = \approx 6$ mA

1. Monitor the collector voltage, with respect to ground, as shown in Figure 4-53. Set $V_C = V_{CE}$ at 3 V by adjusting the V_{BB} voltage. Therefore, since $V_{CC} = 12$ V, Then $V_{RC} = 9$ V and $I_C \approx 6$ mA, from $I_C \approx V_{RC}/RC = 9$ V$/1.5$ k$\Omega = 6$ mA.

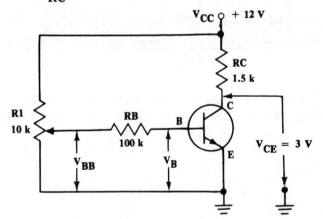

FIGURE 4-53

2. Monitor the V_{BB} and V_B voltages, with respect to ground, where $V_B = V_{BE} = 0.6$ V.

3. Use the measured V_C, V_B, and V_{BB} voltages to calculate:

   a. V_{RC} from $V_{RC} = V_{CC} - V_C$
   b. V_{RB} from $V_{RB} = V_{BB} - V_B$

c. I_C from $I_C = V_{RC}/RC$

d. I_B from $I_B = V_{RB}/RB$

e. Beta from $\beta = I_C/I_B$

f. V_{CB} from $V_{CB} = V_C - V_B$

g. V_{BE} from $V_{BE} = V_B - V_E$

4. Insert the calculated and measured values, as indicated, into Table 4-3.

TABLE 4-3	V_{CE}	V_{BB}	V_B	V_{RC}	V_{RB}	I_C	I_B	Beta	V_{CB}	V_{BE}
CALCULATED	3 V	////	////							
MEASURED				////	////	////	////	////	////	////

Part 2: Saturation and Cutoff Modes of Operation

A. Cutoff Mode: $V_{CE} \approx 12$ V, $V_{RC} \approx 0$ V, and $I_C \approx 0$ mA

1. Monitor the collector voltage, with respect to ground, as shown in Figure 4-54. Set the $V_C = V_{CE}$ voltage to ≈ 12 V by adjusting the V_{BB} voltage. Therefore, since $V_{CE} \approx 12$ V, then $V_{RC} \approx 0$ V and $I_C \approx 0$ mA, from $I_C \approx V_{RC}/RC = 0$ V$/1.5$ k$\Omega = 0$ mA.

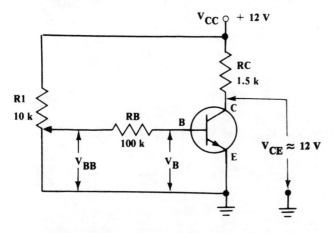

FIGURE 4-54

2. Monitor the V_{BB} and V_B voltages, with respect to ground.

3. Use the measured V_C, V_B, and V_{BB} voltages to calculate:

a. V_{RC} from $V_{RC} = V_{CC} - V_C$

b. V_{RB} from $V_{RB} = V_{BB} - V_B$

c. I_C from $I_C = V_{RC}/RC$

d. I_B from $I_B = V_{RB}/RB$

e. Beta from $\beta = I_C/I_B$

f. V_{CB} from $V_{CB} = V_C - V_B$

g. V_{BE} from $V_{BE} = V_B - V_E$

4. Insert the calculated and measured values, as indicated, into Table 4-4.

TABLE 4-4	V_{CE}	V_{BB}	V_B	V_{RC}	V_{RB}	I_C	I_B	Beta	V_{CB}	V_{BE}
CALCULATED	12 V	/////	/////							
MEASURED				/////	/////	/////	/////	/////	/////	/////

B. Saturation Mode: $V_{CE} \leq 0.5$ V, $V_{RC} \geq 11.5$ V, and I_C (SAT) $\approx$ 8 mA

1. Monitor the collector voltage VC, with respect to ground, as shown in Figure 4-55. Increase the V_{BB} voltage causing I_B and I_C to increase and the voltage V_{CE} to decrease. Continue adjusting until V_{CE} is less than 0.5 V, and where the further increase of V_{BB}, and hence I_B and I_C, does not appreciably lower the V_{CE} (SAT) condition.

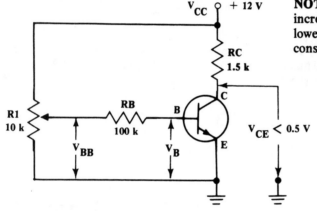

NOTE: Once V_{CE}(SAT) is achieved, continuing to increase I_B lowers V_{CE}(SAT) only slightly, but it does lower beta a great deal, since I_C remains relatively constant.

FIGURE 4-55

2. Monitor V_{BB} and V_{BE} and use the measured V_C, V_{BB}, and V_{BE} voltages to calculate:

a. V_{RC} from $V_{RC} = V_{CC} - V_C$

b. V_{RB} from $V_{RB} = V_{BB} - V_{BE}$

c. I_C from $I_C = V_{RC}/RC$

d. I_B from $I_B = V_{RB}/RB$

e. Beta from $\beta = I_C/I_B$

f. V_{CE} from $V_{CE} = V_C - V_E$

g. V_{CB} from $V_{CB} = V_C - V_B$

h. V_{BE} from $V_{BE} = V_B - V_E$

where: $V_E = 0$ V

3. Insert the calculated and measured values, as indicated, into Table 4-5.

TABLE 4-5	V_{CE}	V_{BB}	V_B	V_{RC}	V_{RB}	I_C	I_B	Beta	V_{CB}	V_{BE}
CALCULATED	/////	/////	/////							
MEASURED				/////	/////	/////	/////	/////	/////	/////

Part 3: Inverse Conditions

1. Connect the transistor in the circuit shown in Figure 4-56, where the collector and emitter leads are interchanged. Adjust the variable resistor R1 and monitor 6 volts at the emitter lead, with respect to ground. Notice, too, that RB has been lowered to 4.7 kΩ in this connection.

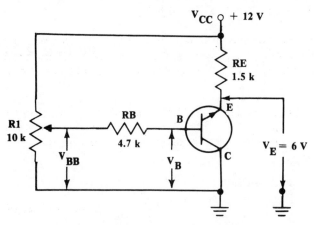

FIGURE 4-56

2. Use the measured V_E, V_B, and V_{BB} voltages to calculate:

   a. V_{RE} from $V_{RE} = V_{CC} - V_E$

   b. V_{RB} from $V_{RB} = V_{BB} - V_B$

   c. I_E from $I_E = V_{RE}/RE$

   d. I_B from $I_B = V_{RB}/RB$

   e. Beta from $\beta = I_E/I_B$

   f. V_{EB} from $V_{EB} = V_E - V_B$

   g. V_{BC} from $V_{BC} = V_B - V_C$ where: $V_C = 0$ V

3. Insert the calculated and measured values, as indicated, into Table 4-6.

NOTE: In the inverse connection, the device might not be forced into cutoff if the breakdown voltage is lower than 12 volts. Monitor V_{CE}, vary V_{BB}, and see if this is true. At the same time, test for V_{CE}(SAT).

TABLE 4-6	V_{CE}	V_{BB}	V_B	V_{RE}	V_{RB}	I_E	I_B	Beta	V_{BC}	V_{EB}
CALCULATED	6 V	/////	/////							
MEASURED				/////	/////	/////	/////	/////	/////	/////

Section II: Constant Collector Current
(To show that the collector current I_C remains relatively constant with V_{CE} change, as long as the device remains in the active linear region.)

Part 1: Varying V_{CE} by Changing the Collector Resistance

A. $RC = 1\ k\Omega$

1. Using a 2N3904 NPN silicon transistor, connect the circuit of Figure 4-57. Set the collector-emitter voltage V_{CE} to 8 volts. Therefore, since $V_{CC} = 12\ V$, then $V_{RC} = 4\ V$ and $I_C = 4\ mA$.

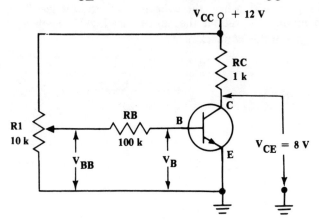

FIGURE 4-57

2. Measure V_{BB} and V_B, with respect to ground, and calculate:

   a. V_{RC} from $V_{RC} = V_{CC} - V_{CE}$

   b. V_{RB} from $V_{RB} = V_{BB} - V_{BE}$

   c. I_C from $I_C = V_{RC}/RC$

   d. I_B from $I_B = V_{RB}/RB$

3. Insert the calculated and measured values, as indicated, into Table **4-7(a)**.

B. $RC = 1.5\ k\Omega$

1. Replace the 1 kΩ collector resistor with a 1.5 kΩ collector resistor, as shown in Figure 4-58. The base current must be kept constant, so the V_{BB} voltage should not differ from that of the previous circuit (Figure 4-57).

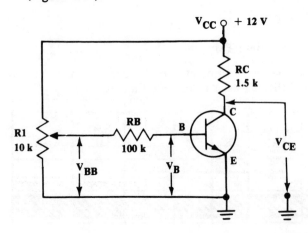

FIGURE 4-58

NOTE: IF RC is 1.5 kΩ and I_C remains at 4 mA, then V_{RC} should equal approximately 6 volts.

2. Measure the V_C, V_{BB}, and VB voltages, with respect to ground. Also, in this circuit, $V_C = V_{CE}$ and $V_B = V_{BE}$. Calculate:

   a. V_{RC} from $V_{RC} = V_{CC} - V_{CE}$ where: $V_C = V_{CE}$

   b. V_{RB} from $V_{RB} = V_{BB} - V_{BE}$

   c. I_C from $I_C = V_{RC}/RC$

   d. I_B from $I_B = V_{RB}/RB$

3. Insert the calculated and measured values, as indicated, into table 4-7(b).

C. RC = 2.2 kΩ

1. Replace the 1.5 kΩ collector resistor with a 2.2 kΩ collector resistor, as shown in Figure 4-59. Again, the base current I_B should be kept constant. This requires that the V_{BB} voltage does not differ from that of the previous RC = 1 kΩ and RC = 1.5 kΩ circuits.

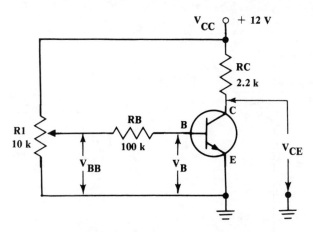

FIGURE 4-59

2. Measure the V_C, V_{BB}, and V_B voltages, with respect to ground. Again, in this circuit, $V_C = V_{CE}$ and $V_B = V_{BE}$. Calculate:

   a. V_{RC} from $V_{RC} = V_{CC} - V_{CE}$

   b. V_{RB} from $V_{RB} = V_{BB} - V_{BE}$

   c. I_C from $I_C = V_{RC}/RC$

   d. I_B from $I_B = V_{RB}/RB$

3. Insert the calculated and measured values, as indicated, into Table 4-7(c).

TABLE 4-7	I_C	V_{CE}	V_{RC}	V_{BB}	V_{BE}	V_{RB}	I_B
(a) RC = 1 kΩ CALCULATED		///////		///////	///////		
MEASURED	///////		///////			///////	///////

TABLE 4-7	I_C	V_{CE}	V_{RC}	V_{BB}	V_{BE}	V_{RB}	I_B
(b) RC = 1.5 kΩ CALCULATED	/////			/////	/////		
MEASURED		/////	/////			/////	/////
(c) RC = 2.2 kΩ CALCULATED		/////		/////	/////		
MEASURED	/////		/////			/////	/////

Part 2: Varying V_{CE} by Varying the Supply Voltage V_{CC}

A. $V_{CC} = 12$ V

1. Using a 2N3904 NPN silicon transistor, construct the circuit of Figure 4-60. Set the V_{CC} voltage to 12 volts by adjusting the R2 resistor and the V_{CE} voltage to 9 volts by adjusting the V_{BB} voltage. Then, $V_{CC} = 12$ V, $V_{CE} = 9$ V, $V_{RC} \approx 3$ V, and $I_C \approx 3$ mA.

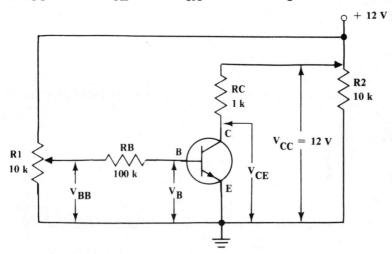

FIGURE 4-60

2. Measure V_{BB} and V_B with respect to ground and calculate:

   a. V_{RC} from $V_{RC} = V_{CC} - V_{CE}$

   b. V_{RB} from $V_{RB} = V_{BB} - V_{BE}$

   c. I_C from $I_C = V_{RC}/RC$

   d. I_B from $I_B = V_{RB}/RB$

3. Insert the calculated and measured values, as indicated, into Table 4-8(a).

B. $V_{CC} = 9$ V

1. Maintain the V_{BB} voltage so that I_B remains constant. Then reduce the V_{CC} voltage to 9 volts, as shown in Figure 4-61.

2. Measure V_C, V_{BB}, and V_B (with respect to ground), and calculate:

   a. V_{RC} from $V_{RC} = V_{CC} - V_C$

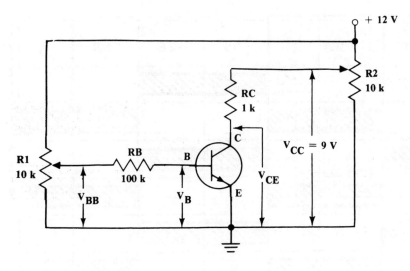

FIGURE 4-61

b. V_{RB} from $V_{RB} = V_{BB} - V_{BE}$

c. I_C from $I_C = V_{RC}/RC$

d. I_B from $I_B = V_{RB}/RB$

3. Insert the calculated and measured values, as indicated, into Table 4-8(b).

C. $V_{CC} = 6$ V

1. Reduce the V_{CC} voltage to 6 volts, as shown in Figure 4-62, and continue to maintain V_{BB} and I_B the same levels as those of the previous two circuits.

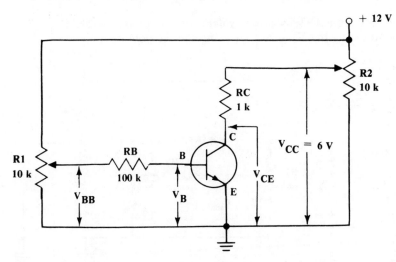

FIGURE 4-62

2. Measure V_C, V_{BB}, and V_B (with respect to ground), and calculate:

a. V_{RC} from $V_{RC} = V_{CC} - V_C$

b. V_{RB} from $V_{RB} = V_{BB} - V_{BE}$

c. I_C from $I_C = V_{RC}/RC$

d. I_B from $I_B = V_{RB}/RB$

3. Insert the calculated and measured values, as indicated, into Table 4-8(c).

TABLE 4-8	V_{CC}	V_{CE}	V_{RC}	I_C	V_{BB}	V_{BE}	V_{RB}	I_B
(a) V_{CC} = 12 V CALCULATED	12 V	9 V			▨	▨		
MEASURED			▨	▨			▨	▨
(b) V_{CC} = 9 V CALCULATED	9 V	▨			▨	▨		
MEASURED			▨	▨			▨	▨
(c) V_{CC} = 6 V CALCULATED	6 V	▨			▨	▨		
MEASURED			▨	▨			▨	▨

Section III: Beta Sensitivity

(To investigate the beta sensitivity of the fixed base versus the fixed emitter current circuit configurations. Three different NPN transistors will be used—preferably low power, medium power, and high power devices.)

Part 1: The Fixed Base Current Circuit

A. Transistor #1 (The 2N3904 or an equivalent low power device).

1. Connect the circuit of Figure 4-63, using a 2N3904 or equivalent low power transistor. Set V_{CC} to 12 volts and then V_{CE} to 6 volts by adjusting the V_{BB} voltage using the R1 potentiometer.

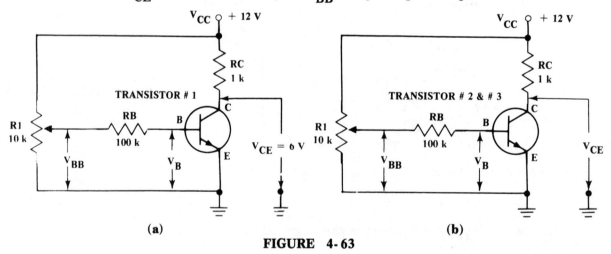

FIGURE 4-63

2. Measure V_{BB} and V_B, with respect to ground, and calculate:

   a. V_{RC} from $V_{RC} = V_{CC} - V_C$ where: $V_{CC} = 12$ V and $V_C = V_{CE} = 6$ V

   b. V_{RB} from $V_{RB} = V_{BB} - V_{BE}$

   c. I_C from $I_C = V_{RC}/RC$

   d. I_B from $I_B = V_{RB}/RB$

e. Beta from $\beta = I_C/I_B$

3. Insert the calculated and measured values, as indicated, into Table 4-9(a). Also, include transistor number to identify device used.

B. Transistor #2 (The 2N3053 or equivalent medium power device.)

1. Remove the first transistor from the circuit and, without changing the circuit voltages, insert the second transistor into the circuit. Measure V_C, V_{BB}, and V_B (with respect to ground), as shown in Figure 4-63(b). Calculate V_{RB}, V_{RC}, I_C, I_B, and beta.

Insert the calculated and measured values, as indicated, into Table 4-9(b). Note the transistor number of the device used.

C. Transistor #3 (The 2N3055 or equivalent higher powered device).

1. Remove the second transistor and again, without changing the circuit voltages, insert the third transistor into the circuit. Measure V_C, V_{BB}, and V_B, with respect to ground, as shown in Figure 4-63(b). Calculate V_{RB}, V_{RC}, I_C, I_B and beta.

3. Insert the calculated and measured values, as indicated, into Table 4-9(c). Note the transistor number of the device used.

TABLE 4-9	V_{CE}	V_{RC}	I_C	V_{BB}	V_{BE}	V_{RB}	I_B	Beta
(a) 2N CALCULATED	6 V			///	///			
MEASURED		///				///	///	///
(b) 2N CALCULATED	///			///	///			
MEASURED		///				///	///	///
(c) 2N CALCULATED	///			///	///			
MEASURED		///				///	///	///

Part 2: The Fixed Emitter Current Circuit

A. Transistor #1 (2N3904 or equivalent low power device).

1. Connect the circuit of Figure 4-64(a), using a 2N3904 or equivalent low power transistor. Set the collector voltage V_C at 9 volts, with respect to ground. Notice that a smaller 10 kΩ RB resistor is used so that the base current can be monitored and the beta calculated. However, the 10 kΩ RB resistor will cause the beta to have slightly more effect on the DC voltage distribution than a zero resistance RB.

2. Measure the V_{BB}, V_B, and V_E voltages, with respect to ground, and calculate:

   a. V_{RC} from $V_{RC} = V_{CC} - V_C$
   b. V_{RE} from $V_{RE} = V_E$
   c. V_{BE} from $V_{BE} = V_B - V_E$
   d. V_{RB} from $V_{RB} = V_{BB} - V_B$

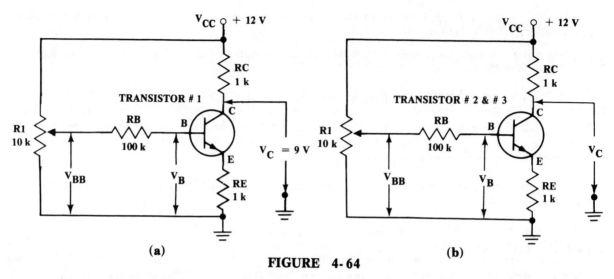

FIGURE 4-64

e. I_C from $I_C = V_{RC}/RC$

f. I_B from $I_B = V_{RB}/RB$

g. I_E from $I_E = V_{RE}/RE$ where: $V_E = V_{RE}$

h. Beta from $\beta = I_C/I_B$

NOTE: An RB of 100 kΩ is needed to obtain reasonably accurate I_B measurements and calculations. However this resistance will effect the DC stability because of beta change.

3. Insert the calculated and measured values, as indicated, into Table 4-10(a). Note the transistor number of the device used.

TABLE 4-10	V_C	V_{BB}	V_B	V_E	V_{RC}	V_{BE}	V_{RB}	I_C	I_B	I_E	Beta
(a) 2N___ CALCULATED	9 V	/////	/////	/////							
MEASURED					/////	/////	/////	/////	/////	/////	/////
(b) 2N___ CALCULATED	/////	/////	/////	/////							
MEASURED					/////	/////	/////	/////	/////	/////	/////
(c) 2N___ CALCULATED	/////	/////	/////	/////							
MEASURED					/////	/////	/////	/////	/////	/////	/////

B. Transistor #2 (The 2N3053 or equivalent medium power device).

1. Remove the first transistor, and without changing the circuit voltages, insert the second transistor into the circuit. Measure V_C, V_E, V_{BB}, and V_B, all with respect to ground as shown in Figure 4.64(b).

2. Calculate V_{RB}, V_{RC}, V_{BE}, I_C, I_B, I_E, and beta.

3. Insert the calculated and measured values, as indicated, into Table 4-10(b). Note the transistor number of the device used.

C. Transistor #3 (The 2N3055 or equivalent larger power device).

1. Remove the second transistor and, again without changing the circuit voltages, insert the third transistor into the circuit. Measure V_C, V_E, V_{BB}, and V_B, all with respect to ground, as shown in Figure 4-64(b).

2. Calculate V_{RB}, V_{RC}, V_{BE}, I_C, I_B, I_E, and beta.

3. Insert the calculated and measured values, as indicated, into Table 4-10(c). Note the transistor number of the device used.

CHAPTER QUESTIONS

1. Why is the base region very thin and lightly doped as compared to the emitter and collector regions?
2. What is the major difference between NPN and PNP transistor devices?
3. Is the TO-92 a plastic or metal package? How about the TO-5 package?
4. What are the four modes of operation into which a transistor can be connected? Which of the four modes is usually connected in error?
5. If the base-collector and base-emitter junctions are both forward biased, the transistor is in what mode of operation?
6. If the base-collector and base-emitter junctions are both reverse biased, the transistor is in what mode of operation?
7. If the base-collector junction is reverse biased and the base-emitter junction is forward biased, the transistor is in what mode of operation?
8. What are the two major drawbacks to the inverse mode of operation?
9. What are the nominal base-emitter diode voltages of silicon and germanium transistors, when operated in the linear mode?
10. BV_{CEo} represents the breakdown voltage of the collector-emitter terminals with the base terminal left open. This is designated by the subscript "o". Therefore, what terminal is left open for the BV_{EBo} junction test? For the BV_{CBo} junction test?
11. Why is the fixed emitter current circuit not as dependent on beta as the fixed base current circuit?
12. Why is the fixed emitter circuit more sensitive to beta change with a base resistor than with no base resistor? Then, why was the base resistor used in the laboratory experiment? (Figure 4-64)
13. Once the transistor enters into saturation, does continuing to increase I_B lower V_{CE}(SAT)? What about beta and why?
14. Did the V_{CE}(SAT) of the circuit of Figure 4-55 appear higher or lower than the V_{CE}(SAT) of the inverse connection of Figure 4-56?
15. From your data and observations, is the transistor truly a constant current source? Explain why or why not, with reference to the collector output.

CHAPTER PROBLEMS

1. Given: $V_{BE} = 0.6$ V, $V_{BB} = 3$ V, and $\beta = 150$

    Find:
    a. V_{RB}
    B. I_B
    C. I_C
    D. V_{RC}
    E. V_{CE}

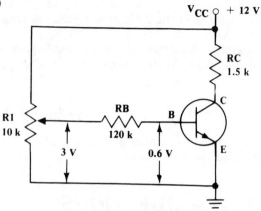

FIGURE 4-65

2. Given: $V_{BE} = 0.6$ V, $V_{CE} = 9$ V, and $\beta = 100$

    Find:
    A. V_{RC}
    B. I_C
    C. I_B
    D. V_{RB}
    E. V_{BB}

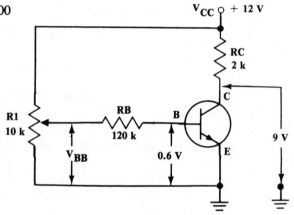

FIGURE 4-66

3. Given: $V_{BE} = 0.6$ V, $V_{CE} = 6.6$ V, and $V_{BB} = 3$ V

    Find:
    A. V_{RC}
    B. I_C
    C. V_{RB}
    D. I_B
    E. Beta

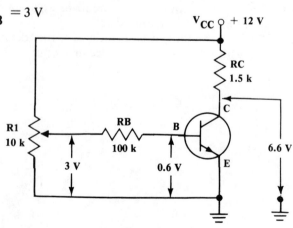

FIGURE 4-67

4. Given: $V_{BE} = 0.6$ V, $V_{BB} = 3$ V, and $\beta = 100$

   Find:

   A. V_{RE}

   B. $I_E \approx I_C$

   C. V_{RC}

   D. V_C

   E. V_{CE}

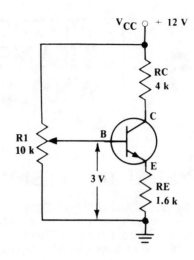

FIGURE 4-68

5. Given: $V_{BE} = 0.6$ V, $V_C = 8$ V, and $\beta = 100$

   Find:

   A. V_{RC}

   B. $I_C \approx I_E$

   C. V_{RE}

   D. V_B

   E. V_{BB}

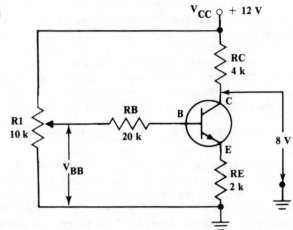

FIGURE 4-69

6. Given: $V_{BE} = 0.6$ V and $V_{BB} = 4$ V

   Find:

   A. V_{RC}

   B. $I_C \approx I_E$

   C. V_{RE}

   D. V_{RB}

   E. I_B

   F. Beta

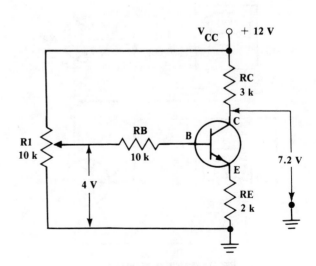

FIGURE 4-70

5 | TRANSISTOR MEASUREMENT AND TESTING TECHNIQUES

GENERAL DISCUSSION

The physical and electrical characteristics of most transistors can be found in manufacturer data sheets, and it is assumed that a new device will operate within those electrical specifications at the time the device is connected into a circuit. Therefore, in most instances, a simple DC voltage check of a transistor in a circuit is all that is necessary to determine whether the device is working or not. However, if the circuit is not working, then the suspected transistor should be removed from the circuit and tested further, because sometimes a perfectly good device can be made to look bad (forced into cutoff or saturation) by another failed component in the circuit.

Transistors can be tested at different levels from go and no-go checks to complete parameter measurements. Go and no-go tests can be made with an ohmmeter, transistor checker, bench measurement methods, or on a curve tracer such as the Tektronix 575. Also, some transistor checkers can be used to measure additional device parameters such as beta and leakage current, and they can be used to find shorts and opens between the junctions of the transistor under test.

Bench measurement techniques using standard laboratory equipment can be used to obtain most basic transistor parameters. However, the amount of work required to obtain data in this manner is time consuming. Curve tracers, on the other hand, are relatively fast and can be used to obtain most of the device parameters found in the manufacturer data sheets.

OHMMETER TEST TECHNIQUES

Since a transistor is effectively two back-to-back diode junctions, the base-emitter junction can be tested as one discrete diode and the base-collector junction can be tested as another discrete diode. As discussed in Chapter 2, a working diode should have a high resistance in one direction and a low resistance in the other direction. Therefore, if the junction under test has a low resistance in both directions, then the junction is probably shorted. If the junction under test has a high resistance in both directions, then the junction is probably open.

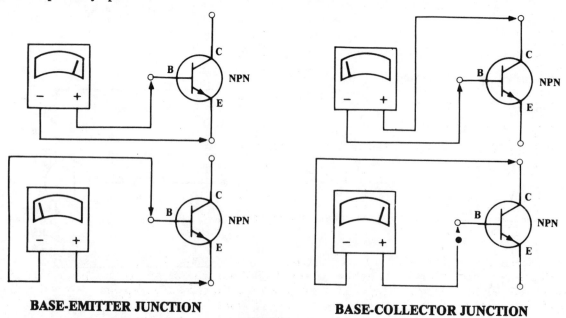

BASE-EMITTER JUNCTION **BASE-COLLECTOR JUNCTION**

FIGURE 5-1

110

Therefore, the technique for measuring either junction is to connect the leads of an ohmmeter across the junction to be tested and take a resistance reading. Then, reverse the ohmmeter leads and take a second resistance reading. The ohmmeter connections for both the base-emitter and base-collector junctions are shown in Figure 5-1.

Ohmmeter junction testing provides a check of a good or bad junction by comparing the resistances of the forward and reverse biased junctions. However, the forward to reverse resistance ratio is relative only, because at each R × 100 Ω, R × 1 kΩ, or R × 100 kΩ meter setting, a different DC current will flow through the non-linear forward biased diode junction providing different resistance measurements.

Ohmmeter junction checking provides only go and no-go tests for junction failure. Hence, a device which passes this go and no-go ohmmeter test can still fail other device parameters such as beta, voltage breakdown and leakage current. Therefore, ohmmeter testing of transistors should only be considered as a first-level check, and further testing must be used to determine whether the device meets the remaining manufacturer specifications.

BENCH MEASUREMENT TECHNIQUES

Bench measurement techniques using standard laboratory equipment can be used also to obtain both the output and input static characteristic curves and, in some instances, the forward and reverse-biased junction breakdown voltages of transistors. However, while the forward-biased base-emitter and base-collector junctions and the reverse biased base-emitter junction measurements can be accomplished with low voltage power supplies, the reverse-biased collector-base junction measurement requires a high voltage power supply. Therefore, with high voltage reverse-biased conditions, the current flow through the junction must be limited to prevent the device from being destroyed while making the test.

OUTPUT STATIC CHARACTERISTIC CURVES PLOTTING POINT-TO-POINT

The output family of characteristic curves, discussed in Chapter 4, providing I_C versus I_B current relationships over a V_{CE} range that extends from zero volts to the breakdown region, can be obtained also by using bench measurement techniques. In order to obtain each individual characteristic curve, the base current I_B is set and the V_{CE} voltage increased from an initial zero volt condition to 12 V, and I_C can be plotted at any of the V_{CE} conditions. Therefore, with base current conditions of 10 μA, 30 μA, 50 μA, 70 μA, and 90 μA, a family of I_C current curves can be plotted similar to those shown in Figure 5-2.

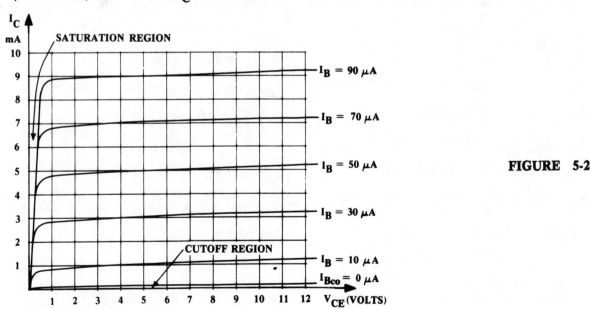

FIGURE 5-2

The circuit used to provide the plotted characteristic curves is shown in Figure 5-3 where: $\beta = 100$, $R_B = 100$ kΩ, and $V_{BE} = 0.6$ V. Therefore, a V_{BB} of 1.6 V will produce an I_B of 10 μA, a V_{BB} of 3.6 V will produce an I_B of 30 μA, a V_{BB} of 5.6 V will produce an I_B of 50 μA, a V_{BB} of 7.6 V will produce an I_B of 70 μA, and a V_{BB} of 9.6 V will produce an I_B of 90 μA. Then, since $\beta = 100$, 10 μA will provide an I_C of 1 mA, 30 μA will provide an I_C of 3 mA, etc., up to 90 μA providing an I_C of 9 mA. The following calculations show I_B at 10 µA through 90 µA and the 1 kΩ collector resistor is used in the indirect solution for I_C.

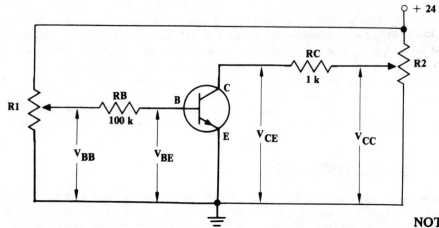

FIGURE 5-3

NOTE: $V_{CE}(SAT) \approx 0.2$ V.

a. $I_B = 10$ µA

$V_{RB} = V_{BB} - V_{BE} = 1.6\text{ V} - 0.6\text{ V} = 1\text{ V}$

$I_B = V_{RB}/RB = 1\text{ V}/100\text{ k}\Omega = 10\text{ µA}$

$I_C = \beta I_B = 100 \times 10\text{ µA} = 1\text{ mA}$

$V_{RC} = I_C R_C = 1\text{ mA} \times 1\text{ k}\Omega = 1\text{ V}$

$V_{CC}(max) = V_{CE} + V_{RC} = 12\text{ V} + 1\text{ V} = 13\text{ V}$

$V_{CC}(min) = V_{CE}(SAT) + V_{RC}$

$\approx 0.2\text{ V} + 1\text{ V} = 1.2\text{ V}$

b. $I_B = 30$ µA

$V_{RB} = V_{BB} - V_{BE} = 3.6\text{ V} - 0.6\text{ V} = 3\text{ V}$

$I_B = V_{RB}/RB = 3\text{ V}/100\text{ k}\Omega = 30\text{ µA}$

$I_C = \beta I_B = 100 \times 30\text{ µA} = 3\text{ mA}$

$V_{RC} = I_C R_C = 3\text{ mA} \times 1\text{ k}\Omega = 3\text{ V}$

$V_{CC}(max) = V_{CE} + V_{RC} = 12\text{ V} + 3\text{ V} = 15\text{ V}$

$V_{CC}(min) = V_{CE}(SAT) + V_{RC}$

$\approx 0.2\text{ V} + 3\text{ V} = 3.2\text{ V}$

c. $I_B = 50$ µA

$V_{RB} = V_{BB} - V_{BE} = 5.6\text{ V} - 0.6\text{ V} = 5\text{ V}$

$I_B = V_{RB}/RB = 5\text{ V}/100\text{ k}\Omega = 50\text{ µA}$

$I_C = \beta I_B = 100 \times 50\text{ µA} = 5\text{ mA}$

$V_{RC} = I_C R_C = 5\text{ mA} \times 1\text{ k}\Omega = 5\text{ V}$

$V_{CC}(max) = V_{CE} + V_{RC} = 12\text{ V} + 5\text{ V} = 17\text{ V}$

$V_{CC}(min) = V_{CE}(SAT) + V_{RC}$

$\approx 0.2\text{ V} + 5\text{ V} = 5.2\text{ V}$

d. $I_B = 70$ µA

$V_{RB} = V_{BB} - V_{BE} = 7.6\text{ V} - 0.6\text{ V} = 7\text{ V}$

$I_B = V_{RB}/RB = 7\text{ V}/100\text{ k}\Omega = 70\text{ µA}$

$I_C = \beta I_B = 100 \times 70\text{ µA} = 7\text{ mA}$

$V_{RC} = I_C R_C = 7\text{ mA} \times 1\text{ k}\Omega = 7\text{ V}$

$V_{CC}(max) = V_{CE} + V_{RC} = 12\text{ V} + 7\text{ V} = 19\text{ V}$

$V_{CC}(min) = V_{CE}(SAT) + V_{RC}$

$\approx 0.2\text{ V} + 7\text{ V} = 7.2\text{ V}$

e. $I_B = 90$ µA

$V_{RB} = V_{BB} - V_{BE} = 9.6\text{ V} - 0.6\text{ V} = 9\text{ V}$

$I_B = V_{RB}/RB = 9\text{ V}/100\text{ k}\Omega = 90\text{ µA}$

$I_C = \beta I_B = 100 \times 90\text{ µA} = 9\text{ mA}$

$V_{RC} = I_C R_C = 9\text{ mA} \times 1\text{ k}\Omega = 9\text{ V}$

$V_{CC}(max) = V_{CE} + V_{RC} = 12\text{ V} + 9\text{ V} = 21\text{ V}$

$V_{CC}(min) = V_{CE}(SAT) + V_{RC}$

$\approx 0.2\text{ V} + 9\text{ V} = 9.2\text{ V}$

INPUT STATIC CHARACTERISTIC CURVE USING POINT-TO-POINT PLOTTING

The input characteristic curve of the base-emitter junction has the same characteristics as that of the discrete diode. Therefore, if V_{CE} is set at 6 volts, to reverse bias the collector-base junction, and V_{BB} is increased slowly from an initial zero volt condition, both I_B and V_{BE} begin to increase rapidly at about 0.5 V and then level off at about 0.6 V. The I_B versus V_{BE} input characteristic curve, along with a representative circuit, is shown in Figure 5-4.

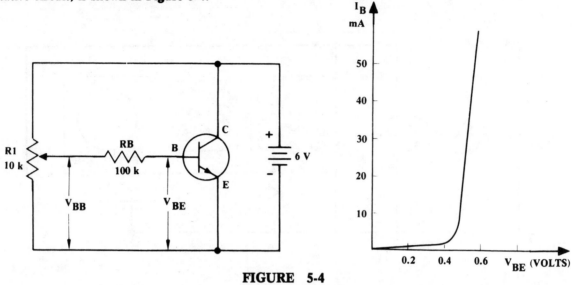

FIGURE 5-4

As shown in Figure 5-4, V_{BE} does change slightly, beyond the knee of the curve, with increased or decreased I_B current, but the change in voltage above or below 0.6 V is rarely greater than a tenth of a volt. Therefore, for the sake of simplicity, the V_{BE} voltage is assumed to be a constant 0.6 V in the following calculations for base currents of 10 µA, 20 µA, 50 µA, and 60 µA.

a. $I_B = \dfrac{V_{BB} - V_{BE}}{RB} = \dfrac{1.6\ V - 0.6\ V}{100\ k\Omega} = \dfrac{1\ V}{100\ k\Omega} = 10\ \mu A$

b. $I_B = \dfrac{V_{BB} - V_{BE}}{RB} = \dfrac{2.6\ V - 0.6\ V}{100\ k\Omega} = \dfrac{2\ V}{100\ k\Omega} = 20\ \mu A$

c. $I_B = \dfrac{V_{BB} - V_{BE}}{RB} = \dfrac{5.6\ V - 0.6\ V}{100\ k\Omega} = \dfrac{5\ V}{100\ k\Omega} = 50\ \mu A$

d. $I_B = \dfrac{V_{BB} - V_{BE}}{RB} = \dfrac{6.6\ V - 0.6\ V}{100\ k\Omega} = \dfrac{6\ V}{100\ k\Omega} = 60\ \mu A$

OUTPUT CHARACTERISTIC CURVES USING SWEPT MEASUREMENTS

Curve tracers like the Tektronix 575 can visually display the entire family of output characteristic curves on a cathode ray tube. Similarly, if the circuit used for bench measurements is modified slightly to include a simply constructed sweep circuit, then each individual output characteristic curve of a transistor can be monitored and measured on an oscilloscope. The schematic diagram and a respresentative response curve are illustrated in Figure 5-5. In the circuit of Figure 5-5, the base current is established by V_{BB}, RB, and $V_{BE} \approx 0.6$ V, and the collector voltage is established by a pulsating half-wave rectified DC voltage. As a result, the collector voltage is constantly being varied between zero volts and the peak voltage of about 9 volts at about 60 times per second. Therefore, the voltage applied to the collector is connected to the horizontal input of the scope to provide the V_{CE} of about 9 volts, and the voltage developed across RS is connected to the vertical amplifier. However, since $I_C = V_{RS}/RS$ and the collector current is proportional to V_{RS}, then I_C can be measured on the oscilloscope. Therefore, the 6.3 V RMS voltage is rectified to a 9 V peak, pulsating DC voltage, that is used to display the V_{CE} versus I_C output characteristic curve for each established base current condition. Hence, if beta of the device is 100 and I_B is set, initially, at 0 μA and then incrementally adjusted in 10 μA steps through 50 μA, for example, then the output characteristic curves will range from about 0 mA through 5 mA. $Vp = V\ RMS\ \sqrt{2} \approx 6.3\ V\ RMS \times 1.414 \approx 9\ Vp$.

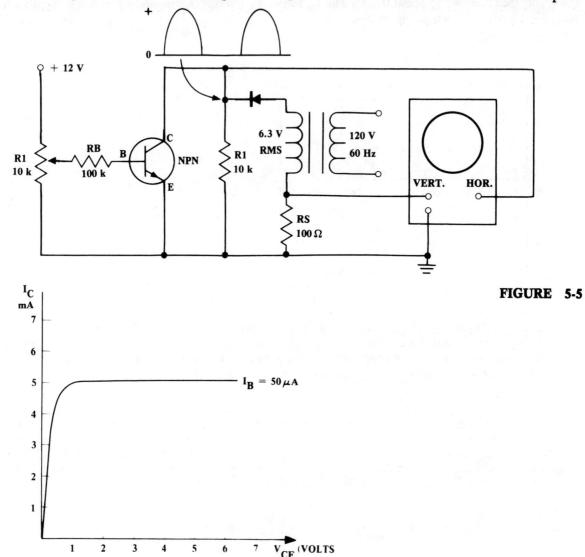

FIGURE 5-5

The procedure for setting up and measuring the output characteristic curves, using the swept measurement technique, is as follows:

1. Turn off all power to the circuit. Locate the 0, 0 origin on the lower left hand side of the oscilloscope grid. Set the vertical amplifier at 100 mV/division and the horizontal amplifier at 1 V/division. This allows V_{CE} to be monitored at about 9 V or nine horizontal divisions and I_C can have a full vertical rise of 10 mA or 1 mA/division, since $I_C = V_{RS}/RS = 100\ mV/100\ \Omega = 1mA$.

2. Turn on the 6.3 V secondary voltage of the circuit but not the power supply. Monitor the half-wave pulses across resistor R1. Set the scope to the horizontal sweep mode and note the left to right horizontal sweep that references V_{CE}. (Using an HP 175 oscilloscope, the horizontal V/cm was set to 1 V/cm.)

3. Turn on the power supply to the circuit and adjust V_{BB} to provide an I_B of 50 μA. Monitor the output characteristic curve. Measure I_C at a V_{CE} of 5 V and calculate beta from $\beta = I_C/I_B$.

4. Reduce I_B, in turn, to 40 μA, 30 μA, 20 μA, and 10 μA and monitor each of the output characteristic curves. Calculate each of the individual betas at the V_{CE} of 6 V.

5. Reduce I_B to 0 μA and monitor $I_C = I_{CEo}$ at $V_{CE} = 6$ V. (I_{CEo} is the leakage current for the device, and it is the collector current when $I_{BCo} = 0.0$ mA.)

6. Representative characteristic curves for the circuit are shown in Figure 5-6. Notice that the active region is all the area from cutoff to the saturated regions, and note that V_{CE}(SAT) is about 0.2 V and I_{CEo} is extremely low.

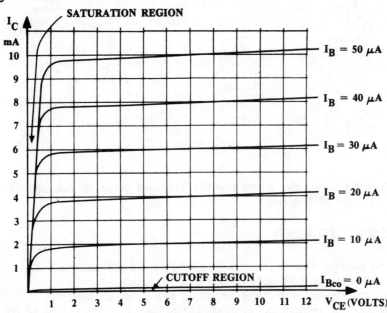

FIGURE 5-6

BASIC CURVE TRACER OPERATION

Swept measurements are much faster than point-to-point plotting of the output characteristic curves, but this method still produces only one output characteristic curve per base current adjustment. This limitation can be resolved by connecting a step generator to the transistor base, where each step represents a 10 μA increment. Then, if the frequency of the step generator is fast enough to make the displayed images on the CRT seem continuous, all the corresponding output characteristic curves can be visually displayed. Both NPN and PNP devices can be tested in this manner.

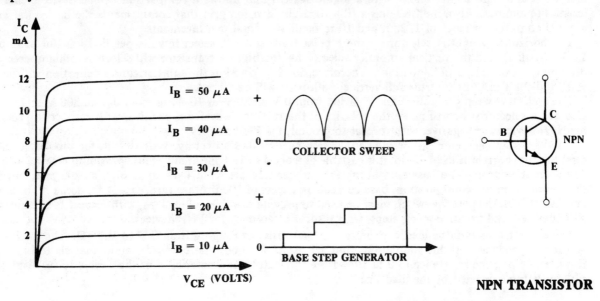

NPN TRANSISTOR

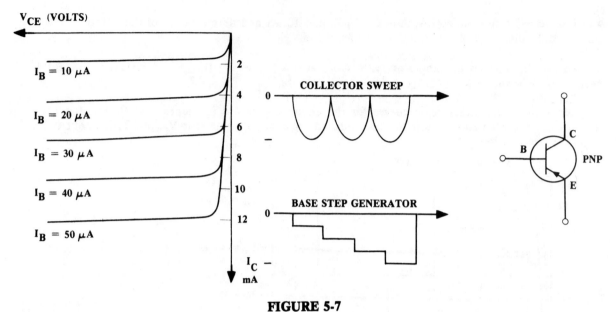

FIGURE 5-7

NOTE: As shown in Figure 5-7, both the collector sweep and the base step generator are positive for the NPN device and they are negative for the PNP device.

THE TEKTRONIX 575 CURVE TRACER

The Tektronix 575 is a completely packaged curve tracer with a CRT, horizontal and vertical amplifiers, and all of the controls necessary to display the full family of transistor characteristic curves. Obviously, the curve tracer is much faster than point-to-point plotting or single sweep bench methods of obtaining characteristic curves, and its wide range of voltage and current control allows for the testing of NPN and PNP transistors — both low and high power. Additionally, the curve tracer is capable of displaying common emitter and common base characteristics, but only the more widely used common emitter characteristics will be investigated.

TEKTRONIX 575 FUNCTIONS

The Tektronix 575 is a complete semiconductor measurement system. It has vertical and horizontal selector switches, a collector sweep generator, a basic step generator, a variable load resistor capability, and a left and right socket (along with a toggle switch) that allows a comparison of one device's characteristics to another. Also, the CRT has a 10 × 10, 1 cm/division grid that complements the horizontal and vertical amplifier switches of 1, 2, 5, and 10 to facilitate visual measurements.

The horizontal amplifier selector for the bipolar transistor will select in volts per division (for example, 1 mV/division), and the vertical amplifier selector for the bipolar transistor will select in milliamperes per division (for example, 1 mA/division). Therefore, for 1 V/division the full horizontal deflection will be 10 volts and for 1 mA/division the full vertical deflection will be 10 mA.

The collector sweep generator can be varied from 0 V to 20 V at 10 A, or from 0 V to 200 V at 1 A. It also has a selector switch to make the polarity of the rectified waveform positive, with respect to ground, for NPN devices, or negative, with respect to ground, for PNP devices.

The base step generator can provide either step current or step voltage sources but, for bipolar transistors, the step current is used. Also, it is capable of from 4 to 12 continuous steps, providing 4 to 12 output characteristic curves. The base current increments per step are from 1 μA to as high as 200 mA, but for the common emitter configuration, base currents in excess of 100 mA are rare, except for some high power devices. Again, the polarity switch must be used to provide positive-going steps, with respect to ground, for NPN devices, and negative-going steps, with respect to ground, for PNP devices.

The curve tracer variable load capability has load resistors ranging from 0 Ω to 100 kΩ. The loads can be used to produce load line simulations that are useful for both static and dynamic analysis techniques. However, setting the resistance to 0 Ω allows the full output characteristics of the device to be displayed without being truncated by the load line.

TRANSISTOR TESTING PROCEDURES FOR A 2N3904

The procedure for displaying the common emitter output characteristic curves on the 575 Tektronix curve tracer for the 2N3904 low power NPN silicon transistor is as follows:

1. Place the 2N3904 device in either the left or right socket, keeping the toggle switch momentarily in the neutral position. Make sure that the switch indicates common emitter (CE).
2. Set the collector sweep voltage supply to the 0-20 V range, and set the polarity switch to positive (+).
3. Adjust the intensity focus controls for a clean, well-contrasted trace.
4. Set the horizontal selector switch to 1 V/DIV and increase the collector sweep voltage to the full 10 V.
5. Set the polarity switch on the base step generator to positive (+), the base step generator to 10 µA/step, the display switch of the base generator to the repetitive postition, and the vertical selector switch to 1 mA/DIV.
6. Set the load resistance to 0 Ω so that the full untruncated characteristic curves are displayed.
7. Throw the toggle switch to the position of the socket that the transistor is in, and display the full family of curves on the grid of the CRT screen. It may be necessary to adjust the number of steps per family, because the base step generator can vary from 4 to 12 recurrent steps per family.

DIRECT CURRENT BETA EVALUATION

The DC beta of a device, with equally spaced characteristic curves will usually vary slightly when the collector current and collector-emitter voltage drop across the device are varied. However, the difference in beta at any number of evaluation points is usually not that dramatic, as long as the device does not approach either saturation or cutoff conditons.

For instance, if the $I_B = 50$ µA characteristic curve of Figure 5-8 is monitored, then the beta at an evaluation point of V_{CE} at 3 V and I_C at 6 mA is about 120. However, the beta at a V_{CE} of 6 V is closer to 124 from beta $= I_C/I_B = 6.2$ mA/50 µA $= 124$. Likewise, slight beta differences can be observed for each characteristic of the device.

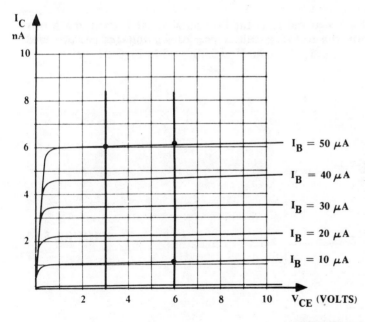

FIGURE 5-8

If the characteristic curves are not equally spaced, then beta should be evaluated over several characteristic curves. In the solved example of Figure 5-8, at 50 µA to 10 µA the corresponding change in I_C (ΔI_C) will be from 6.2 mA to 1.2 mA at $V_{CE} = 6$ V. Both beta determining techniques are shown in Figure 5-8.

Essentially, the DC beta is a single point evaluation of beta from $\beta_{DC} = I_C/I_B$ and the AC beta of $\beta_{AC} = \Delta I_C/\Delta I_B$ is evaluated over several characteristic curves. However, in most instances, the difference between the DC and AC beta values is not important enough to consider one method over the other.

DISPLAY OF LOAD LINE

By setting the dissipation limiting resistance to 1 kΩ, the load line extending from a V_{CE} of 10 volts to an I_C of 10 mA should be observed as shown in Figure 5-9.

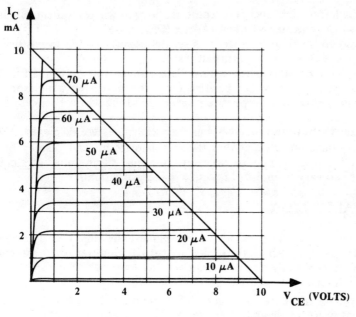

HORZ(VCE) = 1 V/DIV

VERT(I_C) = 1 mA/DIV

STEPS(I_B) = 10 μA/STEP

FIGURE 5-9

V_{CE} BREAKDOWN VOLTAGE OBSERVATION (BV$_{CE}$)

Set the collector sweep voltage to the 0 -200 V range and reset the horizontal volts/division to 5 V/div. Slowly increase the collector sweep voltage until the collector-emitter breakdown voltages are observed. For the 2N3904 device, the breakdown occurs at slightly above 40 V as indicated in Figure 5-10.

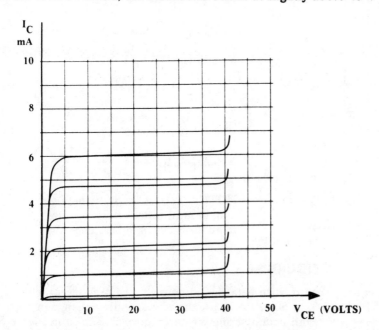

FIGURE 5-10

BV$_{CBo}$ AND BV$_{BEo}$

A discrete diode is tested in the forward and reverse biased mode by connecting the diode across the marked collector and emitter terminals of the curve tracer. Therefore, the individual collector-base and base-emitter junctions can be tested in a similar manner. Remember, however, that if the polarity switch of the peak voltage collector sweep is positive, then the collector voltage is positive.

NOTE: The subscript "o" signifies that the emitter lead is left unconnected.

To determine the BV_{CB_o}, connect the collector lead of the device to the collector terminal of the curve tracer, the base lead of the device to the emitter terminal of the curve tracer, set the horizontal selector to 10 V/div, and set the vertical selector to 10 µA/div to limit current flow through the reverse-biased junction at breakdown conditions. Increase the horizontal voltage and monitor the voltage breakdown conditions. For the 2N3904 the breakdown region is from about 60 V to 80 V.

To observe BV_{EB_o}, connect the emitter lead of the device to the collector terminal of the curve tracer, the base lead of the device to the emitter terminal of the curve tracer, set the horizontal selector to 1 V/div, and set the vertical selector to 1 mA/div. Increase the horizontal voltage and monitor the voltage breakdown conditions. For the 2N3904, the breakdown region is from about 8 V to 10 V.

LABORATORY EXERCISE

OBJECTIVES

To investigate the available techniques for testing transistors — from go and no-go tests to complete parameter measurements. Section I involves ohmmeter testing, Section II involves curve tracer techniques, and Section III involves bench measurement techniques. (Section III is optional if a curve tracer is available for Section II.)

LIST OF MATERIALS AND EQUIPMENT

1. Transistor: 2N3904 or equivalent (one)
2. Diode: 1N4001 or equivalent (one)
3. Resistors (one each except where indicated):
   1 kΩ   10 kΩ (potentiometer — two)   100 kΩ (two)
4. Transformer: 6.3 V RMS SEC (preferred)
5. Oscilloscope (horizontal input capability)
6. Power Supply

PROCEDURE

Section I: To Become Familiar With the Go and No-Go Junction Testing of Transistors Using an Ohmmeter

1. Locate the collector, base, and emitter leads of the device to be tested. Use the data sheets in the appendix if necessary.
2. Set the ohmmeter at the R × 1 kΩ setting and measure the resistance across the base-emitter junction, first in one direction and then in the reverse direction.
3. Repeat the procedure for the collector-base leads of the transistor.

NOTE: The R × 1 kΩ setting could be changed to a lower resistance setting of the ohmmeter for forward biased junction testing, or it could be set to a higher resistance ohmmeter setting for reverse biased junction testing. However, since ohmmeter testing is used to indicate resistance ratios only, it does not matter.

4. Measure the collector-emitter leads of the transistor in both directions with the ohmmeter resistance at the R × 1 kΩ setting.
5. Insert the measured forward and reverse junction resistance for the base-emitter, collector-base, and collector-emitter junctions into Table 5-1.

TABLE 5-1	BASE-EMITTER		COLLECTOR-BASE		COLLECTOR-EMITTER	
	FORWARD	REVERSE	FORWARD	REVERSE	FORWARD	REVERSE
MEASURED						

Section II: To Become Familiar With the General Testing of Transistors on the Curve Tracer (TEK 575)

Part 1: Output Characteristic Curves

1. Connect the 2N3904, or equivalent device, into the left or right socket of the curve tracer. Set the toggle switch in the neutral position and set the load resistor to the 0 Ω position to observe the full untruncated output response curves.
2. Set the horizontal selector knob at 1 V/DIV and the vertical knob at 1 mA/DIV. Set the base current selector knob to 10 μA/STEP.
3. For the NPN device, make sure the polarity switches for the peak voltage collector sweep and the base step generator are both positive. Also, make sure the display switch is set to the repetitive position and the collector sweep is in the 0-20 volt range.
4. Increase the horizontal deflection 10 full divisions, so that the V_{CE} of the device will be 10 volts. Throw the toggle switch in the direction of the device under test and monitor the output characteristic curves on the screen. It may be necessary to increase the repetitivity of the displayed curves or the number of steps being displayed.
5. For each I_B step between 10 μA and 50 μA, plot the I_C current for each of the V_{CE} voltages indicated in Table 5-2. Then, calculate the beta of the device at the V_{CE} voltages of 10 V, 6 V, and 2 V, as indicated in Table 5-2. Insert the values, as indicated, into Table 5-2.

NOTE: If more detail is required, reread the transistor set-up procedure in the text material.

TABLE 5-2 V_{CE}		10 V	8 V	6 V	4 V	2 V	1 V	0.5 V	0 V
I_B	I_C								
10 μA	β		/////		/////		/////	/////	/////
I_B	I_C								
20 μA	β		/////		/////		/////	/////	/////
I_B	I_C								
30 μA	β		/////		/////		/////	/////	/////
I_B	I_C								
40 μA	β		/////		/////		/////	/////	/////
I_B	I_C								
50 μA	β		/////		/////		/////	/////	/////

Part 2: Breakdown Voltage Measurements

A. BV_{CE} Breakdown
 1. The breakdown voltage across the collector-emitter of the transistor is obtained by displaying the output characteristics of the device and increasing the V_{CE} voltage until the breakdown condition occurs, as illustrated in the text material.
 2. Therefore, display the output characteristic curves and switch the horizontal knob to 5 V/DIV, so

that the full deflection capability will be 50 V. Slowly increase the V_{CE} amplitude until breakdown occurs. For voltage peaks greater than 20 V, switch the collector sweep to the 0-200 mode.

B. Breakdown Voltages

1. BV_{CEo} — Measure the collector-emitter breakdown voltage. Connect the collector lead to the collector terminal, the emitter lead to the emitter terminal, and leave the base lead open (unconnected). Set the horizontal selector knob to 10 V/DIV and the vertical selector knob to 10 µA/DIV and increase the collector sweep voltage until the breakdown voltage occurs. Insert the measured value into Table 5-3.

2. BV_{CBo} — Measure the collector-base breakdown voltage. Connect the collector lead to the collector terminal, the base lead to the emitter terminal, and leave the emitter lead open (unconnected). Set the horizontal selector knob to 20 V/DIV and the vertical selector knob to 10 µA/DIV and increase the collector sweep voltage until the breakdown voltage occurs. Insert the measured value into Table 5-3.

3. BV_{EBo} — Measure the emitter-base breakdown voltage. Connect the emitter lead to the collector terminal, the base lead to the emitter terminal, and leave the collector terminal open (unconnected). Set the horizontal selector knob to 1 V/DIV, the vertical knob to 1 mA/DIV, and increase the collector sweep voltage until the breakdown voltage occurs. Insert the measured value into Table 5-3.

NOTE: For further protection, during BV_{CEo} and BV_{CBo} measurements, dial in 100 kΩ.

TABLE 5-3	BV_{CEo}	BV_{CBo}	BV_{EBo}	V_{BE}	V_{BC}
MEASURED					

C. Forward Biased Transistor Characteristics

1. V_{BE} — Measure the forward biased base-emitter junction voltage of the transistor. Connect the base lead to the collector terminal, the emitter lead to the emitter terminal, and leave the collector terminal open (unconnected). Set the horizontal selector knob to 0.1 V/DIV, the vertical selector knob to 1 mA/DIV, and increase the collector sweep voltage until the characteristic curve is shown. This is the input characteristic curve of the device. Insert the measured V_{BE} value, at 1 mA of current flow through the junction, into Table 5-3.

2. V_{BC} — Measure the forward biased transistor base-collector junction voltage in the same manner as the base-emitter voltage was measured. Insert the measured value, at 1 mA of current flow through the junction, into Table 5-3.

NOTE: The individual base-emitter and base-collector junctions of the transistor are measured the same way as a regular diode is measured. However, the base-collector reverse biased junction voltage breakdown is in excess of 60 V, while the reverse biased base-emitter junction voltage is less than 10 V. Therefore, in some instances, the reverse biased base-emitter junction voltage can be used as a reference voltage.

Part 3: Inverse Connection

1. Connect the transistor in the inverse mode, where the collector lead connects to the emitter terminal of the curve tracer, the emitter lead to the collector terminal of the curve tracer, and the base lead to the base terminal of the curve tracer.

2. Set the horizontal indicating knob to 2 V/DIV and the vertical knob to 1 mA/DIV. Then, increase the base current indicating knob to display about 5 characteristic I_B step curves and set I_B in the 50 µA to 100 µA range.

3. Gradually increase the V_{CE} voltage and note that the breakdown voltage occurs at about the voltage of the previously measured BV_{BEo}. Also, monitor the collector current of the third characteristic I_B step curve at a V_{CE} of 6 V and calculate the beta of the inverse connection from: $\beta = I_C/I_B$. Insert these values, as indicated, into Table 5-4.

TABLE 5-4	BREAKDOWN VOLTAGE V_{EC}	I_B	I_C	Beta
MEASURED				////////
CALCULATED	////////	////////	////////	

Section III: (Optional) Bench Techniques Using Standard Swept Measurement Techniques

Part 1: Standard Bench Techniques for Determining Output Characteristic Curves

1. Connect the circuit shown in Figure 5-11.

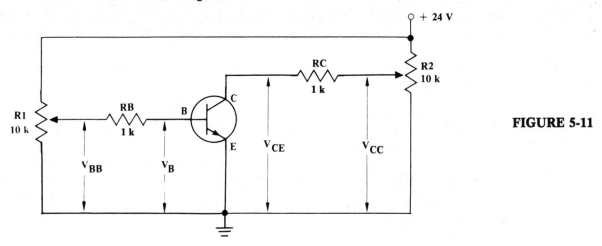

FIGURE 5-11

2. Initially, set V_{BB} at 1.6 V and V_{CE} at 12 V.
   a. Measure V_B and V_{CC} and calculate I_C and I_B from:

   $$I_B = (V_{BB} - V_B)/R_B \quad \text{where:} \quad R_B = 100 \text{ k}\Omega$$

   $$I_C = (V_{CC} - V_C)/R_C \quad \text{where:} \quad R_C = 100 \text{ }\Omega$$

   b. Decrease V_{CE} to 6 V, 3 V, 1.5 V, 1 V, 0.5 V, and 0 V by adjusting R2.

NOTE: Variable resistor R2 controls V_{CC}. Also, RC at 1 kΩ is needed to obtain reasonably accurate I_C readings.

   c. Insert the calculated and measured values, as indicated into Table 5-5.

TABLE 5-5		$V_{BB} = 1.6$ V, $I_B \approx 10$ μA					
MEAS.	V_{CE}	12 V	6 V	3 V	1.5 V	1 V	0.5 V	0 V
MEAS.	V_{CC}							
CALC.	V_{RC}							
CALC.	V_{RB}		////////	////////	////////	////////	////////	////////
CALC.	I_C							
CALC.	I_B		////////	////////	////////	////////	////////	////////
CALC.	Beta							

TABLE 5-5 (continued)		\multicolumn{7}{c}{$V_{BB} = 3.6$ V, $I_B \approx 30$ μA}						
MEAS.	V_{CE}	12 V	6 V	3 V	1.5 V	1 V	0.5 V	0 V
MEAS.	V_{CC}							
CALC.	V_{RC}							
CALC.	V_{RB}		/////	/////	/////	/////	/////	/////
CALC.	I_C							
CALC.	I_B		/////	/////	/////	/////	/////	/////
CALC.	Beta		/////	/////	/////	/////	/////	/////
		\multicolumn{7}{c}{$V_{BB} = 5.6$ V, $I_B \approx 50$ μA}						
MEAS.	V_{CE}	12 V	6 V	3 V	1.5 V	1 V	0.5 V	0 V
MEAS.	V_{CC}							
CALC.	V_{RC}							
CALC.	V_{RB}		/////	/////	/////	/////	/////	/////
CALC.	I_C							
CALC.	I_B		/////	/////	/////	/////	/////	/////
CALC.	Beta		/////	/////	/////	/////	/////	/////

3. Set V_{BB} to 3.6 V and calculate and measure the values called for in Table 5-5. Again, follow the procedure of step 2. The V_{BB} at 5.6 V will provide an I_B of approximately 50 μA.
5. Insert the calculated and measured values, as indicated, into Table 5-5.
6. Use the data obtained in Table 5-5 to graph a set of characteristic curves. Graph each I_C (versus I_B) curve for the incremental V_{CE} voltages of 0 V through 12 V, as indicated in the table. Use the graph form of Figure 5-12.

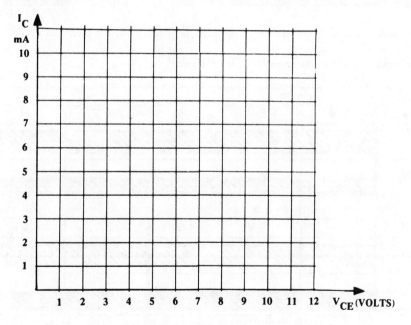

FIGURE 5-12

Part 2: Input Characteristic Curve

1. Connect the circuit shown in Figure 5-13.

NOTE: Use an HP 175 or equivalent oscilloscope.

2. Disconnect all power. Locate a 0,0 (x,y) origin in the lower left hand side of the oscilloscope grid.
3. Set the vertical selector to 100 mV/DIV and the horizontal selector in the sweep mode at 1 V/DIV.

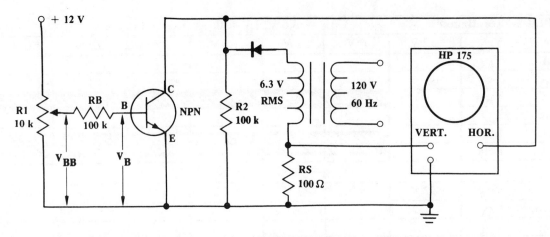

FIGURE 5-13

4. Connect the 120 V, 60 Hz power source and monitor a left to right horizontal deflection on the scope. The deflection for an RMS of 6.3 V should approach 9 V, if signal losses are at a minimum.

NOTE: If the horizonal deflection is not achieved, check for the half-wave pulse across resistor R2. Also, depending of the oscilloscope used, a left to right horizontal deflection or a bottom to top vertical deflection may not be possible.

5. When the V_{CE} horizontal deflection is obtained, connect the V_{CC} power supply and adjust R1 so that about 50 μA of base current exists. Note the characteristic curve as its amplitude is decreased.
6. Since the vertical selector is set to 100 mV/DIV, then each vertical division should indicate 1 mA/DIV since RS = 100 Ω. Set the base current to 30 μA and then to 10 μA, using the standard bench techniques of Part I, step 2.
7. Monitor the I_C currents at each of the V_{CE} conditions indicated in Table 5-6, for each 10 μA, 30 μA, and 50 μA base current condition.
8. Insert the calculated and measured values, as indicated, into Table 5-6.

TABLE 5-6		\multicolumn{6}{c	}{V_{BB} = 1.6 V}
MEAS.	V_{CE}	6 V	3 V	1.5 V	1 V	0.5 V	0 V
MEAS.	V_{RB}						
CALC.	I_B						
MEAS.	I_C						
CALC.	Beta						
\multicolumn{8}{	c	}{V_{BB} = 3.6 V}
MEAS.	V_{CE}	6 V	3 V	1.5 V	1 V	0.5 V	0 V
MEAS.	V_{RB}						
CALC.	I_B						
MEAS.	I_C						
CALC.	Beta						
\multicolumn{8}{	c	}{V_{BB} = 5.6 V}
MEAS.	V_{CE}	6 V	3 V	1.5 V	1 V	0.5 V	0 V
MEAS.	V_{RB}						
CALC.	I_B						
MEAS.	I_C						
CALC.	Beta						

9. Use the data obtained in Table 5-6 to graph a set of characteristic curves. Graph each I_C (versus I_B)

curve for the incremental V_{CE} voltages of 0 V through 10 V indicated in the table. Use the graph form of Figure 5-14

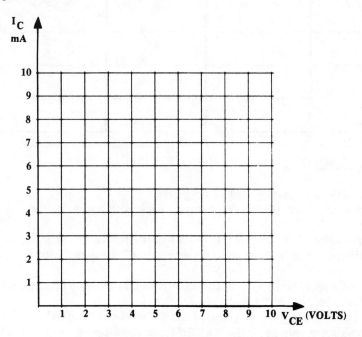

FIGURE 5-14

Part 3: Swept Measurement Procedures for Obtaining Output Characteristic Curves

1. Set V_{CE} to 3 V, as shown in Figure 5-15, so that the collector-base junction is reverse biased. Gradually, increase V_{BB} from an approximate zero volt condition. Calculate I_B for every incremental voltage listed in Table 5-7, beginning with a V_{BE} of 0.4 V.

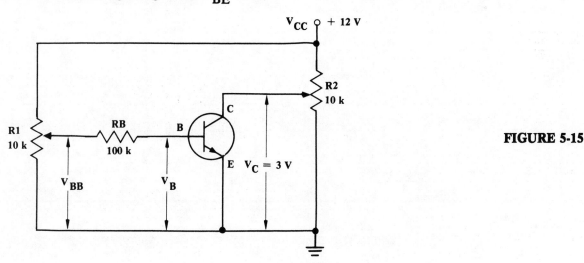

FIGURE 5-15

NOTE: I_B can be measured directly by inserting an ammeter in series with the RB base resistor. However, proper measuring procedure is to use the voltage drop across a known resistor to calculate current flow conditions.

2. Measure V_{BB} and V_B voltage conditions and calculate I_B from $I_B = (V_{BB} - V_{BE})/RB$, for each V_{BE} condition.
3. Insert the calculated and measured values, as indicated, into Table 5-7.

TABLE 5-7					
V_{BE}	0.4 V	0.5 V	0.55 V	0.6 V	0.65 V
V_{BB}					
V_{RB}					
I_B					

LABORATORY QUESTIONS

1. Why is the ohmmeter testing of transistors primarily a go or no-go test?
2. Why does the measured resistance of forward biased junctions change with every R × ohmmeter setting?
3. If a PNP transistor, 2N3906 or equivalent, is to be tested on the curve tracer or by using bench measurement techniques, what major differences in voltage setting are required? Respond with regard to NPN devices.
4. On the curve tracer, test a 2N3906, or equivalent PNP transistor, to gain familiarity with the device.
    a. In what corner is the origin? How does this differ from the 2N3904?
    b. Do the horizontal and vertical selector settings differ greatly from those of the 2N3904?
    c. What is the inverse connected breakdown voltage of the PNP 2N3906 transistor?
    d. Why did the polarity selectors of the peak control collector sweep and base step generator have to be negative?
    e. What are the collector-emitter, breakdown voltages of the 2N3904 and 2N3906 devices?
5. Does the beta of the device increase, although slightly, with an increase in I_C? Use the data tables for the answer and explain why the increase is or is not possible.
6. Give three similarities and two differences between the 2N3904 and 2N3906 transistors.

6 | SINGLE STAGE TRANSISTOR BIASING

GENERAL DISCUSSION

The transistor can be biased in the linear mode through base biasing, collector feedback biasing, universal biasing, and two power supply emitter biasing, as shown in Figure 6-1. Each of these single stage circuits provides linear biasing, where the base-emitter junction is forward biased and the collector-base junction is reverse biased. Too, in these circuits the NPN transistor is shown being used, but PNP transistors can also be used, and only the power supply connections and the polarity direction of the voltage drops would differ.

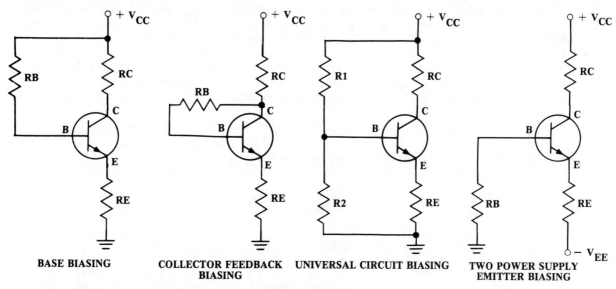

FIGURE 6-1

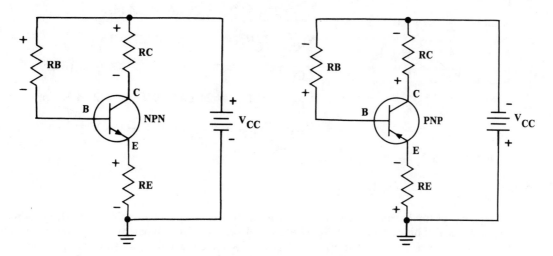

FIGURE 6-2

SECTION I: BASE-BIASED TRANSISTOR CIRCUITS

The base-biased circuit requires only a single power supply and a minimum number of components. However, the drawback to this form of biasing is that the base current is fixed and the collector current and voltage drops around the circuit are too dependent upon the beta of the transistor. Therefore, if the beta of the device changes because of temperature, or a replacement device with a different beta is used, then the voltage drops around the circuit will change. Either an NPN or PNP device can be used, and the use of one or the other determines the power supply connections, as shown in Figure 6-2.

The basic base-biased circuit of Figure 6-3 can be analyzed in terms of the base current I_B or the collector current I_C, when the current gain (beta) and the circuit component values are known. Both approaches to solving the circuit currents and the voltage drops around the circuit are discussed in the examples that follow. However, the analysis of the simple base-biased circuit is so straight-forward that it can be done by inspection. Therefore, it is an excellent circuit for use in introducing Kirchhoff's loop equations to solve for I_B and then for I_C.

1. SOLVING IN TERMS OF I_B

The DC analysis of the simple base-biased circuit, solving in terms of I_B, begins by writing loop equations using the voltage drops across the base resistor V_{RB} and the base-emitter diode V_{BE}. Since $V_{BE} = 0.6$ volts for silicon transistors, V_{RB} is found by subtracting V_{BE} from V_{CC}. Also, since $V_{RB} = I_B R_B$, I_B can be solved from V_{RB}/R_B. Then, if the circuit gain of the device is known, I_C is found from βI_B and V_{RC} is found from $I_C R_C$. V_{CE} is solved by subtracting V_{RC} from V_{CC}.

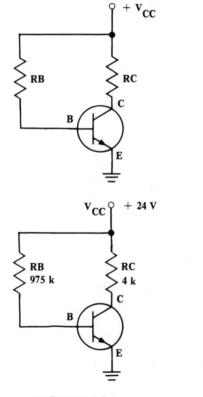

$$V_{CC} = V_{RB} + V_{BE}$$
$$V_{CC} - V_{BE} = V_{RB}$$
$$V_{CC} - V_{BE} = I_B R_B$$
$$\frac{V_{CC} - V_{BE}}{R_B} = I_B$$

Given: $\beta = 100$ and $V_{CC} = 24$ V

$$I_B = \frac{V_{CC} - V_{BE}}{R_B} = \frac{24\text{ V} - 0.6\text{ V}}{975\text{ k}\Omega} = \frac{23.4\text{ V}}{975\text{ k}\Omega} = 24\ \mu A$$

$$I_C = \beta I_B = 100 \times 24\ \mu A = 2.4\text{ mA}$$

$$V_{RC} = I_C R_C = 2.4\text{ mA} \times 4\text{ k}\Omega = 9.6\text{ V}$$

$$V_{CE} = V_{CC} - V_{RC} = 24\text{ V} - 9.6\text{ V} = 14.4\text{ V}$$

$$V_{RB} = I_B R_B = 24\ \mu A \times 975\text{ k}\Omega = 23.4\text{ V}\quad \text{or}$$

$$V_{RB} = V_{CC} - V_{BE} = 24\text{ V} - 0.6\text{ V} = 23.4\text{ V}$$

FIGURE 6-3

2. SOLVING IN TERMS OF I_C

The DC analysis of the base-biased circuit, solving in terms of I_C, also begins by writing loop equations using V_{RB} and V_{BE}. This is because V_{RB} is easily solved, and I_B is fixed by V_{RB}/R_B, regardless of the beta of the device. However, once the $V_{CC} - V_{BE} = I_B R_B$ equation is written, I_C/β is substituted for the base current I_B so that the equation can be solved in terms of I_C.

NOTE: Remember that $I_C = \beta I_B$. Therefore, $I_B = I_C/\beta$

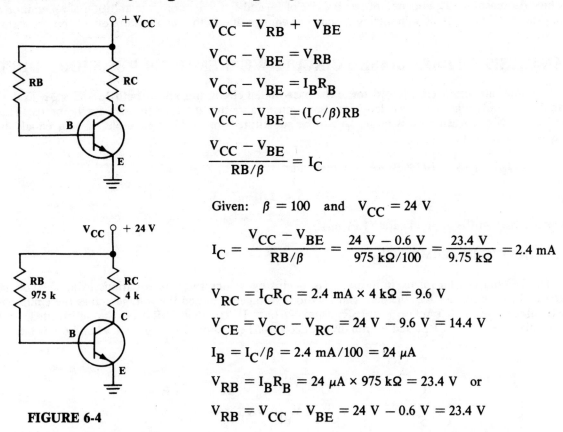

$$V_{CC} = V_{RB} + V_{BE}$$
$$V_{CC} - V_{BE} = V_{RB}$$
$$V_{CC} - V_{BE} = I_B R_B$$
$$V_{CC} - V_{BE} = (I_C/\beta) RB$$
$$\frac{V_{CC} - V_{BE}}{RB/\beta} = I_C$$

Given: $\beta = 100$ and $V_{CC} = 24$ V

$$I_C = \frac{V_{CC} - V_{BE}}{RB/\beta} = \frac{24 \text{ V} - 0.6 \text{ V}}{975 \text{ k}\Omega/100} = \frac{23.4 \text{ V}}{9.75 \text{ k}\Omega} = 2.4 \text{ mA}$$

$$V_{RC} = I_C R_C = 2.4 \text{ mA} \times 4 \text{ k}\Omega = 9.6 \text{ V}$$
$$V_{CE} = V_{CC} - V_{RC} = 24 \text{ V} - 9.6 \text{ V} = 14.4 \text{ V}$$
$$I_B = I_C/\beta = 2.4 \text{ mA}/100 = 24 \text{ }\mu\text{A}$$
$$V_{RB} = I_B R_B = 24 \text{ }\mu\text{A} \times 975 \text{ k}\Omega = 23.4 \text{ V} \text{ or}$$
$$V_{RB} = V_{CC} - V_{BE} = 24 \text{ V} - 0.6 \text{ V} = 23.4 \text{ V}$$

FIGURE 6-4

3. VOLTAGE DISTRIBUTION

The voltage distribution of the circuit is shown in Figure 6-5. It is important to remember that all calculated and measured voltage drops between V_{CC} and ground must equal the applied V_{CC}.

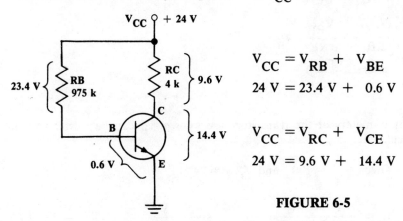

$$V_{CC} = V_{RB} + V_{BE}$$
$$24 \text{ V} = 23.4 \text{ V} + 0.6 \text{ V}$$

$$V_{CC} = V_{RC} + V_{CE}$$
$$24 \text{ V} = 9.6 \text{ V} + 14.4 \text{ V}$$

FIGURE 6-5

BASE-BIASED CIRCUIT WITH ADDITION OF EMITTER RESISTOR

Although the simple base-biased circuit is easy to analyze, as previously stated, it has poor DC stability because the base-emitter junction and the beta of the device vary with temperature change. As shown in the analysis, the beta of the simple base-biased circuit is directly proportional to the collector current I_C, and any increase in I_C causes the voltage drops around the circuit to change accordingly. Hence, if an emitter resistor is added to the circuit, any increase in collector current causes an emitter current increase. And an increased voltage drop across the emitter resistor decreases the voltage drop across V_{RB}, which decreases the I_B and minimizes the rise of I_C and I_E. Therefore, the addition of an emitter resistor improves DC stability by minimizing I_C current change due to the temperature effects on the beta of the device.

ANALYSIS OF BASE-BIASED CIRCUIT WITH EMITTER RESISTOR ADDED

The inclusion of the emitter resistor to the base-biased circuit introduces both I_E and V_{RE} to the circuit analysis calculations. However, loop equations must be written in order to solve for I_C or I_B. Therefore, since $I_C \approx I_E$, when beta is large, I_C can be substituted for I_E. For example, if beta equals 100 and $I_B = 10 \mu A$, then:

$$I_C = \beta I_B = 100 \times 10 \mu A = 1 \text{ mA} \quad \text{and}$$

$$I_E = (\beta + 1)I_B = (100 + 1)10 \mu A = 1.01 \text{ mA}$$

Obviously, 1 mA is approximately 1.01 mA.

1. **SOLVING IN TERMS OF I_B:**

The DC analysis, for solving in terms of I_B, begins by writing the loop equations using the voltage drop across the base resistor RB, the base-emitter diode voltage V_{BE}, and the voltage across the emitter resistor RE. Since $V_{BE} = 0.6$ V, $V_{RB} = I_B R_B$, and $V_{RE} = I_E R_E$, then $I_E R_E$ can be substituted for by the equivalent $(\beta + 1) I_B R_E$, and eventually reduced to $\beta I_B R_E$, since $\beta \approx \beta + 1$ when beta is large.

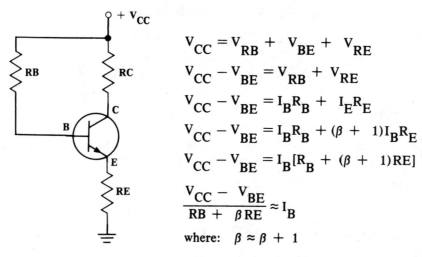

$$V_{CC} = V_{RB} + V_{BE} + V_{RE}$$

$$V_{CC} - V_{BE} = V_{RB} + V_{RE}$$

$$V_{CC} - V_{BE} = I_B R_B + I_E R_E$$

$$V_{CC} - V_{BE} = I_B R_B + (\beta + 1) I_B R_E$$

$$V_{CC} - V_{BE} = I_B [R_B + (\beta + 1) RE]$$

$$\frac{V_{CC} - V_{BE}}{RB + \beta RE} \approx I_B$$

where: $\beta \approx \beta + 1$

FIGURE 6-6

Once the resistor values and the parameters of the transistors are known, I_B can be solved and the voltage drops around the circuit calculated.

Given: $\beta = 100$ and $V_{CC} = 24$ V

$$I_B \approx \frac{V_{CC} - V_{BE}}{RB + \beta RE} = \frac{24 \text{ V} - 0.6 \text{ V}}{875 \text{ k}\Omega + (100 \times 1 \text{ k}\Omega)} = \frac{23.4 \text{ V}}{975 \text{ k}\Omega} = 24 \mu A$$

$$V_{RB} = I_B R_B = 24 \mu A \times 875 \text{ k}\Omega = 21 \text{ V}$$

$$I_C = \beta I_B = 100 \times 24 \mu A = 2.4 \text{ mA}$$

$$V_{RC} = I_C R_C = 2.4 \text{ mA} \times 4 \text{ k}\Omega = 9.6 \text{ V}$$

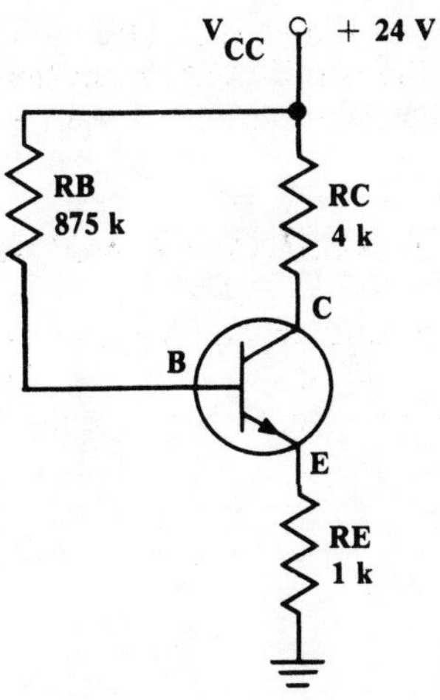

$$V_{RE} = I_E R_E \approx 2.4 \text{ mA} \times 1 \text{ k}\Omega = 2.4 \text{ V}$$

$$V_C = V_{CC} - V_{RC} = 24 \text{ V} - 9.6 \text{ V} = 14.4 \text{ V}$$

$$V_E = V_{RE} = 2.4 \text{ V}$$

$$V_B = V_{RE} + V_{BE} = 2.4 \text{ V} + 0.6 \text{ V} = 3 \text{ V} \quad \text{or}$$

$$V_B = V_{CC} - V_{RB} = 24 \text{ V} - 21 \text{ V} = 3 \text{ V}$$

$$V_{CE} = V_C - V_E = 14.4 \text{ V} - 2.4 \text{ V} = 12 \text{ V}$$

$$V_{CB} = V_C - V_B = 14.4 \text{ V} - 3 \text{ V} = 11.4 \text{ V}$$

FIGURE 6-7

2. SOLVING IN TERMS OF I_C

The DC analysis, in terms of solving for I_C, also begins by writing loop equations using V_{RB}, V_{BE}, and V_{RE}. Then, since $V_{BE} = 0.6$ V, $V_{RB} = I_B R_B$, and $V_{RE} = I_E R_E$, I_B is substituted for by I_C/β and I_E by I_C where, again, $I_C \approx I_E$ is taken into account. Therefore, once I_C is solved, the voltage drops around the circuit can be calculated.

$$V_{CC} = V_{RB} + V_{BE} + V_{RE}$$

$$V_{CC} - V_{BE} = V_{RB} + V_{RE}$$

$$V_{CC} - V_{BE} = I_B R_B + I_E R_E$$

$$V_{CC} - V_{BE} = (I_C/\beta \times RB) + I_C R_E$$

$$V_{CC} - V_{BE} = I_C(RB/\beta + RE)$$

$$\frac{V_{CC} - V_{BE}}{RB/\beta + RE} \approx I_C$$

where: $I_B = I_C/\beta$ and $I_E \approx I_C$

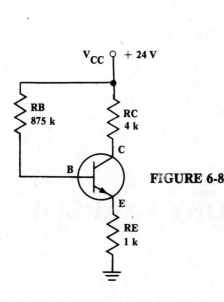

FIGURE 6-8

Given $\beta = 100$ and $V_{CC} = 24$ V

$$I_C \approx \frac{V_{CC} - V_{BE}}{RB/\beta + RE} = \frac{24 \text{ V} - 0.6 \text{ V}}{875 \text{ k}\Omega/100 + 1 \text{ k}\Omega}$$

$$= \frac{23.4 \text{ V}}{8.75 \text{ k}\Omega + 1 \text{ k}\Omega} = \frac{23.4 \text{ V}}{9.75 \text{ k}\Omega} = 2.4 \text{ mA}$$

$$V_{RC} = I_C R_C = 2.4 \text{ mA} \times 4 \text{ k}\Omega = 9.6 \text{ V}$$

$$V_{RE} = I_E RE = 2.4 \text{ mA} \times 1 \text{ k}\Omega = 2.4 \text{ V}$$

$$I_B = I_C/\beta = 2.4 \text{ mA}/100 = 24 \text{ }\mu\text{A}$$

$$V_{RB} = I_B R_B = 24 \text{ }\mu\text{A} \times 875 \text{ k}\Omega = 21 \text{ V}$$

$$V_C = V_{CC} - V_{RC} = 24 \text{ V} - 9.6 \text{ V} = 14.4 \text{ V}$$

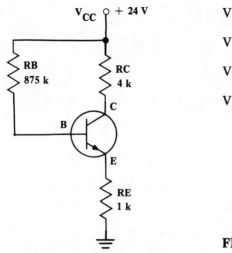

$V_E = V_{RE} = 2.4\ V$

$V_B = V_{RE} + V_{BE} = 2.4\ V + 0.6\ V = 3\ V$

$V_{CE} = V_C - V_E = 14.4\ V - 2.4\ V = 12\ V$

$V_{CB} = V_C - V_B = 14.4\ V - 3\ V = 11.4\ V$

FIGURE 6-9

3. VOLTAGE DISTRIBUTION

The voltage distribution for the circuit is shown in Figure 6-10. Again, it is important to remember that all calculated and measured voltage drops between V_{CC} and ground must equal the applied V_{CC}.

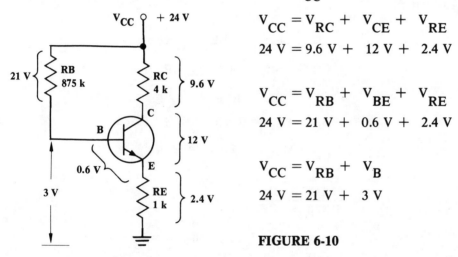

$V_{CC} = V_{RC} + V_{CE} + V_{RE}$
$24\ V = 9.6\ V + 12\ V + 2.4\ V$

$V_{CC} = V_{RB} + V_{BE} + V_{RE}$
$24\ V = 21\ V + 0.6\ V + 2.4\ V$

$V_{CC} = V_{RB} + V_B$
$24\ V = 21\ V + 3\ V$

FIGURE 6-10

SECTION I: LABORATORY EXERCISE

OBJECTIVES
To investigate the DC characteristics of the base-biased circuits.

LIST OF MATERIALS AND EQUIPMENT
1. Transistor: 2N3904 or equivalent
2. Resistors (one each): 1 kΩ   6.8 kΩ   2.2 MΩ
3. Power Supply
4. Voltmeter

PROCEDURE
Part 1: Basic Base-Biased Circuit
1. Connect the base-biased circuit, as shown in Figure 6-11.
2. Measure V_C, the voltage at the collector with respect to ground. In this circuit, $V_C = V_{CE}$.

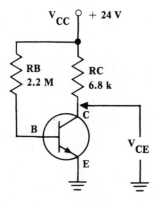

FIGURE 6-11

3. Use the measured V_C and calculate V_{RC} and V_{CE}.

   a. Calculate V_{RC} from: $V_{RC} = V_{CC} - V_C$.

   b. Calculate V_{CE} from: $V_{CE} = V_C - V_E$ where: $V_E = 0$ V

4. Measure V_B, the voltage at the base with respect to ground. In this circuit $V_B = V_{BE}$.

5. Use the measured V_B and calculate V_{RB} and V_{BE}.

   a. Calculate V_{RB} from: $V_{RB} = V_{CC} - V_B$

   b. Calculate V_{BE} from: $V_{BE} = V_B - V_E$ where: $V_E = 0$ V

6. Calculate the collector current I_C and the base current I_B.

   a. Calculate I_C from: $I_C = V_{RC}/RC$

   b. Calculate I_B from: $I_B = V_{RB}/RB$

7. Calculate beta, the current gain of the transistor from $\beta = I_C/I_B$.

8. Insert the calculated and measured values, as indicated, into Table 6-1.

TABLE 6-1	V_C	V_{RC}	V_{CE}	V_B	V_{RB}	V_{BE}	I_C	I_B	Beta
CALCULATED		15 V	9 V		23.33	.67	2.2m	10.6µ	207.5
MEASURED	9 V			.67					

Handwritten annotations above table: 15.5V, 8.8V, 23.7, 2.3 mA, 10.6 µA

Handwritten: ✓ o|< P A+

Part 2: Basic Base-Biased Circuit with Emitter Resistor Added

1. Connect the circuit as shown in Figure 6-12, where a 1 kΩ resistor is added to the basic base-biased circuit. Use the same transistor used in the previous circuit, because the beta of the device will be required in the calculations. If a different device must be used, obtain the beta off the curve tracer or through bench measurement techniques.

2. Calculate I_C, the collector current, from: $I_C = \dfrac{V_{CC} - V_{BE}}{RB/\beta + RE}$

NOTE: $\beta \approx \beta + 1$ when beta is high. Therefore, $I_C \approx I_E$.

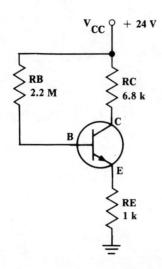

FIGURE 6-12

3. Calculate the voltage drops of V_{RC}, V_{RE}, and V_{RB}. Use these values to calculate the remaining voltage drops around the circuit.

   a. Calculate the voltage drop across the collector resistor R_C from: $V_{RC} = I_C R_C$.

   b. Calculate the voltage drop across the emitter resistor R_E from: $V_{RE} = I_E R_E$.

   c. Calculate the voltage at the collector of the transistor, with respect to ground, from:

   $$V_C = V_{CC} - V_{RC}$$

   d. Calculate the voltage at the base of the transistor, with respect to ground from:

   $$V_B = V_{RE} + V_{BE}$$

   e. Calculate the voltage drop across the base resistor from either:

   $$V_{RB} = V_{CC} - V_B \quad \text{where:} \quad V_B = V_{RE} + V_{BE}$$
   $$V_{RB} = I_B R_B \quad \text{where:} \quad I_B = I_C/\beta$$

4. Measure the DC voltage drops around the circuit.

   a. Measure V_C, the voltage at the collector with respect to ground.

   b. Measure V_E, the voltage at the emitter with respect to ground. In this circuit, $V_E = V_{RE}$.

   c. Measure V_B, the voltage at the base with respect to ground.

NOTE: Good voltage measurement procedure requires that all DC measurements be made with respect to ground. This is because measured values taken at the collector and then the emitter, with respect to ground, will provide the V_{CE} voltage drop across the collector-emitter without having to physically connect the voltmeter across the collector-emitter of the device. The same reasoning holds true for measuring V_{RC} and V_{RB}. Also, taking measurements with respect to ground helps to avoid ground loops that could be caused by a grounded measuring instrument.

5. Use the V_C, V_E and V_B voltage measurement to calculate the measured voltage drops of V_{RC}, V_{RB}, and V_{CE}.

   a. Calculate V_{RC} from: $V_{RC} = V_{CC} - V_C$

   b. Calculate V_{RB} from: $V_{RB} = V_{CC} - V_B$

c. Calculate V_{CE} from: $V_{CE} = V_C - V_E$ where: $V_E = V_{RE}$

6. Insert the calculated and measured values, as indicated, into Table 6-2.

TABLE 6-2	I_C	V_{RC}	V_{RB}	V_{RE}	V_{CE}	V_C	V_B	I_B	Beta
CALCULATED	2.2mA	14.96	21.09	2.2V	▓▓▓	9.04	2.91	10.8μA	203.7
MEASURED	▓▓▓	13.97	21.2V	2.21V	8.2V	10.3V	2.70V	▓▓▓	207.54

SECTION QUESTIONS

1. Is the base-biased circuit a fixed base current or a fixed emitter current circuit? Explain why.
2. What is the main drawback to base-biased circuits?
3. Does DC stability of the circuit increase or decrease with the addition of an emitter resistor? Explain.
4. Why is it a good idea to sum the voltage drops of the circuit between V_{CC} and ground?
5. With reference to Figure 6-12, what voltage measurement, with respect to ground, will allow I_C to be solved directly? How about for solving I_B indirectly?
6. Did beta change with reference to the circuits of Figure 6-11 and Figure 6-12? Why or why not?

SECTION PROBLEMS

1. Given: $\beta = 100$ and $V_{BE} = 0.6$ V

   Find:

   A. I_C 2.6 mA
   B. V_{RC} $(I_C)(R_C) = 13$ 26μA
   C. V_{RE} $(I_C)(R_E)$ 2.6 V
   D. $V_{CE} = 8.4 = (11V - 2.6V) = 8.4V$
   E. V_{RB} $(800K)(26μA) = 20.8$

2. Given: $\beta = 150$ and $V_{BE} = 0.6$ V

   Find:

   A. I_C $(\beta)(I_B)(150)(25μ) = 3.8$ mA
   B. V_{RC} $(3.8)(5k) = 19$
   C. V_{RE} $(I_C)(R_C) = 3.8$
   D. V_{CE} .6V
   E. V_{RB} 19.5V

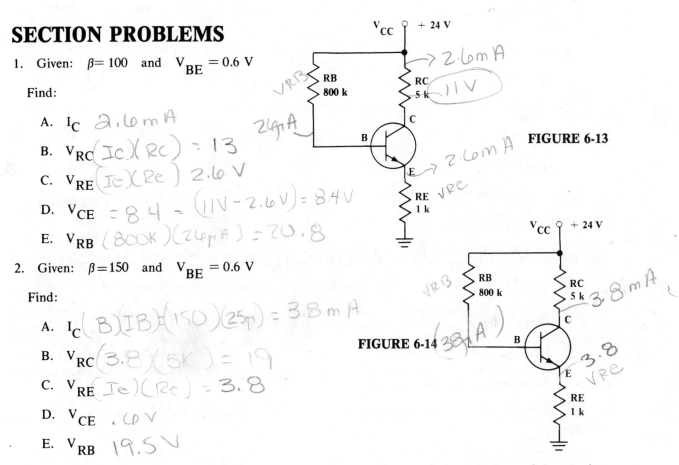

FIGURE 6-13

FIGURE 6-14

NOTE: The problem number one and number two circuits are the same except the beta of the transistor is changed from 100 to 150. This demonstrates clearly the effect beta has on base-biased circuits.

3. Given: $\beta = 100$, $V_{BE} = 0.6$ V, and $V_C = 12$ V

   Find:

   A. V_{RC} 12 V
   B. I_C 2.4 mA
   C. V_{RE} (2.4 mA)(1.5k) 3.6
   D. V_{CE} .6 V
   E. V_B
   F. V_{RB}
   G. R_B

   Reminder: $V_{RC} = V_{CC} - V_C$, $I_E \approx I_C = V_{RC}/RC$

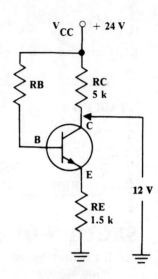

FIGURE 6-15

4. Given: $\beta = 150$, $V_{BE} = 0.6$ V, $V_{RE} = 2.4$ V, and $I_C \approx I_E$

   Find:

   A. I_E
   B. V_{RC}
   C. V_{CE}
   D. V_{RB}
   E. R_B

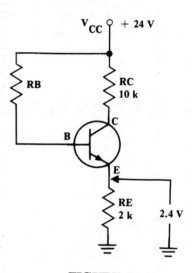

FIGURE 6-16

5. Given: $\beta = 100$, $V_{BE} = 0.6$ V, $V_B = 3.6$ V, and $I_C \approx I_E$

   Find:

   A. V_{RB}
   B. I_B
   C. V_{RE}
   D. V_{RC}
   E. V_{CE}
   F. R_E

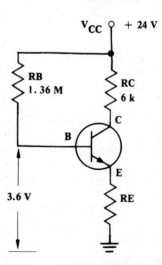

FIGURE 6-17

6. Given: $\beta = 100$ and $V_{BE} = 0.6$ V

Find:

A. I_C
B. V_{RC}
C. V_{RE}
D. V_C
E. V_E
F. V_B
G. V_{RB}
H. V_{CE}

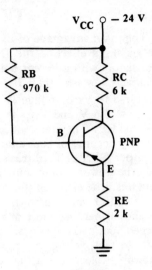

FIGURE 6-18

SECTION II: TWO POWER SUPPLY, EMITTER CURRENT CONTROLLED CIRCUITS

The main advantage of using fixed emitter current control circuits is that the transistor's beta will have only a slight effect on the DC voltage distribution of the circuit. Therefore, in this circuit configuration, transistors with different betas can be substituted for each other as long as the base resistor value is kept relatively low (less than 100 kΩ), to minimize DC voltage distribution changes. In the two power supply circuit, the V_{EE} voltage is distributed across RE, RB, and the base-emitter diode. However, because $V_{BE} \approx 0.6$ V and V_{RB} is rarely greater than 1 V, most of the voltage is developed across the emitter resistor RE. Therefore, since $I_E = V_{RE}/RE$ and beta only effects I_B, then only $V_{RB} = I_B R_B$ changes directly because of beta change.

Either NPN or PNP transistors can be used in the two power supply circuit configuration and, again, only the power supply connections, as shown in Figure 6-19, differ for NPN and PNP devices. Also, since beta has little effect on the voltage distribution in a well designed two power supply circuit, a nominal beta value of 100 can be used arbitrarily even if the actual beta of the device is not known. A beta of 100 can be chosen because most of the betas of low and medium power transistors manufactured today usually exceed 100, and only slight voltage distribution discrepancies can occur in the circuit by assuming a beta of 100. In fact, voltage discrepancies caused by assuming a beta of 100 are less than those caused by the allowable tolerances of the resistors used in constructing the circut.

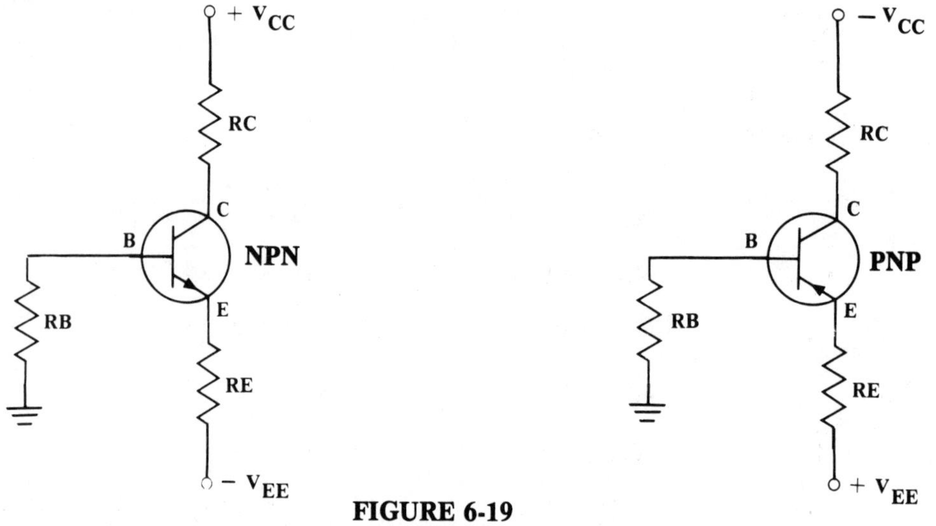

FIGURE 6-19

THE BASIC TWO POWER SUPPLY CIRCUIT

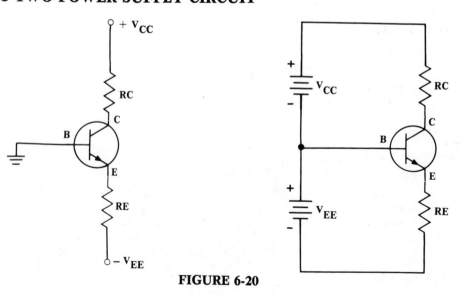

FIGURE 6-20

The basic two power supply circuit is analyzed in terms of the emitter current I_E and, like the basic base-biased circuit, it is easy to solve. The schematic and power supply connections of the circuit are shown in Figure 6-20.

The DC analysis of the simple two power supply circuit begins by writing loop equations in terms of the voltage dropped across the emitter resistor V_{RE} and the base-emitter diode voltage V_{BE}. Therefore, since $V_{BE} \approx 0.6$ V, then V_{RE} is solved by subtracting V_{BE} from V_{EE}, and, because $V_{RE} = I_E R_E$, $I_E = V_{RE}/RE$. Also, $I_C \approx I_E$ and $V_{RC} = I_C R_C$. Therefore, $V_C = V_{CC} - V_{RC}$, $V_E = V_B - V_{BE}$, and $V_{CE} = V_C - V_E$.

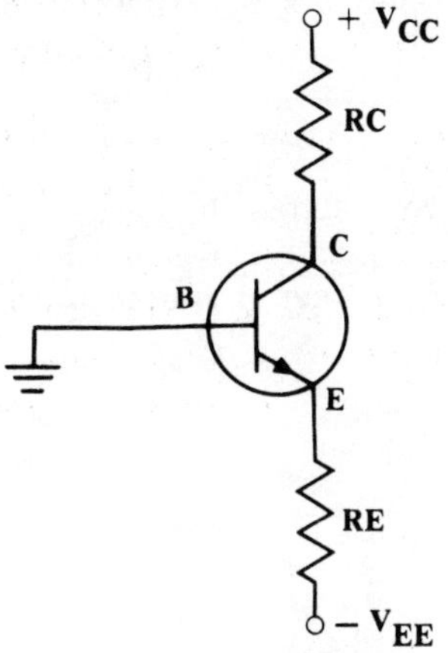

$$V_{EE} = V_{BE} + V_{RE}$$
$$V_{EE} - V_{BE} = V_{RE}$$
$$V_{EE} - V_{BE} = I_E R_E$$
$$(V_{EE} - V_{BE})/R_E = I_E$$

FIGURE 6-21

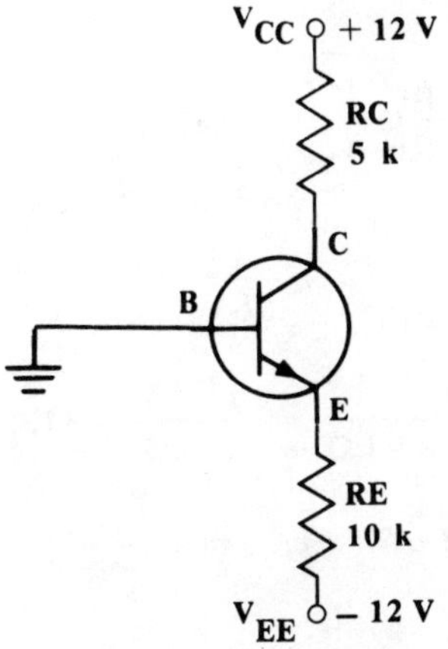

Given: $\beta = 100$, $V_{CC} = 12$ V, and $V_{EE} = -12$ V

$$I_E = \frac{V_{EE} - V_{BE}}{RE} = \frac{|12\text{ V}| - 0.6\text{ V}}{10\text{ k}\Omega} = \frac{11.4\text{ V}}{10\text{ k}\Omega} = 1.14\text{ mA}$$

$$V_{RC} = I_C R_C \approx I_E R_C = 1.14\text{ mA} \times 5\text{ k}\Omega = 5.7\text{ V}$$

$$V_C = V_{CC} - V_{RC} = 12\text{ V} - 5.7\text{ V} = 6.3\text{ V}$$

$$V_E = V_B - V_{BE} = 0\text{ V} - 0.6\text{ V} = -0.6\text{ V}$$

$$V_{RE} = V_E - V_{EE} = -0.6\text{ V} - (-12\text{ V}) = -0.6\text{ V} + 12\text{ V} = 11.4\text{ V}$$

$$V_{CE} = V_C - V_E = 6.3\text{ V} - (-0.6\text{ V}) = 6.3\text{ V} + 0.6\text{ V} = 6.9\text{ V}$$

$$V_{CB} = V_C - V_B = 6.3\text{ V} - 0\text{ V} = 6.3\text{ V}$$

NOTE: Mathematically $-(-1) = (+1)$, therefore, $-(-12\text{ V}) = 12\text{ V}$ and $-(-0.6\text{ V}) = 0.6\text{ V}$.

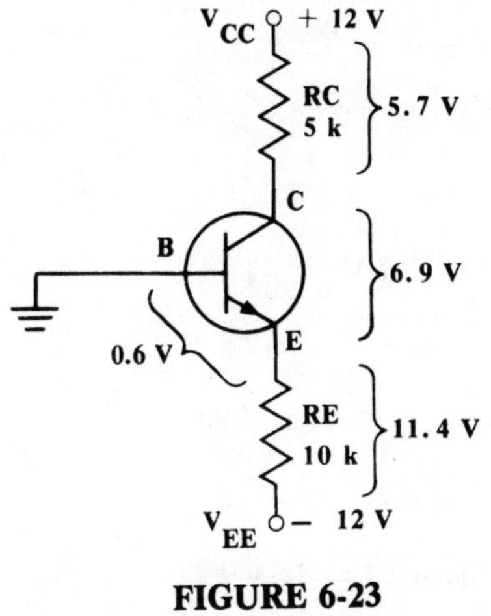

FIGURE 6-22

$$V_{CC} = V_{RC} + V_{CB}$$
$$12\text{ V} = 5.7\text{ V} + 6.3\text{ V}$$

$$V_{EE} = V_{BE} + V_{RE}$$
$$12\text{ V} = 0.6\text{ V} + 11.4\text{ V}$$

$$V_{CC} - V_{EE} = 12\text{ V} - (-12\text{ V}) = 12\text{ V} + 12\text{ V} = 24\text{ V}$$

$$V_{CC} - V_{EE} = V_{RC} + V_{CE} + V_{RE}$$
$$24\text{ V} = 5.7\text{ V} + 6.9\text{ V} + 11.4\text{ V}$$

FIGURE 6-23

The voltage distribution of the circuit is shown in Figure 6-23. Remember that the voltage drops between V_{CC} and ground must equal V_{CC}, between ground and V_{EE} must equal V_{EE}, and between V_{CC} and V_{EE} must equal the algebraic difference between the V_{CC} and V_{EE} voltages.

THE BASIC TWO POWER SUPPLY CIRCUIT WITH BASE RESISTOR ADDED

Adding a base resistor to the basic two power supply circuit is required if the circuit is to be used as an amplifier, where the AC signal is applied to the base lead. Then, the base resistor must be taken into account in determining the base current and the V_{RB} voltage drop. The circuit is analyzed in terms of I_B and then I_E.

1. SOLVING IN TERMS OF THE BASE CURRENT I_B

The DC analysis begins by writing the loop equations in terms of the voltage drops across the base resistor RB, the base-emitter diode, and the emitter resistor RE. Since $V_{RB} = I_B R_B$, $V_{RE} = I_E R_E$ and $I_E = (\beta + 1)I_B$. Then substitutions are made to solve for I_B. Also, since $\beta \approx \beta + 1$, then β is used in the final equations.

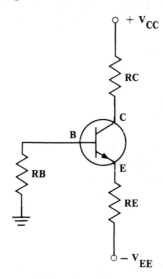

FIGURE 6-24

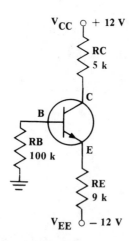

FIGURE 6-25

$$V_{EE} = V_{RB} + V_{BE} + V_{RE}$$

$$V_{EE} = V_{BE} + V_{RB} + V_{RE}$$

$$V_{EE} = V_{BE} + I_B R_B + I_E R_E$$

$$V_{EE} = V_{BE} + I_B[R_B + (\beta + 1)RE]$$

$$V_{EE} - V_{BE} = I_B[R_B + (\beta + 1)R_E]$$

$$I_B = \frac{V_{EE} - V_{BE}}{R_B + (\beta + 1)R_E} \approx \frac{V_{EE} - V_{BE}}{R_B + \beta R_E}$$

where: $\beta \approx \beta + 1$

Given: $\beta = 100$, $V_{CC} = 12$ V, and $V_{EE} = -12$ V

$$I_B = \frac{V_{EE} - V_{BE}}{R_B + \beta R_E} = \frac{12\text{ V} - 0.6\text{ V}}{100\text{ k}\Omega + (100 \times 9\text{ k}\Omega)} = \frac{11.4\text{ V}}{1\text{ M}\Omega} = 11.4\ \mu\text{A}$$

$$I_E = (\beta + 1)I_B \approx \beta I_B = 100 \times 11.4\ \mu\text{A} = 1.14\text{ mA}$$

$$I_C = \beta I_B = 100 \times 11.4\ \mu\text{A} = 1.14\text{ mA}$$

$$V_{RB} = I_B R_B = 11.4\ \mu\text{A} \times 100\text{ k}\Omega = 1.14\text{ V}$$

$$V_{RE} = I_E R_E = 1.14\text{ mA} \times 9\text{ k}\Omega = 10.26\text{ V}$$

$$V_{RC} = I_C R_C = 1.14\text{ mA} \times 5\text{ k}\Omega = 5.7\text{ V}$$

$$V_C = V_{CC} - V_{RC} = 12\text{ V} - 5.7\text{ V} = 6.3\text{ V}$$

$$V_B = \text{Ground} - V_{RB} = 0.0\text{ V} - 1.14\text{ V} = -1.14\text{ V}$$

$$V_E = V_B - V_{BE} = -1.14\text{ V} - 0.6\text{ V} = -1.74\text{ V}$$

$$V_{RE} = V_E - V_{EE} = -1.74\text{ V} - (-12\text{ V}) = 10.26\text{ V}$$

$$V_{CB} = V_C - V_B = 6.3\text{ V} - (-1.14\text{ V}) = 7.44\text{ V}$$

$$V_{CE} = V_C - V_E = 6.3\text{ V} - (-1.74\text{ V}) = 8.04\text{ V}$$

2. SOLVING IN TERMS OF THE EMITTER CURRENT I_E

In analyzing the two power supply circuit, in terms of I_E, the loop equations are similar to those for the I_B analysis except the substitutions are made in terms of I_E rather than I_B.

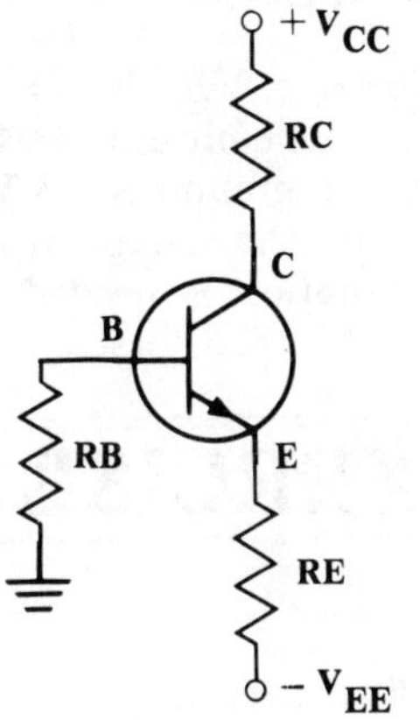

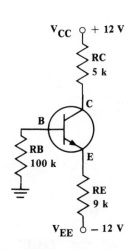

FIGURE 6-26

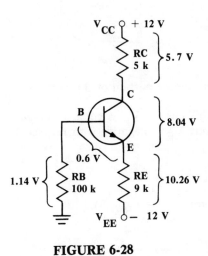

FIGURE 6-27

$$V_{EE} = V_{RB} + V_{BE} + V_{RE}$$
$$V_{EE} - V_{BE} = V_{RB} + V_{RE}$$
$$V_{EE} - V_{BE} = I_B R_B + I_E R_E$$
$$V_{EE} - V_{BE} = (I_E R_B/[\beta + 1]) + I_E R_E$$
$$V_{EE} - V_{BE} = I_E(R_B/[\beta + 1] + R_E)$$
$$I_E = \frac{V_{EE} - V_{BE}}{RB/(\beta + 1) + RE} \approx \frac{V_{EE} - V_{BE}}{RB/\beta + RE}$$

Given: $\beta = 100$, $V_{CC} = 12$ V, and $V_{EE} = -12$ V

$$I_E = \frac{V_{EE} - V_{BE}}{RB/\beta + RE} = \frac{12\text{ V} - 0.6\text{ V}}{100\text{ k}\Omega/100 + 9\text{ k}\Omega} = \frac{11.4\text{ V}}{10\text{ k}\Omega} = 1.14\text{ mA}$$

$I_B = I_E/\beta + 1 \approx I_E/\beta = 1.14$ mA$/100 = 11.4\ \mu$A

$I_C = \beta I_B = 100 \times 11.4\ \mu$A $= 1.14$ mA

$V_{RB} = I_B R_B = 11.4\ \mu$A $\times 100$ k$\Omega = 1.14$ V

$V_{RE} = I_E R_E = 1.14$ mA $\times 9$ k$\Omega = 10.26$ V

$V_{RC} = I_C R_C = 1.14$ mA $\times 5$ k$\Omega = 5.7$ V

$V_C = V_{CC} - V_{RC} = 12$ V $- 5.7$ V $= 6.3$ V

$V_B = $ Ground $- V_{RB} = 0.0$ V $- 1.14$ V $= -1.14$ V

$V_E = V_B - V_{BE} = -1.14$ V $- (0.6$ V$) = -1.74$ V

$V_{RE} = V_E - V_{EE} = -1.74$ V $-(-12$ V$) = 10.26$ V

$V_{CB} = V_C - V_B = 6.3$ V $-(-1.14$ V$) = 7.44$ V

$V_{CE} = V_C - V_E = 6.3$ V $-(-1.74$ V$) = 8.04$ V

$$V_{CC} = V_{RC} + V_{CB} - V_{RB}$$
$$12\text{ V} = 5.7\text{ V} + 7.44\text{ V} - 1.14\text{ V}$$

$$V_{EE} = V_{RB} + V_{BE} + B_{RE}$$
$$12\text{ V} = 1.14\text{ V} + 0.6\text{ V} + 10.26\text{ V}$$

$$V_{CC} - V_{EE} = 12\text{ V} - (-12\text{ V}) = 12\text{ V} + 12\text{ V} = 24\text{ V}$$
$$V_{CC} - V_{EE} = V_{RC} + V_{CE} + V_{RE} = 5.7\text{ V} + 8.04\text{ V} + 10.26\text{ V} = 24\text{ V}$$

FIGURE 6-28

3. VOLTAGE DISTRIBUTION

The voltage distribution for the two power supply circuit with an emitter resistor is shown in Figure 6-28. Again, remember that the voltage drops between V_{CC} and ground must equal V_{CC}, between ground and V_{EE} must equal V_{EE}, and between V_{CC} and V_{EE} must equal the $V_{CC} - (-V_{EE}) = |V_{CC}| + |V_{EE}|$.

NOTE: With respect to ground, the V_{CC} voltage is 12 V and V_C is 6.3 V. However, V_{EE} is −12 V, V_E is −1.74 V, and V_B is −1.14 V. Therefore, the voltage drop from the collector to the emitter is 8.04 V, the voltage from the collector to the base is 7.44 V, and the voltage from the collector to ground is 6.3 V. Too, do not become confused by plus and minus power supply connections. Simply solve the voltage drops first and then reference the voltages with respect to ground. In case further explanation is needed, review Chapter 1, where two power supply connections were first introduced using only two resistors.

SECTION II: LABORATORY EXERCISE

OBJECTIVES

To investigate the DC characteristics of the two power supply emitter controlled circuit.

LIST OF MATERIALS AND EQUIPMENT

1. Transistor: 2N3904 or equivalent
2. Resistors (one each except where indicated)   4.7 kΩ   10 kΩ (two)
3. Power Supplies
4. Voltmeter

PROCEDURE

Part 1: Basic Two Power Supply Emitter Controlled Circuit

1. Connect the two power supply circuit as shown in Figure 6-29

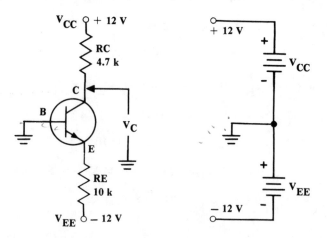

FIGURE 6-29

NOTE: The two power supply connections are further illustrated in the theory portion of this section.

2. Measure V_E, the voltage at the emitter with respect to ground. In this circuit, V_E is equal in magnitude to V_{BE}. Therefore, $|V_E| = |V_{BE}|$, and $V_{BE} \approx 0.6$ V. However, V_E is negative with respect to ground.

3. Use the measured V_E value and calculate V_{RE} from: $V_{RE} = V_E - V_{EE}$

NOTE: Since $V_{EE} = V_{BE} + V_{RE}$, then V_{RE} can also be solved from: $I_E = V_{RE}/RE$

4. Calculate the emitter current I_E from: $I_E = V_{RE}/RE$.

5. Measure V_C, the voltage at the collector, with respect to ground. V_C is positive with respect to ground.

142

6. Use the measured V_C value and calculate V_{RC} from: $V_{RC} = V_{CC} - V_C$
7. Calculate the collector current from: $I_C = V_{RC}/R_C$
8. Calculate alpha from: $a = I_C/I_E$

NOTE: $I_E \approx I_C$ and unless precise voltage measurements are made and the emitter and collector resistors are known exactly, then alpha cannot be accurately calculated. Hence the equation $a = I_C/I_E$ should provide a nominal alpha value of about 0.99, but a 10% error in measurement can provide greater than unity calculations, which is obviously in error.

9. Insert the calculated and measured values, as indicated, into Table 6-3.

TABLE 6-3	V_E	V_{RE}	I_E	V_C	I_C	V_{RC}	V_{BE}	alpha
CALCULATED		11.5	1.15 mA		2.54 mA	11.08	.6 V	2.20
MEASURED	.5 V			.02				

Part 2: Basic Two Power Supply Emitter Controlled Circuit with Base Resistor Added

1. Connect the circuit of Figure 6-30, where a 10 kΩ base resistor has been added to the basic two power supply circuit.

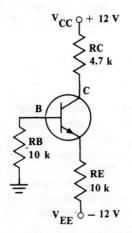

FIGURE 6-30

2. Calculate I_C, the collector current from:

$$I_C \approx \frac{V_{EE} - V_{BE}}{(R_B/\beta) + R_E}$$ where: $V_{BE} \approx 0.6$ V and β is known

$I_C = \dfrac{-12V - 0.6V}{(10K/100) + 10K} = \dfrac{-12.6}{100 + 10K} = \dfrac{10100}{}$ 1.24 mA

1.17 mA

NOTE: $I_C \approx I_E$ and use $\beta = 100$ if the beta is not known, since beta change effects only V_{RB} directly.

3. Calculate all the DC voltage drops around the circut.
   a. Calculate V_{RE} from: $V_{RE} = I_E R_E$ $V_{RE} = (1.24mA)(10K) = 12.4$
   b. Calculate V_{RC} from: $V_{RC} = I_C R_C$ $(1.24mA)(4.7K) = 5.828$
   c. Calculate I_B from: $I_B \approx I_E/\beta$ $1.24m/100 = 12.6 \mu A$

E B C

d. Calclate V_{RB} from: $V_{RB} = I_B R_B$

e. Calculate the single point transistor voltages with respect to ground.

$V_B = -V_{RB}$

$V_E = -(V_{RB} + V_{BE})$ or $V_E = V_{EE} - V_{RE}$

$V_C = V_{CC} - V_{RC}$

f. Calculate V_{CE} from: $V_{CE} = V_C - V_E$

g. Calculate V_{CB} from: $V_{CB} = V_C - V_B$

4. Measure the DC voltage drops around the circuit. Include all resistors and the transistor.

NOTE: Make all measurements with respect to ground. However, V_B, V_E, and V_{EE} will all be negative voltages. Therefore, use these values along with the measured V_C to calculate V_{RB}, V_{RC}, V_{RE}, V_{CE}, and V_{CB} "measured" values.

5. Insert the calcuated and measured values, as indicated, into Table 6-4.

TABLE 6-4	I_E	I_C	I_B	V_{RE}	V_{RC}	V_{RB}	V_B	V_E	V_C	beta	V_{CE}	V_{CB}
CALCULATED	1.32	1.25	7.27	13.2	5.28	.07	.07	.67	6.72		6.05	6.05
MEASURED				11.5	5.36	.06	.06	.71	5.3		7.42	6.75

SECTION QUESTIONS

1. What is the main advantage of fixed emitter current biasing?
2. What would be the effect on the voltage distribution of the circuit if all of the transistors were given a beta of 100 instead of their actual betas? Would the difference in the distributed voltages be only minimal?
3. The base resistor value in Section II was given at 100 kΩ to demonstrate clearly the beta effect on the circuit. If the base resistor value were lowered to 10 kΩ, would the effect of beta change, through substituted transistor, cause larger or smaller distribution changes? Why? Explain!
4. Why is using beta more practical than using alpha in circuit calculations and measurements?
5. Why is it easier, in plus and minus two power supply circuits, to solve for the voltage drops first, before referencing the circuit voltages with respect to ground? Is there another approach that is less confusing?

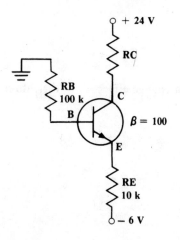
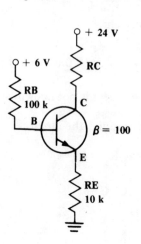

FIGURE 6-31

6. In the circuits of Figure 6-31, are the voltage drops identical for the similar components? With reference to plus and minus voltage supplies, does ground location effect the voltage drops across the circuit resistors? Why or why not?

SECTION PROBLEMS

1. Given: $\beta = 100$ and $V_{BE} = 0.6$ V

   Find:

   a. I_E

   b. V_{RE}

   c. I_B

   d. V_{RB}

   e. V_{RC}

   f. V_{CE}

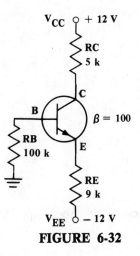

FIGURE 6-32

2. Given: $\beta = 200$ and $V_{BE} = 0.6$ V

   Find:

   a. I_C

   b. V_{RC}

   c. I_B

   d. V_{RB}

   e. V_{RE}

   f. V_{CE}

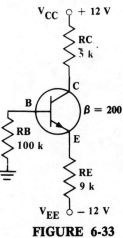

FIGURE 6-33

NOTE: With reference to problems one and two, the circuit remains the same and only the betas are changed from 100 to 200. This demonstrates the effect of beta on two power supply emitter controlled circuits.

3. Given: $\beta = 150$ and $V_{BE} = 0.6$ V

   Find:

   a. $I_C \approx I_E$

   b. V_{RB}

   c. V_{RE}

   d. V_E

   e. V_{CE}

   f. R_E

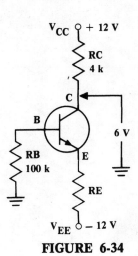

FIGURE 6-34

145

4. Given: $\beta = 50$ and $V_{BE} = 0.6$ V

   Find:

   a. I_E

   b. V_{RE}

   c. V_{RB}

   d. V_{RC}

   e. V_C

   f. V_E

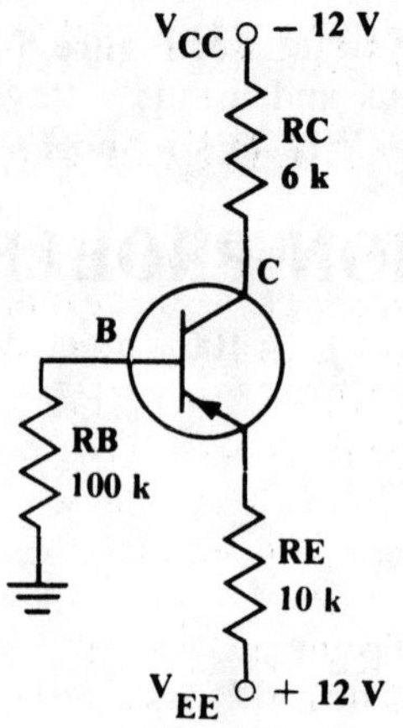

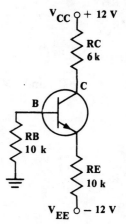

FIGURE 6-35

5. Given: $\beta = 100$ and $V_{BE} = 0.6$ V

   Find:

   a. I_E

   b. V_{RE}

   c. V_{RB}

   d. V_E

   e. V_{CE}

FIGURE 6-36

SECTION III: UNIVERSAL CIRCUIT BIASING

GENERAL DISCUSSION

The main advantage of universal circuit biasing is that it uses only a single power supply to establish fixed emitter current conditions. Therefore, like the two power supply circuit, the effect of beta is minimized but, because the voltage distribution is now from only one source, a much smaller voltage is dropped across the RE resistor. This causes I_E to be slightly more dependent on beta than it was in the two power supply circuit, where V_{RE} and RE are the main factors in controlling I_E. Therefore, in the universal circuits, such as those shown in Figure 6-37, $I_E = V_{RE}/RE$ and V_{RE} is only $V_{BE} \approx 0.6$ V lower than V_{R2}. Hence, the voltage divider network of R1 and R2 is a major factor in determining I_E.

For instance, if $V_{R2} = 4$ V, then $V_{RE} = 3.4$ V, when $V_{BE} = 0.6$ V. Since $I_E = V_{RE}/RE$ and $I_C \approx I_E$, then the circuit voltage distribution is solved easily by using approximation techniques. The NPN and PNP devices used in universal circuit configurations are shown in Figure 6-37 where, again, only voltage polarities differ.

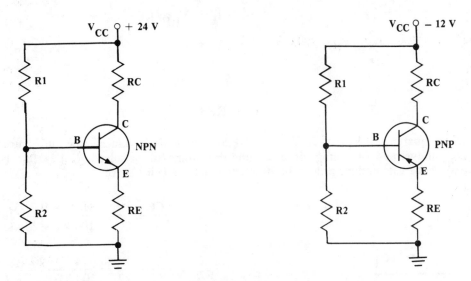

FIGURE 6-37

DIRECT CURRENT CIRCUIT ANALYSIS TECHNIQUES

Universal circuits can be analyzed by either approximation or exact methods. However, approximation techniques use far fewer steps than exact circuit analysis techniques, and the calculated versus measured results of approximation techniques are almost always more exact than the allowable tolerance of the circuit resistors used to construct the actual circuit.

NOTE: Approximation techniques are valid only if the circuit is well designed — where substituting a transistor with another transistor with a different beta should cause little change to the direct current voltage distribution of the circuit. On the other hand, exact techniques can provide the direct current analysis necessary to determine whether good design is employed since the transistor's base current loading is taken into account, which shows up in the voltage distribution of the circuit.

DIRECT CURRENT CIRCUIT ANALYSIS APPROXIMATION TECHNIQUES

Most small and medium power transistors have betas in excess of 100, and in a well-designed circuit a transistor with a beta of 100 should have little effect on the DC distribution of the circuit. Therefore, in practice, a beta of 100 can be safely assumed in the analysis of all well-designed universal biased circuits.

NOTE: Unless a transistor is bad or incorrectly connected, designing and analyzing assuming a beta of 100 is practical and should be well within the allowable voltage discrepancies caused by the tolerances of the resistors used in the construction of the circuit.

In using approximation techniques to analyze the universal circuit of Figure 6-38, the voltage drops across R1 and R2 are solved without regard to the current loading effect of the transistor's base current. Then, V_{RE} is solved from $V_{RE} = V_{R2} - V_{BE}$ and I_E is solved from $I_E = V_{RE}/RE$. Since $I_E \approx I_C$, then $V_{RC} = I_C R_C$, $V_C = V_{CC} - V_{RC}$, and V_{CE} is solved from $V_{CE} = V_C - V_E$, where $V_E = V_{RE}$. The equations necessary for the analysis are shown in conjunction with Figure 6-38.

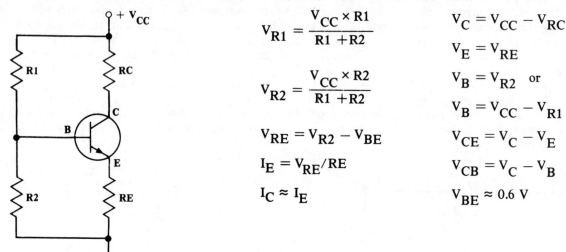

$$V_{R1} = \frac{V_{CC} \times R1}{R1 + R2}$$

$$V_{R2} = \frac{V_{CC} \times R2}{R1 + R2}$$

$$V_{RE} = V_{R2} - V_{BE}$$

$$I_E = V_{RE}/RE$$

$$I_C \approx I_E$$

$$V_C = V_{CC} - V_{RC}$$

$$V_E = V_{RE}$$

$$V_B = V_{R2} \text{ or}$$

$$V_B = V_{CC} - V_{R1}$$

$$V_{CE} = V_C - V_E$$

$$V_{CB} = V_C - V_B$$

$$V_{BE} \approx 0.6 \text{ V}$$

FIGURE 6-38

Using these equations, if $V_{CC} = 24$ V, $\beta = 100$, and the resistor values shown in Figure 6-39 are given, then the DC voltages and DC voltage drops around the circuit can be solved. Notice that the beta of 100 is not required in the calculations unless the base current I_B is being solved.

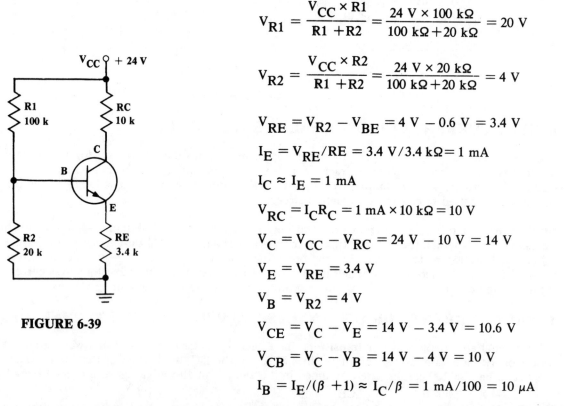

FIGURE 6-39

$$V_{R1} = \frac{V_{CC} \times R1}{R1 + R2} = \frac{24 \text{ V} \times 100 \text{ k}\Omega}{100 \text{ k}\Omega + 20 \text{ k}\Omega} = 20 \text{ V}$$

$$V_{R2} = \frac{V_{CC} \times R2}{R1 + R2} = \frac{24 \text{ V} \times 20 \text{ k}\Omega}{100 \text{ k}\Omega + 20 \text{ k}\Omega} = 4 \text{ V}$$

$$V_{RE} = V_{R2} - V_{BE} = 4 \text{ V} - 0.6 \text{ V} = 3.4 \text{ V}$$

$$I_E = V_{RE}/RE = 3.4 \text{ V}/3.4 \text{ k}\Omega = 1 \text{ mA}$$

$$I_C \approx I_E = 1 \text{ mA}$$

$$V_{RC} = I_C R_C = 1 \text{ mA} \times 10 \text{ k}\Omega = 10 \text{ V}$$

$$V_C = V_{CC} - V_{RC} = 24 \text{ V} - 10 \text{ V} = 14 \text{ V}$$

$$V_E = V_{RE} = 3.4 \text{ V}$$

$$V_B = V_{R2} = 4 \text{ V}$$

$$V_{CE} = V_C - V_E = 14 \text{ V} - 3.4 \text{ V} = 10.6 \text{ V}$$

$$V_{CB} = V_C - V_B = 14 \text{ V} - 4 \text{ V} = 10 \text{ V}$$

$$I_B = I_E/(\beta + 1) \approx I_C/\beta = 1 \text{ mA}/100 = 10 \text{ }\mu\text{A}$$

The distributed voltage drops showing three paths are shown in Figure 6-40. Remember, the sum of the distributed voltages between V_{CC} and ground, regardless of the path taken, must always equal the applied V_{CC}.

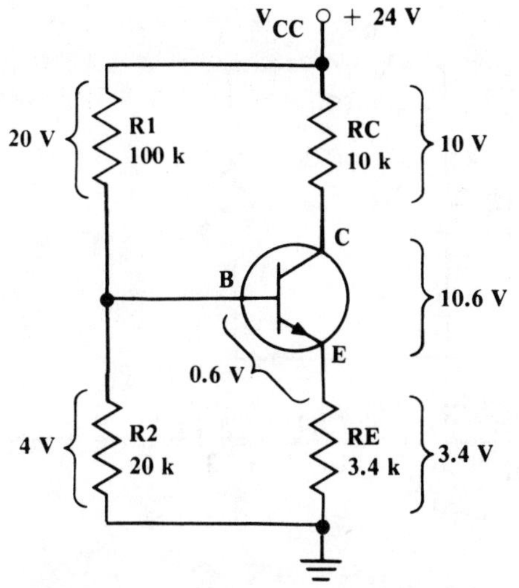

FIGURE 6-40

$$V_{CC} = V_{R1} + V_{R2}$$
$$24\ V = 20\ V + 4\ V$$

$$V_{CC} = V_{RC} + V_{CE} + V_{RE}$$
$$24\ V = 10\ V + 10.6\ V + 3.4\ V$$

$$V_{CC} = V_{R1} + V_{BE} + V_{RE}$$
$$24\ V = 20\ V + 0.6\ V + 3.4\ V$$

DIRECT CURRENT CIRCUIT ANALYSIS EXACT TECHNIQUES

In using approximation techniques for the DC analysis of the circuit of Figure 6-39, it is assumed that the transistor has a beta of 100 and the effect of the base-current load in a well-designed circuit is too minimal to be included in the calculations. However, to analyze the circuit using exact techniques, the base current loading effect is considered, which causes the voltage drop across R1 to increase and the voltage drop across R2 to decrease. Hence, the remaining circuit voltage drops will change and the exact values can be solved by using Thevenin's equivalent theorem in the calculations.

The circuit of Figure 6-41 has the same circuit parameters as those of Figure 6-39. However, in using exact analysis techniques, the effective base resistance is reflected in series with the 3.4 kΩ emitter resistance. Once V_{RE} is known, the distributed voltages are found. Figure 6-42(a) shows the voltage distribution for the circuit of Figure 6-41, and Figure 6-42(b) shows the Thevenin's equivalent circuit.

Given: $\beta = 100$ and
$V_{CC} = 24\ V$

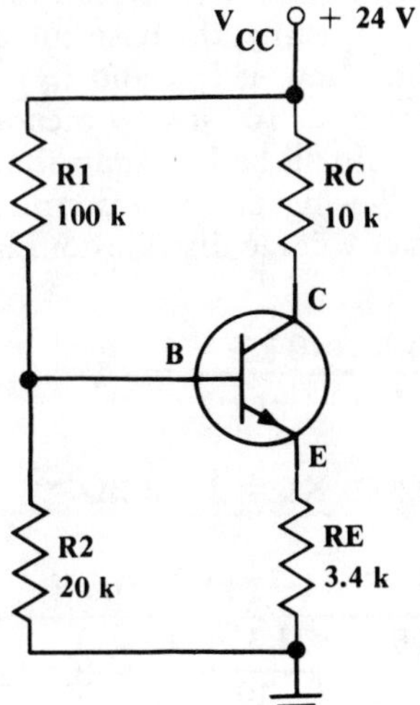

FIGURE 6-41

$V_{TH} = V_{R2} - V_{BE} = 4\ V - 0.6\ V = 3.4\ V$,

where: $V_{R2} = \dfrac{24\ V \times 20\ k\Omega}{100\ k\Omega + 20\ k\Omega} = 4\ V$

$R_{TH} = \dfrac{R1 \ /\!/\ R2}{(\beta + 1)} = \dfrac{100\ k\Omega \ /\!/\ 20\ k\Omega}{(100 + 1)} \approx \dfrac{16.67\ \Omega}{101} \approx 165\ \Omega$

$I_E = \dfrac{V_{TH} - V_{BE}}{R_{TH} + R_E} = \dfrac{4\ V - 0.6\ V}{165\ \Omega + 3.4\ k\Omega} = \dfrac{3.4\ V}{3565\ \Omega} = 0.9537\ mA$

$V_{RE} = I_E R_E = 0.9537\ mA \times 3.4\ k\Omega = 3.2426\ V$

$I_B = I_E/(\beta + 1) = 0.9537\ mA/101 = 9.443\ \mu A$

$I_C = \beta I_B = 100 \times 9.443\ \mu A = 0.9443\ mA$

$V_{RC} = I_C R_C = 0.9433\ mA \times 10\ k\Omega = 9.443\ V$

$V_C = V_{CC} - V_{RC} = 24\ V - 9.443\ V = 14.557\ V$

$V_E = V_{RE} = 3.2426\ V$

$V_B = V_{RE} + V_{BE} = 3.2426\ V + 0.6\ V = 3.8426\ V$

$V_{CE} = V_C - V_E = 14.557\ V - 3.2426\ V = 11.314\ V$

$V_{R2} = V_B = 3.8426\ V \approx 3.84\ V$

$V_{R1} = V_{CC} - V_{R2} = 24\ V - 3.8426\ V \approx 20.16\ V$

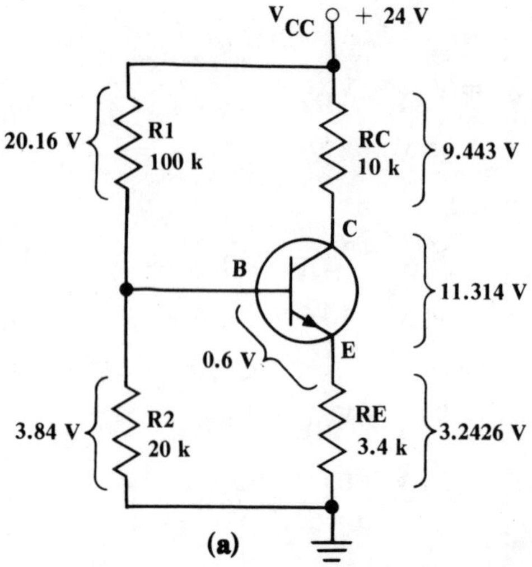

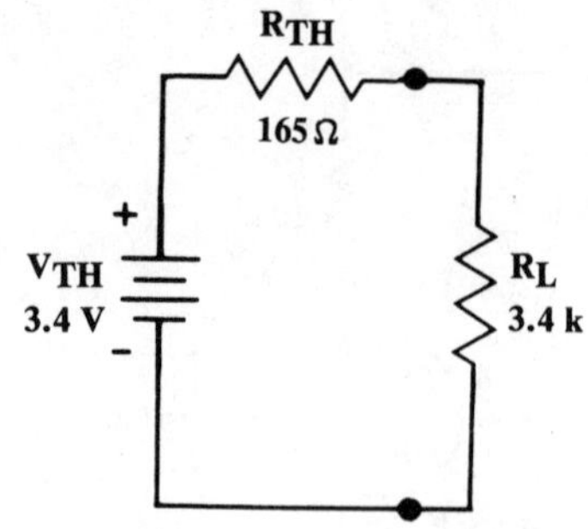

FIGURE 6-42

DESIGN CONSIDERATIONS

The comparison of exact to approximate DC analysis techniques shows only a slight voltage distribution change caused by base current loading. Therefore, the circuit is considered "well designed" because the distributed voltages are well within the voltage distribution variations that could be caused by using resistors with a tolerance of 5 percent. Since most circuits use 10 percent tolerance resistors, then it is definitely a well-designed circuit. At this point, it might be appropriate to consider what constitutes a good design and why certain resistor values are chosen.

To start with, good design depends on the circuit application. For instance, low power signal voltage applications use collector currents of from 1 mA to 10 mA, medium power from 5 mA to 30 mA, and high power from 30 mA and up. Since the circuits considered in this manual have dealt primarily with low-power, signal voltage, then a collector current of 1 mA is a reasonable choice.

For a low-power, signal voltage circuit using universal circuit form with a collector current of 1 mA, if we assume a beta of 100, then the base current will be 10 μA. Therefore, the current flow in the R1 and R2 resistors of the voltage divider must be a minimum of 10 times greater, or 100 μA. (Recall that in the previously analyzed circuit of Figure 6-41, the I_{R1} was about 200 μA. Consequently, the base current of 10 microamps was only 5 percent of the available 200 μA of "bleeder current" flow of I_{R1} and I_{R2}, and the overall voltage distribution effect was minimal.) With the bleeder current set at 100 μA by increasing R1 and R2 to 200 kΩ and 40 kΩ, respectively, the voltage distribution effect will still be less than that caused by the allowable tolerance of 10 percent resistors, as shown in the following solved example. Again, Thevenin's equivalent theorem is used in the analysis to determine the exact voltage distribution, as shown in Figure 6-44.

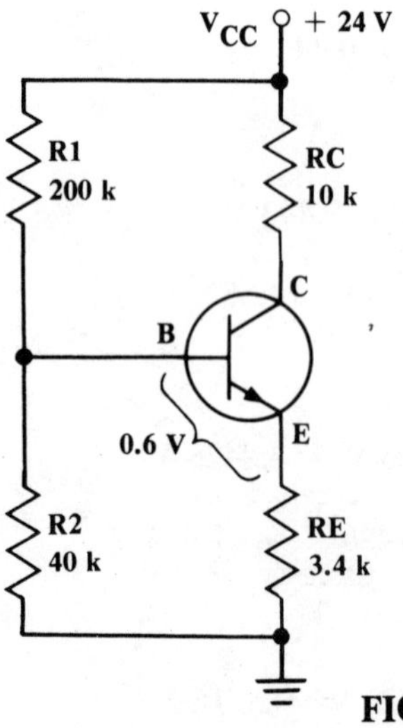

$$V_{TH} = \frac{V_{CC} \times R2}{R1 + R2} = \frac{24 \text{ V} \times 40 \text{ k}\Omega}{200 \text{ k}\Omega + 40 \text{ k}\Omega} = 4 \text{ V}$$

$$R_{TH} = R1 \mathbin{/\mkern-5mu/} R2 = 200 \text{ k}\Omega \mathbin{/\mkern-5mu/} 40 \text{ K}\Omega = 33.33 \text{ k}\Omega$$

$$I_E = \frac{V_{TH} - V_{BE}}{R_{TH}/(\beta + 1) + R_E} = \frac{4 \text{ V} - 0.6 \text{ V}}{\dfrac{33.33 \text{ k}\Omega}{101} + 3.4 \text{ k}\Omega}$$

$$= \frac{3.4 \text{ V}}{330 \ \Omega + 3400 \ \Omega} = \frac{3.4 \text{ V}}{3730 \ \Omega} = 911.5 \ \mu\text{A}$$

$$V_{RE} = I_E R_E = 911.5 \ \mu\text{A} \times 3.4 \text{ k}\Omega \approx 3.1 \text{ V}$$

$$I_B = I_E/(\beta + 1) = 911.5 \ \mu\text{A}/101 = 9.025 \ \mu\text{A}$$

FIGURE 6-43

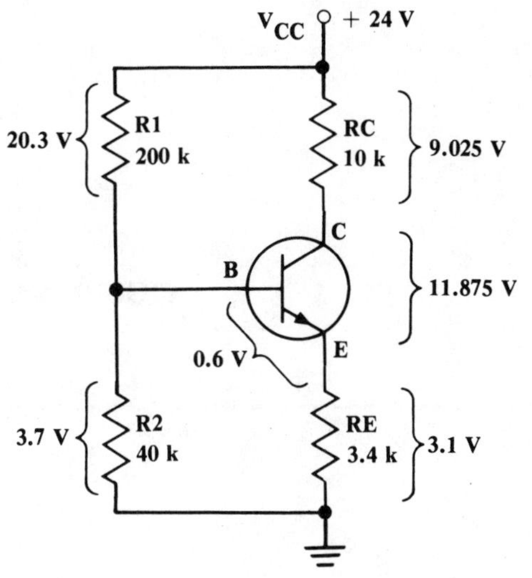

$$I_C = \beta I_B = 100 \times 9.025 \ \mu A = 902.5 \ \mu A$$

$$V_{RC} = I_C R_C = 902.5 \ \mu A \times 10 \ k\Omega = 9.025 \ V$$

$$V_C = V_{CC} - V_{RC} = 24 \ V - 9.025 \ V = 14.957 \ V$$

$$V_E = V_{RE} \approx 3.1 \ V$$

$$V_B = V_{RE} + V_{BE} = 3.1 \ V + 0.6 \ V = 3.7 \ V$$

$$V_{CE} = V_C - V_E = 14.95 \ V - 3.1 \ V = 11.857 \ V$$

$$V_{R2} = V_B = 3.7 \ V$$

$$V_{R1} = V_{CC} - V_{R2} = 24 \ V - 3.7 \ V = 20.3 \ V$$

FIGURE 6-44

For a circuit such as that of Figure 6-43, the bleeder current should be a minimum of 10 times greater than I_B, and a good ratio of R1 to R2 is 5 to 1 where, for instance, R1 = 100 kΩ and R2 = 20 kΩ. The reason for this 5 to 1 ratio is so the voltage drop across the emitter resistor can be moderately low, but not too low when compared with the V_{BE} of about 0.6 V. Therefore, if $V_{R2} = 4$ V and $V_{BE} = 0.6$ V, then $V_{RE} = 3.4$ V, allowing the remaining 20.6 V to be distributed across the collector-emitter of the transistor and the collector resistor RC.

Therefore, if $V_{RE} = 3.4$ V and I_E is 1 mA, then RE should be 3.4 kΩ. Likewise, if I_E is 2 mA, then RE must be 1.7 kΩ. However, if it is assumed that RE = 3.4 kΩ and $I_E = 1$ mA, then for a beta of 100, $I_B = 10 \ \mu A$ and R1 and R2 should have a minimum 100 μA of current flow to satisfy the minimum 10 to 1 ratio. Hence, R1 + R2 at 240 kΩ is the maximum resistance value selected, but lowering the value of the R1 + R2 total would improve the circuit stability because of beta change. However, a higher than 20 to 1 bleeder current to I_B ratio, which is achieved when R1 is 100 kΩ and R2 is 20 kΩ, is usually not necessary.

The value of R_C is determined by I_E, and when $I_E = 1$ mA, then RC at 10 kΩ will drop 10 V, or about half the remaining 20.6 V. Therefore, the start of any design can begin with the amount of voltage drop available across the emitter resistor RE and the amount of emitter current required for the circuit application. Then, the remaining circuit current, voltage drops, and resistor values can be selected. Further reasons for selecting resistor values, other than for the DC requirements, are discussed in the alternating current circuit analysis section of this book.

SECTION III: LABORATORY EXERCISE

OBJECTIVES
To investigate the DC characteristics of universal circuit biasing.

LIST OF MATERIALS AND EQUIPMENT
1. Transistor: 2N3904 or equivalent
2. Resistors (one each except where indicated): 3.3 kΩ 10 kΩ 22 kΩ 100 kΩ
3. Power Supply
4. Voltmeter

PROCEDURE
Part 1: Basic Universal Circuit Analysis Techniques
1. Connect the circuit shown in Figure 6-45.

2. Measure V_C, the voltage at the collector of the transistor, with respect to ground. Then, calculate V_{RC}, the voltage drop across the collector resistor from: $V_{RC} = V_{CC} - V_C$.

3. Measure V_E, the voltage drop across the emitter resistor from: $V_{RE} = V_E$.

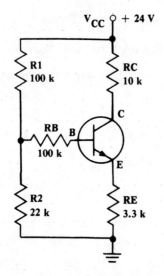

FIGURE 6-45

4. Measure V_B, the voltage at the base of the transistor with respect to ground. Also, measure V_{R2}, the voltage drop across Resistor R2, which is also the voltage at the junction of resistors R1 and R2, with respect to ground.

5. Use the V_B and V_{R2} voltage measurements to calculate the voltage drops across:

   a. V_{R1} from: $V_{R1} = V_{CC} - V_{R2}$ 24-3

   b. V_{RB} from: $V_{RB} = V_{R2} - V_B$

   c. V_{BE} from: $V_{BE} = V_B - V_E$ where: $V_E = V_{RE}$ $VBE = .6$

6. Use the calculated V_{RB}, V_{RE}, and V_{RC} voltage drops to calculate the base, emitter and collector currents. Calculate:

   a. I_B from: $I_B = V_{RB}/RB$ .4/100k

   b. I_E from: $I_E = V_{RE}/RE$

   c. I_C from: $I_C = V_{RC}/RC$

7. Use the calculated I_C and I_B currents to calculate both the beta and alpha current gains of the device.

   a. Calculate beta from: $\beta = I_C/I_B$

   b. Calculate alpha from: $a = \beta/(\beta + 1)$.

8. Insert the calculated and measured values, as indicated, into Table 6-5.

NOTE: Accurately calculated base, emitter, and collector currents require that the circuit resistors be measured precisely.

VR1 = 24.3

TABLE 6-5	V_C	V_{RC}	V_E	V_{RE}	V_{CE}	V_{R2}	V_B	V_{R1}	V_{RB}	I_B	I_C	I_E	beta	alpha
CALCULATED		11.4		3.2	9.3			19.8	.4	4µ		.14m	24.25	.940
MEASURED	12.6		3.2			4.2	3.8							
	11.7		3.2	9.3				24.3	.3					

152

Part 2: Predicting Voltage Drops of Universal Circuit

1. Connect the circuit as shown in Figure 6-46.

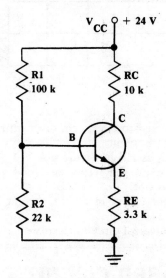

FIGURE 6-46

2. Calculate all the DC voltage drops around the circuit.

   a. Calculate $V_{R1} = \dfrac{V_{CC} \times R1}{R1 + R2}$ $\quad \dfrac{24 \times 100K}{100K + 22K} = \dfrac{2.4^4}{122K} = 19.67$

   b. Calculate $V_{R2} = \dfrac{V_{CC} \times R2}{R1 + R2}$ $\quad \dfrac{24 \times 22K}{122K} = 4.32$

   c. Calculate $V_{RE} = V_{R2} - V_{BE}$ $\quad 4.32 - .6 = 3.72$

   d. Calculate $I_E = V_{RE}/RE$ $\quad 3.72/3.3K = 1.12\,mA$

   e. Calculate $I_C \approx I_E$ $\quad 1.12\,mA$

   f. Calculate $V_{RC} = I_C R_C$ $\quad 1.12\,mA / 10K = .11$

   g. Solve for the single point transistor voltages, with respect to ground.

   4.3V
   $V_B = V_{R2}$

   3.6 $V_E = V_{RE}$

   $V_C = V_{CC} - V_{RC}$ $\quad 24 - 13 = 11$

   h. Calculate $V_{CE} = V_C - V_E$ $\quad 11 - 3.6 = 7.4$

   i. Calculate $V_{CB} = V_C - V_B$ $\quad 11 - 4.3 = 6.7$

3. Measure the DC voltage drops around the circuit. Include all the resistor and transistor measurements.

NOTE: Make all measurements with respect to ground to avoid possible ground loops. Use the measured values to calculate the remaining voltage drops.

4. Insert the calculated and measured values, as indicated, into Table 6-6.

TABLE 6-6	V_{R1}	V_{R2}	V_{RE}	V_{RC}	V_{CE}	V_{CB}	V_{BE}	V_B	V_C	V_E	I_E
CALCULATED	19.67	4.32	3.72	11	7.4	6.7	≈ 0.6 V	4.3	11	3.6	1.12 mA
MEASURED	20	4.2	3.6	13	7.5	6.7	7.6	4.2	11.7	3.6	

SECTION QUESTIONS

1. Does the practice of using a beta of 100, instead of the actual transistor beta, cause anything but minimal voltage differences in the analysis of universal biased circuits? Explain.
2. What is the analysis advantage of using a 100 kΩ base resistor such as RB in Figure 6-45?
3. Why does the method of measuring voltage drops, with respect to ground, represent good voltage measurement technique? Is another method better? Explain why or why not.
4. Why does solving for beta, instead of alpha, provide less chance for errors in circuit current calculations?
5. Are exact analysis techniques really necessary for the analysis of universal circuits? Explain.

SECTION IV: THE PNP UNIVERSAL-BIASED CIRCUIT

NPN and PNP transistor circuits are solved in a similar manner, despite the power supply polarities of the single stage PNP circuit being opposite to that of the single stage NPN stage. In fact, if the resistor values, the transistor betas, and power supply voltages are equal, then the voltage drops across the resistors of respective circuits will be equal. This is illustrated and solved for in Figure 6-47. Notice, however, that because the power supply polarities are opposite, the single point voltages of the circuit will differ, in regard to polarity, with respect to ground.

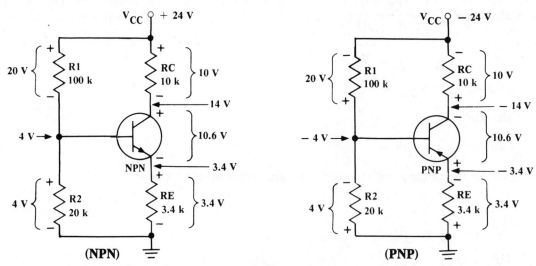

FIGURE 6-47

$$V_{R1} = \frac{V_{CC} \times R1}{R1 + R2} = \frac{24\text{ V} \times 100\text{ k}\Omega}{100\text{ k}\Omega + 20\text{ k}\Omega} = 20\text{ V} \qquad V_{R1} = \frac{V_{CC} \times R1}{R1 + R2} = \frac{24\text{ V} \times 100\text{ k}\Omega}{100\text{ k}\Omega + 20\text{ k}\Omega} = 20\text{ V}$$

$$V_{R2} = \frac{V_{CC} \times R2}{R1 + R2} = \frac{24\text{ V} \times 20\text{ k}\Omega}{100\text{ k}\Omega + 20\text{ k}\Omega} = 4\text{ V} \qquad V_{R2} = \frac{V_{CC} \times R2}{R1 + R2} = \frac{24\text{ V} \times 20\text{ k}\Omega}{100\text{ k}\Omega + 20\text{ k}\Omega} = 4\text{ V}$$

$$V_{RE} = V_{R2} - V_{BE} = 4\text{ V} - 0.6\text{ V} = 3.4\text{ V} \qquad V_{RE} = V_{R2} - V_{BE} = 4\text{ V} - 0.6\text{ V} = 3.4\text{ V}$$

(NPN)	(PNP)
$I_E = V_{RE}/RE = 3.4\text{ V}/3.4\text{ k}\Omega = 1\text{ mA}$	$I_E = V_{RE}/RE = 3.4\text{ V}/3.4\text{ k}\Omega = 1\text{ mA}$
$V_{RC} = I_C R_C = 1\text{ mA} \times 10\text{ k}\Omega = 10\text{ V}$	$V_{RC} = I_C R_C = 1\text{ mA} \times 10\text{ k}\Omega = 10\text{ V}$
$V_C = V_{CC} - V_{RC} = 24\text{ V} - 10\text{ V} = 14\text{ V}$	$V_C = V_{CC} - V_{RC} = -24\text{ V} + 10\text{ V} = -14\text{ V}$
$V_E = V_{RE} = 3.4\text{ V}$	$V_E = -V_{RE} = -3.4\text{ V}$
$V_B = V_{R2} = 4\text{ V}$	$V_B = -V_{R2} = -4\text{ V}$
$V_{CE} = V_C - V_E = 14\text{ V} - 3.4\text{ V} = 10.6\text{ V}$	$V_{CE} = V_{EC} = V_E - V_C = -3.4\text{ V} - (-14\text{ V})$ $= -3.4\text{ V} + 14\text{ V} = 10.6\text{ V}$

FIGURE 6-47 CALCULATIONS

Obviously, the only difference in the two circuits is the polarity of the voltage drops, which effect the single point voltages of the circuit with regard to voltage drops or voltage rises. For instance, in the single point voltage calculations of the PNP circuit the V_{CC} is at -24 V, and all the voltages between V_{CC} and ground are voltage rises. Therefore, the single point voltage of the collector is solved by adding the V_{RC} to the V_{CC} (10 V + -24 V). Likewise, the V_E voltage can be solved by adding the V_{CE} to the collector voltage.

INVERTED PNP AND NPN CIRCUITS

Inverting the circuits, or turning them "upside down", is nothing more than moving the ground around the circuit, and then referencing the voltages of the circuit with respect to ground. In Figure 6-48, both the PNP and NPN circuits are inverted. Notice that the voltage drops of the circuits are identically equal to the previous circuit connections of Figure 6-47, and only single point voltages of the circuit differ. However, notice that the V_{BE}, of both the NPN and PNP circuits, must be forward biased. Therefore, while the NPN circuit looks different than the PNP circuit, they are actually analyzed in the same way.

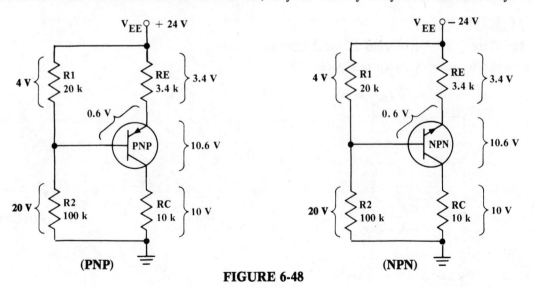

FIGURE 6-48

$$V_{R1} = \frac{V_{CC} \times R1}{R1 + R2} = \frac{24\text{ V} \times 20\text{ k}\Omega}{20\text{ k}\Omega + 100\text{ k}\Omega} = 4\text{ V} \qquad V_{R1} = \frac{V_{CC} \times R1}{R1 + R2} = \frac{24\text{ V} \times 20\text{ k}\Omega}{20\text{ k}\Omega + 100\text{ k}\Omega} = 4\text{ V}$$

$$V_{R2} = \frac{V_{CC} \times R2}{R1 + R2} = \frac{24\text{ V} \times 100\text{ k}\Omega}{20\text{ k}\Omega + 100\text{ k}\Omega} = 20\text{ V} \qquad V_{R2} = \frac{V_{CC} \times R2}{R1 + R2} = \frac{24\text{ V} \times 100\text{ k}\Omega}{20\text{ k}\Omega + 100\text{ k}\Omega} = 20\text{ V}$$

$V_{RE} = V_{R1} - V_{BE} = 4\text{ V} - 0.6\text{ V} = 3.4\text{ V}$ $\qquad V_{RE} = V_{R1} - V_{BE} = 4\text{ V} - 0.6\text{ V} = 3.4\text{ V}$

$I_E = V_{RE}/RE = 3.4\text{ V}/3.4\text{ k}\Omega = 1\text{ mA}$ $\qquad I_E = V_{RE}/RE = 3.4\text{ V}/3.4\text{ k}\Omega = 1\text{ ma}$

$V_{RC} = I_C R_C = 1\text{ mA} \times 10\text{ k}\Omega = 10\text{ V}$ $\qquad V_{RC} = I_C R_C = 1\text{ mA} \times 10\text{ k}\Omega = 10\text{ V}$

$V_C = V_{RC} = 10\text{ V}$ $\qquad V_C = = V_{RC} = -10\text{ V}$

$V_E = V_{EE} - V_{RE} = 24\text{ V} - 3.4\text{ V} = 20.6\text{ V}$ $\qquad V_E = V_{EE} - V_{RE} = -24\text{ V} + 3.4\text{ V} = -20.6\text{ V}$

$V_B = V_{EE} - V_{R1} = 24\text{ V} - 4\text{ V} = 20\text{ V}$, or $\qquad V_B = V_{EE} - V_{R1} = -24\text{ V} + 4\text{ V} = -20\text{ V}$, or

$V_B = V_E - V_{BE} = 20.6\text{ V} - 0.6\text{ V} = 20\text{ V}$ $\qquad V_B = V_E - V_{BE} = -20.6\text{ V} + 0.6\text{ V} = -20\text{ V}$

$V_{CE} = V_{EC} = V_E - V_C$ $\qquad V_{CE} = V_C - V_E$

$\qquad = 20.6\text{ V} - 10\text{ V} = 10.6\text{ V}$ $\qquad\qquad = -10\text{ V} - (-20.6\text{ V}) = 10.6\text{ V}$

(PNP)   **FIGURE 6-48 CALCULATIONS**   **(NPN)**

SECTION IV: LABORATORY EXERCISE

OBJECTIVES

To investigate the DC characteristics of universal circuit biasing.

LIST OF MATERIALS AND EQUIPMENT

1. Transistor: 2N3906 or equivalent
2. Resistors (one each except where indicated):   3.3 kΩ 10 kΩ 22 kΩ 100 kΩ
3. Power Supply
4. Voltmeter

PROCEDURE

Part 1: The PNP Universal Biased Circuit

1. Connect the circuit as shown in Figure 6-49.

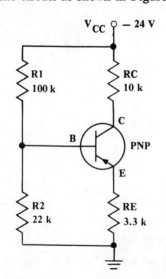

FIGURE 6-49

2. Calculate all the DC voltage drops around the circuit.

    a. Calculate $V_{R1} = \dfrac{V_{CC} \times R1}{R1 + R2}$

b. Calculate $V_{R2} = \dfrac{V_{CC} \times R2}{R1 + R2}$ $\dfrac{-24 \times 22K}{122K} = -4.32$

c. Calculate $V_{RE} = V_{R2} - V_{BE}$ where: $V_{RE} = V_E$ $-4.32 - -.6 = -3.72$

d. Calculate $I_E = V_{RE}/RE$ $3.72/3.3K = 1.12\,mA$

e. Calculate $I_C \approx I_E$

f. Calculate $V_{RC} = I_C R_C$ $1.12/10K = 11.2$

g. Solve for the single point transistor voltages, with respect to ground.

$V_B = -V_{R2}$

$V_E = -V_{RE}$

$V_C = V_{CC} - V_{RC}$ $-24 - 11.2$

h. Calculate $V_{CE} = V_{EC} = V_E - V_C$ 3.7

i. Calculate $V_{CB} = V_{BC} = V_B - V_C$ 4.2

3. Measure the DC voltage drops around the circuit. Include all the resistor and transistor measurements.

NOTE: Make all measurements with respect to ground to avoid possible ground loops. Use the measured values to calculate the remaining voltage drops.

4. Insert the calculated and measured values, as indicated, into Table 6-7.

TABLE 6-7	V_{R1}	V_{R2}	V_{RE}	V_{RC}	V_{CE}	V_{CB}	V_{BE}	V_B	V_C	V_E	I_E
CALCULATED	-19.67	-4.32	3.72	11.2	9.1	8.5	≈0.6 V	4.32	12.8	3.7	1.12 mA
MEASURED	20	-4.1	3.6	12.7	7.9	7.1	.6	4.2	11.3	3.5	/////

Part 2: The Inverted "Upside Down" PNP Universal Circuit

1. Connect the circuit as shown in Figure 6-50.

same as pt 1

FIGURE 6-50

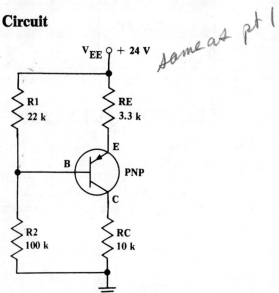

2. Calculate all the DC voltage drops around the circuit.

   a. Calculate $V_{R1} = \dfrac{V_{CC} \times R1}{R1 + R2}$

   b. Calculate $V_{R2} = \dfrac{V_{CC} \times R2}{R1 + R2}$

   c. Calculate $V_{RE} = V_{R1} - V_{BE}$

   d. Calculate $I_E = V_{RE}/RE$

   e. Calculate $I_C \approx I_E$

   f. Calculate $V_{RC} = I_C R_C$

   g. Solve for the single point transistor voltages, with respect to ground.

   $$V_B = V_{CC} - V_{R1}$$
   $$V_E = V_{CC} - V_{RE}$$
   $$V_C = V_{RC}$$

   h. Calculate $V_{CE} = V_{EC} = V_E - V_C$

   i. Calculate $V_{CB} = V_{BC} = V_B - V_C$

3. Measure the DC voltage drops around the circuit. Include all the resistor and transistor measurements.

NOTE: Make all measurements with respect to ground to avoid the possible ground loops. Use the measured values to calculate the remaining voltage drops.

4. Insert the calculated and measured values, as indicated, into Table 6-8.

TABLE 6-8	VR1	VR2	VRE	VRC	VCE	VCB	VBE	VB	VC	VE	IE
CALCULATED							≈ 0.6 V				
MEASURED											/////

SECTION QUESTIONS

1. Were the voltage drops of the PNP circuits of Figures 6-49 and 6-50 similar? How about the single point voltage drops of the circuit, such as V_C, V_E, and V_B?

2. In the "upside-down" PNP circuit of Figure 6-48, the base voltage is solved at 20 V and the emitter voltage at 20.6 V. Is the base-emitter junction forward biased? Explain! How about the VBE of the similar, but opposite polarity connected, NPN circuit of Figure 6-48?

SECTION PROBLEMS

1. Given: $\beta = 100$ and $V_{BE} = 0.6$ V

   Find:
   a. V_{R1}
   b. V_{R2}
   c. V_{RC}
   d. V_{RE}
   e. V_C
   f. V_{CE}

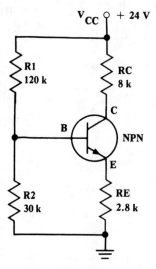

FIGURE 6-51

2. Given: $\beta = 100$ and $V_{BE} = 0.6$ V

   Find:
   a. V_{R1}
   b. V_{R2}
   c. V_{RE}
   d. V_{RC}
   e. V_E
   f. V_C
   g. $V_{CE} = V_{EC}$

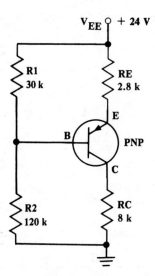

FIGURE 6-52

3. Given: $\beta = 100$ and $V_{BE} = 0.6$ V

   Find:
   a. V_{R1}
   b. V_{R2}
   c. V_{RC}
   d. V_{RE}
   e. V_C
   f. V_E
   g. V_{CE}

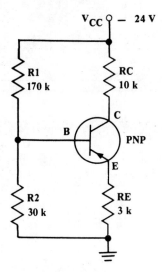

FIGURE 6-53

4. Given: $\beta = 100$, $V_{BE} = 0.6$ V, and $V_C = 15$ V

   Find:

   a. V_{R3}
   b. I_C
   c. V_{R4}
   d. V_B
   e. V_{R1}
   f. R1

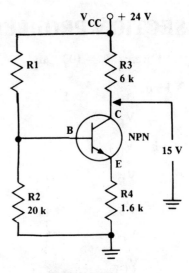

FIGURE 6-54

5. Given: $\beta = 100$, $V_{BE} = 0.6$ V, and $V_C = -8$ V

   Find:

   a. V_{R1}
   b. V_{R2}
   c. V_{R4}
   d. I_E
   e. V_{R3}
   f. V_E
   g. V_B
   h. V_{CE}

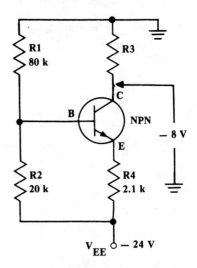

FIGURE 6-55

SECTION V: COLLECTOR FEEDBACK-BIASED CIRCUIT

Collector feedback biased circuits are analyzed in the same way as the base-biased circuits and, like the approximation methods shown for base-biased circuits, $\beta \approx \beta + 1$ and $I_E \approx I_C$. Therefore, writing and solving the loop equations for the collector feedback biased circuit of Figure 6-56 produces the following:

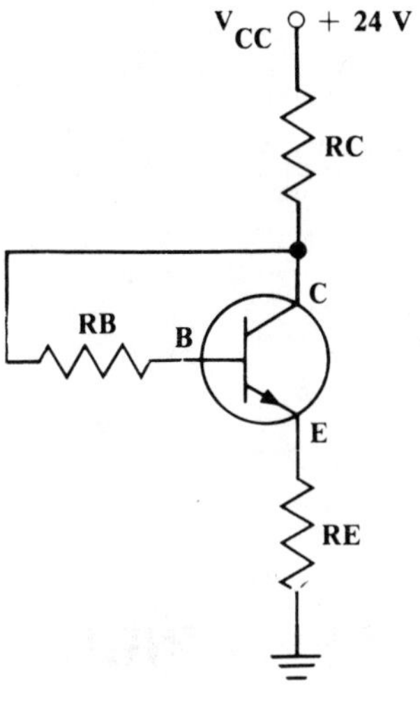

$$V_{CC} = V_{RC} + V_{RB} + V_{BE} + V_{RE}$$

$$V_{CC} - V_{BE} = V_{RC} + V_{RB} + V_{RE}$$

$$V_{CC} - V_{BE} = I_C R_C + I_B R_B + I_E R_E$$

$$V_{CC} - V_{BE} \approx (I_C \times RC) + (I_C/\beta) R_B + I_C R_E$$

where: $I_E \approx I_C$

$$V_{CC} - V_{BE} \approx I_C (RC + RB/\beta + RE)$$

$$I_C \approx \frac{V_{CC} - V_{BE}}{RC + RB/\beta + RE}$$

FIGURE 6-56

If beta = 100 and the circuit resistor values are as shown in Figure 6-57, then I_C and the circuit voltage drops can be solved as follows:

$$I_C \approx \frac{V_{CC} - V_{BE}}{RC + RB/\beta + RE}$$

$$= \frac{24\ V - 0.6\ V}{5\ k\Omega + 300\ k\Omega/100 + 1\ k\Omega} = \frac{23.4\ V}{9\ k\Omega} = 2.6\ mA$$

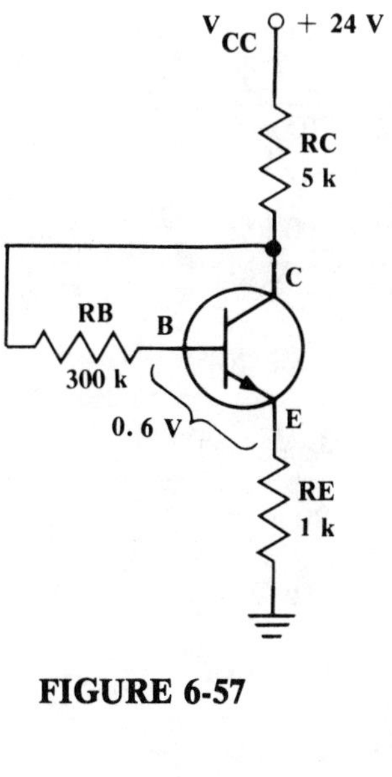

$$V_{RC} = I_C R_C = 2.6\ mA \times 5\ k\Omega = 13\ V$$

$$V_{RE} = I_E R_C = 2.6\ mA \times 1\ k\Omega = 2.6\ V$$

$$I_B = I_C/\beta = 2.6\ mA/100 = 26\ \mu A$$

$$V_{RB} = I_B R_B = 26\ \mu A \times 300\ k\Omega = 7.8\ V$$

$$V_C = V_{CC} - V_{RC} = 24\ V - 13\ V = 11\ V$$

$$V_E = V_{RE} = 2.6\ V$$

$$V_B = V_{RE} + V_{BE} = 2.6\ V + 0.6\ V = 3.2\ V$$

$$V_{CE} = V_C - V_E = 11\ V - 2.6\ V = 8.4\ V$$

$$V_{CB} = V_C - V_B = 11\ V - 3.2\ V = 7.8\ V$$

$$I_E \approx I_C = 2.6\ mA$$

where $\beta = \beta + 1 = 100$

FIGURE 6-57

The voltage distribution, using approximation techniques, is shown in Figure 6-58.

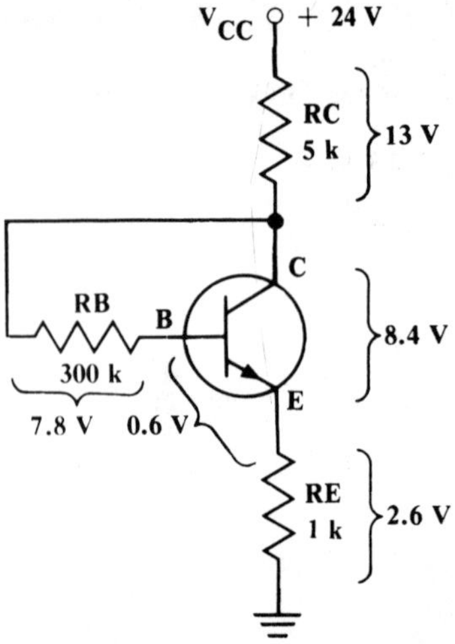

$V_{CC} = V_{RC} + R_{CE} + V_{RE}$

$24\text{ V} = 13\text{ V} + 8.4\text{ V} + 2.6\text{ V}$

$V_{CC} = V_{RC} + V_{RB} + V_{BE} + V_{RE}$

$24\text{ V} = 13\text{ V} + 7.8\text{ V} + 0.6\text{ V} + 2.6\text{ V}$

$V_{CC} = V_{RC} + R_{RB} + V_B$

$24\text{ V} = 13\text{ V} + 7.8\text{ V} + 3.2\text{ V}$

FIGURE 6-58

SECTION VI: ANALYZING CIRCUITS WITH VARIOUS CONFIGURATIONS

In introducing direct current circuit analysis, circuit configurations generally are kept as straight forward as possible. However, it is necessary to develop awareness with regard to circuits that have been turned around or changed slightly from what is normally expected, where some thought has to be given before proceeding with the analysis. Therefore, in this section, a variety of circuit configurations are given and then analyzed. For instance, in the circuit of Figure 6-59, a PNP transistor is used and connected so that the base-emitter diode junction is forward biased.

NOTE: All betas are 100 and $V_{BE} = 0.6\text{ V}$ for the circuits that follow.

A.

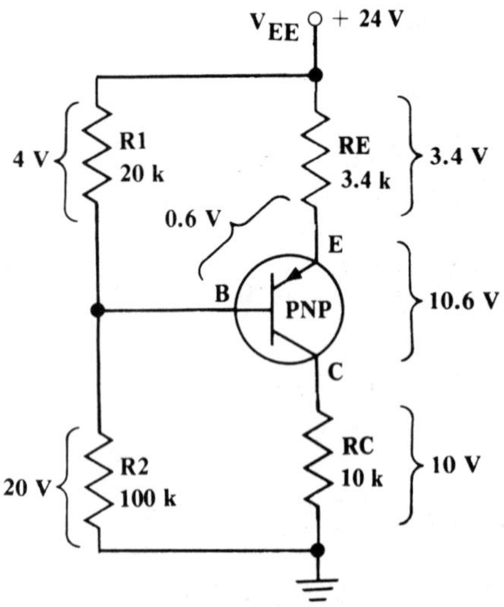

$V_{R1} = \dfrac{V_{EE} \times R1}{R1 + R2} = \dfrac{24\text{ V} \times 20\text{ k}\Omega}{20\text{ k}\Omega + 100\text{ k}\Omega} = 4\text{ V}$

$V_{R2} = \dfrac{V_{EE} \times R2}{R1 + R2} = \dfrac{24\text{ V} \times 100\text{ k}\Omega}{20\text{ k}\Omega + 100\text{ k}\Omega} = 20\text{ V}$

$V_B = V_{EE} - V_{R1} = 24\text{ V} - 4\text{ V} = 20\text{ V}$

$V_E = V_B + V_{BE} = 20\text{ V} + 0.6\text{ V} = 20.6\text{ V}$

$V_{RE} = V_{EE} - V_E = 24\text{ V} - 20.6\text{ V} = 3.4\text{ V}$ or

$V_{RE} = V_{R1} - V_{BE} = 4\text{ V} - 0.6\text{ V} = 3.4\text{ V}$

$I_E = V_{RE}/RE = 3.4\text{ V}/3.4\text{ k}\Omega = 1\text{ mA}$

$V_{RC} = I_C R_C = 1\text{ mA} \times 10\text{ k}\Omega = 10\text{ V}$

$V_C = V_{RC} = 10\text{ V}$

$V_{CE} = V_{EC} = V_E - V_C = 20.6\text{ V} - 10\text{ V} = 10.6\text{ V}$

FIGURE 6-59

B.

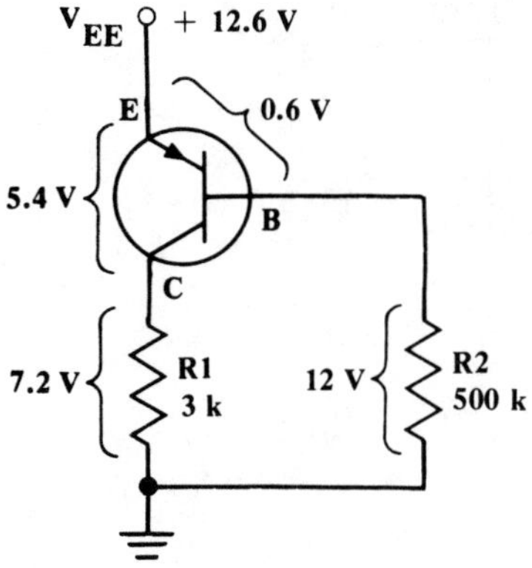

FIGURE 6-60

$V_{R2} = V_{EE} - V_{BE} = 12.6 \text{ V} - 0.6 \text{ V} = 12 \text{ V}$

$I_B = V_{R2}/R2 = 12 \text{ V}/500 \text{ k}\Omega = 24 \text{ μA}$

$I_C = \beta I_B = 100 \times 24 \text{ μA} = 2.4 \text{ mA}$

$V_{R1} = I_C R1 = 2.4 \text{ mA} \times 3 \text{ k}\Omega = 7.2 \text{ V}$

$V_C = V_{R1} = 7.2 \text{ V}$

$V_{EE} = V_E = 12.6 \text{ V}$

$V_{CE} = V_{EC} = V_{EE} - V_C = 12.6 \text{ V} - 7.2 \text{ V} = 5.4 \text{ V}$

C.

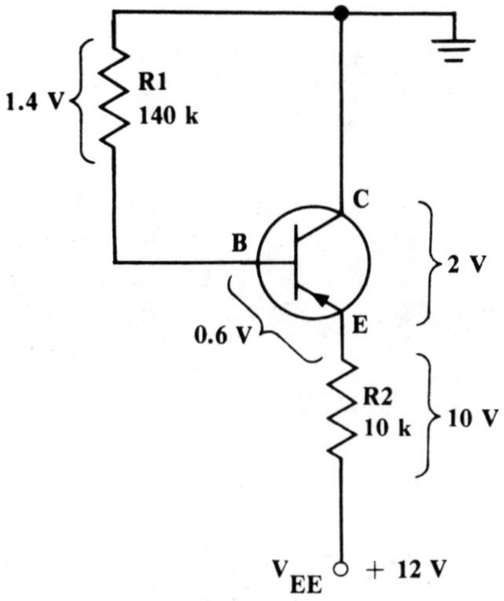

FIGURE 6-61

$I_E \approx \dfrac{V_{EE} - V_{BE}}{(R1/\beta) + R2} = \dfrac{12 \text{ V} - 0.6 \text{ V}}{(140 \text{ k}\Omega/100) + 10 \text{ k}\Omega} = \dfrac{11.4 \text{ V}}{1.4 \text{ k}\Omega + 10 \text{ k}\Omega} = 1 \text{ mA}$

$V_{RE} = I_E R2 = 1 \text{ mA} \times 10 \text{ k}\Omega = 10 \text{ V}$

$I_B = I_E/\beta + 1 \approx I_B/\beta = 1 \text{ mA}/100 = 10 \text{ μA}$

$V_{R1} = I_B R1 = 10 \text{ μA} \times 140 \text{ k}\Omega = 1.4 \text{ V}$

$V_E = V_{EE} - V_{R2} = 12 \text{ V} - 10 \text{ V} = 2 \text{ V}$

$V_B = V_E - V_{BE} = 2 \text{ V} - 0.6 \text{ V} = 1.4 \text{ V}$ or

$V_B = V_{R1} = 1.4 \text{ V}$

$V_{CE} = V_{EC} = V_E - V_C = 2 \text{ V} - 0 \text{ V} = 2 \text{ V}$

D.

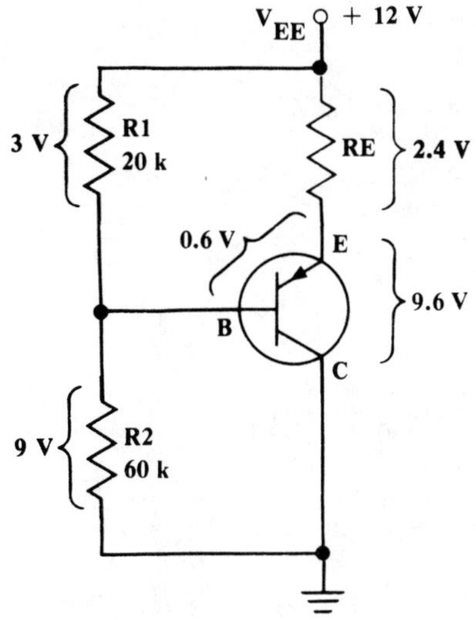

FIGURE 6-62

$V_{R1} = \dfrac{V_{EE} \times R1}{R1 + R2} = \dfrac{12 \text{ V} \times 20 \text{ k}\Omega}{20 \text{ k}\Omega + 60 \text{ k}\Omega} = 3 \text{ V}$

$V_{R2} = \dfrac{V_{EE} \times R2}{R1 + R2} = \dfrac{12 \text{ V} \times 60 \text{ k}\Omega}{20 \text{ k}\Omega + 60 \text{ k}\Omega} = 9 \text{ V}$

$V_B = V_{EE} - V_{R1} = 12 \text{ V} - 3 \text{ V} = 9 \text{ V}$ or

$V_B = V_{R2} = 9 \text{ V}$

$V_E = V_B + V_{BE} = 9 \text{ V} + 0.6 \text{ V} = 9.6 \text{ V}$

$V_{RE} = V_{EE} - V_E = 12 \text{ V} - 9.6 \text{ V} = 2.4 \text{ V}$, or

$V_{RE} = V_{R1} - V_{BE} = 3 \text{ V} - 0.6 \text{ V} = 2.4 \text{ V}$

$V_{CE} = V_{EC} = V_E - V_C = 9.6 \text{ V} - 0 \text{ V} = 9.6 \text{ V}$

E.

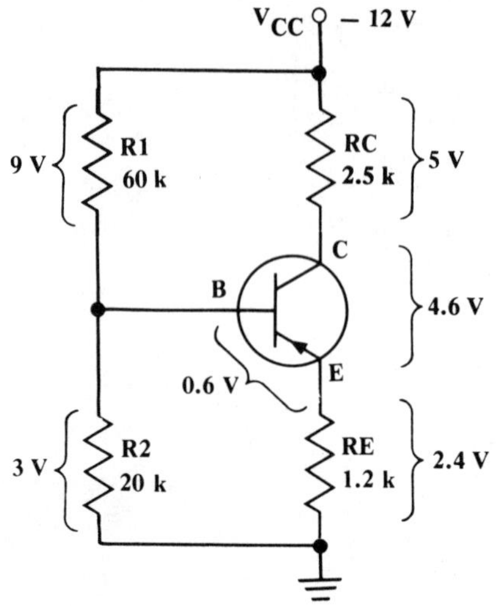

FIGURE 6-63

$$V_{R1} = \frac{V_{CC} \times R1}{R1 + R2} = \frac{12\text{ V} \times 60\text{ k}\Omega}{60\text{ k}\Omega + 20\text{ k}\Omega} = 9\text{ V}$$

$$V_{R2} = \frac{V_{CC} \times R2}{R1 + R2} = \frac{12\text{ V} \times 20\text{ k}\Omega}{60\text{ k}\Omega + 20\text{ k}\Omega} = 3\text{ V}$$

$V_B = V_{CC} - V_{R1} = -3\text{ V}$

$V_E = V_B - V_{BE} = -3\text{ V} - (-0.6\text{ V}) = -2.4\text{ V}$

$V_{RE} = V_E = 2.4\text{ V}$

$I_E = V_{RE}/RE = 2.4\text{ V}/1.2\text{ k}\Omega = 2\text{ mA}$

$V_{RC} = I_C R_C \approx 2\text{ mA} \times 2.5\text{ k}\Omega = 5\text{ V}$

$V_C = V_{CC} - V_{RC} = -12\text{ V} + 5\text{ V} = -7\text{ V}$

$V_{CE} = V_C - V_E = -7\text{ V} - (-2.4\text{ V}) = 4.6\text{ V}$

$V_{EC} = V_E - V_C = -2.4\text{ V} - (-7\text{ V}) = 4.6\text{ V}$

F.

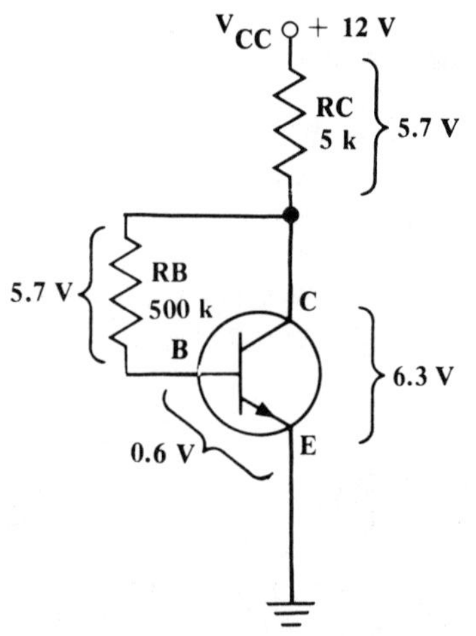

FIGURE 6-64

$$I_C \approx \frac{V_{CC} - V_{BE}}{RB/\beta + RC} = \frac{12\text{ V} - 0.6\text{ V}}{500\text{ k}\Omega/100 + 5\text{ k}\Omega} = \frac{11.4\text{ V}}{10\text{ k}\Omega} = 1.14\text{ mA}$$

$V_{RC} = I_C R_C \approx 1.14\text{ mA} \times 5\text{ k}\Omega = 5.7\text{ V}$

$V_C = V_{CC} - V_{RC} = 12\text{ V} - 5.7\text{ V} = 6.3\text{ V}$

$I_B = I_C/\beta = 1.14\text{ mA}/100 = 11.4\text{ }\mu\text{A}$

$V_{RB} = I_B R_B = 11.4\text{ }\mu\text{A} \times 500\text{ k}\Omega = 5.7\text{ V}$

$V_{CE} = V_C - V_E = 6.3\text{ V} - 0\text{ V} = 6.3\text{ V}$

$V_{CC} = V_{RC} + V_{RB} + V_{BE}$
$\quad\quad = 5.7\text{ V} + 5.7\text{ V} + 0.6\text{ V} = 12\text{ V}$

$V_{CC} = V_{RC} + V_{CE} = 5.7\text{ V} + 6.3\text{ V} = 12\text{ V}$

SECTION VII: EXACT TECHNIQUES REVIEWED

When the universal circuit is well designed it is the simplest circuit configuration to analyze. However, when a base resistor is inserted into the circuit, as in Figure 6-65, then the voltage drop V_{RB} is not easily obtained unless Thevenin's equivalent theorem is used. Therefore, the exact analysis techniques previously used in this chapter are again used to analyze the circuit of Figure 6-65 but, in this case, the R3 resistance, along with R_{TH}, is reflected in series with the emitter resistance to solve for both the emitter and base currents.

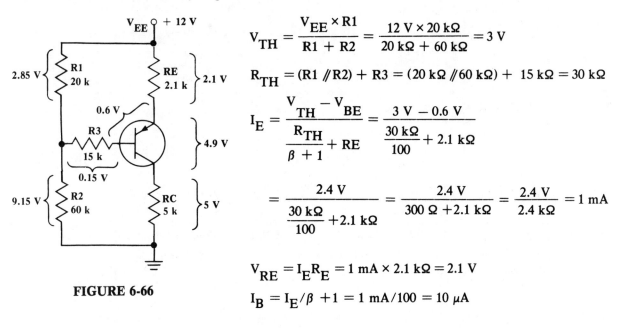

$$V_{TH} = \frac{V_{CC} \times R2}{R1 + R2} = \frac{12\,V \times 20\,k\Omega}{60\,k\Omega + 20\,k\Omega} = \frac{240\,V}{80} = 3\,V$$

$$R_{TH} = (R1 // R2) + R3 = \frac{60\,k\Omega \times 20\,k\Omega}{60\,k\Omega + 20\,k\Omega} + 15\,k\Omega$$

$$= \frac{1200\,k\Omega}{80\,k\Omega} + 15\,k\Omega = 15\,k\Omega + 15\,k\Omega = 30\,k\Omega$$

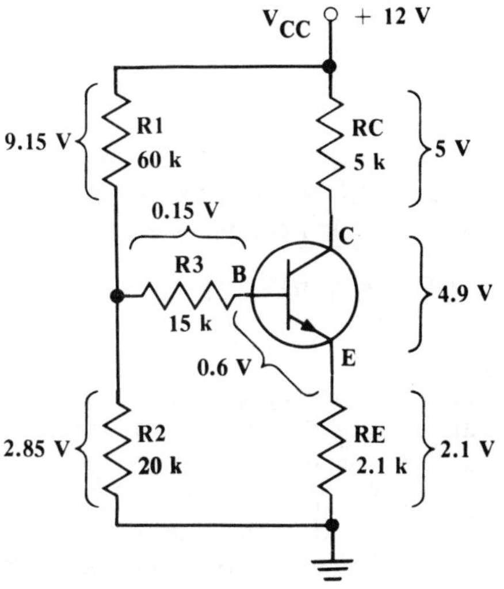

$$I_E = \frac{V_{TH} - V_{BE}}{\frac{R_{TH}}{\beta + 1} + RE} = \frac{3\,V - 0.6\,V}{\frac{30\,k\Omega}{100} + 2.1\,k\Omega}$$

$$= \frac{2.4\,V}{300\,\Omega + 2.1\,k\Omega} = \frac{2.4\,V}{2.4\,k\Omega} = 1\,mA$$

$$V_{RE} = I_E R_E = 1\,mA \times 2.1\,k\Omega = 2.1\,V$$

$$V_B = V_{RE} + V_{BE} = 2.1\,V + 0.6\,V = 2.7\,V$$

$$I_B = I_E/\beta + 1 = 1\,mA/100 = 10\,\mu A$$

$$V_{RB} = I_B R3 = 10\,\mu A \times 15\,k\Omega = 0.15\,V$$

$$V_{R2} = V_B + V_{RB} = 2.7\,V + 0.15\,V = 2.85\,V$$

$$V_{R1} = V_{CC} - V_{R2} = 12\,V - 2.85\,V = 9.15\,V$$

$$V_{RC} = I_C R_C = 1\,mA \times 5\,k\Omega = 5\,V$$

$$V_C = V_{CC} - V_{RC} = 12\,V - 5\,V = 7\,V$$

$$V_{CE} = V_C - V_E = 7\,V - 2.1\,V = 4.9\,V$$

FIGURE 6-65

NOTE: Rough approximation techniques could also be used to analyze this circuit if the V_{R3} voltage is ignored (initially), the approximate I_E current solved, then I_B, and so on. This technique, however, does not take into account extremely large values of R3.

$$V_{TH} = \frac{V_{EE} \times R1}{R1 + R2} = \frac{12\,V \times 20\,k\Omega}{20\,k\Omega + 60\,k\Omega} = 3\,V$$

$$R_{TH} = (R1 // R2) + R3 = (20\,k\Omega // 60\,k\Omega) + 15\,k\Omega = 30\,k\Omega$$

$$I_E = \frac{V_{TH} - V_{BE}}{\frac{R_{TH}}{\beta + 1} + RE} = \frac{3\,V - 0.6\,V}{\frac{30\,k\Omega}{100} + 2.1\,k\Omega}$$

$$= \frac{2.4\,V}{\frac{30\,k\Omega}{100} + 2.1\,k\Omega} = \frac{2.4\,V}{300\,\Omega + 2.1\,k\Omega} = \frac{2.4\,V}{2.4\,k\Omega} = 1\,mA$$

$$V_{RE} = I_E R_E = 1\,mA \times 2.1\,k\Omega = 2.1\,V$$

$$I_B = I_E/\beta + 1 = 1\,mA/100 = 10\,\mu A$$

FIGURE 6-66

$$V_{RB} = I_B R_B = 10 \; \mu A \times 15 \; k\Omega = 0.15 \; V$$

$$V_E = V_{CC} - V_{RE} = 12 \; V - 2.1 \; V = 9.9 \; V$$

$$V_B = V_E - V_{BE} = 9.9 \; V - 0.6 \; V = 9.3 \; V$$

$$V_{R2} = V_B - V_{R3} = 9.3 \; V - 0.15 \; V = 9.15 \; V$$

$$V_{R1} = V_{CC} - V_{R2} = 12 \; V - 9.15 \; V = 2.85 \; V$$

NOTE: To avoid confusion, solve all the V_{TH} voltages with reference to the voltage drop across the emitter resistor. Then, solve for the single point voltages of the circuit, with respect to ground. The circuits of Figure 6-65 and Figure 6-66 have identical component values to allow a comparison.

SECTION VIII: STABILITY CONSIDERATIONS AND CIRCUIT CHOICE

The choice of one circuit configuration over another is directly related to the degree of direct current stability required. For instance, the beta parameter of transistors can be very different from device to device and it will vary with temperature change. Hence, the base-biased circuit, which is very dependent on beta, is the least stable. The two power supply circuit, which is the least dependent on beta, is the most stable. In between, with regard to DC stability, are the collector feedback circuits, which are more stable than base-biased circuit, and the universal circuits, which provide emitter control and only require a single power supply. A brief comparison of the circuit collector currents, as the beta is changed from 100 to 200 to 50, is given for the base-biased, collector feedback biased, universal biased, and two power supply circuits that follow.

BASE BIASED CIRCUIT

In the base-biased circuit, the base resistor RB is reflected into the emitter by the beta when solving for the collector current I_C. And, because the base resistor values are inherently large, even with RE as large as 2 kΩ, the effect of beta on the I_C calculations, and the associated voltage drops around the circuit, can be dramatic, as shown in Figure 6-67.

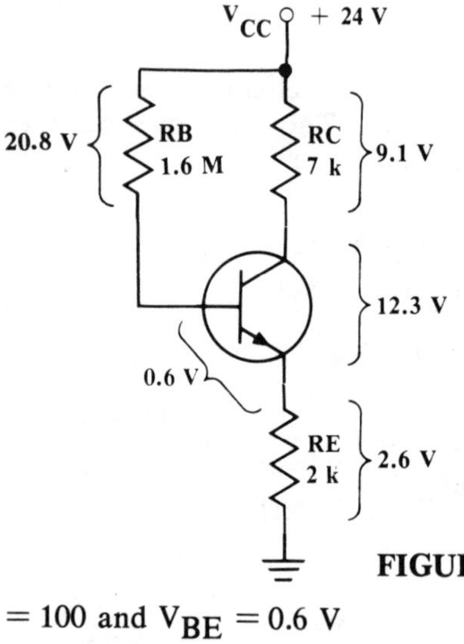

FIGURE 6-67(a)

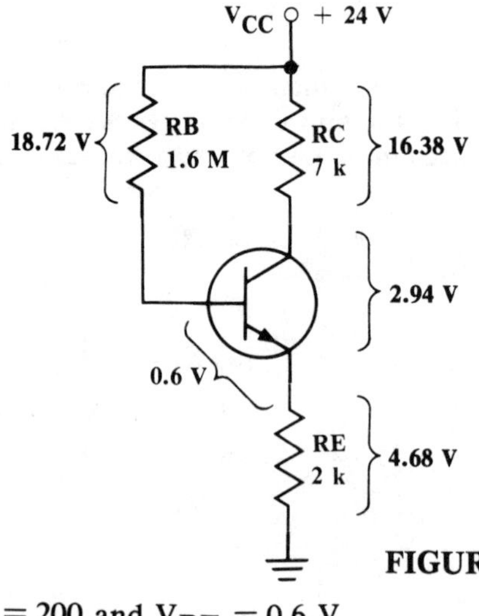

FIGURE 6-67(b)

Beta = 100 and V_{BE} = 0.6 V

$$I_C = \frac{V_{CC} - V_{BE}}{RB/\beta + RE} = \frac{24 \; V - 0.6 \; V}{1.6 \; M\Omega/100 + 2 \; k\Omega}$$

$$= \frac{23.4 \; V}{16 \; k\Omega + 2 \; k\Omega} = \frac{23.4 \; V}{18 \; k\Omega} = 1.3 \; mA$$

Beta = 200 and V_{BE} = 0.6 V

$$I_C = \frac{V_{CC} - V_{BE}}{RB/\beta + RE} = \frac{24 \; V - 0.6 \; V}{1.6 \; M\Omega/200 + 1 \; k\Omega}$$

$$\frac{23.4 \; V}{8 \; k\Omega + 2 \; k\Omega} = \frac{23.4 \; V}{10 \; k\Omega} = 2.34 \; mA$$

$V_{RC} = 9.1$ V $V_{RE} = 2.6$ V $V_C = 14.9$ V $V_{RC} = 16.38$ $V_{RE} = 4.68$ V $V_C = 7.62$ V

$V_{CE} = 12.3$ V $V_{RB} = 20.8$ V $V_B = 3.2$ V $V_{CE} = 2.94$ V $V_{RB} = 18.72$ V $V_B = 5.28$ V

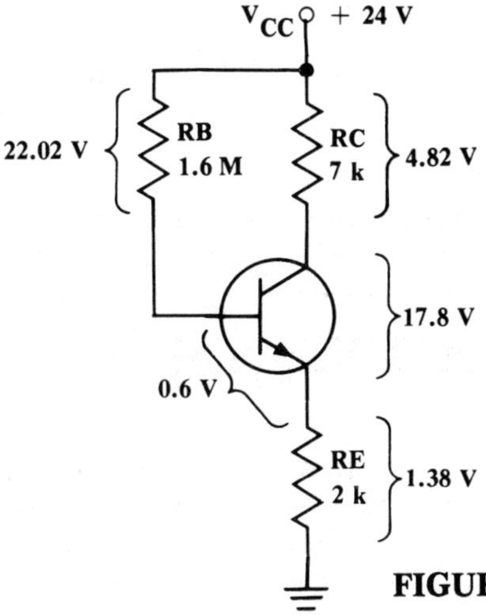

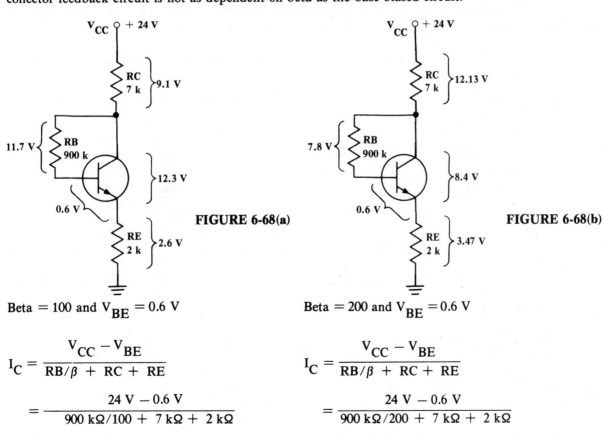

Beta = 50 and $V_{BE} = 0.6$ V

$$I_C = \frac{V_{CC} - V_{BE}}{RB/\beta + RE} = \frac{24\text{ V} - 0.6\text{ V}}{1.6\text{ M}\Omega/50 + 2\text{ k}\Omega}$$

$$= \frac{23.4\text{ V}}{32\text{ k}\Omega + 2\text{ k}\Omega} = \frac{23.4\text{ V}}{34\text{ k}\Omega} = 0.668\text{ mA}$$

$V_{RC} = 4.82$ V $V_{RE} = 1.38$ V $V_C = 19.18$ V

$V_{CE} = 17.8$ V $V_{RB} = 22.02$ V $V_B = 1.98$ V

FIGURE 6-67(c)

FIGURE 6-67

COLLECTOR FEEDBACK BIASED CIRCUIT

Connecting the base resistor between the base and the collector leads includes RC, effectively, into the calculations. Therefore, the base resistor RB does not have to be as large to obtain 1.3 mA of collector current, when beta equals 100. Therefore, beta changes to 200 and then to 50 will not cause as large a collector current change as the base-biased circuit. Hence, as shown and solved for in Figure 6-68, the collector feedback circuit is not as dependent on beta as the base-biased circuit.

FIGURE 6-68(a)

Beta = 100 and $V_{BE} = 0.6$ V

$$I_C = \frac{V_{CC} - V_{BE}}{RB/\beta + RC + RE}$$

$$= \frac{24\text{ V} - 0.6\text{ V}}{900\text{ k}\Omega/100 + 7\text{ k}\Omega + 2\text{ k}\Omega}$$

FIGURE 6-68(b)

Beta = 200 and $V_{BE} = 0.6$ V

$$I_C = \frac{V_{CC} - V_{BE}}{RB/\beta + RC + RE}$$

$$= \frac{24\text{ V} - 0.6\text{ V}}{900\text{ k}\Omega/200 + 7\text{ k}\Omega + 2\text{ k}\Omega}$$

$$= \frac{23.4 \text{ V}}{9 \text{ k}\Omega + 7 \text{ k}\Omega + 2 \text{ k}\Omega} = \frac{23.4 \text{ V}}{18 \text{ k}\Omega} = 1.3 \text{ mA} \qquad = \frac{23.4 \text{ V}}{4.5 \text{ k}\Omega + 7 \text{ k}\Omega + 2 \text{ k}\Omega} = \frac{23.4 \text{ V}}{13.5 \text{ k}\Omega} = 1.73 \text{ mA}$$

$V_{RC} = 9.1 \text{ V}$ $V_{RE} = 2.6 \text{ V}$ $V_C = 14.9 \text{ V}$ $V_{RC} = 12.13 \text{ V}$ $V_{RE} = 3.47 \text{ V}$ $V_C = 19.87 \text{ V}$

$V_{CE} = 12.3 \text{ V}$ $V_{RB} = 11.7 \text{ V}$ $V_B = 3.2 \text{ V}$ $V_{CE} = 8.4 \text{ V}$ $V_{RB} = 7.8 \text{ V}$ $V_B = 4.07 \text{ V}$

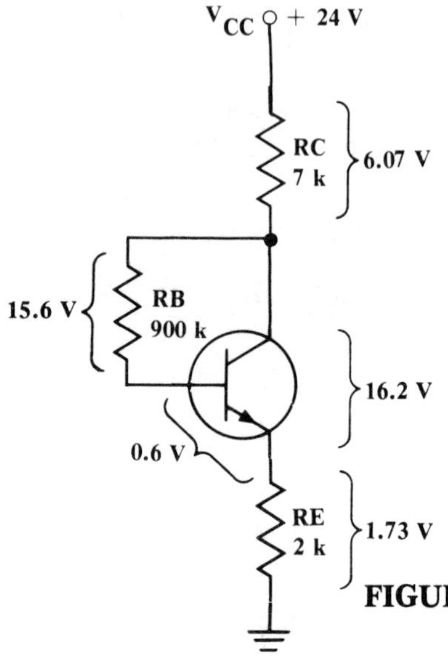

Beta $= 50$ and $V_{BE} = 0.6$ V

$$I_C = \frac{V_{CC} - V_{BE}}{RB/\beta + RC + RE} = \frac{24 \text{ V} - 0.6 \text{ V}}{900 \text{ k}\Omega/50 + 7 \text{ k}\Omega + 2 \text{ k}\Omega}$$

$$= \frac{23.4 \text{ V}}{18 \text{ k}\Omega + 7 \text{ k}\Omega + 2 \text{ k}\Omega} = \frac{23.4 \text{ V}}{27 \text{ k}\Omega} = 0.867 \text{ mA}$$

$V_{RC} = 6.07 \text{ V}$ $V_{RE} = 1.73 \text{ V}$ $V_C = 11.87 \text{ V}$

$V_{CE} = 16.2 \text{ V}$ $V_{RB} = 15.6 \text{ V}$ $V_B = 1.93 \text{ V}$

FIGURE 6-68(c)

FIGURE 6-68

TWO POWER SUPPLY EMITTER CURRENT CONTROL BIASING

Once emitter current control techniques are employed the effect of beta is minimized, even with large beta variations. Additionally, the base resistor value does not have to be too large (about 30 kΩ), so its effect on the emitter current will be minimal, regardless of the beta changes from 100 to 200 to 50. Therefore, the voltage distribution change with beta change will also be minimal as shown in Figure 6-69.

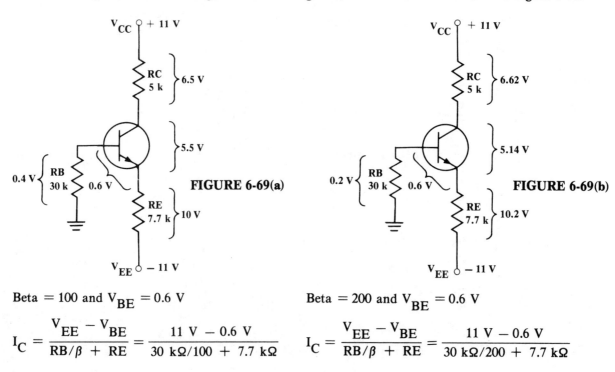

FIGURE 6-69(a)

FIGURE 6-69(b)

Beta $= 100$ and $V_{BE} = 0.6$ V

$$I_C = \frac{V_{EE} - V_{BE}}{RB/\beta + RE} = \frac{11 \text{ V} - 0.6 \text{ V}}{30 \text{ k}\Omega/100 + 7.7 \text{ k}\Omega}$$

Beta $= 200$ and $V_{BE} = 0.6$ V

$$I_C = \frac{V_{EE} - V_{BE}}{RB/\beta + RE} = \frac{11 \text{ V} - 0.6 \text{ V}}{30 \text{ k}\Omega/200 + 7.7 \text{ k}\Omega}$$

$$= \frac{10.4 \text{ V}}{300 \text{ }\Omega + 7.7 \text{ k}\Omega} = \frac{10.4 \text{ V}}{8 \text{ k}\Omega} = 1.3 \text{ mA} \qquad = \frac{10.4 \text{ V}}{150 \text{ }\Omega + 7.7 \text{ k}\Omega} = \frac{10.4 \text{ V}}{7.85 \text{ k}\Omega} = 1.325 \text{ mA}$$

$V_{RC} = 6.5 \text{ V} \qquad V_{RB} \approx 0.4 \text{ V} \qquad V_E = -1 \text{ V} \qquad V_{RC} = 6.62 \text{ V} \qquad V_{RB} = 0.2 \text{ V} \qquad V_E = -0.8 \text{ V}$

$V_{RE} = 10 \text{ V} \qquad V_{CE} = 5.5 \text{ V} \qquad V_C = 4.5 \text{ V} \qquad V_{RE} = 10.2 \text{ V} \qquad V_{CE} = 5.18 \text{ V} \qquad V_C = 4.38 \text{ V}$

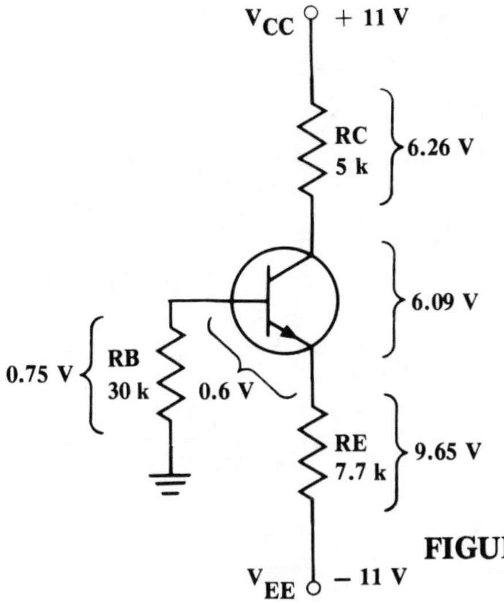

Beta = 50 and $V_{BE} = 0.6$ V

$$I_C = \frac{V_{EE} - V_{BE}}{R_B/\beta + R_E} = \frac{11 \text{ V} - 0.6 \text{ V}}{30 \text{ k}\Omega/50 + 7.7 \text{ k}\Omega}$$

$$= \frac{10.4 \text{ V}}{600 \text{ }\Omega + 7.7 \text{ k}\Omega} = \frac{10.4 \text{ V}}{8.3 \text{ k}\Omega} = 1.253 \text{ mA}$$

$V_{RC} = 6.26 \text{ V} \qquad V_{RB} = 0.75 \text{ V} \qquad V_E = -1.35 \text{ V}$

$V_{RE} = 9.65 \text{ V} \qquad V_{CE} = 6.09 \text{ V} \qquad V_C = 4.74 \text{ V}$

FIGURE 6-69(c)

FIGURE 6-69

UNIVERSAL CIRCUIT

The universal circuit is a single power supply emitter current controlled circuit, which is not nearly as stable with beta change as is the two power supply circuit, because of the lower voltage distributed across the emitter resistor. However, it is more stable than either the base-biased or the collector feedback circuits. Also, because the effect of the beta change from 100 to 200 to 50 must be calculated, Thevenin's equivalent circuit will be used, as shown in Figure 6-70.

Beta = 100 and $V_{BE} = 0.6$ V

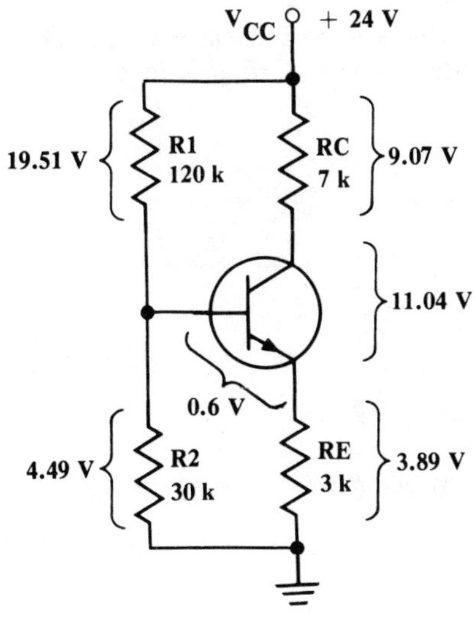

$$V_{TH} = \frac{V_{CC} \times R2}{R1 + R2} = \frac{24 \text{ V} \times 30 \text{ k}\Omega}{30 \text{ k}\Omega + 120 \text{ k}\Omega} = 4.8 \text{ V}$$

$$R_{TH} = R1 \parallel R2 = \frac{R1 \times R2}{R1 + R2} = \frac{120 \text{ k}\Omega \times 30 \text{ k}\Omega}{120 \text{ k}\Omega + 30 \text{ k}\Omega} = 24 \text{ k}\Omega$$

$$I_E = \frac{V_{TH} - V_{BE}}{R_{TH}/\beta + R_E} = \frac{4.8 \text{ V} - 0.6 \text{ V}}{24 \text{ k}\Omega/100 + 3 \text{ k}\Omega}$$

$$= \frac{4.2 \text{ V}}{24 \text{ V} + 3 \text{ k}\Omega} = \frac{4.2 \text{ V}}{3240 \text{ }\Omega} = 1.296 \text{ mA}$$

$V_{RC} = 9.07 \text{ V} \qquad V_{R1} = 19.51 \text{ V} \qquad V_C = 14.93 \text{ V}$

$V_{RE} = 3.89 \text{ V} \qquad V_{R2} = 4.49 \text{ V} \qquad V_{CE} = 11.04 \text{ V}$

FIGURE 6-70(a)

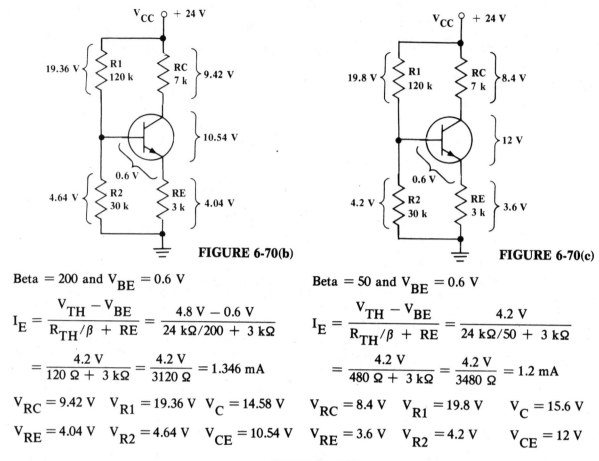

FIGURE 6-70(b)   FIGURE 6-70(c)

Beta = 200 and V_{BE} = 0.6 V

$$I_E = \frac{V_{TH} - V_{BE}}{R_{TH}/\beta + R_E} = \frac{4.8\ V - 0.6\ V}{24\ k\Omega/200 + 3\ k\Omega}$$

$$= \frac{4.2\ V}{120\ \Omega + 3\ k\Omega} = \frac{4.2\ V}{3120\ \Omega} = 1.346\ mA$$

$V_{RC} = 9.42\ V$ $V_{R1} = 19.36\ V$ $V_C = 14.58\ V$

$V_{RE} = 4.04\ V$ $V_{R2} = 4.64\ V$ $V_{CE} = 10.54\ V$

Beta = 50 and V_{BE} = 0.6 V

$$I_E = \frac{V_{TH} - V_{BE}}{R_{TH}/\beta + R_E} = \frac{4.2\ V}{24\ k\Omega/50 + 3\ k\Omega}$$

$$= \frac{4.2\ V}{480\ \Omega + 3\ k\Omega} = \frac{4.2\ V}{3480\ \Omega} = 1.2\ mA$$

$V_{RC} = 8.4\ V$ $V_{R1} = 19.8\ V$ $V_C = 15.6\ V$

$V_{RE} = 3.6\ V$ $V_{R2} = 4.2\ V$ $V_{CE} = 12\ V$

FIGURE 6-70

SECTION IX: REFLECTED RESISTANCE CIRCUIT ANALYSIS

Direct current approximation analysis ignores the effect of beta in the analysis of universal circuits, and Thevenin's equivalent provides exact analysis by taking into account the effect of beta. A third approach to DC circuit analysis is reflected resistance techniques, which **solves** for the voltage drops of the circuit by reflecting the emitter resistance into the base in parallel with R2, as shown in Figure 6-71(a). This approach is not as accurate as the Thevenin approach, but it provides a good indicator of the effect beta has on the distributed voltages of the circuit, and it is an excellent method to use in the analysis of complex circuitry.

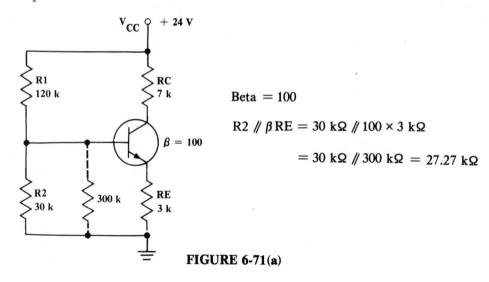

Beta = 100

R2 // β RE = 30 kΩ // 100 × 3 kΩ

= 30 kΩ // 300 kΩ = 27.27 kΩ

FIGURE 6-71(a)

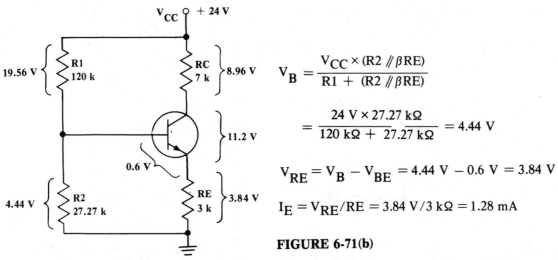

$$V_B = \frac{V_{CC} \times (R2 \,/\!/\, \beta RE)}{R1 + (R2 \,/\!/\, \beta RE)}$$

$$= \frac{24\text{ V} \times 27.27\text{ k}\Omega}{120\text{ k}\Omega + 27.27\text{ k}\Omega} = 4.44\text{ V}$$

$$V_{RE} = V_B - V_{BE} = 4.44\text{ V} - 0.6\text{ V} = 3.84\text{ V}$$

$$I_E = V_{RE}/RE = 3.84\text{ V}/3\text{ k}\Omega = 1.28\text{ mA}$$

FIGURE 6-71(b)

Essentially, the reflected resistance technique, with reference to the circuit of Figure 6-71, is to reflect the emitter resistance of 3 kΩ into the base circuit by multiplying it by the beta of the device and then placing this reflected resistance in parallel with R2. Then, the voltage divider equation is used to solve for V_{R1} and V_{R2} in parallel with beta times RE.

A comparison of values derived for the universal circuit using approximation, Thevenin's, and reflected resistance analysis techniques is given in Table 6-9. This table shows the relative variations in calculated values for a device with a beta of 100.

TABLE 6-9	V_{R1}	V_{R2}	V_{RE}	V_{RC}	V_{CE}	I_C
APPROXIMATION	19.2 V	4.8 V	4.2 V	9.8 V	12.8 V	1.4 mA
THEVENIN'S	19.8 V	4.2 V	3.6 V	8.4 V	12 V	1.2 mA
REFLECTED	19.56 V	4.44 V	3.84 V	8.98 V	11.2 V	1.28 mA

CHAPTER PROBLEMS

1. Given: $\beta = 100$ and $V_{BE} = 0.6$ V

   Solve for V_{R3}.

2. Given: $\beta = 100$ and $V_{BE} = 0.6$ V

   Solve for V_{R1}

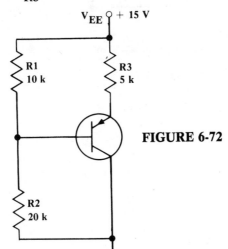

FIGURE 6-72

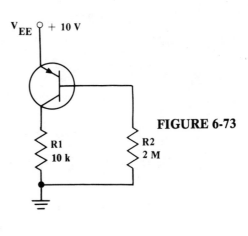

FIGURE 6-73

3. Given: $\beta = 100$ and $V_{BE} = 0.6$ V
   Solve for V_{CE}.

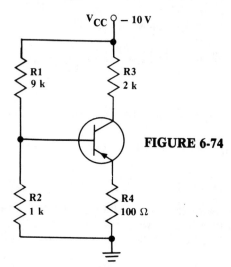

FIGURE 6-74

4. Given: $\beta = 100$, $V_{BE} = 0.6$ V, and $V_{CE} = 10.4$ V
   Solve for R3.

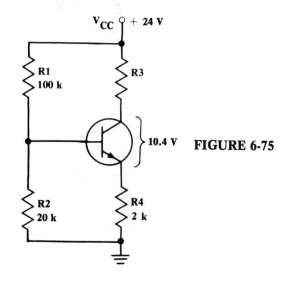

FIGURE 6-75

5. Given: $\beta = 100$ and $V_{BE} = 0.6$ V
   Solve for I_E.

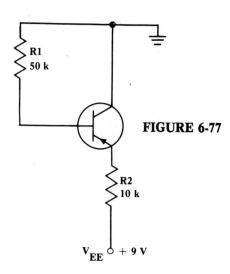

FIGURE 6-77

6. Given: $\beta = 100$ and $V_{BE} = 0.6$ V
   Solve for V_{R3}.

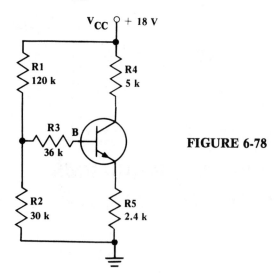

FIGURE 6-78

7 | INTRODUCTION TO ALTERNATING CURRENT AMPLIFIER CIRCUIT ANALYSIS
Using Base-Biased Amplifier Circuits

GENERAL DISCUSSION

The transistor can be connected in three useful alternating current configurations (common emitter, common collector, and common base) to provide amplification, current gain, impedance transfer, or power gain. Quality voltage amplification occurs when an input voltage is enlarged without changing the shape of the original wave. For instance, if 10 Vp-p is monitored at the output of the amplifier, and the input voltage is 200 mVp-p, then the voltage gain (amplification) is 50. Current gain is the ability to control a larger output current with a smaller input current and, since transistors are current gain devices, the current flow in the base lead is always much smaller than the current flow in the emitter and collector leads.

Where the input signal is applied and where the output signal is taken off determines whether the circuit configuration is common emitter, common collector, or common base. For instance, for the common emitter configuration, the input signal is applied to the base lead, the output signal is taken off the collector lead, and the emitter lead is common. For the common collector configuration, the input signal is applied to the base lead, the output is taken off the emitter lead, and the collector lead is common. For the common base configurations, the input is applied to the emitter lead, the output is taken off the collector lead, and the base lead is common. In other words, the lead not being used for either the input or the output is common and identifies the configuration.

The common emitter, common collector, and common base configurations using the base-biased transistor circuit are shown in Figure 7-1. Notice that all three configurations are DC biased similarly, and the emitter, collector, and base bypass capacitors provide the respective AC grounds that aid in identifying the common terminal of the three terminal transistor device. The main characteristics of each of these circuit configurations is that the common emitter circuit provides both voltage and current gain, the common collector circuit provides current gain only, and the common base circuit provides voltage gain only.

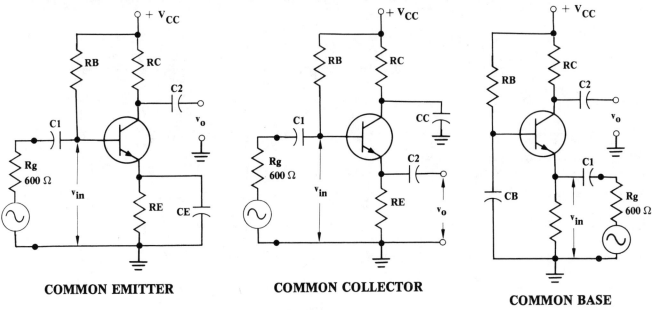

COMMON EMITTER   **COMMON COLLECTOR**   **COMMON BASE**

FIGURE 7-1

173

The main reason for using the base-biased circuit to introduce alternating current parameters is because it is the simplest of all configurations and the mathematical complexity can be kept at a minimum in solving the AC parameters. The base-biased common emitter circuits using both NPN and PNP devices are shown in Figure 7-2. Notice that they are identical except for the polarity differences. In this chapter, the common emitter base-biased circuit will be connected in both NPN and PNP configurations, while only the NPN device will be used for the common collector and common base configurations.

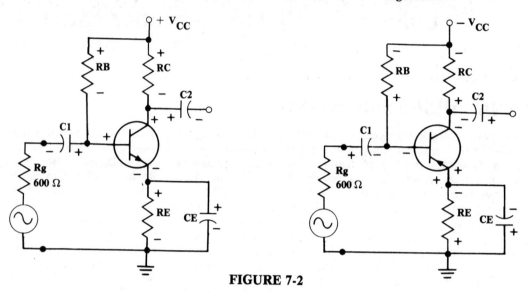

FIGURE 7-2

NOTE: With reference to Figure 7-2, the polarity of the capacitors must be opposite in the NPN and PNP circuits. Recall that the straight line on the capacitor symbol denotes the positive polarity and the curved line denotes the negative polarity.

SECTION I: THE COMMON-EMITTER, NPN, BASE-BIASED CIRCUIT

The analysis of circuits usually begins with the DC analysis and then proceeds to the AC analysis because of the interelating dependence of the AC parameters on the DC conditions of the circuit. The base-biased circuit to be analyzed is shown in Figure 7-3.

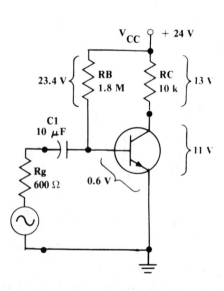

Given: $\beta = 100$

$$I_C \approx \frac{V_{CC} - V_{BE}}{R_B/\beta} = \frac{24\text{ V} - 0.6\text{ V}}{1.8\text{ M}\Omega/100} = \frac{23.4\text{ V}}{18\text{ k}\Omega} = 1.3\text{ mA}$$

$$V_{RC} = I_C R_C = 1.3\text{ mA} \times 10\text{ k}\Omega = 13\text{ V}$$

$$V_C = V_{CC} - V_{RC} = 24\text{ V} - 13\text{ V} = 11\text{ V}$$

$$V_{CE} = V_C - V_E = 11\text{ V} - 0\text{ V} = 11\text{ V}$$

$$I_B = I_C/\beta = 1.3\text{ mA}/100 = 13\text{ }\mu\text{A}$$

$$V_{RB} = I_B R_B = 13\text{ }\mu\text{A} \times 1.8\text{ M}\Omega = 23.4\text{ V, or}$$

$$V_{RB} = V_{CC} - V_B = 24\text{ V} - 0.6\text{ V} = 23.4\text{ V}$$

where: $V_B = V_{BE} = 0.6\text{ V}$

FIGURE 7-3

DIRECT CURRENT CIRCUIT ANALYSIS OF THE BASE-BIASED CIRCUIT

The direct current circuit analysis begins by solving for the collector current using the formula developed in Chapter 6. Once the collector current is calculated, the voltage drops around the base-biased circuit are easily derived, as shown with the equations of Figure 7-3.

ALTERNATING CURRENT CIRCUIT ANALYSIS OF THE BASE-BIASED CIRCUIT

Once the collector current I_C and the DC voltage drops around the circuit are known, then the AC parameters can be found. For the basic base-biased circuit, the voltage gain will be high and the input impedance moderately low, and both of these AC parameter conditions are directly related to beta, the device current gain, and the base-emitter-diode resistance r_e.

BASE EMITTER DIODE RESISTANCE r_e

The base-emitter-diode resistance is solved from $r_e \approx 26 \text{ mV}/I_E$ which is similar to discrete diode analysis where $r_j \approx 26 \text{ mV}/I_F$. Hence, if $I_E = 1$ mA, then $r_e = 26 \text{ mV}/1 \text{ mA} = 26 \, \Omega$, and if $I_E = 2$ mA then $r_e \approx 26 \text{ mV}/2 \text{ mA} = 13 \, \Omega$. For the circuit of Figure 7-3, which has an I_E of 1.3 mA, the base-emitter-diode resistance is solved from:

$$r_e \approx 26 \text{ mV}/I_E = 26 \text{ mV}/1.3 \text{ mA} = 20 \, \Omega$$

NOTE: The theoretical AC resistance of r_e is solved from $r_e \approx 26 \text{ mV}/I_E$. However, in practice, r_e can vary over a two to one range, from device to device, and with temperature. In this chapter, the effect of the two to one r_e change will be analyzed, and then minimized, through the circuit technique of swamping.

VOLTAGE GAIN A_v

The voltage gain of amplifier circuits is determined by the ratio of the output signal voltage to the input signal voltage or $A_v = v_o/v_{in}$. Therefore, if the input signal voltage to a basic single stage circuit is 20 mVp-p and the output signal voltage is 10 Vp-p, then the voltage gain of the stage is 500 solved from:

$$A_v = v_o/v_{in} = 10 \text{ Vp-p}/20 \text{ mVp-p} = 10 \text{ Vp-p}/(20 \times 10^{-3} \text{ Vp-p}) = 500$$

Stated differently, if the input signal applied to the base of the common emitter stage is 20 mVp-p, then the output signal voltage at the collector with respect to ground is 10 Vp-p, when $A_v = 500$. The input and output peak-to-peak voltage swing of a basic single stage circuit is illustrated in Figure 7-4.

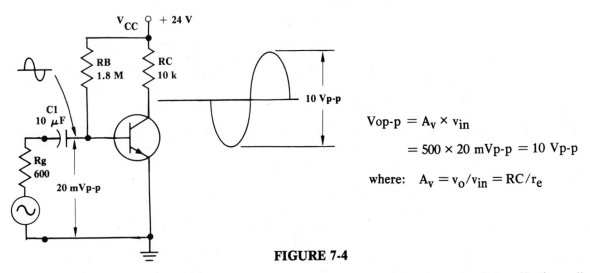

FIGURE 7-4

The voltage gain of the circuit of Figure 7-4 can also be solved from the ratio of the effective collector resistance to the effective emitter resistance. For example, if the effective collector resistance is 10 kΩ and the effective emitter resistance is 20 Ω, then:

$$A_v = RC/r_e = 10 \text{ k}\Omega/20 \text{ }\Omega = 500$$

The reason $A_v = RC/r_e$ is because approximately the same signal current value flows in the emitter and collector leads. Therefore:

$$A_v = V_o/V_{in} = V_{RC}/V_{r_e} = i_c RC/I_E r_e \approx RC/r_e \quad \text{where: } i_c \approx I_E$$

For instance, 20 mVp-p of input signal at the base is developed across the r_e of 20 Ω and $i_e = 1$ mAp-p from:

$$i_e = V_{in}/r_e = V_{r_e}/r_e = 20 \text{ mVp-p}/20 \text{ }\Omega = 1 \text{ mAp-p}$$

Therefore, solving for A_v using the $i_e \approx I_C = 1$ mAp-p value:

$$A_v = V_o/V_{in} = 10 \text{ Vp-p}/20 \text{ mVp-p} \approx \frac{1 \text{ mAp-p} \times 10 \text{ k}\Omega}{1 \text{ mAp-p} \times 20 \text{ }\Omega} = \frac{10 \text{ k}\Omega}{20 \text{ }\Omega} = 500$$

INPUT IMPEDANCE Zin

The input impedance of the circuit is the impedance that the signal generator sees at the base of the single stage amplifier, with respect to ground. The input impedance to the circuit of Figure 7-4 is calculated from:

$$Z_{in} = RB \mathbin{/\mkern-5mu/} \beta r_e = 1.8 \text{ M}\Omega \mathbin{/\mkern-5mu/} 100 \times 20 \text{ }\Omega = 1.8 \text{ M}\Omega \mathbin{/\mkern-5mu/} 2 \text{ k}\Omega \approx 1998 \text{ }\Omega$$

In this equation, the base-emitter-diode resistance of 20 Ω is shown reflected into the base by the beta of the device, which is then seen in parallel with the RB of 1.8 MΩ.

The effect of the base-emitter-diode resistance being reflected into the base by the beta of the transistor can be solved by knowing the current gain capabilities of the transistor and Ohm's law. Since, $r_e = 20$ Ω and $v_{in} = 20$ mVp-p, then the signal current in the emitter is 1 mAp-p. Since the beta of the transistor is about 100, then the signal current at the base is 10 μAp-p from:

$$i_b = i_c/(\beta + 1) \approx i_e/\beta = 1 \text{ mAp-p}/100 = 10 \text{ }\mu\text{Ap-p}$$

Therefore, the input signal voltage of 20 mVp-p produces 10 μAp-p of base signal current, which is exactly the same amount of signal current produced by applying 20 mVp-p to a 2 kΩ load. So, the effective emitter resistance reflected into the base (Zin[base]) is solved from:

$$i_e = v_{r_e}/r_e = 20 \text{ mVp-p}/20 \text{ }\Omega = 1 \text{ mAp-p} \quad \text{where: } v_{in} = v_{r_e} = v_e$$

$$i_b = i_e/(\beta + 1) = 1 \text{ mAp-p}/(100 + 1) \approx 1 \text{ mAp-p}/100 = 10 \text{ }\mu\text{Ap-p}$$

$$Z_{in}(base) = v_{in}/i_b = 20 \text{ mVp-p}/10 \text{ }\mu\text{Ap-p} = 2 \text{ k}\Omega$$

Stated differently, because $v_{in} = v_{r_e} = 20$ mVp-p and the base current is 100 times less than the emitter current, the base-emitter-diode resistance reflected into the base must be 100 times large than r_e through simple Ohm's law application. Therefore, Zin(base) can also be calculated from:

$$Z_{in}(base) = (\beta + 1)r_e \approx \beta r_e = 100 \times 20 \text{ }\Omega = 2 \text{ k}\Omega$$

The equivalent circuit for the input impedance of the stage is shown in Figure 7-5, where both Zin(base) and the base resistor RB are seen going to "AC" ground. Also, although r_e is reflected into the base in "series", the signal generator connected to the junction of RB and zin(base) "sees" them in parallel.

Notice, also, that the 1.8 MΩ base resistor connected to the positive terminal of the DC power supply is considered an AC ground in the equivalent circuit. The reason is that the internal resistance of most regulated supplies is less than an ohm to maintain a constant output voltage under load or no-load conditions. Therefore, less than an ohm separates the positive and negative terminals of the supply and either terminal is considered at "AC" ground. Consequently, the signal generator "looking" into the input of the stage "sees" RB and βr_e in parallel, and connected to "AC" ground, as shown in Figure 7-5.

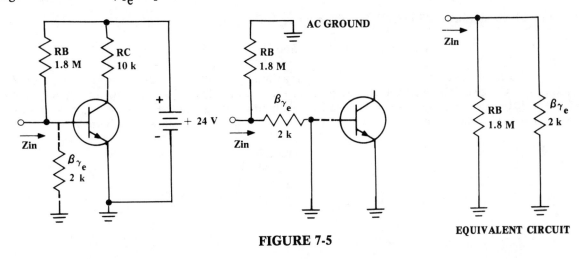

FIGURE 7-5

NOTE: Practical equivalent circuits will be discussed in detail in Section VII of this chapter.

MAXIMUM PEAK-TO-PEAK VOLTAGE SWING

Maximum peak-to-peak output voltage swing is the ability of the amplifier to produce the largest possible output voltage swing without distortion being caused — either by clipping or alpha crowding. The maximum peak-to-peak output voltage swing for the basic base-biased circuit is determined by either $2V_{CE}$ or $2I_C R_E$, whichever, is smaller.

For instance, if $V_{CC} = 24$ V and RC = 6 kΩ, as shown in Figure 7-6, then $I_C(MAX) = 4$ mA from:

$$I_C(MAX) = V_{CC}/RC = 24 \text{ V}/6 \text{ k}\Omega = 4 \text{ mA}$$

Therefore, V_{CC} and $I_C(MAX)$ determine the extreme conditions of the load line and the operating Q points can be located anywhere on the load line, simply by varying I_C and the corresponding V_{CE} voltages. In the following examples, the maximum peak-to-peak voltages will be solved for Q point V_{CE} conditions of 12 volts, 18 volts, and 6 volts, as shown in Figure 7-6.

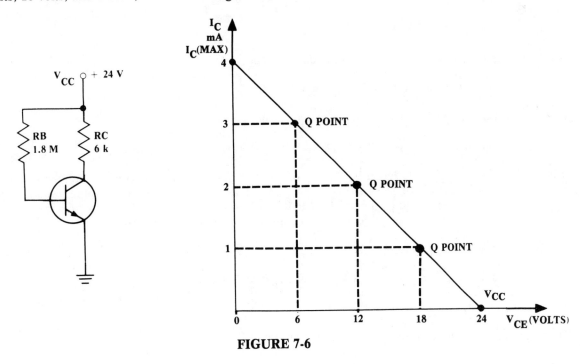

FIGURE 7-6

177

A. $I_C = 2$ mA and $V_{CE} = 12$ V

When I_C is established at 2 mA, V_{RC} will equal 12 V from:

$$V_{RC} = I_C R_C = 2 \text{ mA} \times 6 \text{ k}\Omega = 12 \text{ V} \quad \text{and } V_{CE} = 12 \text{ V}.$$

Then, if alpha crowding is temporarily ignored, the maximum peak-to-peak voltage swing will be 24 Vp-p, because the AC signal voltage at the output rides on the collector DC voltage (Q operating point) of 12 V and the positive excursion can go from 12 V to 24 V without clipping and the negative voltage swing can go from 12 V to 0 V without clipping. The ideal Vop-p(max) of 24 Vp-p is shown in Figure 7-7.

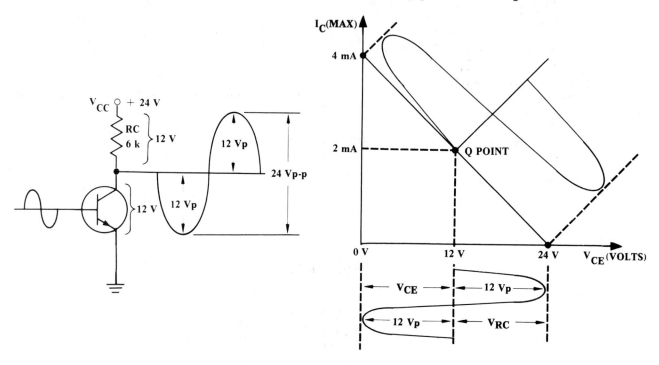

FIGURE 7-7

B. $I_C = 1$ mA and $V_{CE} = 18$ V

When I_C is lowered to 1 mA, V_{RC} will equal 6 V from:

$$I_C R_C = 1 \text{ mA} \times 6 \text{ k}\Omega = 6 \text{ V} \quad \text{and } V_{CE} = 18 \text{ V}.$$

Then, the maximum peak-to-peak voltage swing will be:

$$\text{Vop-p(max)} = 2I_C R_C = 2 \times 1 \text{ mA} \times 6 \text{ k}\Omega = 12 \text{ Vp-p}$$

It is 12 Vp-p because the AC signal voltage at the output rides on the collector DC voltage Q operating point of 18 V and the positive excursion can only go from 18 V to 24 V without clipping, even though the negative-going excursion could go the full 18 V to 0 V. Therefore, $2I_C R_C = 2 \times 1$ mA $\times 6$ k$\Omega = 12$ Vp-p is the maximum output voltage possible, without clipping, as shown in Figure 7-8.

C. $I_C = 3$ mA and $V_{CC} = 6$ V

When I_C is increased to 3 mA, then V_{RC} will equal 18 V from:

$$V_{RC} = I_C R_C = 3 \text{ mA} \times 6 \text{ k}\Omega = 18 \text{ V} \quad \text{and } V_{CE} = 6 \text{ V}.$$

Then the maximum peak-to-peak output voltage swing will be:

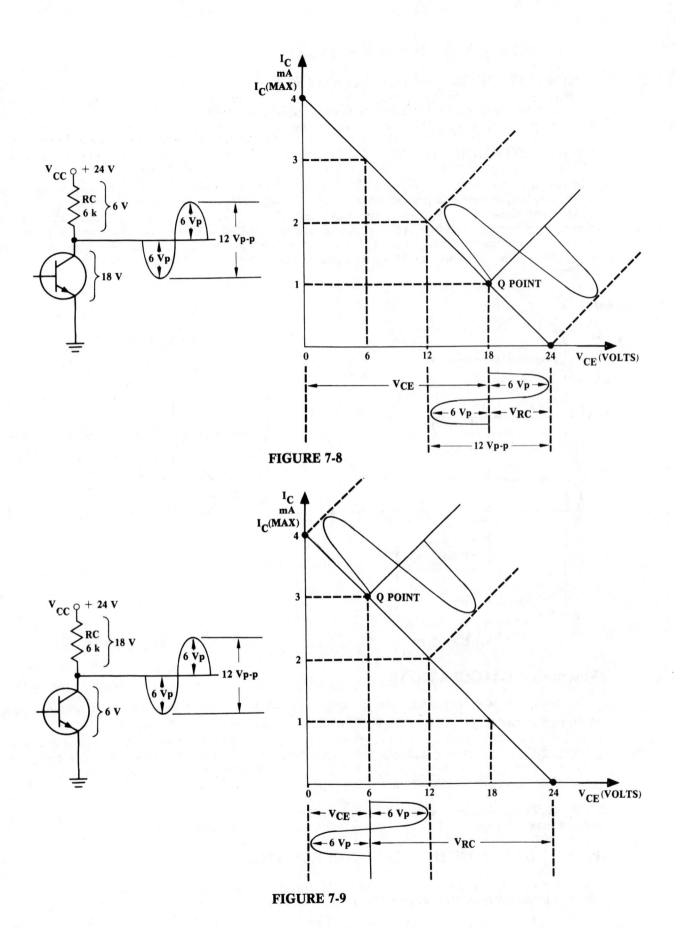

FIGURE 7-8

FIGURE 7-9

$$V_{op\text{-}p}(max) = 2V_{CE} = 2 \times 6\ V = 12\ V_{p\text{-}p}$$

It will be 12 Vp-p because the signal voltage riding on the collector DC voltage Q operating point of 6 V DC can only have a negative-going excursion of 6 V, which in turn limits the positive-going excursion to 6 volts, even though the positive-going excursion can swing about 18 V. Therefore, the $2V_{CE} = 12$ Vp-p is the maximum ideal output voltage swing, without clipping, as shown in Figure 7-9.

ALPHA CROWDING

Ideally, the Vop-p(max) for a V_{CE} of 12 V (Figure 7-7) is 24 Vp-p but, in fact, it will only swing about 23 Vp-p, or less, because of alpha crowding. Essentially, alpha crowding occurs in the region approaching $I_C(SAT)$ and causes a relatively clean sine wave to be rounded off. The "rounding" occurs because the current gain begins to decrease as the value of I_C approaches $I_C(SAT)$, where the characteristic curves are no longer in the linear region. $V_{CE}(SAT)$, which rarely goes below 0.2 V DC, also limits the maximum Vop-p excursions. The Vop-p of 23 Vp-p, shown in Figure 7-10, is limited primarily by alpha crowding distortion.

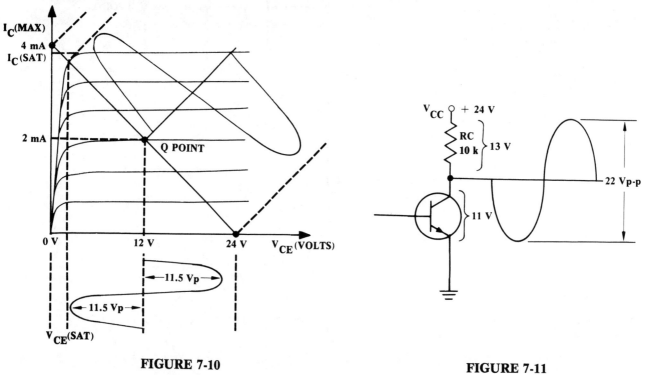

FIGURE 7-10   **FIGURE 7-11**

Vop-p(max) CALCULATIONS

The peak-to-peak output voltage swing, ignoring $V_{CE}(SAT)$, is determined by the smaller of $2V_{CE}$ or $2I_CR_C$. Applying the $2V_{CE}$ or $2I_CR_C$ formula to the circuit of Figure 7-3 provides:

$$2V_{CE} = 2 \times 11\ V = 22\ V_{p\text{-}p}$$

$$2I_CR_C = 2 \times 1.3\ mA \times 10\ k\Omega = 26\ V_{p\text{-}p}$$

Therefore, 22 Vp-p is the smaller and establishes the ideal maximum peak-to-peak output voltage condition, as shown in Figure 7-11. This ideal condition ignores $V_{CE}(SAT)$ and alpha crowding.

POWER GAIN TO THE COLLECTOR RESISTOR

Power gain is the ratio of the power out of the circuit in relation to the power into the circuit or, in equation form, PG = Po/Pin. Power gain can be written also as A_p, similar to the A_v for voltage gain and the A_i for current gain. For the basic base-biased circuit under discussion, $Po = Vo^2/RC$ and $Pin = Vin^2/Rin$. Then, taking the ratio, transposing, and substituting A_v^2 for $(Vo/Vin)^2$, where $A_v = Vo/Vin$, and Zin for Rin, provides:

$$PG = A_p = \frac{Po}{Pin} = \frac{Vo^2/RC}{Vin^2/Rin} = \frac{Vo^2 \times Rin}{Vin^2 \times RC} = A_v^2 \times \frac{Rin}{RC} = A_v^2 \times \frac{Zin}{RC} \quad \text{where:} \quad Rin = Zin$$

Therefore, the power gain of the basic base-biased circuit is solved from:

$$PG = \frac{A_v^2 \times Zin}{RC} \approx \frac{500^2 \times 1998\,\Omega}{10\,k\Omega} = 49,950$$

NOTE: Power gain is also found in the literature as A_p.

Also, the power gain converted to dB is solved from:

$$PG(dB) = 10 \log 49{,}950 = 46.98\ dB$$

Another method to solve for power gain from $PG = Po/Pin$, if both the RMS power out and the RMS power in for the circuit is calculated, is to take the ratio. For instance, if the input signal is 10 mV RMS, then the output voltage is 5 V RMS, when $A_v = 500$. A PG versus dB table is given in the appendix.

$$Po = \frac{V\ RMS^2}{RC} = \frac{5\ V^2}{10\ k\Omega} = \frac{25\ V}{10\ k\Omega} = 2.5 \times 10^{-4}\ W = 2.5\ mW$$

$$Pin = \frac{V\ RMS^2}{Zin} = \frac{10\ mV^2}{1998\ \Omega} = \frac{100 \times 10^{-6}}{1.998\ \Omega \times 10^3} = 50.05 \times 10^{-9}\ W = 50.05\ nW$$

$$PG = Po/Pin = \frac{2.5 \times 10^{-3}}{5.005 \times 10^{-8}} = \frac{2.5 \times 10^{-3} \times 10^8}{5.005} = \frac{250{,}000}{5.005} = 49{,}950$$

CURRENT GAIN

The current gain of the circuit is solved by dividing the power gain by the voltage gain. Therefore:

$$A_i = PG/A_v = 49{,}950/500 = 99.9$$

NOTE: If the effect of the 1.8 MΩ RB resistor is not included in the calculations, then Zin would equal 2 kΩ, PG would equal 50,000, and A_i would equal 100.

THE 180° PHASE SHIFT OF THE COMMON EMITTER AMPLIFIER

In the common collector circuit, the positive-going excursion of the input signal at the base is processed to the output emitter in phase with the input signal. Likewise, for the common base circuit, a positive-going excursion of the input signal at the emitter is processed to the output collector in phase with the input signal. However, in the common emitter circuit, a positive-going excursion of the input signal at the base will be negative-going at the collector output, or it will be 180° out of phase with the input signal.

With reference to the common emitter circuit, a positive-going signal voltage at the input base causes the base current to rise which, in turn, causes the I_C current to rise. Then, since $I_C = \beta \times I_B$, as the I_C current rises, the voltage drop across the collector resistor V_{RC} rises, and the collector voltage V_C falls proportionately, with respect to ground. Therefore, a positive-going signal at the input base of the common emitter stage will produce a negative-going signal at the collector output, while a negative-going signal at the base input will produce a positive-going signal at the collector output. The signal process between input and output for the common collector, common base, and common emitter circuits is shown in Figure 7-12, where the signal voltages are referenced with respect to ground.

THE OUTPUT IMPEDANCE Zo

The output impedance of the common emitter stage circuit is equal, effectively, to the collector resistor

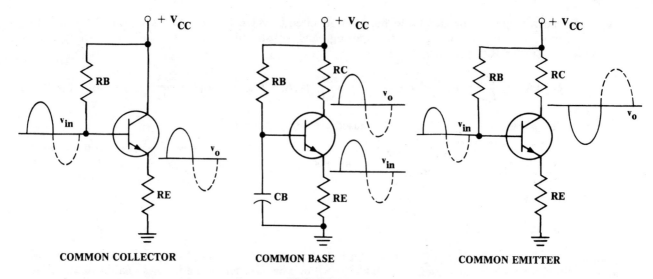

FIGURE 7-12

RC and for the circuit of Figure 7-13, Zo ≈ RC = 10 kΩ. Essentially, the output impedance is the impedance looking back into the collector of the device. Therefore, RC is seen in parallel with the collector-base-junction resistance of the device which, for most low power devices is greater than 1 MΩ and too large to be significant in the calculations. Therefore, Zo ≈ RC = 10 kΩ.

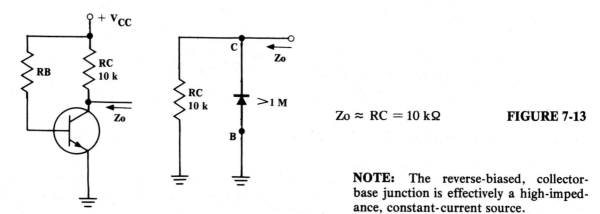

Zo ≈ RC = 10 kΩ   **FIGURE 7-13**

NOTE: The reverse-biased, collector-base junction is effectively a high-impedance, constant-current source.

ADDITION OF EMITTER RESISTOR

The addition of an emitter resistor to the basic base-biased circuit will increase both the DC and AC stability of the circuit. For instance, the base-emitter-diode resistance r_e is unpredictable and can vary over a 2 to 1 range. However, the effect of r_e change can be minimized by having some unbypassed emitter resistance in series with r_e, as shown in Figure 17-14.

THE DIRECT CURRENT CIRCUIT ANALYSIS

Adding the 1 kΩ emitter resistor to the basic base-biased circuit requires that resistance value of RB be lowered to 1.7 MΩ to maintain the same I_E of 1.3 mA. Then, as with the basic circuit, once I_C is solved by using the formula developed in Chapter 6, the voltage drops around the circuit can be solved. Too, all voltage drops between V_{CC} and ground must equal the applied V_{CC} as shown in the calculations with Figure 7-14.

$$I_C \approx \frac{V_{CC} - V_{BE}}{RB/\beta + RE} = \frac{24\,V - 0.6\,V}{(1.7\,M\Omega/100) + 1\,k\Omega} = \frac{23.4\,V}{17\,k\Omega + 1\,k\Omega} = 1.3\,mA$$

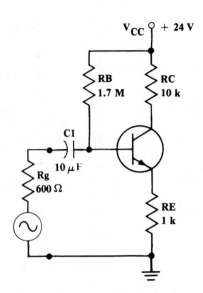
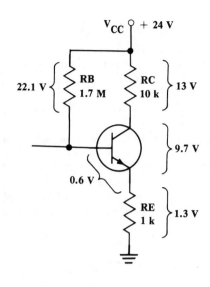

FIGURE 7-14

$V_{RC} = I_C R_C = 1.3 \text{ mA} \times 10 \text{ k}\Omega = 13 \text{ V}$

$V_{RE} = I_E R_E \approx 1.3 \text{ mA} \times 1 \text{ k}\Omega = 1.3 \text{ V}$

$V_C = V_{CC} - V_{RC} = 24 \text{ V} - 13 \text{ V} = 11 \text{ V}$

$V_B = V_E + V_{BE} = 1.3 \text{ V} + 0.6 \text{ V} = 1.9 \text{ V}$

$V_E = V_{RE} = 1.3 \text{ V}$

$V_{RB} = V_{CC} - V_B = 24 \text{ V} - 1.9 \text{ V} = 22.1 \text{ V}$

$I_B = V_{RB}/RB = 22.1 \text{ V}/1.7 \text{ M}\Omega = 13 \text{ }\mu\text{A}$ or,

$I_B = I_C/\beta = 1.3 \text{ mA}/100 = 13 \text{ }\mu\text{A}$

$V_{CE} = V_C - V_E = 11 \text{ V} - 1.3 \text{ V} = 9.7 \text{ V}$

$V_{CC} = V_{RC} + V_{CE} + V_{RE}$

$24 \text{ V} = 13 \text{ V} + 9.7 \text{ V} + 1.3 \text{ V}$

$V_{CC} = V_{RB} + V_{BE} + V_{RE}$

$24 \text{ V} = 22.1 \text{ V} + 0.6 \text{ V} + 1.3 \text{ V}$

ALTERNATING CURRENT CIRCUIT ANALYSIS

The emitter resistor of the basic base-biased circuit can be fully bypassed by a large valued capacitor, be unbypassed, or have only part of the total emitter resistance bypassed. All three types of circuits are shown in Figure 7-15. The fully bypassed circuit provides high voltage gain but has a low input impedance and is susceptible to instability. The fully unbypassed circuit is stable, has high input impedance, but it has low voltage gain. The partially unbypassed circuit provides a good compromise and is the most widely used.

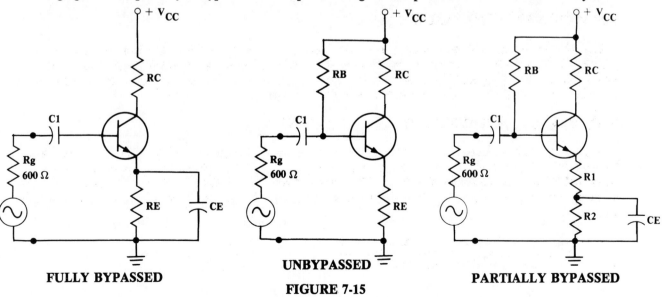

FULLY BYPASSED **UNBYPASSED** **PARTIALLY BYPASSED**

FIGURE 7-15

VOLTAGE GAIN

Voltage gain for the circuits of Figure 7-15 is determined by the ratio of the effective collector resistance to the effective emitter resistance.

1. **Fully-Bypassed, Emitter-Resistance Circuit**

Connecting a large-valued capacitor across the total emitter resistor provides a circuit that is solved much like the base-biased circuit, where only the base-emitter-diode resistance r_e is effectively left unbypassed. Therefore, the effect of the 100 µF CE capacitor is to place an effective AC ground at the emitter junction because at 1 kHz, which is considered the mid-band frequency, X_C is calculated at 1.59 Ω, as shown in Figure 7-16. Also shown, is the effect of voltage gain which can vary 2 to 1 if r_e varies over a 2 to 1 range from the theoretical $r_e = 26$ mV/I_E to a possible 52 mV/I_E condition.

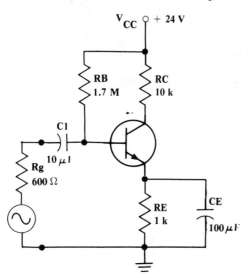

A. $A_V = RC/r_e = 10$ kΩ/20 Ω $= 500$

where: $r_e = 26$ mV/1.3 mA $= 20$ Ω

B. $A_V = RC/r_e = 10$ kΩ/40 Ω $= 250$

where: $r_e = 52$ mV/1.3 mA $= 40$ Ω

C. $X_C = \dfrac{1}{2\pi f C} = \dfrac{0.159}{1 \text{ kHz} \times 100 \text{ } \mu F}$

$= \dfrac{0.159}{10^3 \times 10^{-4}} = 1.59$ Ω

FIGURE 7-16

2. **The Unbypassed, Emitter-Resistance Circuit**

When capacitor CE is removed from the circuit, the resistance of r_e is then seen in series with the 1 kΩ emitter resistor. Therefore, if r_e again varies from 20 Ω to 40 Ω, the inclusion of the series 1 kΩ resistor RE will minimize the effect of r_e change on voltage gain. The voltage gain for an r_e of 20 Ω and then 40 Ω is shown in the following calculations.

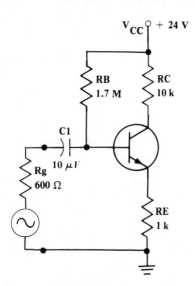

A. $A_V = \dfrac{RC}{r_e + RE} = \dfrac{10 \text{ k}\Omega}{20 \text{ }\Omega + 1 \text{ k}\Omega} = 9.8$

where: $r_e = 20$ Ω

B. $A_V = \dfrac{RC}{r_e + RE} = \dfrac{10 \text{ k}\Omega}{40 \text{ }\Omega + 1 \text{ k}\Omega} = 9.62$

where: $r_e = 40$ Ω

NOTE: If r_e is not considered in the calculations, then $A_V \approx RC/RE = 10$ kΩ/1 kΩ $= 10$

FIGURE 7-17

3. The Partially-Bypassed, Emitter-Resistance Circuit

The partially unbypassed emitter resistor is used in most circuits to attain voltage gain, where the unbypassed portion of the emitter resistance is normally selected to be at least 10 times larger than the theoretical $r_e = 26\text{ mV}/I_E$ resistance. Therefore, since the theoretical $r_e = 26\text{ mV}/I_E = 26\text{ mV}/1.3\text{ mA} = 20\text{ }\Omega$, then R1 should be a minimum 200 Ω as shown in Figure 7-18. Also, note that R2 equals 800 Ω, which makes the series total to R1 and R2 equal to the previous RE of 1 kΩ (200 Ω + 800 Ω = 1 kΩ), so that the DC voltage distribution will remain the same. The voltage gain for an r_e of 20 Ω and of 40 Ω is shown in the calculations included with Figure 7-18, where the difference in voltage gains when r_e goes from 20 Ω to 40 Ω is still within the 10% tolerance of the circuit resistors.

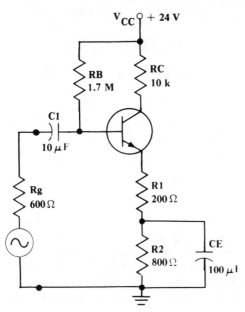

A. $A_v = \dfrac{RC}{r_e + R1} = \dfrac{10\text{ k}\Omega}{20\text{ }\Omega + 200\text{ }\Omega} = 45.45$

where: $r_e = 20\text{ }\Omega$

B. $A_v = \dfrac{RC}{r_e + R1} = \dfrac{10\text{ k}\Omega}{40\text{ }\Omega + 200\text{ }\Omega} = 41.67$

where: $r_e = 40\text{ }\Omega$

NOTE: R1 is usually selected to be $\geq 10 r_e$.

FIGURE 7-18

INPUT IMPEDANCE

The input impedance of the fully bypassed, partially bypassed, and unbypassed emitter resistor circuits is almost directly dependent on the beta of the device and on the unbypassed emitter resistance.

1. Fully-Bypassed, Emitter-Resistance Circuit

The input impedance of the fully bypassed emitter resistance is solved in the same manner as the basic base-biased circuit because the 100 μF capacitor has a capacitive reactance of about 1.59 Ω which effectively places an AC ground at the emitter terminal. Therefore, as shown in Figure 7-19, the Zin for the fully bypassed emitter resistance circuit is about 1998 Ω.

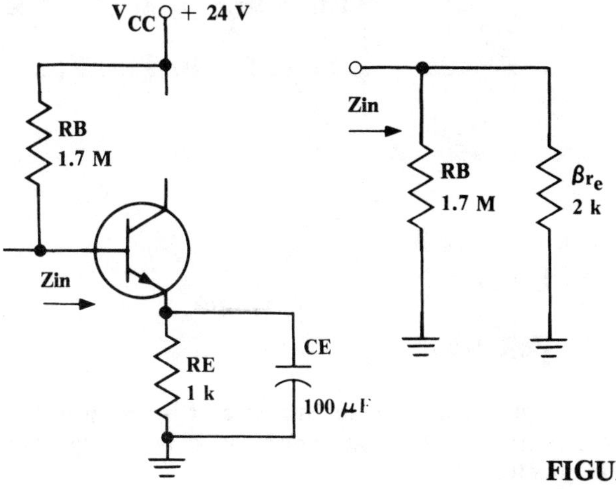

$I_E = 1.3\text{ mA}$ and $r_e = 20\text{ }\Omega$

$Zin = \beta r_e \mathbin{/\!/} RB = (100 \times 20\text{ }\Omega) \mathbin{/\!/} 1.7\text{ M}\Omega$

$= 2\text{ k}\Omega \mathbin{/\!/} 1.7\text{ M}\Omega \approx 1998\text{ }\Omega$

NOTE: If the effect of X_{CE} is included in the calculations, Zin $\approx$ 2004 Ω at 1 kHz.

FIGURE 7-19

2. The Unbypassed, Emitter-Resistance Circuit

Unbypassing the emitter resistance allows the total emitter resistance to be in series with r_e. Therefore, the emitter resistance reflected back into the base will be large, and the base resistance RB in parallel will have an effect in determining Zin, as shown in Figure 7-20.

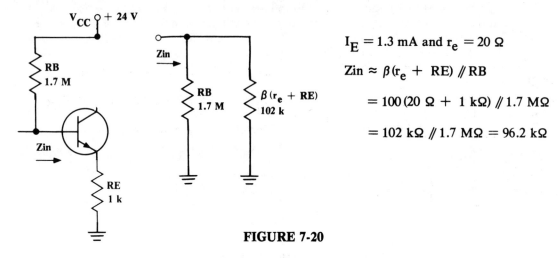

$I_E = 1.3$ mA and $r_e = 20\ \Omega$

$Zin \approx \beta(r_e + RE) \mathbin{/\mkern-5mu/} RB$

$= 100(20\ \Omega + 1\ k\Omega) \mathbin{/\mkern-5mu/} 1.7\ M\Omega$

$= 102\ k\Omega \mathbin{/\mkern-5mu/} 1.7\ M\Omega = 96.2\ k\Omega$

FIGURE 7-20

3. The Partially-Bypassed, Emitter-Resistance Circuit

Partially bypassing the emitter resistance allows the unbypassed portion of the resistance to be in series with r_e. Essentially, capacitor CE places an AC ground at the junction of R1 and R2 which causes R1 to be unbypassed and R2 to be fully bypassed. Therefore, the total emitter resistance is $r_e + R1$ which, when reflected into the base by beta, provides $\beta(r_e + R1)$. Therefore, the emitter resistance reflected into the base will be moderately high and will be only slightly effected by the RB of 1.7 MΩ, as shown in the following calculations.

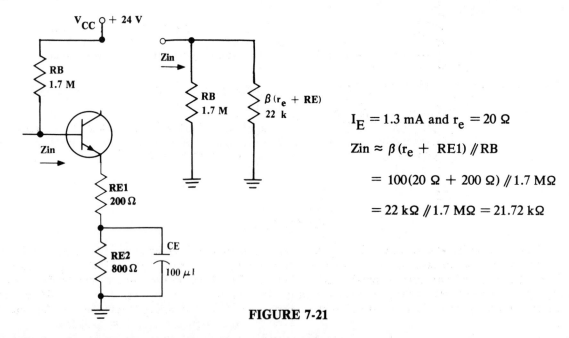

$I_E = 1.3$ mA and $r_e = 20\ \Omega$

$Zin \approx \beta(r_e + RE1) \mathbin{/\mkern-5mu/} RB$

$= 100(20\ \Omega + 200\ \Omega) \mathbin{/\mkern-5mu/} 1.7\ M\Omega$

$= 22\ k\Omega \mathbin{/\mkern-5mu/} 1.7\ M\Omega = 21.72\ k\Omega$

FIGURE 7-21

MAXIMUM PEAK-TO-PEAK VOLTAGE SWING

The maximum peak-to-peak voltage swing for the bypassed emitter resistor circuit is solved in the same manner as the basic base-biased circuit, but once a portion of the emitter resistor is left unbypassed, the maximum peak-to-peak voltage swing condition is effected.

1. **The Fully-Bypassed, Emitter-Resistance Circuit**

The maximum peak-to-peak voltage swing for the bypassed emitter resistor circuit is solved from $2V_{CE}$ or $2I_C R_C$, whichever is smaller, as shown in the calculations with Figure 7-22.

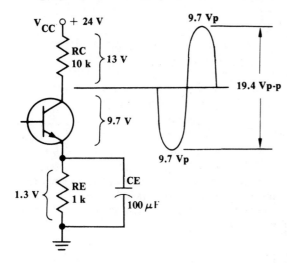

$2V_{CE} = 2 \times 9.7 \text{ Vp} = 19.40 \text{ Vp-p}$

$2I_C R_C = 2 \times 1.3 \text{ mA} \times 10 \text{ k}\Omega = 26 \text{ Vp-p}$

Therefore, Vop-p(max) = 19.4 Vp-p

FIGURE 7-22

2. **The Unbypassed, Emitter-Resistance Circuit**

The maximum peak-to-peak voltage swing for the unbypassed emitter resistor is solved from $2I_C R_C$ or $2V_{CE} \times RC/(RC + RE)$, whichever is smaller. Therefore:

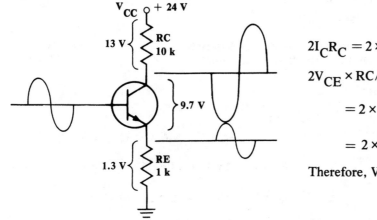

$2I_C R_C = 2 \times 1.3 \text{ mA} \times 10 \text{ k}\Omega = 26 \text{ Vp-p}$

$2V_{CE} \times RC/(RC + RE)$

$= 2 \times 9.7 \text{ V} \times 10 \text{ k}\Omega/(10 \text{ k}\Omega + 1 \text{ k}\Omega)$

$= 2 \times 9.7 \text{ V} \times 10 \text{ k}\Omega/11 \text{ k}\Omega = 17.6 \text{ Vp-p}$

Therefore, Vop-p(max) = 17.6 Vp-p

FIGURE 7-23

The reason Vop-p(max) is less than $2V_{CE}$ where RE is unbypassed is because the input signal to the base of an unbypassed emitter resistor circuit is processed across the emitter in phase with the input signal but 180° out of phase with the output signal at the collector. Then, because the voltage gain of the circuit is low, the peak-to-peak input voltage is relatively large, and the positive-going signal at the emitter, riding on the DC emitter voltage, is seen opposing the negative-going output voltage that rides on the collector voltage.

For instance, if Vop-p is calculated at about 17.6 Vp-p, the output signal riding on the DC collector voltage of 11 V DC will have a maximum positive excursion to about 19.8 V and a negative excursion to about 2.2 V. Likewise, the approximate input signal riding on the 1.3 V DC of the emitter will have a positive excursion to about 2.2 V and a negative excursion to about 0.4 V. Therefore, any further increase of Vinp-p or Vop-p will cause the signals to overlap and distortion to occur. See Figure 7-24.

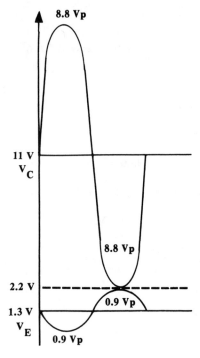

$V_{in}\text{p-p} = V_o\text{p-p}/A_v = 17.64\ V\text{p-p}/9.8 = 1.8\ V\text{p-p}$

$V_o(\text{peak}) = V_o\text{p-p}/2 = 17.64/2 = 8.8\ V(\text{peak})$

$V_{in}(\text{peak}) = V_{in}\text{p-p}/2 = 1.8\ V\text{p-p}/2 = 0.9\ V(\text{peak})$

$V_C(\text{low}) = V_C - V_o(\text{peak}) = 11\ V - 8.8\ V = 2.2\ V$

NOTE: 2.2 V is the imaginary line that intersects the opposing waves, since the input wave rides on the 1.3 V DC and the positive-going wave touches 2.2 V at the same time as the negative-going Vop-p wave. Vop-p at 17 Vp-p rides on 11 V DC.

FIGURE 7-24

Stated in other terms, any signal voltage appearing at the emitter of the common emitter stage is in opposition to Vop-p = 2 V_{CE} and subtracts from the 2V_{CE}. Therefore, if:

$2V_{CE} = 2 \times 9.7\ V = 19.4\ V\text{p-p}$ and $V_E\text{p-p} = 1.8\ V\text{p-p}$, then

$V_o\text{p-p} = 2V_{CE}\text{p-p} - V_E\text{p-p} = 19.4\ V\text{p-p} - 1.8\ V\text{p-p} = 17.6\ V\text{p-p}$

3. The Partially-Bypassed, Emitter-Resistance Circuit

The maximum peak-to-peak voltage swing for the partially bypassed emitter resistor is solved from $2I_C R_C$ or $2V_{CE} \times RC/(RC + R1)$, whichever is smaller. Therefore:

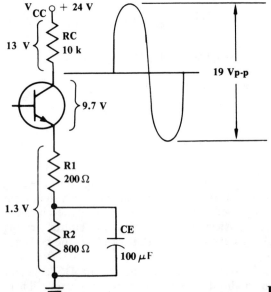

$2I_C R_C = 2 \times 1.3\ \text{mA} \times 10\ k\Omega = 26\ V\text{p-p}$

$2V_{CE} \times RC/(RC + R1)$

$= 2 \times 9.7\ V \times 10\ k\Omega/(10\ k\Omega + 200\ \Omega)$

$= 2 \times 9.7\ V \times 10\ k\Omega/10.2\ k\Omega = 19\ V\text{p-p}$

Therefore, $V_o\text{p-p(max)} = 19\ V\text{p-p}$

FIGURE 7-25

Hence, the 19 Vp-p riding on the collector voltage of $V_C = 11\ V$ will provide positive-going and negative-going excursions of 9.5 V, and it is limited by the positive-going and negative-going input signal of approximately 200 mV(peak). The calculations are shown in conjunction with Figure 7-26.

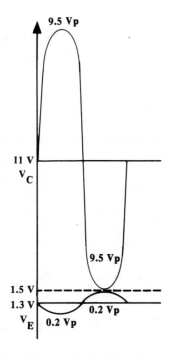

$V_{in}\text{p-p} = V_o\text{p-p}/A_v = 19 \text{ Vp-p}/45.45 = 418 \text{ mVp-p}$

$V_{in}(\text{peak}) = V_{in}\text{p-p}/2 = 418 \text{ mVp-p}/2 \approx 0.2 \text{ V(peak)}$

$V_o(\text{peak}) = V_o\text{p-p}/2 = 19 \text{ Vp-p}/2 = 9.5 \text{ V(peak)}$

$V_C(\text{low}) = V_C - V_o(\text{peak}) = 11 \text{ V} - 9.5 \text{ V(peak)} = 1.5 \text{ V}$

$V_E(\text{high}) = V_E + V_{in}(\text{peak}) = 1.3 \text{ V} + 0.2 \text{ V(peak)} = 1.5 \text{ V}$

FIGURE 7-26

Also, Vop-p(max) can be solved from:

$V_o\text{p-p}(\text{max}) = 2V_{CE}\text{p-p} - V_E\text{p-p} = 19.4 \text{ Vp-p} - 0.4 \text{ Vp-p} = 19 \text{ Vp-p}.$

Therefore, for the partially bypassed emitter resistor circuit, $2V_{CE} = 19.4$ Vp-p and $2V_{CE} \times RC/(RC + R1) = 19$ Vp-p. Hence, $2 V_{CE} = 2 V_{CE} \times RC/(RC + R1)$, or the approximate $2V_{CE}$, will be used in determining one of the two calculated possibilities of Vop-p(max) for the partially bypassed emitter resistance — $2I_CR_C$ being the other.

NOTE: A good rule to follow when deciding whether to use $2V_{CE}$ or the more involved $2V_{CE} \times RC/(RC + R1)$ formula, is to use the more involved formula only when the unbypassed emitter resistance is greater than 10% of the effective load resistance. In all other cases use $2V_{CE}$, since the circuit component tolerance, in addition to the visual measurement of Vop-p(max), can easily contribute to a 10% error in the calculated versus measured values.

SUMMARY OF PARTIALLY BYPASSED CIRCUIT

The most widely used of the three circuits analyzed is the partially bypassed circuit, because the unbypassed portion of the emitter resistance provides adequate "swamping" of the base-emitter-diode resistance r_e and, at the same time, provides moderate input impedance and voltage gain. Both A_v and Z_{in} parameters were "traded off" to improve circuit stability.

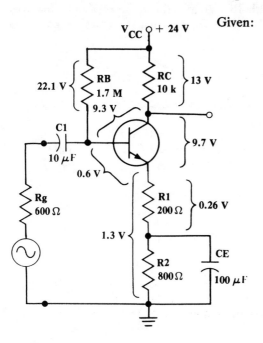

Given: $I_E \approx I_C = 1.3 \text{ mA}$ $V_{CE} = 9.7 \text{ V}$ $V_{RC} = 13 \text{ V}$

$V_{RE} = 1.3 \text{ V}$ $V_{R2} = 1.04 \text{ V}$ $V_{R1} = 0.26 \text{ V}$

$V_C = 11 \text{ V}$ $V_E = 1.3 \text{ V}$ $V_B = 1.9 \text{ V}$

$A_v = \dfrac{RC}{r_e + R1} = \dfrac{10 \text{ k}\Omega}{20 \text{ }\Omega + 200 \text{ }\Omega} = \dfrac{10 \text{ k}\Omega}{220 \text{ }\Omega} = 45.45$

$Z_{in} = RB \mathbin{/\mkern-5mu/} \beta(r_e + R1) = 1.7 \text{ M}\Omega \mathbin{/\mkern-5mu/} 100 \times 220 \text{ }\Omega \approx 21.72 \text{ k}\Omega$

$Z_o = RC = 10 \text{ k}\Omega$

$PG = A_v^2 \times Z_{in}/RC = 45.45^2 \times 21.72 \text{ k}\Omega/10 \text{ k}\Omega = 4486.7$

$V_o\text{p-p}(\text{max}) \approx 2V_{CE} = 2 \times 9.7 \text{ V} = 19.4 \text{ Vp-p}$, or

$2I_CR_C = 2 \times 1.3 \text{ mA} \times 10 \text{ k}\Omega = 26 \text{ Vp-p}$, therefore,

The appoximate Vop-p(max) is 19 Vp-p.   **FIGURE 7-27**

NOTE: Because of the slight degeneration caused by the effect of R1 on gain, the actual Vop-p = 19 Vp-p.

ADDITION OF LOAD RESISTOR

Adding a load resistor to the common emitter base-biased circuit, as shown and solved in Figure 7-28, lowers the voltage gain, power gain, and maximum peak-to-peak voltage swing, but it does not alter the DC voltage distribution or the input impedance of the circuit. The following circuit analysis summary is given and this is then followed by the more detailed analysis.

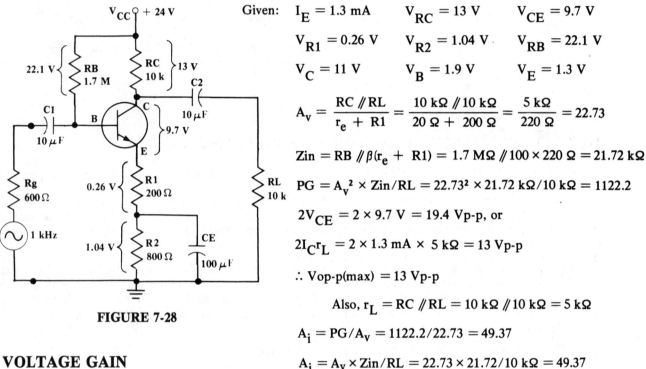

Given: $I_E = 1.3$ mA $V_{RC} = 13$ V $V_{CE} = 9.7$ V

$V_{R1} = 0.26$ V $V_{R2} = 1.04$ V $V_{RB} = 22.1$ V

$V_C = 11$ V $V_B = 1.9$ V $V_E = 1.3$ V

$$A_v = \frac{RC \parallel RL}{r_e + R1} = \frac{10\ k\Omega \parallel 10\ k\Omega}{20\ \Omega + 200\ \Omega} = \frac{5\ k\Omega}{220\ \Omega} = 22.73$$

$Z_{in} = RB \parallel \beta(r_e + R1) = 1.7\ M\Omega \parallel 100 \times 220\ \Omega = 21.72\ k\Omega$

$PG = A_v^2 \times Z_{in}/RL = 22.73^2 \times 21.72\ k\Omega/10\ k\Omega = 1122.2$

$2V_{CE} = 2 \times 9.7$ V $= 19.4$ Vp-p, or

$2I_C r_L = 2 \times 1.3$ mA $\times 5\ k\Omega = 13$ Vp-p

$\therefore$ Vop-p(max) $= 13$ Vp-p

Also, $r_L = RC \parallel RL = 10\ k\Omega \parallel 10\ k\Omega = 5\ k\Omega$

$A_i = PG/A_v = 1122.2/22.73 = 49.37$

$A_i = A_v \times Z_{in}/RL = 22.73 \times 21.72/10\ k\Omega = 49.37$

FIGURE 7-28

VOLTAGE GAIN

The voltage gain of either the unloaded or loaded circuit is solved by taking the ratio of the effective collector resistance to the effective emitter resistance. Therefore, the emitter resistance remains at $r_e + R1$ but, from the collector terminal "looking outward", RC is seen connected to AC ground and, because RL is also connected to ground, RL is in parallel with RC. Again the value of X_{C2} is low with regard to either RL or RC, at a mid-band frequency of 1 kHz (15.9 Ω), so it can be disregarded in the calculations.

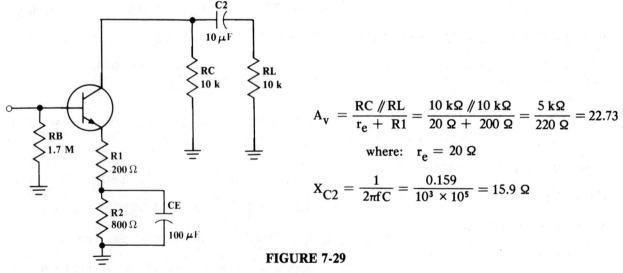

$$A_v = \frac{RC \parallel RL}{r_e + R1} = \frac{10\ k\Omega \parallel 10\ k\Omega}{20\ \Omega + 200\ \Omega} = \frac{5\ k\Omega}{220\ \Omega} = 22.73$$

where: $r_e = 20\ \Omega$

$$X_{C2} = \frac{1}{2\pi fC} = \frac{0.159}{10^3 \times 10^5} = 15.9\ \Omega$$

FIGURE 7-29

THE INPUT IMPEDANCE Zin

The input impedance is not effected by the load resistance. Therefore, Zin, for both loaded and unloaded partially bypassed circuits, is solved in the same manner at approximately 22.72 kΩ.

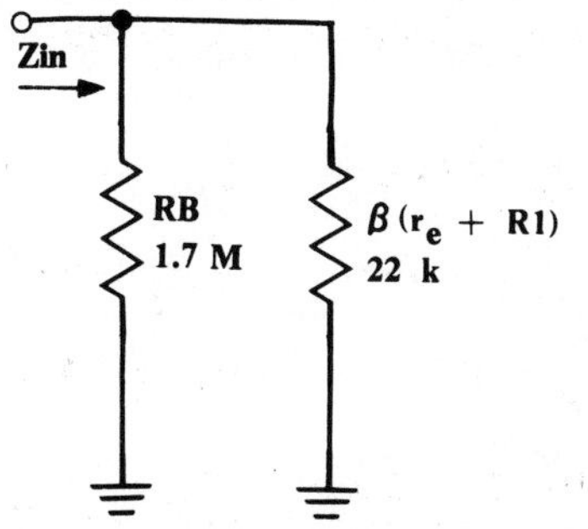

$$Zin = RB \;//\; \beta(r_e + R1)$$
$$= 1.7\,M \;//\; 100(20\,\Omega + 200\,\Omega)$$
$$= 1.7\,M\Omega \;//\; 22\,k\Omega = 21.72\,k\Omega$$

where: $r_e = 20\,\Omega$

FIGURE 7-30

OUTPUT IMPEDANCE OF LOADED COMMON EMITTER CIRCUITS

The output impedance of any common emitter circuit is approximately equal to the collector resistance only. Therefore, connecting a load resistor does not change the output impedance of the circuit, regardless of the value of RL. Hence, Zo remains equal to RC, or Zo = RC = 10 kΩ.

A good analogy to this circuit parameter situation is a signal generator which has a specified output impedance, where the generator is specified by that output impedance, regardless of the load value. The only exception is when the generator is deliberately "bridged" to make the output impedance lower. And this is done by connecting a resistor in parallel with the signal generator prior to the load being connected. The Zo equal to RC equal to 10 kΩ is shown in Figure 7-31.

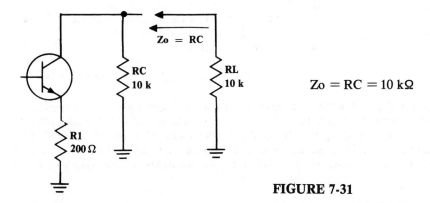

$Zo = RC = 10\,k\Omega$

FIGURE 7-31

MAXIMUM PEAK-TO-PEAK VOLTAGE SWING

The maximum peak-to-peak out voltage swing for the loaded common emitter stage is solved from approximately $2V_{CE}$ or $2I_C r_L$, whichever is smaller. Therefore:

$$2V_{CE} = 2 \times 9.7\,V = 19.4\,Vp\text{-}p$$

$$2I_C r_L = 2 \times 1.3\,mA \times (10\,k\Omega \;//\; 10\,k\Omega) = 2 \times 1.3\,mA \times 5\,k\Omega = 13\,Vp\text{-}p$$

Therefore, Vop-p(max) = 13 Vp-p. Again, $r_L = RC \;//\; RL = 10\,k\Omega \;//\; 10\,k\Omega = 5\,k\Omega$

GRAPHICAL ANALYSIS OF THE LOADED Vop-p(max)

A. Unloaded Conditions

Previously in this chapter, Figures 7-6, 7-7, 7-8, and 7-9 were used to illustrate the non-loaded, basic, base-biased circuit, where the DC load line was established by V_{CC} and $I_C(MAX)$. The Vop-p(max) was determined by where the operating point was located on the load line, with regard to either V_{CC} or $I_C(MAX)$.

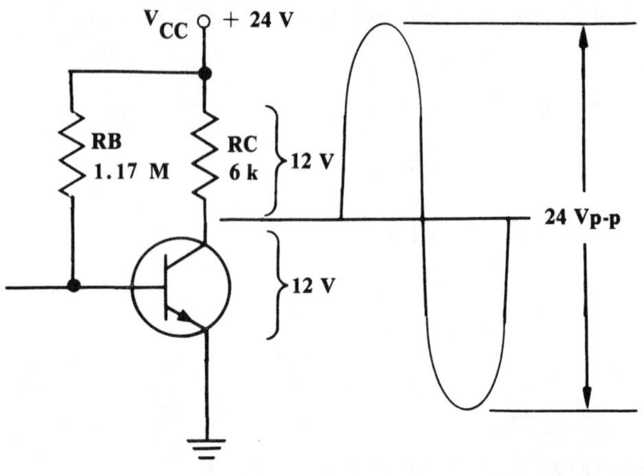

$$I_C = \frac{V_{CC} - V_{BE}}{R_B/\beta} = \frac{24\ V - 0.6\ V}{1.17\ M\Omega/100} = 2\ mA$$

$$I_C(SAT) = V_{CC}/R_C = 24\ V/6\ k\Omega = 4\ mA$$

$$V_{RC} = I_C R_C = 2\ mA \times 6\ k\Omega = 12\ V$$

$$V_{CE} = V_{RC} = 24\ V - 12\ V = 12\ V,\ \text{therefore}$$

$$V_C = V_{CE} = 12\ V$$

$$V_{CC} = V_{CE} + I_C R_C$$

$$24\ V = 12\ V + (2\ mA \times 6\ k\Omega)$$

$$24\ V = 12\ V + 12\ V$$

FIGURE 7-32

The cutoff voltage on the DC load line is solved from:

$$V_{CO} = V_{CE} + I_C R_C = 12\ V + (2\ mA \times 6\ k\Omega) = 24\ V$$

and the maximum on the DC load line is solved from:

$$I_C(MAX) = I_C + V_{CE}/R_C = 2\ mA + 12\ V/6\ k\Omega = 2\ mA + 2\ mA = 4\ mA$$

Therefore, the DC load line is established by the extreme cutoff and saturation conditions of the transistor, as shown in Figure 7-34.

B. Loaded Conditions

Loading the circuit through a coupling capacitor does not change the DC voltage distributions or the operating point established by V_{CE} and I_C. However, the load resistor RL is seen in parallel with the collector resistor RC and the AC load will provide an AC load line. Therefore, Vop-p(max) is solved from $2V_{CE}$ or $2I_C r_L$, whichever is smaller, and the graphical analysis is shown in Figure 7-33. Again, since V_{CE} and I_C remain the same, both the DC and AC load lines must go through the operating point.

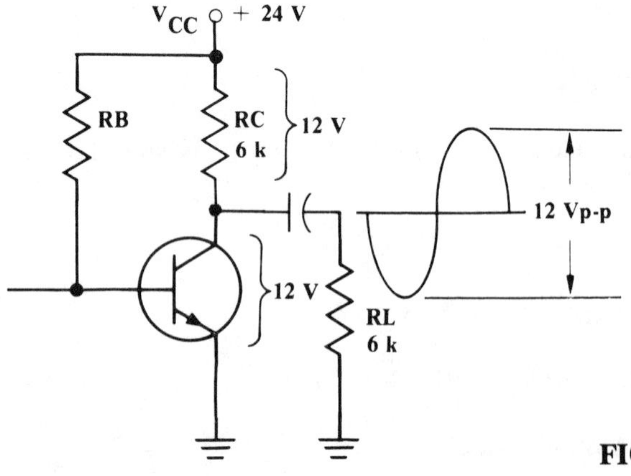

$$2V_{CE} = 2 \times 12\ V = 24\ Vp\text{-}p$$

$$2I_C r_L = 2 \times 2\ mA \times (6\ k\Omega\ //\ 6\ k\Omega)$$

$$= 2 \times 2\ mA \times 3\ k\Omega = 12\ Vp\text{-}p$$

$$V_{CO} = V_{CE} + I_C r_L$$

$$18\ V = 12\ V + (2\ mA \times 3\ k\Omega)$$

$$18\ V = 12\ V + 6\ V$$

FIGURE 7-33

The cutoff voltage on the AC load line is solved from:

$$V_{CO} = V_{CE} + I_C r_L = 12\ V + 2\ mA(6\ k\Omega\ //\ 6\ k\Omega) = 12\ V + 6\ V = 18\ V$$

and the maximum current on the AC load line is solved from:

$$I_{CO} = I_C + V_{CE}/r_L = 2\text{ mA} + 12\text{ V}/3\text{ k}\Omega = 2\text{ mA} + 4\text{ mA} = 6\text{ mA}$$

Therefore, the AC load line is established by the extreme cutoff and maximum current conditions as determined by the AC load r_L. The DC and AC load lines are shown in Figure 7-34.

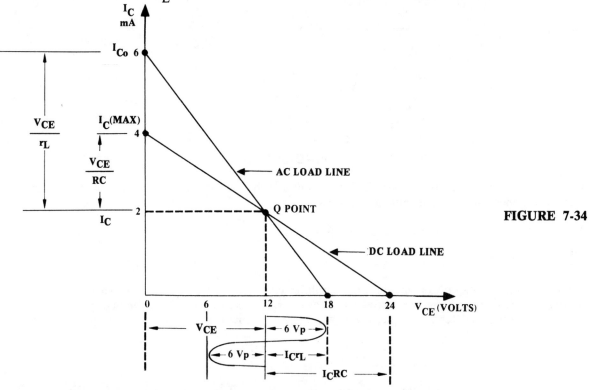

FIGURE 7-34

MEASURING THE ALTERNATING CURRENT PARAMETERS

Measuring the AC parameters of voltage gain and input impedance are accomplished by making specific measurements, and then using these measurements to "calculate" the measured voltage gain. The maximum peak-to-peak measurement, on the other hand, is a direct measurement that is visually monitored on the oscilloscope.

VOLTAGE GAIN

If the collector resistor of an unloaded base-biased circuit is 10 kΩ and the effective emitter resistance, including the r_e, is 250 ohms, then the voltage gain of the stage is solved from:

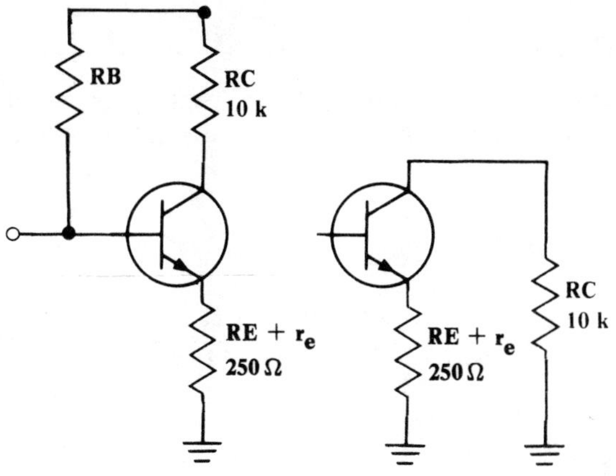

$$A_V = \frac{RC}{r_e + RE} = \frac{10\text{ k}\Omega}{25\text{ }\Omega + 225\text{ }\Omega} = \frac{10\text{ k}\Omega}{250\text{ }\Omega} = 40$$

where: $r_e = 25\text{ }\Omega$

FIGURE 7-35

193

Therefore, if 100 mVp-p of input signal is applied, the output voltage at the collector, with respect to ground, would be solved from:

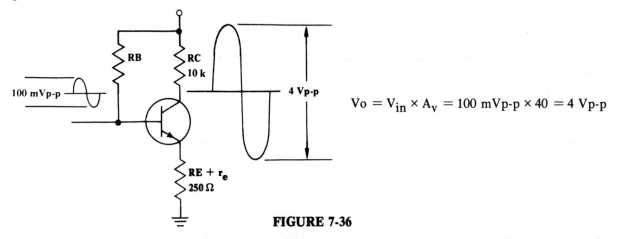

$V_o = V_{in} \times A_v = 100 \text{ mVp-p} \times 40 = 4 \text{ Vp-p}$

FIGURE 7-36

Then, if the process is reversed, where both the input and the output voltages are measured with respect to ground, the measured voltage gain can be solved from:

$A_v = V_o/V_{in} = 4 \text{ Vp-p}/100 \text{ mVp-p} = 40$

INPUT IMPEDANCE

The input impedance of any single stage amplifier can be solved by using a known series resistor and comparing the signal voltage drop across the known series resistor to the signal voltage at the input of the circuit. The series resistor technique, along with the Zin formula, is shown in Figure 7-37.

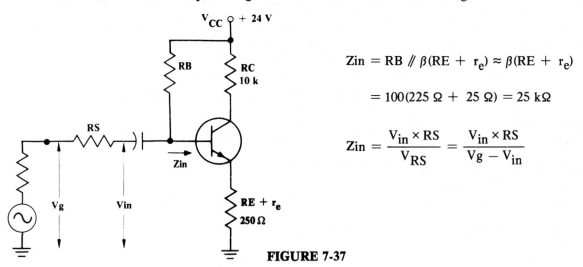

$Z_{in} = RB \mathbin{/\mkern-5mu/} \beta(RE + r_e) \approx \beta(RE + r_e)$

$= 100(225 \text{ }\Omega + 25 \text{ }\Omega) = 25 \text{ k}\Omega$

$Z_{in} = \dfrac{V_{in} \times RS}{V_{RS}} = \dfrac{V_{in} \times RS}{V_g - V_{in}}$

FIGURE 7-37

For instance, if Zin is 25 kΩ and a known 15 kΩ series resistor is used, then a generator voltage of 400 mVp-p would produce 250 mVp-p of voltage drop across Zin. Using the voltage divider equation:

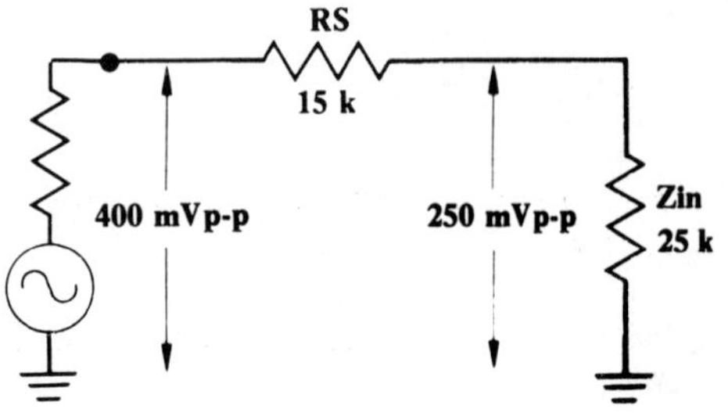

$V_{Zin} = \dfrac{V_g \times Z_{in}}{RS + Z_{in}} = \dfrac{400 \text{ mVp-p} \times 25 \text{ k}\Omega}{15 \text{ k}\Omega + 25 \text{ k}\Omega} = 250 \text{ mVp-p}$

FIGURE 7-38

If the process is reversed, where both Vg and Vin are measured, and RS is known, then V_{RS} can be calculated from:

$$V_{RS} = Vg - Vin = 400 \text{ mVp-p} - 250 \text{ mVp-p} = 150 \text{ mVp-p}$$

Then, since the same signal current flows through RS and Zin, the ratio of V_{RS} to V_{Zin} equals the ratio of RS to Zin.

$$\frac{V_{RS}}{Vin} = \frac{i \times RS}{i \times Zin} = \frac{RS}{Zin} \quad \text{or} \quad \frac{V_{RS}}{Vin} = \frac{RS}{Zin}$$

Therefore, solving for Zin by transposing, and then subtracting the measured values of Vg and Vin for V_{RS}, provides the Zin of 25 kΩ.

$$Zin = \frac{RS \times Vin}{V_{RS}} = \frac{RS \times Vin}{Vg - Vin} = \frac{15 \text{ k}\Omega \times 250 \text{ mVp-p}}{400 \text{ mVp-p} - 250 \text{ mVp-p}} = \frac{15 \text{ k}\Omega \times 250 \text{ mVp-p}}{150 \text{ mVp-p}} = 25 \text{ k}\Omega$$

MAXIMUM PEAK-TO-PEAK OUTPUT VOLTAGE SWING

The maximum peak-to-peak output voltage swing is a visual measurement, where the output signal is monitored on an oscilloscope and the Vin is increased until the sine wave begins to distort. A measurement that will show the extreme boundaries is to overdrive the input so that clipping of both the positive and the negative excursions occur. However, unless the positive-going and negative-going waves are equal (symmetrical), and alpha crowding is not a factor, the maximum peak-to-peak voltage swing should be well within these extreme boundaries.

NOTE: The visual indication of alpha is first seen on the oscilloscope as the clean sine wave begins to round off. However, only the use of a distortion analyzer and practical percentage of signal distortion can provide hard guidelines as to how much rounding off must occur before the amount of alpha crowding is too much. A rule of thumb in regard to alpha crowding is to increase the Vop-p to the calculated value and, if alpha crowding distortion looks to be too high, use a distortion analyzer, or monitor the DC voltage at the collector, where a distortion greater than 5% will cause the DC voltage to shift.

SECTION I: LABORATORY EXERCISE

OBJECTIVES
Investigate the DC and AC parameters of the base-biased, NPN, common emitter, amplifier circuit.

LIST OF MATERIALS AND EQUIPMENT
1. Transistor: 2N3904 or equivalent
2. Resistors (one each except where indicated): 330 Ω 680 Ω 1 kΩ (2) 4.7 kΩ
   6.8 kΩ 47 kΩ 100 kΩ 2.2 MΩ
3. Capacitors (10-35 V): 10 µF (two) 100 µF (one)
4. Power Supply (24 Volt)
5. Voltmeter
6. Signal Generator
7. Oscilloscope

PROCEDURE
Part 1: The Common-Emitter, NPN Circuit
1. Connect the circuit with the NPN transistor, as shown in Figure 7-39.

2. Measure the voltages at the collector, base, and emitter of the transistor, with respect to ground. Insert the measured V_C, V_E, and V_B voltages, as indicated, into Table 7-1.

3. Calculate the circuit voltage drops of V_{RC}, V_{RE}, V_{RB}, V_{CE}, and V_{BE}, and the circuit currents of I_C and I_B.

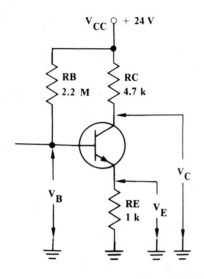

FIGURE 7-39

a. Calculate the collector-resistor voltage drop from: $V_{RC} = V_{CC} - V_C$
b. Calculate the emitter-resistor voltage drop from: $V_{RE} = V_E$
c. Calculate the base-resistor voltage drop from: $V_{RB} = V_{CC} - V_B$
d. Calculate the voltage drop across the collector-emitter of the device from: $V_{CE} = V_C - V_E$
e. Calculate the voltage drop across the base emitter of the device from: $V_{BE} = V_B - V_E$
f. Calculate the collector current from: $I_C = V_{RC}/RC$
g. Calculate the base current from: $I_B = V_{RB}/RB$
h. Calculate the current gain of the device from: $\beta = I_C/I_B$

4. Insert the calculated and measured values, as indicated, into Table 7-1.

TABLE 7-1	V_C	V_B	V_E	V_{RC}	V_{RE}	V_{RB}	V_{CE}	V_{BE}	I_C	I_B	Beta
CALCULATED											
MEASURED											

Part 2: The Common-Emitter, NPN, Base-Biased Circuit

1. Connect the common emitter circuit as shown in Figure 7-40. Notice that the collector resistor RC has been changed to 6.8 kΩ and the signal generator has been added.

2. Calculate the collector current IC, using the calculated current gain (beta) and the base-emitter-diode voltage value obtained in the previous circuit. Calculate from:

$$I_C = \frac{V_{CC} - V_{BE}}{(RB/\beta) + RE} \quad \text{where;} \quad V_{BE} \approx 0.6 \text{ V}$$

3. Calculate the DC voltage drops around the circuit. Include all resistors and transistors.

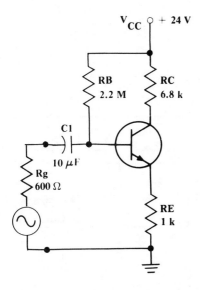

FIGURE 7-40

a. Calculate the voltage drop across the collector resistor RC from:

$$V_{RC} = I_C R_C \quad \text{where:} \quad RC = 6.8 \text{ k}\Omega$$

b. Calculate the voltage drop across the emitter resistor RE from:

$$V_{RE} = I_E R_E \approx I_C R_E \quad \text{where:} \quad RE = 1 \text{ k}\Omega$$

c. Calculate the voltage at the collector of the device from: $V_C = V_{CC} - V_{RC}$

d. Calculate the voltage drop across the device from:

$$V_{CE} = V_C - V_E \quad \text{where:} \quad V_E = V_{RE} \quad \text{or}$$
$$V_{CE} = V_{CC} - (V_{RC} + V_{RE})$$

e. Calculate the voltage drop across the base resistor from:

$$V_{RB} = V_{CC} - (V_{BE} + V_{RE}), \quad \text{where:} \quad V_B = V_{BE} + V_{RE}, \quad \text{or}$$
$$V_{RB} = (I_C/\beta)RB \quad \text{where} \quad I_B = I_C/\beta$$

4. Measure the DC voltage drops around the circuit.

   a. Measure V_{RE}, the voltage drop across the emitter resistor.

   b. Measure V_C, the voltage at the collector of the device, with respect to ground.

   c. Measure V_B, the voltage at the base of the device, with respect to ground.

NOTE: Good voltage measurement technique requires that all DC measurements are made with respect to ground. This helps to avoid ground loops that could be caused by the measuring instruments that might also be grounded. For instance, a measured value at the collector and then another at the emitter, taken with respect to ground, will provide the V_{CE} voltage drop without having to go across, physically, the collector-emitter of the device. In most instances, the measured and calculated values should be close.

   d. Use the V_C, V_B, and V_E voltage measurements to calculate, instead of measuring directly, the voltage drops of V_{RB}, V_{RC}, and V_{CE}. Calculate:

   V_{RB} from: $V_{RB} = V_{CC} - V_B$

VRC from: $V_{RC} = V_{CC} - V_C$

VCE from: $V_{CE} = V_C - V_E$

5. Insert the calculated and measured values, as indicated, into Table 7-2.

TABLE 7-2	I_C	V_{RC}	V_{RE}	V_C	V_{CE}	V_{RB}	V_B	I_B	Beta
CALCULATED									
MEASURED	▨							▨	▨

6. **VOLTAGE GAIN:**

   a. Calculate the mid-frequency voltage gain from:

   $A_v = RC/(RE + r_e)$ where: $r_e = 26 \text{ mV}/I_E$ and $I_E \approx I_C$

   b. Measure the mid-frequency voltage gain using an input signal voltage of 100 mVp-p at a frequency of 1 kHz. Measure Vo at the collector with respect to ground. Measure both the Vin and the Vo with an oscilloscope, as shown in Figure 7-41.

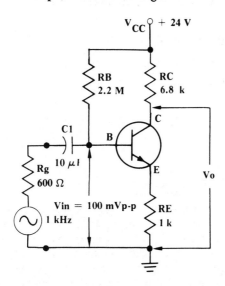

FIGURE 7-41

   c. Use the measured Vop-p to calculate the measured voltage gain from:

   $A_v = V_o/V_{in}$ where: $V_{in} = 100 \text{ mVp-p}$

NOTE: It should be obvious that if A_v is calculated and Vinp-p is known, Vop-p could be calcuated from: $V_{op-p} = A_v \times V_{in}$.

7. **INPUT IMPEDANCE:**

   a. Calculate the input impedance to the circuit from:

   $Z_{in} = RB \,//\, \beta(RE + r_e)$ where: $RB = 2.2 \text{ M}\Omega$, $RE = 1 \text{ k}\Omega$, and $r_e = 26 \text{ mV}/I_E$

NOTE: At low frequencies, the DC and AC betas are relatively close and only at frequencies much beyond 1 MHz do they begin to differ. Use a curve tracer and compare the static (DC beta) and dynamic

(AC beta) betas. Unless wide differences are found between the DC and AC betas, the DC beta can be used in all calculations.

b. Measure the input impedance. Use a known 100 kΩ series resistor and an input signal of 100 mVp-p at a frequency of 1 kHz, as shown in Figure 7-42.

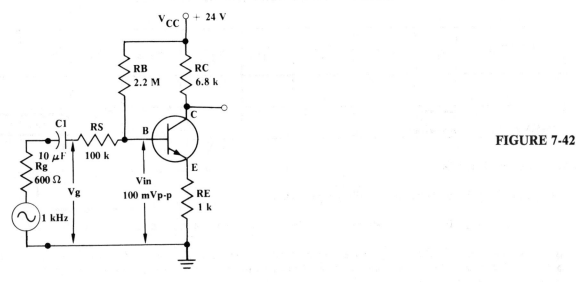

FIGURE 7-42

1. Adjust the generator voltage and measure 100 mVp-p of input signal at the base of the device, hence to the circuit.

2. Calculate the input impedance Zin from:

$$\text{Zin} = \frac{\text{Vin} \times \text{RS}}{V_{RS}} = \frac{\text{Vin} \times \text{RS}}{\text{Vg} - \text{Vin}} \quad \text{where:} \quad \text{RS} = 100 \text{ k}\Omega \text{ and Vin} = 100 \text{ mVp-p}$$

NOTE: The technique of solving the input impedance using a known series resistor value involves adjusting the signal generator voltage so that Vin = 100 mVp-p. A sample problem using the known series resistor technique is given in the theory section of this section.

8. MAXIMUM PEAK-TO-PEAK OUTPUT VOLTAGE SWING:

The maximum peak-to-peak output voltage signal, taken off the collector of the device in this circuit, rides on the DC voltage of the collector terminal. Therefore, the positive-going excursion is processed across the collector resistance and it is limited by $I_C R_C$ (in this case), and the negative-going excursion is processed by the device and limited to V_{CE} and any form of degeneration caused by the unbypassed emitter resistance. Therefore, the approximate maximum peak-to-peak signal voltage swing at the collector of the device is either $2I_C R_C$ or $2V_{CE}$, whichever is smaller. (If the degenerative effect of RE is included to provide more exactness, $2V_{CE} \times RC/(RC + RE)$. Therefore:

a. Calculate the maximum peak-to-peak output voltage swing from $2I_C R_C$ or $2V_{CE} \times RC/(RC + RE)$, whichever is smaller.

b. Monitor the collector output and increase the 1 kHz input signal level until clipping or distortion occurs because of alpha crowding. This is the maximum peak-to-peak volage swing.

NOTE: Remove RS before making the Vop-p(max) measurement.

9. POWER GAIN TO THE COLLECTOR RESISTOR:

a. Calculate the power gain to the collector resistor from: $PG = A_v^2 \times \text{Zin}/RC$ where: $RC = 6.8 \text{ k}\Omega$

b. Convert power gain to decibel form from: PG(db) = 10 log PG

NOTE: Power gain is also written as A_p, but in this text PG will be used. Also, power gain is normally construed as being the power gain to the load. However, in this circuit, the collector resistor is the load resistor. Also, use the measured A_v and Z_{in} values in the calculations to find power gain.

10. CIRCUIT CURRENT GAIN:

Use the calculated power gain and voltage gain to calculate current gain from: $A_i = PG/A_v$.

11. Insert the calculated and measured values, as indicated, into Table 7-3.

TABLE 7-3	A_v	Vop-p for Vin = 100 mVp-p	Zin	Vop-p MAX	POWER GAIN	PG dB	A_i
CALCULATED		//////					
MEASURED					//////	//////	//////

Part 3: The Partially-Bypassed, Emitter-Resistor Circuit

1. Connect the partially bypassed emitter resistor in the base-biased, common-emitter circuit, as shown in Figure 7-43.

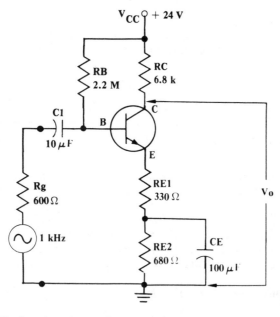

FIGURE 7-43

2. Notice that the 1 kΩ resistor has been substituted for by two series resistors that total 1010 Ω. However, the additional 10 ohms is well within the 10% tolerance of 10% resistors (or even 5% in this instance), and should have only minimal effect on the DC biasing. Monitor the voltages around the circuit to prove the minimal effect.

3. Measure V_C, V_B, V_E, and V_{RE2} and calculate the remaining voltage drops of the circuit.

   a. V_{RB} from: $V_{RB} = V_{CC} - V_B$

   b. V_{RC} from: $V_{RC} = V_{CC} - V_C$

c. V_{RE1} from: $V_{RE1} = V_E - V_{RE2}$

d. V_{CE} from: $V_{CE} = V_C - V_E$

e. I_C from: $I_C = V_{RC}/RC$

f. I_B from: $I_B = V_{RB}/RB$

g. Beta from: $\beta = I_C/I_B$

4. Insert the calculated and measured values, as indicated, into Table 7-4.

TABLE 7-4	V_C	V_B	V_E	V_{RE2}	V_{RB}	V_{RC}	V_{RE1}	V_{CE}	I_C	I_B	Beta
CALCULATED	///	///	///	///							
MEASURED					///	///	///	///	///	///	///

5. VOLTAGE GAIN:

   a. Calculate the mid-frequency voltage gain from:

   $$A_v = RC/(RE + r_e) \quad \text{where:} \quad RE = 330\ \Omega \quad \text{and} \quad r_e = 26\ \text{mV}/I_E$$

   b. Measure the mid-frequency voltage gain using an input voltage of 100 mVp-p at a frequency of 1 kHz. Measure the Vop-p at the collector, with respect to ground, and use Vop-p to calculate the measured voltage gain from:

   $$A_v = Vo/Vin \quad \text{where:} \quad Vin = 100\ \text{mVp-p}$$

NOTE: The effect of the unpredictable r_e is minimized by the 330 Ω emitter resistor RE, which also provides a gain that is not too low. RE1 is called a "swamping" resistor because it minimizes or "swamps" the effect of the variable resistance of r_e, and the gain of the stage is determined by RC/RE1, approximately, and only slightly by r_e.

6. INPUT INPEDANCE:

   a. Calculate the input impedance to the stage from:

   $$Zin = RB \mathbin{/\mkern-5mu/} \beta(RE1 + r_e) \quad \text{where;} \quad RE1 = 330\ \Omega \quad \text{and} \quad r_e = 26\ \text{mV}/IE$$

   b. Measure the input impedance. Use a known 47 kΩ resistor and an input signal level of 100 mVp-p at 1 kHz, as shown in Figure 7-44. Calculate Zin from:

   $$Zin = \frac{Vin \times RS}{V_{RS}} = \frac{Vin \times RS}{Vg - Vin}$$

7. MAXIMUM PEAK-TO-PEAK OUTPUT VOLTAGE SWING

   a. Calculate the maximum peak-to-peak output voltage swing from $2I_C R_C$ or $2V_{CE}$, whichever is smaller. More accuracy can be achieved by including the effect of RE1 from $2V_{CE}[RC/(RC + RE1)]$. However, when the ratio of RC to the unbypassed portion of RE is greater than 10 to 1, the percent error is less than 10% and the effect of RE1 is usually ignored in the initial analysis. However, if $2V_{CE}$ is the smaller of the two parameters and the calculated value of $2V_{CE}$ is larger than measured, including the effect of RE1 will narrow the discrepancy.

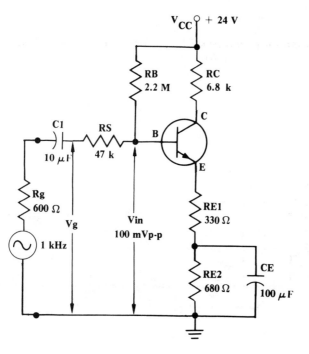

FIGURE 7-44

b. Monitor the collector output with an oscilloscope and increase the 1 kHz input signal level until clipping or distortion, from alpha crowding, occurs.

NOTE: Since RE is partially bypassed, some negative feedback occurs because of RE1. However, because the ratio of RC to RE1 is greater than 10 (about 20), the choice of using $2V_{CE}$ or $2V_{CE}RC/(RC + RE1)$ depends on how accurate the calculated to measured value need be. Also, this is only a concern when $2V_{CE}$ is smaller than $2I_C R_C$ and, since 10% resistors are almost always used, the extra attempt for accuracy may be strictly pencil pushing. Remember to remove RS for this measurement.

8. POWER GAIN:

   a. Calculate the power gain of the circuit from: $PG = A_v^2 \times Zin/RC$

   b. Convert PG to dB from: $PG(db) = 10 \log PG$

NOTE: Use the measured A_v and Zin values in the power gain calculations.

9. CIRCUIT CURRENT GAIN:

   Use the calculated power and voltage gains to calculate current gain from: $A_i = PG/A_v$

10. Insert the calculated and measured values, as indicated, into Table 7-5.

TABLE 7-5	A_v	Vop-p for Vin = 100 mVp-p	Zin	Vop-p MAX	POWER GAIN	PG dB	A_i
CALCULATED		/////					
MEASURED					/////	/////	/////

Part 4: The Loaded Amplifier

1. Load the partially bypassed emitter resistor circuit with a 10 kΩ load resistor, as shown in Figure 7-45.

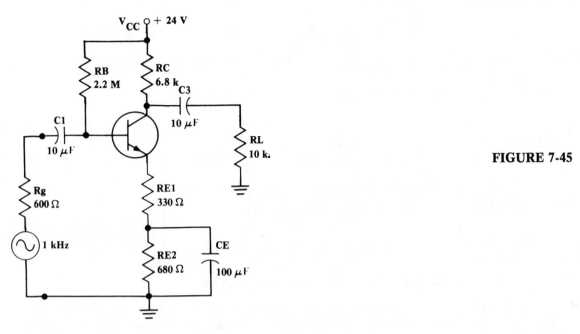

FIGURE 7-45

2. Measure the DC voltage drops around the circuit to demonstrate that the DC voltages of the circuit have not been effected by the circuit change. Insert the measured DC voltages into Table 7-6.

TABLE 7-6	V_C	V_B	V_E	V_{RC}	V_{RB}	V_{RE1}	V_{RE2}	V_{CE}	I_B	I_C	Beta
MEASURED											

3. **VOLTAGE GAIN:**

   a. Calculate the mid-frequency voltage gain from: $A_v = \dfrac{RC \parallel RL}{r_e + RE1}$ where: $r_e = 26$ mV/I_E and $I_E \approx I_C$

   b. Measure the mid-frequency voltage gain using an input voltage of 100 mVp-p at a frequency of 1 kHz. Measure the output voltage across the load resistor, which is the same as measuring the output voltage at the collector with respect to ground. Calculate the measured voltage gain from:

   $A_v = Vo/Vin$ where: $Vin = 100$ mVp-p

4. **INPUT IMPEDANCE:**

   a. The calculated input impedance of this circuit is similar to that of the unloaded, partially-bypassed, emitter-resistor circuit and need not be recalculated.

   b. Measure the input impedance to verify the statement of Part a. Again, use the known 47 kΩ series resistor technique to measure the input impedance.

5. **MAXIMUM PEAK-TO-PEAK OUTPUT VOLTAGE SWING:**

   a. Calculate the maximum peak-to-peak output voltage swing from approximately $2V_{CE}$ or $2I_C r_L$, whichever is smaller. Again: $r_L = RC \parallel RL$.

b. Monitor the output voltage swing and increase the 1 kHz input signal level until clipping or distortion, because of alpha crowding, occurs.

6. POWER GAIN TO THE LOAD:

   a. Calculate the power gain of the circuit from: $PG = A_V^2 \times Z_{in}/R_L$ where: $R_L = 10\ k\Omega$

   b. Convert the power gain to db using 10 log PG. (Use the measured values of Z_{in} and A_V.)

7. Insert the calculated and measured values, as indicated, into Table 7-7.

TABLE 7-7	A_V	Vop-p for Vin = 100 mVp-p	Zin	Vop-p MAX	POWER GAIN	PG dB
CALCULATED		/////	/////			
MEASURED					/////	/////

Part 5: The Fully-Bypassed, Emitter-Resistor Circuit

1. Connect the common emitter circuit as shown in Figure 7-46. Notice that the 100 µF CE capacitor bypasses the entire emitter resistance. Therefore, the DC biasing of the circuit should be identical to that of the previous circuit.

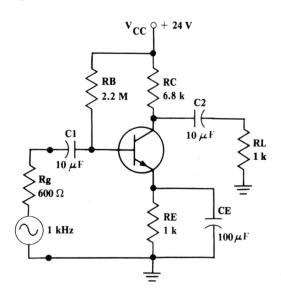

FIGURE 7-46

2. Measure the DC voltage drops around the circuit to verify that the C2 and CE capacitors do not effect the DC voltage distribution of the circuit. Also, find I_C from V_{RC}/RC or current meter measurement.

3. Insert the "measured" DC voltages of the circuit, as indicated, into Table 7-8, as reference for the AC analysis of the circuit.

TABLE 7-8	V_C	V_B	V_E	V_{RB}	V_{RC}	V_{CE}	I_C
MEASURED							

NOTE: Unless capacitors C2 and CE are leaky, the values of Table 7-8 and Table 7-6 should be identical.

VOLTAGE GAIN:

a. Calculate the mid-frequency voltage gain from:

$$A_V = \frac{RC \,\|\, RL}{r_e} \quad \text{where:} \quad r_e \approx 26 \text{ mV}/I_E, \; I_E \approx I_C, \text{ and } RL = 1 \text{ k}\Omega$$

b. Measure the mid-frequency voltage gain using an input voltage of 30 mVp-p at 1 kHz. Measure the Vop-p at the collector, with respect to ground, and use the Vop-p to calculate the measured voltage gain from:

$$A_v = V_o/V_{in} \quad \text{where:} \quad V_{in} = 30 \text{ mVp-p}$$

NOTE: The X_C of the 100 μF capacitor CE is about 1.6 ohms of capacitive reactance at 1 kHz, and r_e can be measured anywhere between 26 mV/I_E to 52 mV/I_E, depending on the device used. Therefore, the voltage gain of the fully-bypassed, emitter resistor is unpredictable, at best.

c. Use the measured A_v to calculate the actual r_e from: $r_e = r_L/A_v$

5. INPUT IMPEDANCE:

a. Calculate the input impedance to the circuit from: $Z_{in} = RB \,\|\, \beta r_e$ where: $RB = 2.2 \text{ M}\Omega$

NOTE: Use the r_e calculated in Step 4(c) and the beta given in Table 7-8.

b. Measure the input impedance. Use a known series resistor of 1 kΩ at an input signal level of 40 mVp-p at 1 kHz, as shown in Figure 7-47. Measure Vin and Calculate Zin from:

$$Z_{in} = V_{in} \times \frac{RS}{V_{RS}} = \frac{V_{in} \times RS}{V_g - V_{in}}$$

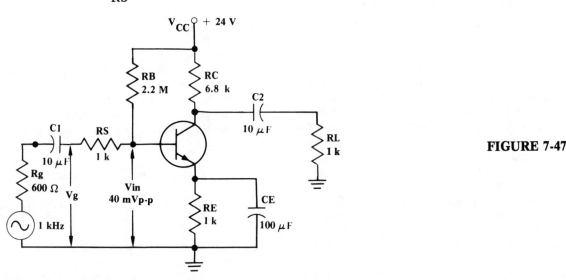

FIGURE 7-47

NOTE: Adjust the generator voltage for a Vin of 40 mVp-p after the circuit is operative because the signal loss across the internal resistance Rg of the generator will be significant as a result of the low impedance load of RS in series with Zin. Remember, measurements are critical because any error in reading Vin is shifted to V_{RS} and the ratio of error is multiplied.

6. MAXIMUM PEAK-TO-PEAK VOLTAGE SWING:

a. Calculate the maximum peak-to-peak output voltage swing from $2I_C r_L = 2I_C(RC \,\|\, RL)$ or $2V_{CE}$, whichever is smaller.

b. Monitor the collector-output and increase the 1 kHz input signal level until clipping or distortion, as a result of alpha crowding, occurs.

NOTE: $2 I_C r_L$ will be very small because $r_L = RC \parallel RL$ and $RL = 1$ kHz.

7. **POWER GAIN TO THE LOAD RESISTOR:**

   a. Calculate the power gain to the resistor load of the circuit from: $PG = A_v^2 \times Zin/RL$

   b. Convert PG to db from 10 log PG.

NOTE: Use the measured A_v and Zin values in the calculations.

8. **CIRCUIT CURRENT GAIN:**

   Use the calculated power and voltage gains to calculate current gain from: $A_i = PG/A_v$

9. Insert the calculated and measured values, as indicated, into Table 7-9.

TABLE 7-9	A_v	Vop-p for Vin = 30 mVp-p	r_e	Zin	Vop-p MAX	POWER GAIN	PG dB	A_i
CALCULATED								
MEASURED						/////	/////	/////

SECTION QUESTIONS

1. What is the main advantage, with regard to voltage gain, of the partially-bypassed, emitter-resistor circuit? How about the totally-bypassed, emitter-resistor circuit?
2. Which of the three circuits discussed in this section has the highest input impedance?
3. Does the DC voltage distribution of the circuit have an effect on the maximum peak-to-peak output voltage swing? Explain how DC can be a factor.
4. What does the 180° phase shift between the input and output signals mean? Explain.
5. Does a circuit with a higher input impedance increase the power gain of the circuit?
6. Does the theoretical output impedance of the common emitter circuit change with load change?
7. Why is the peak-to-peak output voltage signal slightly smaller when any portion of the emitter resistor is unbypassed?
8. Does loading a single stage amplifier circuit effect the input impedance? How about the maximum peak-to-peak output voltage swing?
9. Did changing the collector resistor from 4.7 kΩ to 6.8 kΩ change the voltage drop across the emitter resistor? The 4.7 kΩ collector resistor was given in the circuit of Figure 7-39 and the 6.8 kΩ collector resistor was given in the circuit of Figure 7-40.
10. Is the circuit current gain (A_i) almost equal to the device current gain (beta) for the fully-bypassed, emitter-resistor circuit? How about the partially-bypassed and unbypassed, emitter-resistor circuits?
11. Why is the load resistor seen in parallel with the collector resistor when computing voltage gain?
12. Can r_e be approximated in the partially bypassed, emitter-resistor circuit? Explain if it can or cannot, with regard to gain.

SECTION PROBLEMS

1. Given: $\beta = 100$ and $V_{BE} = 0.6$ V

   Find:

   a. I_C
   b. V_{RC}
   c. V_{RE}
   d. V_{CE}
   e. V_{RB}
   f. r_e
   g. A_v
   h. Zin
   i. PG
   j. Vop-p(max)

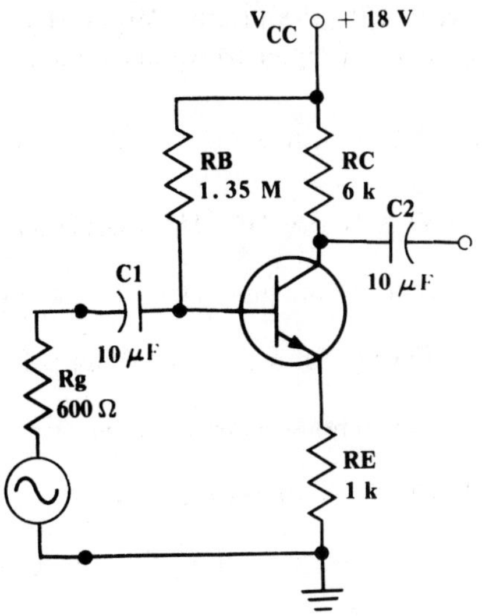

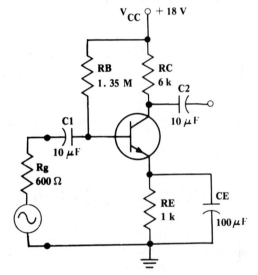

FIGURE 7-48

2. Given: $\beta = 100$ and $V_{BE} = 0.6$ V

   Find:

   a. r_e
   b. A_v
   c. Zin
   d. PG
   e. Vop-p(max)

FIGURE 7-49

3. Given: $\beta = 100$ and $V_{BE} = 0.6$ V

   Find:

   a. r_e
   b. A_v
   c. Zin
   d. PG
   e. Vop-p(max)

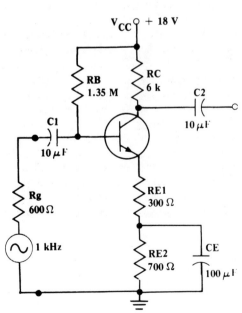

FIGURE 7-50

207

4. Given: $\beta = 100$ and $V_{BE} = 0.6$ V

   Find:

   a. r_e
   b. A_v
   c. Zin
   d. PG
   e. Vop-p(max)

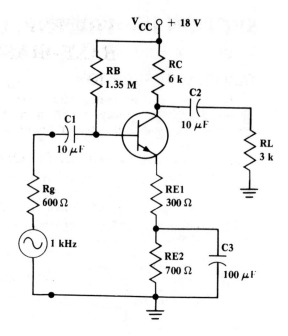

FIGURE 7-51

SECTION II: THE PNP, COMMON-EMITTER, BASE-BIASED CIRCUIT

GENERAL DISCUSSION

The analysis of the common emitter circuit using the PNP transistor is solved the same as that circuit using the NPN device except the DC polarities of the circuit differ. Therefore, it is only necessary to solve for the partially-bypassed, emitter circuit to show the AC circuit parameter similarities.

DIRECT CURRENT CIRCUIT ANALYSIS

The DC analysis for the common emitter circuit using the PNP device begins by solving for the collector current I_C, and then solving for the remaining voltage drops around the circuit. Therefore, the circuit currents and voltages are solved like those of the NPN circuit, but the single point voltages of the circuit will be negative with respect to ground. As a result, V_C will be -11 V, V_E will be -1.3 V, and V_B will be -1.9 V, all with respect to ground. The DC analysis for the circuit of Figure 7-52 is as follows:

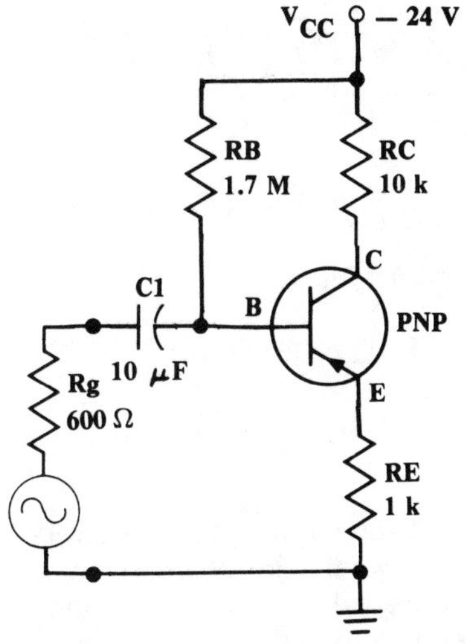

$$I_C \approx \frac{V_{CC} - V_{BE}}{R_B/\beta + R_E} = \frac{24\text{ V} - 0.6\text{ V}}{1.7\text{ M}\Omega/100 + 1\text{ k}\Omega}$$

$$= \frac{23.4\text{ V}}{17\text{ k}\Omega + 1\text{ k}\Omega} = 1.3\text{ mA}$$

$$V_{RC} = I_C R_C = 1.3\text{ mA} \times 10\text{ k}\Omega = 13\text{ V}$$

$$V_{RE} = I_E R_E \approx 1.3\text{ mA} \times 1\text{ k}\Omega = 1.3\text{ V}$$

$$V_C = V_{CC} - V_{RC} = -24\text{ V} + 13\text{ V} = -11\text{ V}$$

$$V_B = V_E + V_{EB} = -1.3\text{ V} - 0.6\text{ V} = -1.9\text{ V}$$

$$V_E = \text{Ground} - V_{RE} = 0.0\text{ V} - 1.3\text{ V} = -1.3\text{ V}$$

$$V_{RB} = V_{CC} - V_B = 24\text{ V} - 1.9\text{ V} = 22.1\text{ V}$$

$$I_B = V_{RB}/R_B = 22.1\text{ V}/1.7\text{ M}\Omega = 13\text{ }\mu\text{A} \quad \text{or,}$$

$$I_B = I_C/\beta = 1.3\text{ mA}/100 = 13\text{ }\mu\text{A}$$

$$V_{CE} = V_{EC} = V_E - V_C = -1.3\text{ V} - (-11\text{ V}) = 9.7\text{ V}$$

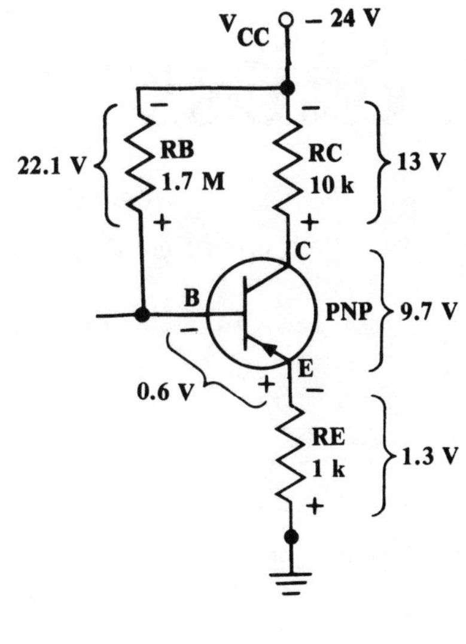

$$V_{CC} = V_{RC} + V_{CE} + V_{RE}$$

$$|24\text{ V}| = 13\text{ V} + 9.7\text{ V} + 1.3\text{ V}$$

$$V_{CC} = V_{RB} + V_{BE} + V_{RE}$$

$$|24\text{ V}| = 22.1\text{ V} + 0.6\text{ V} + 1.3\text{ V}$$

FIGURE 7-52

ALTERNATING CURRENT CIRCUIT ANALYSIS

The AC parameters of voltage gain, input impedance, power gain, and maximum peak-to-peak output voltage of the PNP circuit are solved identically to those of the NPN circuits. The calculations are shown in conjunction with Figure 7-53.

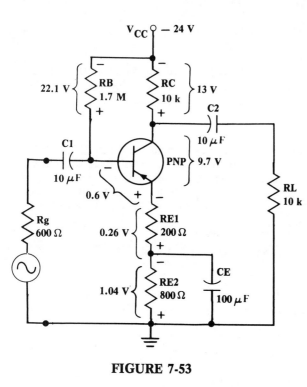

FIGURE 7-53

Given:

$I_E = 1.3$ mA $\quad V_{RC} = 13$ V $\quad V_{CE} = 9.7$ V

$V_{R1} = 0.26$ V $\quad V_{R2} = 1.04$ V $\quad V_{RB} = 22.1$ V

$V_C = -11$ V $\quad V_B = -1.9$ V $\quad V_E = -1.3$ V

$$A_v = \frac{RC \,/\!/\, RL}{r_e + R1} = \frac{10\text{ k}\Omega \,/\!/\, 10\text{ k}\Omega}{20\ \Omega + 200\ \Omega} = \frac{5\text{ k}\Omega}{220\ \Omega} = 22.73$$

$\text{Zin} = RB \,/\!/\, \beta\,(r_e + R1)$

$\quad = 1.7\text{ M}\Omega \,/\!/\, 100 \times 220\ \Omega = 21.72\text{ k}\Omega$

$PG = A_v^2 \times \text{Zin}/RL = 22.73^2 \times 21.72\text{ k}\Omega / 10\text{ k}\Omega = 1122$

$PG(dB) = 10 \log PG = 10 \log 1122 = 30.5$ dB

$Z_o = RC = 10\text{ k}\Omega$

$V_{op\text{-}p(max)} = 2V_{CE}$ or $2I_C r_L$, whichever is smaller.

$\quad 2V_{CE} = 2 \times 9.7\text{ V} = 19.4$ Vp-p

$\quad 2I_C r_L = 2 \times 1.3\text{ mA} \times 5\text{ k}\Omega = 13$ Vp-p

Therefore, $V_{op\text{-}p(max)} = 2I_C r_L = 13$ Vp-p

SUMMARY

The PNP and NPN circuits are solved in the same way and only the polarity of the connecting capacitors need be changed, because of the opposite DC requirements. Therefore, voltage gain, input impedance, power gain, maximum peak-to-peak output voltage swing, and output impedance will be identical as long as the parameters of the respective PNP and NPN devices are equal. The analysis for the similar NPN circuit was shown in conjunction with Figure 7-28.

SECTION II: LABORATORY EXERCISE

OBJECTIVE

To investigate the DC and AC parameters of the base-biased, common-emitter, PNP circuit configuration to see how they compare with those of a similar NPN circuit configuration.

LIST OF MATERIALS AND EQUIPMENT

1. Transistor: 2N3906 or equivalent PNP
2. Resistors (one each): 330 Ω   680 Ω   6.8 kΩ   10 kΩ   47 kΩ   2.2 MΩ
3. Capacitors (15-35 V): 10 μF (two)   100 μF (one)
4. Power Supply: 24 volt
5. Voltmeter
6. Signal Generator
7. Oscilloscope

PROCEDURE

1. Connect the PNP transistor into the base-biased, common-emitter circuit configuration of Figure 7-54, which has a partially-bypassed, emitter resistor and a minus (−) V_{CC} power supply voltage. Remember that the partially-bypassed, emitter resistor provides swamping of the base-emitter-diode resistance re, and that the power supply connection of the PNP device is exactly opposite to that of the NPN device.

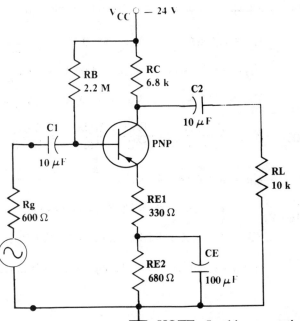

FIGURE 7-54

NOTE: In this connection, the reference ground is the positive voltage.

2. Measure the voltages at the collector, base, and emitter terminals of the transistor and the junction of RE1 and RE2, with respect to ground. Insert the measured V_C, V_E, V_B, and V_{RE2} voltages, as indicated, into Table 7-10.

3. Calculate the circuit voltage drops of V_{RC}, V_{RE1}, V_{RB}, V_{CE}, and V_{BE} and the circuit currents of I_C and I_B from:

   a. $V_{RC} = V_{CC} - V_C$
   b. $V_{RE1} = V_E - V_{RE2}$
   c. $V_{RB} = V_{CC} - V_B$
   d. $V_{CE} = V_C - V_E$

   e. $V_{BE} = V_B - V_E$
   f. $I_C = V_{RC}/RC$
   g. $I_B = V_{RB}/RB$
   h. $\beta = I_C/I_B$

4. Insert the calculated and measured values, as indicated, into Table 7-10.

TABLE 7-10	V_C	V_B	V_E	V_{RE2}	V_{RC}	V_{RE1}	V_{RB}	V_{CE}	V_{BE}	I_C	I_B	Beta
CALCULATED												
MEASURED												

6. **VOLTAGE GAIN:**

   a. Calculate the mid-frequency voltage gain from:

   $$A_v = \frac{r_L}{r_e + RE1} = \frac{RC \parallel RL}{r_e + RE1} \quad \text{where: } r_e \approx 26\,mV/I_E \text{ and } r_L \approx RC \parallel RL$$

   b. Measure the mid-frequency voltage gain using an input voltage of 100 mVp-p at a frequency of

211

1 kHz. Measure Vo at the collector, with respect to ground, and use the Vin and Vo measurements to calculate the "measured" voltage gain from:

$$A_v = Vo/Vin \quad \text{where:} \quad Vin = 100 \text{ mVp-p}$$

7. **INPUT IMPEDANCE:**

   a. Calculate the input impedance to the circuit from:

   $$Zin = RB \mathbin{/\mkern-6mu/} \beta(r_e + RE1) \quad \text{where:} \quad RE1 = 330 \text{ }\Omega \quad \text{and} \quad r_e \approx 26 \text{ mV}/I_E$$

   b. Measure the input impedance. Use a known 47 kΩ resistor and an input signal level of 100 mVp-p at 1 kHz, as shown in Figure 7-55. Calculate Zin from:

   $$Zin = \frac{Vin \times RS}{V_{RS}} = \frac{Vin \times RS}{Vg - Vin} \quad \text{where:} \quad Vin = 100 \text{ mVp-p}$$

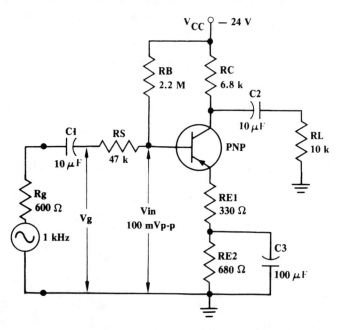

FIGURE 7-55

8. **MAXIMUM PEAK-TO-PEAK OUTPUT VOLTAGE SWING:**

   a. Calculate the maximum peak-to-peak output voltage swing from $2I_C r_L$ or $2V_{CE}$, whichever is smaller. Again, more accuracy can be achieved by including the slight effect of the 330 Ω resistor RE1 and solving from $2V_{CE} \times r_L/(r_L + RE1)$

   b. Monitor the collector output and increase the 1 kHz input signal level until clipping or distortion, caused by alpha crowding, occurs.

NOTE: Remove the RS resistor for this measurement.

9. **POWER GAIN TO THE LOAD:**

   a. Calculate the power gain to the load resistor of the circuit, using both the measured A_v and Zin values. Calculate from:

   $$PG = A_v^2 \times Zin/RL$$

   b. Convert to dB from 10 log PG.

10. Insert the calculated and measured values, as indicated, into Table 7-11.

TABLE 7-11	A_v	Vop-p for Vin = 100 mVp-p	Zin	Vop-p MAX	POWER GAIN	PG dB
CALCULATED		/////////				
MEASURED					/////////	/////////

SECTION QUESTIONS

1. Do the AC circuit parameters differ because a device is PNP as compared to NPN? Compare the PNP circuit parameters of Figure 7-53 with those of the NPN circuit of Figure 7-28.
2. Were there any other major differences between using PNP and NPN devices, aside from the polarity connections of the power supply and circuit capacitors?
3. What AC parameters would be affected if bypass capacitor C3 were removed? Name three and explain why.

SECTION PROBLEMS

1. Given: Beta = 150 and V_{BE} = 0.6 V.

   Find:
   - a. I_C
   - b. V_{RC}
   - c. V_{RE1}
        V_{RE2}
   - d. V_{CE}
   - e. V_{RB}
   - f. V_C
   - g. V_E
   - h. V_B
   - i. r_e
   - j. A_v
   - k. Zin
   - l. PG
   - m. A_i
   - n. Vop-p(max)

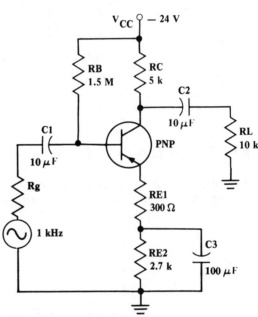

FIGURE 7-56

2. Given: Beta = 100 and V_{BE} = 0.6 V

   Find:
   - a. I_C
   - b. V_{RC}
   - c. V_{RE1}
        V_{RE2}
   - d. V_{CE}
   - e. V_{RB}
   - f. V_C
   - g. V_E
   - h. V_B
   - i. r_e
   - j. A_v
   - k. Zin
   - l. PG
   - m. A_i
   - n. Vop-p(max)

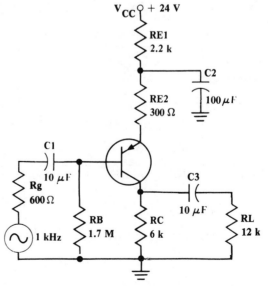

FIGURE 7-57

SECTION III: THE COMMON-COLLECTOR, BASE-BIASED CIRCUIT

GENERAL DISCUSSION

The main function of the common collector circuit is to use the current gain of the transistor to increase the load resistance. In this circuit, as shown in Figure 7-58, a load resistance can be reflected into the base by beta and "seen" in parallel with the base resistor RB. As a result, the circuit increases the load resistance from r_L to Zin. Additionally, because the signal is applied to the base and taken off the emitter, the voltage gain of the circuit will be slightly less than unity.

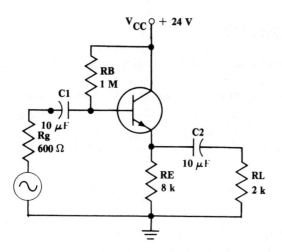

FIGURE 7-58

DIRECT CURRENT CIRCUIT ANALYSIS

The direct current circuit analysis begins by solving for the collector current I_C and then solving the remaining circuit currents and voltage drops. The calculations are shown in conjunction with Figure 7-59.

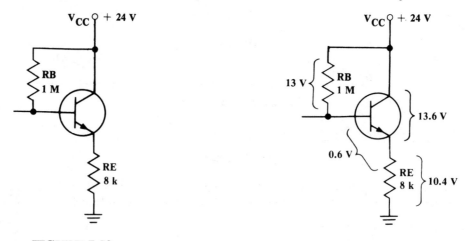

FIGURE 7-59 FIGURE 7-60

$$I_C = \frac{V_{CC} - V_{BE}}{RB/\beta + RE} = \frac{24\text{ V} - 0.6\text{ V}}{1\text{ M}\Omega/100 + 8\text{ k}\Omega} = \frac{23.4\text{ V}}{10\text{ k}\Omega + 8\text{ k}\Omega} = 1.3\text{ mA}$$

$$V_{RE} = I_E R_E = 1.3\text{ mA} \times 8\text{ k}\Omega = 10.4\text{ V}$$

$$V_{CE} = V_{CC} - V_{RE} = 24\text{ V} - 10.4\text{ V} = 13.6\text{ V}$$

$$I_B = I_C/\beta = 1.3\text{ mA}/100 = 13\text{ }\mu\text{A} \quad \text{where:} \quad I_C \approx I_E$$

$$V_{RB} = I_B R_B = 13 \ \mu A \times 1 \ M\Omega = 13 \ V$$

$$V_B = V_{CC} - V_{RB} = 24 \ V - 13 \ V = 11 \ V \quad \text{or,}$$

$$V_B = V_{RE} + V_{BE} = 10.4 \ V + 0.6 \ V = 11 \ V$$

$$V_{CC} = V_{CE} + V_{RE} \qquad\qquad V_{CC} = V_{RB} + V_{BE} + V_{RE}$$

$$24 \ V = 13.6 \ V + 10.4 \ V \qquad 24 \ V = 13 \ V + 6 \ V + 10.4 \ V$$

ALTERNATING CURRENT CIRCUIT ANALYSIS

The alternating current circuit parameters to be solved for the common-collector circuit are the voltage gain, input impedance, power gain, and maximum peak-to-peak voltage swing.

VOLTAGE GAIN

The voltage gain of the common collector circuit is slightly less than unity gain because a small amount of signal voltage is dropped across the base-emitter-diode resistance as the signal is being processed from the input base to the emitter output.

A. The Unloaded Circuit

The voltage gain is the ratio of voltage out to voltage in and the voltage out is solved using the voltage divider equation. There is no signal phase shift between input and output.

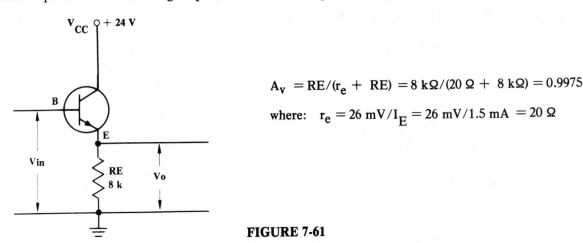

$$A_v = RE/(r_e + RE) = 8 \ k\Omega/(20 \ \Omega + 8 \ k\Omega) = 0.9975$$

where: $r_e = 26 \ mV/I_E = 26 \ mV/1.5 \ mA = 20 \ \Omega$

FIGURE 7-61

Therefore, if 4 Vp-p of input signal is applied, the voltage developed across the unloaded RE resistor is 3.99 Vp-p, and about 0.01 Vp-p of signal voltage is dropped across the 20 Ω base-emitter diode resistance.

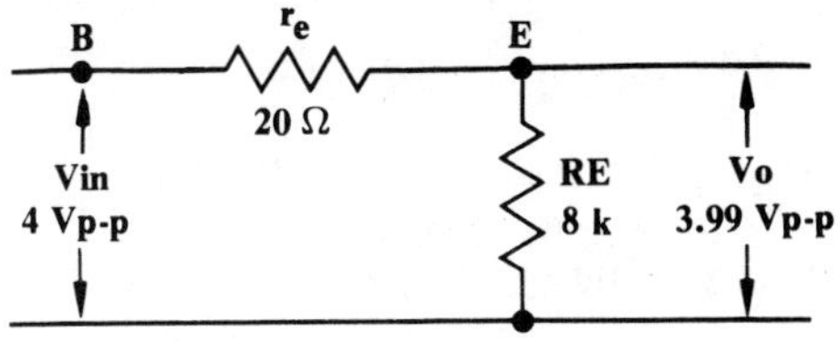

$$Vo = \frac{Vin \times RE}{r_e + RE} = \frac{4 \ Vp\text{-}p \times 8 \ k\Omega}{20 \ \Omega + 8 \ k\Omega} = 3.99 \ Vp\text{-}p, \quad \text{also}$$

$$Vo = A_v \times Vin = 0.9975 \times 4 \ Vp\text{-}p = 3.99 \ Vp\text{-}p$$

FIGURE 7-62

B. The Loaded Circuit

If a load resistor of 2 kΩ is added to the circuit, then the AC load is the parallel or RE ∥ RL = 1.6 kΩ.

The voltage gain is:

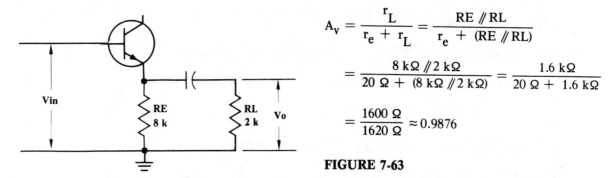

$$A_v = \frac{r_L}{r_e + r_L} = \frac{RE \mathbin{/\mkern-5mu/} RL}{r_e + (RE \mathbin{/\mkern-5mu/} RL)}$$

$$= \frac{8\,k\Omega \mathbin{/\mkern-5mu/} 2\,k\Omega}{20\,\Omega + (8\,k\Omega \mathbin{/\mkern-5mu/} 2\,k\Omega)} = \frac{1.6\,k\Omega}{20\,\Omega + 1.6\,k\Omega}$$

$$= \frac{1600\,\Omega}{1620\,\Omega} \approx 0.9876$$

FIGURE 7-63

Therefore, if a 4 Vp-p input signal is applied to the circuit, the voltage across the load is 3.975 Vp-p.

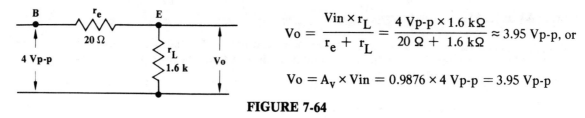

$$Vo = \frac{Vin \times r_L}{r_e + r_L} = \frac{4\,Vp\text{-}p \times 1.6\,k\Omega}{20\,\Omega + 1.6\,k\Omega} \approx 3.95\,Vp\text{-}p,\text{ or}$$

$$Vo = A_v \times Vin = 0.9876 \times 4\,Vp\text{-}p = 3.95\,Vp\text{-}p$$

FIGURE 7-64

INPUT IMPEDANCE

The input impedance to the common collector circuit changes with load resistance because the output is taken off the emitter. This is shown in the following unloaded and loaded circuit calculations.

A. The Unloaded Circuit

The input impedance of the unloaded circuit is solved by reflecting the effective emitter resistance in the base in parallel with the base resistor RB. The effective emitter resistance is r_e in series with RE.

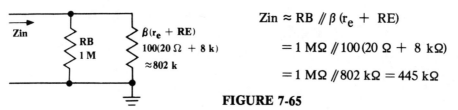

$$Zin \approx RB \mathbin{/\mkern-5mu/} \beta(r_e + RE)$$

$$= 1\,M\Omega \mathbin{/\mkern-5mu/} 100(20\,\Omega + 8\,k\Omega)$$

$$= 1\,M\Omega \mathbin{/\mkern-5mu/} 802\,k\Omega = 445\,k\Omega$$

FIGURE 7-65

B. The Loaded Circuit

The input impedance of the loaded stage is solved by reflecting the load resistance into the base in parallel with the base resistor RB. The effective reflected AC load resistance is r_e in series with the parallel RE and RL resistors.

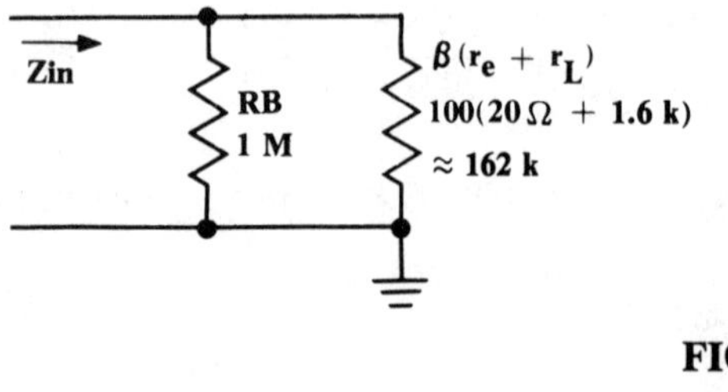

$$Zin = RB \mathbin{/\mkern-5mu/} \beta(r_e + r_L) = RB \mathbin{/\mkern-5mu/} \beta[r_e + (RE \mathbin{/\mkern-5mu/} RL)]$$

$$= 1\,M\Omega \mathbin{/\mkern-5mu/} 100[20\,\Omega + (8\,k\Omega \mathbin{/\mkern-5mu/} 2\,k\Omega)]$$

$$= 1\,M\Omega \mathbin{/\mkern-5mu/} 100(20\,\Omega + 1.6\,k\Omega)$$

$$= 1\,M\Omega \mathbin{/\mkern-5mu/} 162\,k\Omega \approx 139.4\,k\Omega$$

FIGURE 7-66

POWER GAIN

Power gain is calculated differently for the unloaded and loaded circuits because the input impedance changes in each circuit. Also, the load for the unloaded circuit is RE while the load for the loaded circuit is RL.

A. The Unloaded Circuit

The power gain for the unloaded circuit has RE for the load. Therefore:

$PG = A_v^2 \times Zin/RE = 0.9975^2 \times 445\ k\Omega/8\ k\Omega \approx 55.38$

$PG(dB) = 10 \log PG = 10 \log 55.38 = 17.43\ dB$

B. The Loaded Circuit

The power gain for the loaded circuit has RL for the load. Therefore:

$PG = A_v^2 \times Zin/RL = 0.9876^2 \times 139.4\ k\Omega/2\ k\Omega \approx 67.99$

$PG(dB) = 10 \log PG = 10 \log 67.99 = 18.32\ dB$

CURRENT GAIN

The unloaded circuit current gain is solved from: $A_i = PG/A_v = 55.38/0.9975 = 55.52$

The loaded circuit current gain is solved from: $A_i = PG/A_v = 67.99/0.9876 = 68.84$

MAXIMUM PEAK-TO-PEAK OUTPUT VOLTAGE SWING

A. The Unloaded Circuit

The maximum peak-to-peak output voltage swing for the unloaded circuit is solved from $2V_{CE}$ or $2I_E R_E$, whichever is smaller.

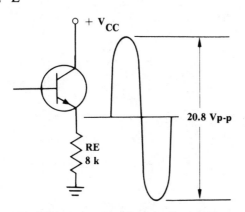

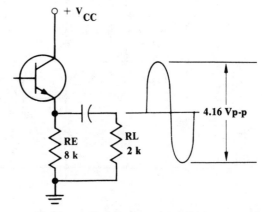

$2V_{CE} = 2 \times 13.6\ V = 27.2\ Vp\text{-}p$

$2I_E R_E = 2 \times 1.3\ mA \times 8\ k\Omega = 20.8\ Vp\text{-}p$

Therefore, since $2I_E R_E$ is the smaller,

$Vop\text{-}p(max) = 2I_E R_E = 20.8\ Vp\text{-}p$

FIGURE 7-67

$2V_{CE} = 2 \times 13.6\ V = 27.2\ Vp\text{-}p$

$2I_E r_L = 2 \times 1.3\ mA \times 1.6\ k\Omega = 4.16\ Vp\text{-}p$

Therefore, since $2I_E r_L$ is the smaller:

$Vop\text{-}p(max) = 2I_E r_L = 4.16\ Vp\text{-}p$

FIGURE 7-68

B. The Loaded Circuit

The maximum peak-to-peak output voltage swing for the loaded circuit is solved from $2V_{CE}$ or $2I_E r_L$, whichever is smaller.

SUMMARY

The main function of the common collector circuit is to provide impedance transfer. For instance, if a 1 kΩ load resistor is connected to a common emitter stage which has a collector resistor of 10 kΩ, the gain will decrease by a factor greater than 10. However, if the 1 kΩ load is "buffered" by a common collector stage, where the 1 kΩ load is multiplied by the beta of 100 so that the reflected load is 100 kΩ, then the loading effect of the 100 kΩ resistance on the 10 kΩ collector resistor will be less than 10 percent.

SECTION III: LABORATORY EXERCISE

OBJECTIVES

Investigate the DC and AC parameters of the NPN, base-biased, common-collector circuit.

LIST OF MATERIALS AND EQUIPMENT

1. Transistor: 2N3904 or eqivalent NPN
2. Resistors (one each):   1 kΩ   3.3 kΩ   100 kΩ   470 kΩ
3. Capacitors (15-35 V):   10 μF (two)
4. Power Supply: 24 V
5. Voltmeter
6. Signal Generator
7. Oscilloscope

PROCEDURE

1. Connect the NPN transistor into the base-biased, common-collector circuit as shown in Figure 7-69. Notice that the collector terminal is connected directly to the V_{CC}, which saves the use of two components — a collector resistor and an associated bypass capacitor.

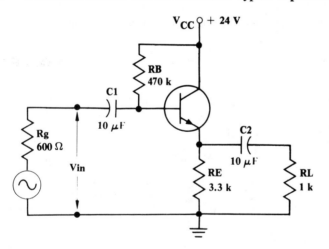

FIGURE 7-69

2. Measure the voltage at the collector, base, and emitter terminals of the transistor, with respect to ground. Insert the measured V_C, V_B, and V_E voltages into Table 7-12.

3. Calculate the circuit voltages of V_{RE}, V_{RB}, V_{BE}, and V_{CE}, along with the circuit currents of I_B and I_C. Calculated from:

    a. $V_{RE} = V_E$ e. $I_E = V_{RE}/RE$

    b. $V_{RB} = V_{CC} - VB$ f. $I_B = V_{RB}/RB$

    c. $V_{BE} = V_B - VE$ g. Beta $= I_C/I_B \approx I_E/I_B$

    d. $V_{CE} = V_{CC} - VE$ where $V_C = V_{CC}$

4. Insert the calculated and measured values, as indicated, into Table 7-12.

TABLE 7-12	V_C	V_B	V_E	V_{RE}	V_{RB}	V_{BE}	V_{CE}	I_E	I_B	Beta
CALCULATED	/////	/////	/////							
MEASURED				/////	/////	/////	/////	/////	/////	/////

5. **VOLTAGE GAIN:**

   a. Calculate the mid-frequency voltage gain from:

   $$A_v = r_L/(r_e + r_L) \quad \text{where:} \quad r_L = RE \parallel RL \quad \text{and} \quad r_e \approx 26 \text{ mV}/I_E$$

   b. Measure the mid-frequency voltage gain using an input voltage of 1 Vp-p at a frequency of 1 kHz. Measure Vo at the emitter with respect to ground, as shown in Figure 7-70. Calculate the measured voltage gain from:

   $$A_v = Vo/Vin \quad \text{where:} \quad Vin = 1 \text{ Vp-p}$$

NOTE: Vo will be very close in value to the input signal voltage, because only a small amount of signal is lost across the base-emitter-diode resistance.

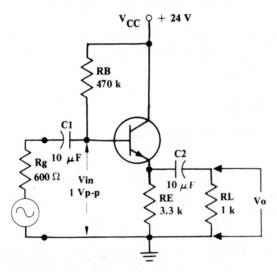

FIGURE 7-70

6. **INPUT IMPEDANCE:**

   a. Calculate the input impedance to the stage from:

   $$Zin = RB \parallel \beta(r_e + r_L) \quad \text{where:} \quad r_L = RE \parallel RL \quad \text{and} \quad r_e \approx 26 \text{ mV}/I_E$$

   b. Measure the input impedance. Use a known 100 kΩ resistor and an input signal level of 1 Vp-p at 1 kHz, as shown in Figure 7-71. Calculate Zin from:

   $$Zin = \frac{Vin \times RS}{V_{RS}} = \frac{Vin \times RS}{Vg - Vin}$$

7. **MAXIMUM PEAK-TO-PEAK OUTPUT VOLTAGE SWING:**

   a. Calculate the maximum peak-to-peak output voltage swing from $2I_E r_L$ or $2V_{CE}$, whichever is smaller.

   b. Monitor the output voltage and increase the 1 kHz input signal level until clipping or distortion, caused by alpha crowding, occurs.

NOTE: Remove RS for this measurement.

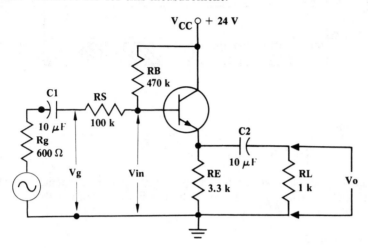

FIGURE 7-71

8. POWER GAIN TO THE LOAD:

   a. Calculate the power gain to the load resistor using both the measured A_v and Zin values. Calculate from:

   $$PG = A_v^2 \times Zin/RL$$

   b. Convert the power gain to dB from: $PG(dB) = 10 \log PG$

9. Calculate the current gain of the loaded circuit from: $A_i = PG/A_v$

10. Insert the calculated and measured values, as indicated, into Table 7-13.

TABLE 7-13	A_v	Vo for Vin = 1 Vp-p	Vop-p MAX	Zin	POWER GAIN	PG dB	A_i
CALCULATED		/////					
MEASURED					/////	/////	/////

SECTION QUESTIONS

1. What is the advantage of connecting the collector lead of the transistor directly to V_{CC}, instead of using a collector resistor and an associated capacitor?
2. Is there any advantage to increasing the I_E of the circuit to improve the voltage gain?
3. Why is the voltage gain of the common collector circuit always less than unity gain? See Figure 7-70.
4. Would the output voltage increase if the 1 kΩ load resistor is removed? If it does, by how much?
5. Is the power gain and current gain of the circuit directly proportional in the common collector circuit? Why?
6. The load does not effect input impedance of the common emitter and common base circuits. Why, then, does the changing of load resistors effect the imput impedance of the common collector circuit?
7. Since the load resistor of Figure 7-71 is 1 kΩ and the emitter resistor is 3.3 kΩ, would lowering the value of RE to 1 kΩ increase the maximum peak-to-peak voltage swing? Why or why not?

SECTION PROBLEMS

1. Given: beta = 100   and $V_{BE} = 0.6$ V

   Find:

   a. I_C h. r_e
   b. V_{RE} i. A_v
   c. V_{CE} j. Z_{in}
   d. V_{RB} k. PG
   e. V_B l. A_i
   f. V_C m. $V_{op\text{-}p(max)}$
   g. V_E

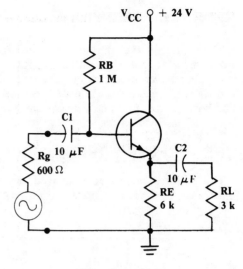

FIGURE 7-72

2. Given: Beta = 100   and $V_{BE} = 0.6$ V

   Find:

   a. I_C h. r_e
   b. V_{RE} i. A_v
   c. V_{CE} j. Z_{in}
   d. V_{RB} k. PG
   e. V_B l. A_i
   f. V_C m. $V_{op\text{-}p(max)}$
   g. V_E

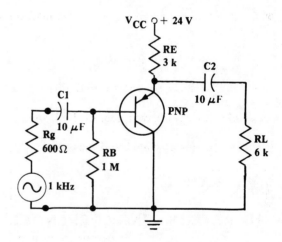

FIGURE 7-73

SECTION IV: THE COMMON-BASE, BASE-BIASED CIRCUIT

GENERAL DISCUSSION

The signal for the common base circuit is applied to the emitter and taken off the collector. Also, the input impedance at the emitter terminal of the circuit is very low, because the signal generator "looks into" the base-emitter-diode resistance, which is calculated at $r_e = 26\ mV/I_E$. Therefore, if a larger input impedance is needed, an emitter resistor, in series with the base-emitter-diode resistance, can be used. The basic circuit and one with an emitter resistor in series with the base-emitter-diode resistance will be discussed in this section.

THE DIRECT CURRENT CIRCUIT ANALYSIS

The direct current circuit analysis for the common base circuit is exactly like that of the common emitter circuit and need not be repeated. The DC voltage drops of the circuit are shown in Figure 7-74.

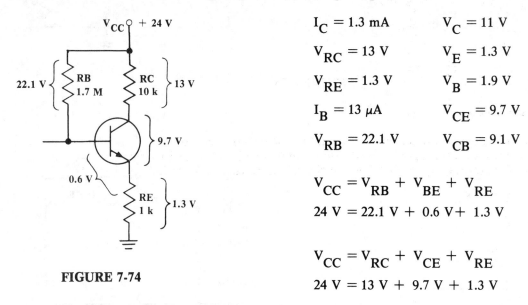

$I_C = 1.3\ mA \qquad V_C = 11\ V$

$V_{RC} = 13\ V \qquad V_E = 1.3\ V$

$V_{RE} = 1.3\ V \qquad V_B = 1.9\ V$

$I_B = 13\ \mu A \qquad V_{CE} = 9.7\ V$

$V_{RB} = 22.1\ V \qquad V_{CB} = 9.1\ V$

$V_{CC} = V_{RB} + V_{BE} + V_{RE}$
$24\ V = 22.1\ V + 0.6\ V + 1.3\ V$

$V_{CC} = V_{RC} + V_{CE} + V_{RE}$
$24\ V = 13\ V + 9.7\ V + 1.3\ V$

FIGURE 7-74

THE ALTERNATING CURRENT CIRCUIT ANALYSIS

The alternating current circuit parameters to be solved for the common base circuit are voltage gain, input impedance, power gain, and maximum peak-to-peak output voltage swing.

VOLTAGE GAIN

The voltage gain of the common base circuit is solved like that of the common emitter circuit, where the voltage gain is the ratio of the effective collector resistance to the effective emitter resistance. The effective emitter resistance can be r_e or $r_e + RE1$, depending on where the input signal from the generator is connected directly to the emitter, as shown in Figure 7-75, or to a split emitter resistance, as shown in Figure 7-76.

A. No Load Conditions

The effect of connecting the signal generator to the emitter terminal of the transistor is similar to that of the fully bypassed emitter resistor of the common emitter circuit because, in both circuit configurations, the effective emitter resistance is r_e, alone. Essentially, the input signal applied to the emitter is developed equally across r_e and RE which, to the input signal, are in parallel. However, only the signal current developed by Vin and r_e flows in the collector, providing $A_v = RC/r_e$. Again, the possibility to have r_e range from $26\ mV/I_E$ to $52\ mV/I_E$ causes the voltage gain of this connection to be unpredictable.

NOTE: if $r_e = 52\ mV/I_E$, then $r_e = 40\ \Omega$ and A_v will be 250. Also, RE = 1 kΩ is not included in the A_v

calculations because only the signal current flow of r_e flows in RC.

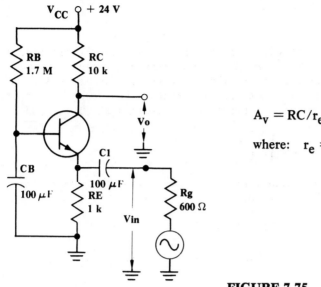

$A_v = RC/r_e = 10\ k\Omega/20\ \Omega = 500$

where: $r_e = 26\ mV/I_E = 26\ mV/1.3\ mA = 20\ \Omega$

FIGURE 7-75

B. Series Emitter Resistors

Splitting the 1 kΩ emitter resistance into 200 Ω and 800 Ω resistors and connecting the signal generator at their junction causes the input signal to be developed across RE2 and r_e + RE1 equally. However, only the signal current flow through r_e + RE1 flows in the collector. Hence, $A_v = RC/(r_e + RE1)$.

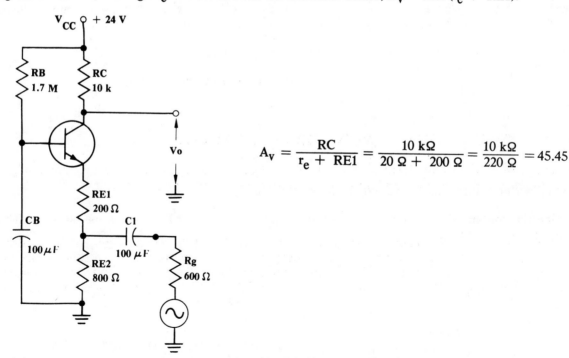

$$A_v = \frac{RC}{r_e + RE1} = \frac{10\ k\Omega}{20\ \Omega + 200\ \Omega} = \frac{10\ k\Omega}{220\ \Omega} = 45.45$$

FIGURE 7-76

NOTE: If r_e is 52 mV/I_E, then $r_e = 40\ \Omega$ and A_v only changes to 42.67.

C. Loaded Conditions

Solving the series emitter resistance connection for the loaded circuit:

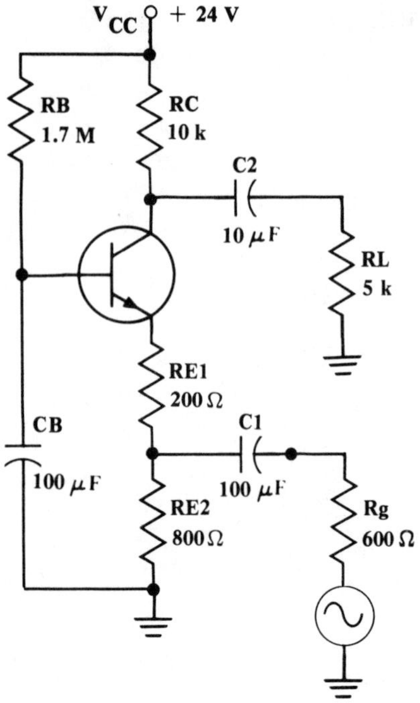

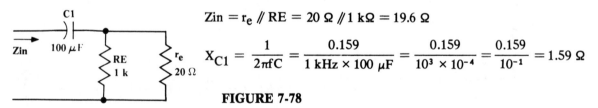

FIGURE 7-77

$$A_v = \frac{RC \parallel RL}{r_e + RE1} = \frac{10\text{ k}\Omega \parallel 5\text{ k}\Omega}{20\ \Omega + 200\ \Omega} \approx \frac{3333\ \Omega}{220\ \Omega} = 15.15$$

INPUT IMPEDANCE

Connecting the input signal to the emitter terminal of the common base circuit produces a low input impedance because the signal is looking directly into the base-emitter-diode resistance in parallel with the emitter resistor resistance. However, the input impedance can be increased by using some of the emitter resistance in series with r_e, which is equivalent to the partially-emitter-bypassed, common-emitter circuit.

A. Zin (Input signal connected directly to the emitter terminal.)

$$Z_{in} = r_e \parallel RE = 20\ \Omega \parallel 1\text{ k}\Omega = 19.6\ \Omega$$

$$X_{C1} = \frac{1}{2\pi f C} = \frac{0.159}{1\text{ kHz} \times 100\ \mu F} = \frac{0.159}{10^3 \times 10^{-4}} = \frac{0.159}{10^{-1}} = 1.59\ \Omega$$

FIGURE 7-78

B. Zin (Input signal applied to split 1 kΩ emitter resistance.)

Splitting the emitter resistor RE, while maintaining the full 1 kΩ for DC distribution purposes, increases the input impedance.

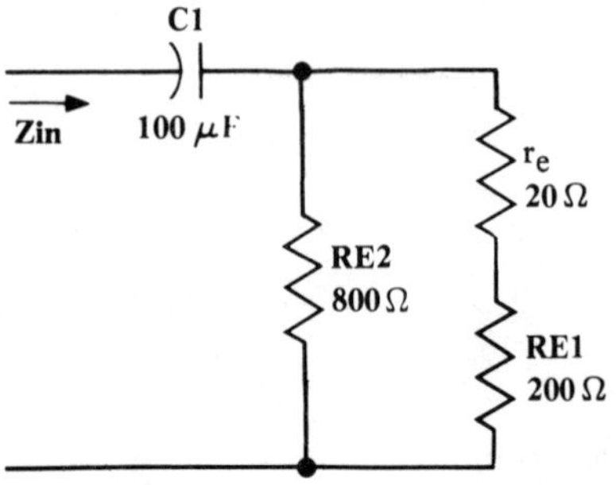

$$Z_{in} = (r_e + RE1) \parallel RE2$$
$$= (20\ \Omega + 200\ \Omega) \parallel 800\ \Omega = 220\ \Omega \parallel 800\ \Omega = 172.55\ \Omega$$

FIGURE 7-79

POWER GAIN

The power gain of the common base circuit is solved much like that of the common emitter and common collector circuits. However, in the no-load circuit, Av = PG because the input and output signal currents are approximately equal.

A. POWER GAIN WHEN $Z_{in} = r_e \parallel RE$

$$PG = A_v{}^2 \times Z_{in}/RC = 500^2 \times 19.6\ \Omega/10\ k\Omega = 490$$

$$PG(dB) = 10 \log PG = 10 \log 490 = 26.9\ dB$$

NOTE: If the effect of RE is ignored, Z_{in} would equal 20 Ω and PG = 500.

B. POWER GAIN WHEN $Z_{in} = (r_e + RE1) \parallel RE2$

1. No load conditions

$$PG = A_v{}^2 \times Z_{in}/RC = 45.45^2 \times 172.55\ \Omega/10\ k\Omega = 35.64$$

$$PG(dB) = 10 \log PG = 10 \log 35.64 = 15.63\ dB$$

2. Loaded conditions

$$PG = A_v{}^2 \times Z_{in}/RL = 15.15^2 \times 172.55\ \Omega/5\ k\Omega = 7.92$$

$$PG(dB) = 10 \log PG = 10 \log 7.92 = 8.99\ dB$$

CURRENT GAIN

The current gain of the common base circuit is unity if there are no signal losses to the circuit, and the current gain decreases as loaded conditions exist. Therefore, since $PG = A_v A_i$, then $A_i = PG/A_v$

A. **No-Load Conditions**

$$A_i = PG/A_v = 490/500 = 0.98$$

NOTE: If PG equalled 500, where Z_{in} = 20 Ω, then A_i would equal one.

B. **Increased Z_{in} with No-Load Conditions**

$$A_i = PG/A_v = 35.64/45.45 = 0.78$$

C. **Increased Z_{in} with Loaded Conditions**

$$A_i = PG/A_v = 7.92/15.15 = 0.523$$

MAXIMUM PEAK-TO-PEAK OUTPUT VOLTAGE SWING

The maximum peak-to-peak output voltage swing for the common-base, no-load circuit is solved from $2I_C R_C$ or $2V_{CB}$, whichever is smaller. The $2V_{CB}$ condition exists because the base lead is at AC ground. Hence, the input signal is developed across the base-emitter-diode resistance and the series emitter resistance, and the output is taken off the collector, with respect to ground. For the loaded circuit, Vop-p(max) equals $2V_{CB}$ or $2I_C r_L$, whichever is smaller. The output for the loaded condition is taken of the load which is in parallel with the collector resistor.

NOTE: The $2V_{CB}$ condition exists because capacitor CB places an AC ground at the base terminal and, since the negative-going excursion of the output signal cannot go below AC ground, the signal is limited to the DC voltage at the base. The $2V_{CE}$ limiting condition for the grounded-emitter, common-emitter circuit is similar to this condition, where the AC ground at the emitter limits the negative excursion of the output wave.

A. **No-Load Conditions**

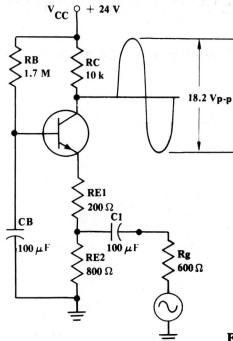

$2I_C R_C = 2 \times 1.3 \text{ mA} \times 10 \text{ k}\Omega = 26 \text{ Vp-p}$

$2V_{CB} = 2 \times 9.1 \text{ V} = 18.2 \text{ Vp-p}$

$V_{op\text{-}p(max)} = 18.2 \text{ Vp-p}$

FIGURE 7-80

B. **Loaded Conditions**

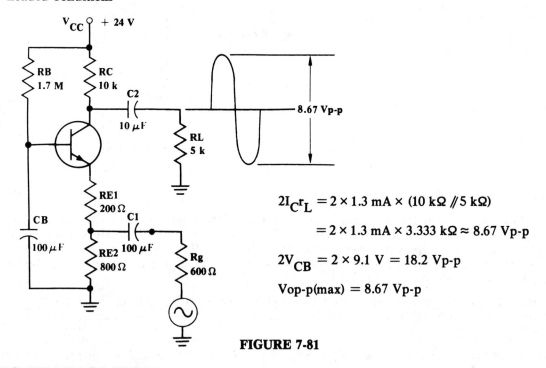

$2I_C r_L = 2 \times 1.3 \text{ mA} \times (10 \text{ k}\Omega \text{ // } 5 \text{ k}\Omega)$

$= 2 \times 1.3 \text{ mA} \times 3.333 \text{ k}\Omega \approx 8.67 \text{ Vp-p}$

$2V_{CB} = 2 \times 9.1 \text{ V} = 18.2 \text{ Vp-p}$

$V_{op\text{-}p(max)} = 8.67 \text{ Vp-p}$

FIGURE 7-81

CIRCUIT PHASE SHIFT

There is no phases shifts between the input and output signal waveforms because a positive-going input wave causes V_{BE} to decrease, I_B to decrease, I_C to decrease, V_{RC} to decrease, and the voltage at the collector, with respect to ground, to increase. Recall from the common emitter circuit that a positive-going input wave causes a positive-going I_C and also causes the signal voltage at the collector to decrease as V_{RC} increased.

SECTION IV: LABORATORY EXERCISE

OBJECTIVES
Investigate the DC and AC parameters of the base-biased, common-base circuit configuration.

LIST OF MATERIALS AND EQUIPMENT
1. Transistor: 2N3904 or equivalent
2. Resistors (one each): 330 Ω 470 Ω 680 Ω 4.7 kΩ 10 kΩ 1.5 MΩ
3. Capacitors: 10 µF (one) 100 µF (two)
4. Power Supply
5. Voltmeter
6. Signal Generator
7. Oscilloscope

PROCEDURE
1. Connect common base configuration circuit shown in Figure 7-82.

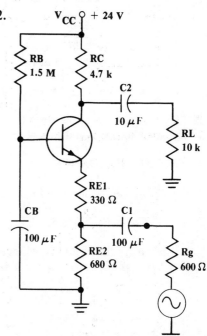

FIGURE 7-82

NOTE: The 100 µF capacitor connected between base and ground provides an AC ground at the base terminal, but it will not effect the DC voltage distribution of the circuit unless CB is defective (leaky). Also, connecting the signal generator to the junction of RE1 and RE2 provides the same function as the partially-bypassed, emitter resistor of the common emitter circuit, because RE1 "swamps" the effect of r_e, and therefore controls the gain of the circuit.

2. Measure the voltages at the collector, base, and emitter terminals of the transistor and at the junction of RE1 and RE2, all with respect to ground. Insert the measured V_C, V_E, V_B, and V_{RE2} voltages into Table 7-14.

3. Calculate the circuit voltage drops of V_{RC}, V_{RE}, V_{RB}, V_{CE}, and V_{BE}, and the circuit currents of I_B and I_C. Calculate from:

a. $V_{RC} = V_{CC} - V_C$ d. $V_{CE} = V_C - V_E$ g. $I_B = V_{RB}/RB$

b. $V_{RB} = V_{CC} - V_B$ e. $V_{BE} = V_B - V_E$ h. $Beta = I_C/I_B$

c. $V_{RE1} = V_E - V_{RE2}$ f. $I_C = V_{RC}/RC$

4. Insert the calculated values, as indicated, into Table 7-14.

TABLE 7-14	V_C	V_B	V_E	V_{RE2}	V_{RC}	V_{RB}	V_{RE1}	V_{CE}	V_{BE}	I_C	I_B	Beta
CALCULATED	//////	//////	//////	//////								
MEASURED					//////	//////	//////	//////	//////	//////	//////	//////

5. **VOLTAGE GAIN**

   a. Calculate the mid-frequency voltage gain from: $A_v = \dfrac{RC \,/\!/\, RL}{RE1 + r_e}$ where: $r_e \approx 26\ mV/I_E$

NOTE: The 330 Ω of RE1 in series with the effective r_e swamps its effect and the gain of the circuit is primarily controlled by the r_L to RE1 ratio, and only slightly by the effect of the approximate r_e, which is in series with RE1.

   b. Measure the mid-frequency voltage gain using an input voltage of 100 mVp-p at a frequency of 1 kHz. Measure the Vo at the collector with respect to ground and use Vop-p to calculate:

   $A_v = Vo/Vin$ where: $Vin = 100\ mVp\text{-}p$

NOTE: Again, measure the Vin only after the circuit is connected, because the impedance is still relatively low and the signal loss across the internal resistance of the signal generator will be significant.

6. **INPUT IMPEDANCE**

   a. Calculate the input impedance to the circuit from:

   $Zin = RE2 \,/\!/\, (RE1 + r_e)$ where: $r_e \approx 26\ mV/IE$ and $RE1 = 330\ \Omega$

   b. Measure the input impedance. Use a known 470 Ω resistor and an input signal level of 100 mVp-p at 1 kHz, as shown in Figure 7-83.

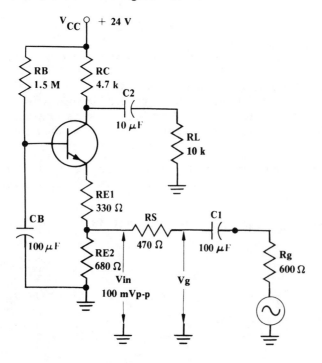

FIGURE 7-83

c. Calculate Zin from:

$$Zin = \frac{Vin \times RS}{V_{RS}} = \frac{Vin \times RS}{VG - Vin} \quad \text{where:} \quad Vin = 100 \text{ mVp-p}$$

7. MAXIMUM PEAK-TO-PEAK VOLTAGE SWING

a. Calculate the maximum peak-to-peak output voltage swing from $2I_C r_L$ or $2V_{CB}$, whichever is smaller.

b. Monitor the collector output and increase the 1 kHz input signal level until clipping or distortion, because of alpha crowding, occurs.

NOTE: RE1 does not provide degeneration and the approximate peak-to-peak excursion is approximately $2V_{CB}$, since the signal voltage off the collector is measured with respect to ground. RS is removed for this measurement.

8. POWER GAIN

a. Calculate the power gain to the load of the circuit from: $PG = A_v^2 \times Zin/RL$

b. Convert the power gain to dB from $PG(dB) = 10 \log PG$

NOTE: Use the A_v and Zin values in the power gain calculations.

9. Insert the calculated and measured values, as indicated, into Table 7-15.

TABLE 7-15	A_v	Vo for Vin = 100 mVp-p	Zin	Vop-p MAX	POWER GAIN	PG (db)
CALCULATED		/////				
MEASURED					/////	/////

SECTION QUESTIONS

1. With reference to the common base circuit, what is the main drawback to connecting the input signal directly into the emitter of the transistor?
2. Is the current gain of the common base circuit ever greater than unity?
3. Why is the voltage gain of the common base circuit about equal to a similar biased common emitter circuit having an equal load resistor?

SECTION PROBLEMS

1. Given: Beta = 100 and $V_{BE} = 0.6$ V

   Find:

   a. I_C i. r_e
   b. V_{RC} j. A_v
   c. V_{RE} k. Zin
   d. V_{CE} l. PG
   e. V_{RB} m. Vop-p(max)
   f. V_C
   g. V_B
   h. V_E

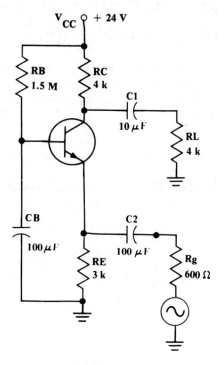

FIGURE 7-84

2. Given: Beta = 100 and $V_{BE} = 0.6$ V

   Find:

   a. I_C i. r_e
   b. V_{RC} j. A_v
   c. V_{RE} k. Zin
   d. V_{CE} l. PG
   e. V_{RB} m. Vop-p(max)
   f. V_C
   g. V_B
   h. V_E

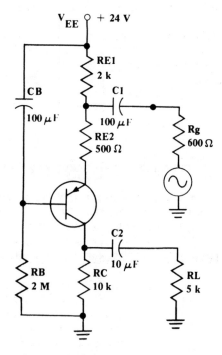

FIGURE 7-85

SECTION V: EQUIVALENT CIRCUITS

GENERAL DISCUSSION

In the earlier years of the developing transistor, many designers used hybrid parameters as a guide to designing and analyzing small signal transistor amplifiers. However, as the specifications on transistors continued to improve and circuit designers became more knowledgeable, the use of approximation techniques became the more accepted approach. However, the early influence of hybrid parameters is still reflected in data sheets, where the small signal characteristics of transistors continue to be given in "h" parameters, as they are often referred to. In this section, common emitter, common collector, and common base circuits will be analyzed using simplified equivalent circuit techniques.

HYBRID PARAMETER NOTATION

The four hybrid parameters of importance in the analysis of small signal amplifier circuits are h_{fe}, h_{ie}, h_{oe}, and h_{re}. However, in practice, only h_{fe} and h_{ie} are widely used.

1. h_{fe} is the current gain of the device and is equal to beta. "h" stands for hybrid, "e" stands for the common emitter configuration, and "f" stands for forward current.
2. h_{ie} is the input impedance at the base of the transistor and is approximately equal to βr_e. Again "h" stands for hybrid, "e" stands for common emitter configuration, and "i" stands for input.
3. h_{oe} is the output admittance, which is the inverse of output resistance of the device proper, not includ- in the collector biasing resistor. Both "h" and "e" are as previously described and "o" signifies output.
4. h_{re} is the reverse voltage ratio and, except for use in "h" parameters or in very high frequency applications, it is rarely used. The "r" stands for reverse. h_{re} is also called the voltage feedback ratio.

OBTAINING h PARAMETER DATA

Data books list the small signal common emitter characteristics of h_{fe}, h_{ie}, h_{oe}, and h_{re} giving minimum and maximum device values. Characteristic values of the 2N3904, taken from data sheets, are given below, along with nominal values that will be used in the calculations of circuit voltage gain, input impedance, and current gain.

CHARACTERISTICS	SYMBOL	MIN	MAX	NOMINAL VALUES
Current Gain	h_{fe}	100	400	$\approx$ 200
Input Impedance	h_{ie}	1 kΩ	10 kΩ	$\approx$ 4 kΩ
Voltage Feedback Ratio	h_{re}	0.5×10^{-4}	8×10^{-4}	$\approx 4 \times 10^{-4}$
Output Admittance	h_{oe}	1 μMho	40 μMho	$\approx$ 10 μMho

THE HYBRID EQUIVALENT CIRCUIT

The equivalent circuit for the basic base-biased circuit is given in Figure 7-86. Notice that the equivalent circuit consists of h_{ie}, the input impedance of the device proper (which is essentially equal to βr_e), and the output circuit consists of a constant current source $h_{fe}i_B$, which equals $\beta i_B = I_C$. In the input circuit, h_{re} is given but, because its effect on the calculated value is almost non-existent, it will be dropped in the simplified equivalent circuit. Too, at mid-band frequencies, X_C approaches zero ohms of capacitive reactance and it is not included. Also, because RB at 1.6 MΩ is so large, it normally is not shown in the hybrid circuit. However, while it is shown in the circuit, it is not used in the calculations.

VOLTAGE GAIN: Using the nominal values established for the 2N3904, the voltage gain is solved from:

$$A_v = \frac{h_{fe} R_C}{h_{ie} + (h_{ie}h_{oe} - h_{re}h_{fe})} = \frac{200 \times 3 \text{ k}\Omega}{4 \text{ k}\Omega + (4 \text{ k}\Omega \times 10 \text{ }\mu\text{Mho} - 4 \times 10^{-4} \times 200)} =$$

$$A_V = \frac{600 \text{ k}\Omega}{4 \text{ k}\Omega + (40 \times 10^{-3} - 80 \times 10^{-3})} = \frac{600 \text{ k}\Omega}{4 \text{ k}\Omega - (40 \times 10^{-3})} \approx \frac{600 \text{ k}\Omega}{4 \text{ k}\Omega} = 150$$

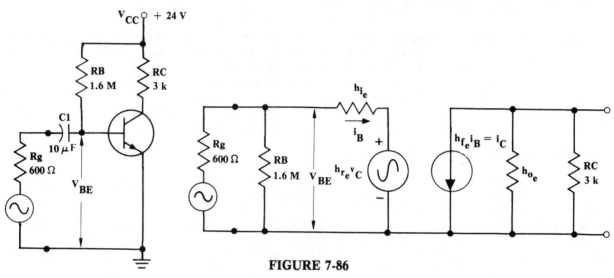

FIGURE 7-86

INPUT IMPEDANCE: The input impedance, again using the nominal 2N3904 values, is solved from:

$$Z_{in} = \frac{h_{ie} + (h_{ie}h_{oe} - h_{re}h_{fe})}{1 + h_{oe}RC} = \frac{4 \text{ k}\Omega + (40 \times 10^{-3})}{1 + (10 \times 10^{-6} \times 3 \text{ k}\Omega)} = \frac{4 \text{ k}\Omega}{1 + (30 \times 10^{-3})} = \frac{4 \text{ k}\Omega}{1.03} = 3883.5 \text{ }\Omega$$

NOTE: RB is not included in the calculations because, at 1.6 MΩ, it is too large to have any effect.

CURRENT GAIN: The current gain is solved from:

$$A_i = \frac{h_{fe}}{1 + h_{oe}RC} = \frac{200}{1 + (30 \times 10^{-3})} = \frac{200}{1.03} = 194$$

THE SIMPLIFIED EQUIVALENT CIRCUIT

The simplified equivalent circuit drops the h_{re} and h_{oe} transistor characteristics from the formula in solving for voltage gain, input impedance, and current gain.

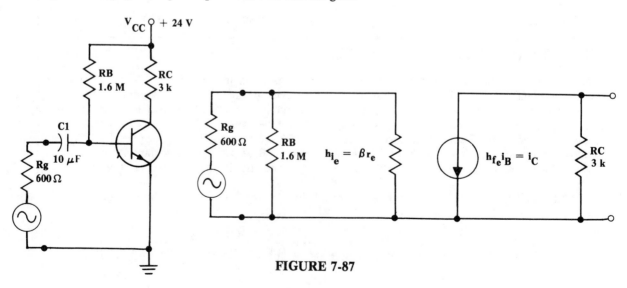

FIGURE 7-87

1. **CURRENT GAIN:** The current gain for the simplified circuit is:

   $h_{f_e} = 200$. In standard notation, $h_{f_e} = \text{Beta} = 200$

2. **INPUT INPEDANCE:** The input impedance for the simplified circuit is:

   $h_{i_e} = \beta r_e = 200 \times 20\ \Omega = 4\ k\Omega$ where: $r_e = h_{i_e}/\beta = 4\ k\Omega/200 = 20\ \Omega$

3. **VOLTAGE GAIN:** The voltage gain for the simplifed circuit is solved from:

   $A_v = h_{f_e} R C / h_{i_e} = 200 \times 3\ k\Omega / 4\ k\Omega = 600\ k\Omega / 4\ k\Omega = 150$

In standard notation:

$$A_v = h_{f_e} RC / h_{i_e} = \beta RC / \beta r_e = RC / r_e = 3\ k\Omega / 20\ \Omega = 150$$

APPROXIMATION AND SIMPLIFIED EQUIVALENT CIRCUIT COMPARISON

Approximation techniques solve for the voltage gain, input impedance, current gain, and power gain without solving for collector or base signal currents. Hence, solving the same circuit using equivalent circuits takes into account the signal current, which gives the solved problem a whole new perspective. In analyzing the circuit of Figure 7-88, both the approximation and equivalent circuit methods of solution are used to provide a comparison — to show that both methods give similar results.

Given: $\text{Beta} = h_{f_e} = 200$ $r_e = 20\ \Omega$

$\text{Vin} = 40\ \text{mVp-p}$ $V_{BE} = 0.6\ V$

$I_C = \dfrac{V_{CC} - V_{BE}}{RB/\beta} = \dfrac{11\ V - 0.6\ V}{1.6\ M\Omega/200} = \dfrac{10.4\ V}{8\ k\Omega} = 1.3\ mA$

$r_e = 26\ mV/I_E = 26\ mV/1.3\ mA = 20\ \Omega$

$V_{RC} = I_C R_C = 1.3\ mA \times 3\ k\Omega = 3.9\ V$

$V_{CE} = V_{CC} - V_{RC} = 11\ V - 3.9\ V = 7.1\ V$

$V_{RB} = V_{CC} - V_{BE} = 11\ V - 0.6\ V = 10.4\ V$

$I_B = V_{RB}/RB = 10.4\ V / 1.6\ M\Omega = 6.5\ \mu A$, or

$I_B = I_C/\beta = 1.3\ mA / 200 = 6.5\ \mu A$

A. Approximation Techniques

$Z_{in} = R_{in} = RB\ //\ \beta r_e = 1.6\ M\Omega\ //\ (200 \times 20\ \Omega)$
$= 1.6\ M\Omega\ //\ 4\ k\Omega \approx 4\ k\Omega$

$A_v = RC/r_e = 3\ k\Omega / 20\ \Omega = 150$

$V_o = \text{Vin} \times A_v = 40\ \text{mVp-p} \times 150 = 6\ \text{Vp-p}$

$A_p = PG = A_v^2 \times Z_{in}/RC = 150^2 \times 4\ k\Omega/3\ k\Omega = 30\ k\Omega$

$A_i = PG/A_v = 30{,}000/150 = 200$

$\text{Vin} = V_o/A_v = 6\ \text{Vp-p}/150 = 40\ \text{mVp-p}$

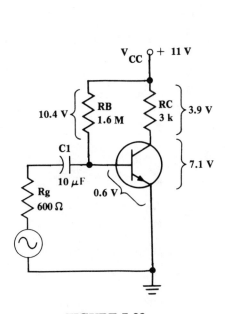

FIGURE 7-88

B. Equivalent Circuit Techniques

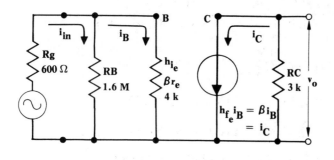

$i_B = V_{in}/R_{in} \approx 40 \text{ mVp-p}/4 \text{ k}\Omega = 10 \text{ μAp-p}$

$i_C = h_{f_e} i_B = \beta i_B = 200 \times 10 \text{ μAp-p} = 2 \text{ mAp-p}$

$V_o = i_C R_C = 2 \text{ mAp-p} \times 3 \text{ k}\Omega = 6 \text{ Vp-p}$

$r_e = R_{in}/h_{f_e} = Z_{in}/\beta = 4 \text{ k}\Omega/200 = 20 \text{ }\Omega$

$A_V = h_{f_e} R_C / h_{ie} = \beta R_C / h_{ie}$
$\quad = (200 \times 3 \text{ k}\Omega)/4 \text{ k}\Omega = 150$

$A_i = h_{f_e} = \beta = 200$

$A_p = PG = A_V A_i = 150 \times 200 = 30 \text{ k}\Omega$

$PG(dB) = 10 \log 30{,}000 = 44.77 \text{ dB}$

FIGURE 7-89

Graphical Analysis

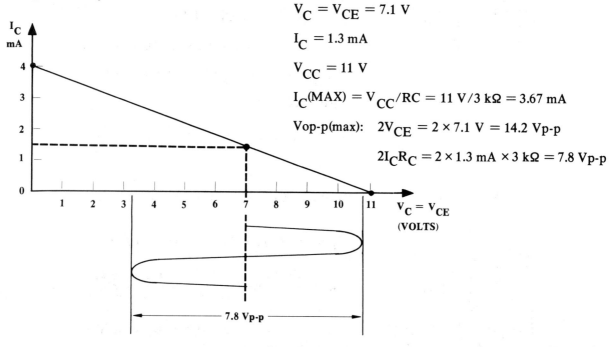

$V_C = V_{CE} = 7.1 \text{ V}$

$I_C = 1.3 \text{ mA}$

$V_{CC} = 11 \text{ V}$

$I_C(\text{MAX}) = V_{CC}/R_C = 11 \text{ V}/3 \text{ k}\Omega = 3.67 \text{ mA}$

Vop-p(max): $2V_{CE} = 2 \times 7.1 \text{ V} = 14.2 \text{ Vp-p}$

$\qquad\qquad 2I_C R_C = 2 \times 1.3 \text{ mA} \times 3 \text{ k}\Omega = 7.8 \text{ Vp-p}$

FIGURE 7-90

PRACTICAL EQUIVALENT CIRCUIT APPLICATION

If simplified circuit are used, then it becomes obvious that none of the four hybrid h_{f_e}, h_{r_e}, h_{ie} and h_{oe} parameters obtained from the data sheets are actually necessary for circuit design or analysis. In fact, only h_{f_e}, which is the current gain or beta, is a required parameter for practical circuit analysis, and beta can be easily measured. However, the hybrid-parameter, equivalent-circuit model does lead nicely into practical equivalent circuit analysis, which will be shown for the previously analyzed common emitter, common collector, and common base circuits.

THE COMMON EMITTER CIRCUIT

The circuit of Figure 7-91 differs only slightly from the common emitter circuit of Figure 7-28. Therefore, only a quick summary is necessary to cover the DC and AC analysis.

A. Approximation Approach

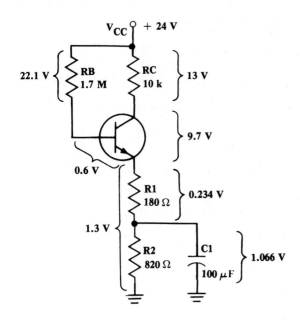

$$I_C = \frac{V_{CC} - V_{BE}}{RB/\beta + RE} = \frac{24\ V - 0.6\ V}{1.7\ M\Omega/100 + 1\ k\Omega}$$

$$= \frac{23.4\ V}{17\ k\Omega + 1\ k\Omega} = 1.3\ mA$$

$V_{RC} = I_C RC = 1.3\ mA \times 10\ k\Omega = 13\ V$

$V_{R1} = I_E R1 = 1.3\ mA \times 180\ \Omega = 0.234\ V$

$V_{R2} = I_E R2 = 1.3\ mA \times 820\ \Omega = 1.066\ V$

$V_{RB} = I_B RB = I_C RB = 13\ \mu A \times 1.7\ M\Omega = 22.1\ V$

$V_C = V_{CC} - V_{RC} = 24\ V - 13\ V = 11\ V$

$V_E = V_{R1} + V_{R2} = I_E(R1 + R2)$

$\quad = 1.3\ mA \times 1\ k\Omega = 1.3\ V$

$V_B = V_E + V_{BE} = 1.3\ V + 0.6\ V = 1.9\ V$

$V_{CE} = V_C - V_E = 11\ V - 1.3\ V = 9.7\ V$

FIGURE 7-91

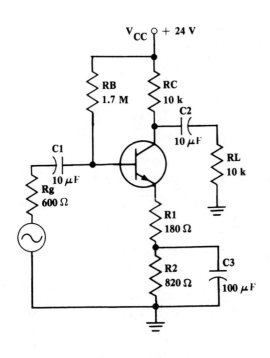

Given: Beta $= 100$ and Vin $= 400$ mVp-p

$r_e = 26\ mV/I_E = 26\ mV/1.3\ mA = 20\ \Omega$

$A_V = \dfrac{RC\ //\ RL}{r_e + R1} = \dfrac{10\ k\Omega\ //\ 10\ k\Omega}{20\ \Omega + 180\ \Omega} = \dfrac{5\ k\Omega}{200\ \Omega} = 25$

$Vo = A_v \times Vin = 25 \times 400\ mVp\text{-}p = 10\ Vp\text{-}p$

$Zin = RB\ //\ \beta(r_e + R1) = 1.7\ M\Omega\ //\ 100(20\ \Omega + 180\ \Omega)$

$\quad = 1.7\ M\Omega\ //\ (100 \times 200\ \Omega) = 1.7\ M\Omega\ //\ 20\ k\Omega \approx 20\ k\Omega$

$PG = A_v{}^2 \times Zin/RL = 25^2 \times 20\ k\Omega/10\ k\Omega = 1250$

$A_i = PG/A_v \approx 1250/25 = 50$

$Vop\text{-}p(max) = 2I_C r_L = 2 \times 1.3\ mA \times 5\ k\Omega = 13\ Vp\text{-}p$

$Vin(max) = Vop\text{-}p(max)/A_v = 13\ Vp\text{-}p/25 = 520\ mVp\text{-}p$

FIGURE 7-92

B. Equivalent Circuit

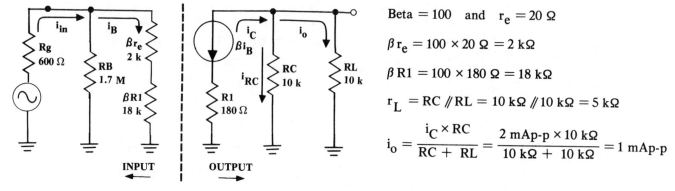

FIGURE 7-93

NOTE: Circuit current calculations using Norton's Theorem will be further investigated in Chapter 8.

$Zin = RB \;//\; \beta(r_e + R1) = 1.7\,M\Omega \;//\; 100(20\,\Omega + 180\,\Omega) = 1.7\,M\Omega \;//\; 20\,k\Omega \approx 20\,k\Omega$

$i_{in} \approx Vin/Zin = 400\,mVp\text{-}p/20\,k\Omega = 20\,\mu Ap\text{-}p$

$i_B = Vin/\beta(r_e + R1) = 400\,mVp\text{-}p/20\,k\Omega = 20\,\mu Ap\text{-}p$

$i_C = \beta i_B = 100 \times 20\,\mu Ap\text{-}p = 2\,mAp\text{-}p$

$Vo = i_C r_L = 2\,mAp\text{-}p \times 5\,k\Omega = 10\,Vp\text{-}p$, or

$Vo = i_o RL = 1\,mAp\text{-}p \times 10\,k\Omega = 10\,Vp\text{-}p$

$A_v = Vo/Vin = 10\,Vp\text{-}p/400\,mVp\text{-}p = 25$, or

$A_v = \beta r_L/Zin = (100 \times 5\,k\Omega)/20\,k\Omega = 500\,k\Omega/20\,k\Omega = 25$

where: $\beta r_L/Zin = \beta r_L/\beta(r_e + R1) = r_L/(r_e + R1)$

$A_i = i_o/i_{in} = 1\,mAp\text{-}p/20\,\mu Ap\text{-}p = 50$, or

$A_i = \dfrac{Vo/RL}{Vin/Zin} = \dfrac{10\,Vp\text{-}p/10\,k\Omega}{400\,mVp\text{-}p/20\,k\Omega} = 50$, or

$A_i = A_v \times Zin/RL = 25 \times 20\,k\Omega/10\,k\Omega = 50$ where: $Zin \approx Zin(base) = 20\,k\Omega$

$A_p = PG = A_v A_i = 25 \times 50 = 1250$

$PG(dB) = 10 \log 1250 = 30.97\,dB$

$Vop\text{-}p(max) = 2I_C r_L = 2 \times 1.3\,mA \times 5\,k\Omega = 13\,Vp\text{-}p$

THE COMMON BASE CIRCUIT

The direct current analysis of the common base circuit is identical to that of the common emitter circuit and need not be repeated. Again, the alternating current analysis for the slightly modified circuit of Figure 7-94 is summarized for convenience, and compared to the equivalent circuit approach to show the similarities.

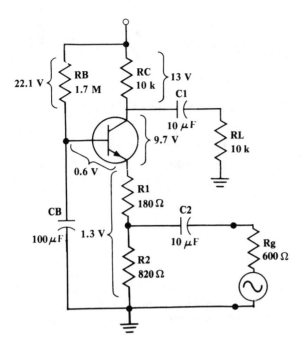

$r_e = 26\text{ mV}/IE = 26\text{ mV}/1.3\text{ mA} = 20\text{ }\Omega$

$A_v = \dfrac{RC\text{ // }RL}{r_e + R1} = \dfrac{10\text{ k}\Omega\text{ // }10\text{ k}\Omega}{20\text{ }\Omega + 180\text{ }\Omega} = \dfrac{5\text{ k}\Omega}{200\text{ }\Omega} = 25$

$Vo = A_v \times Vin = 25 \times 400\text{ mVp-p} = 10\text{ Vp-p}$

$Zin = (r_e + R1)\text{ // }R2 = (20\text{ }\Omega + 180\text{ }\Omega)\text{ // }820\text{ }\Omega = 160\text{ }\Omega$

$PG = A_v^2 \times Zin/RL = 25^2 \times 160\text{ }\Omega/10\text{ k}\Omega = 10$

$A_i = PG/A_v = 10/25 = 0.4$

$Vop\text{-}p(max) = 2I_C r_L = 2 \times 1.3\text{ mA} \times 5\text{ k}\Omega = 13\text{ Vp-p}$

$Vin(max) = Vop\text{-}p(max)/A_v = 13\text{ Vp-p}/25 = 520\text{ mVp-p}$

FIGURE 7-94

EQUVALENT CIRCUIT

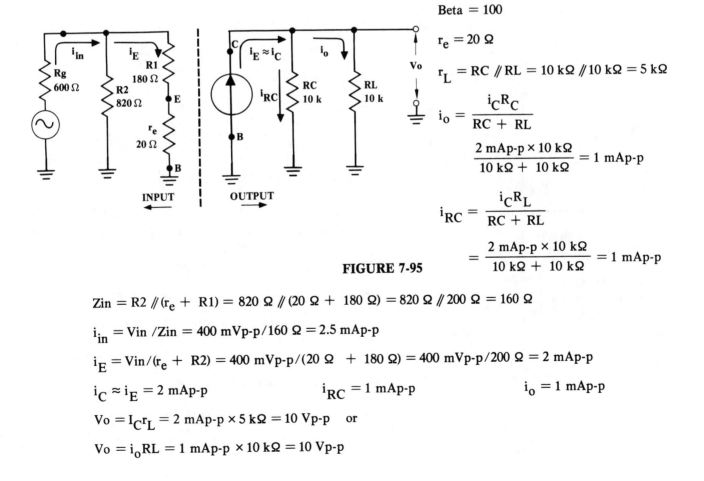

Beta = 100

$r_e = 20\text{ }\Omega$

$r_L = RC\text{ // }RL = 10\text{ k}\Omega\text{ // }10\text{ k}\Omega = 5\text{ k}\Omega$

$i_o = \dfrac{i_C R_C}{RC + RL}$

$\dfrac{2\text{ mAp-p} \times 10\text{ k}\Omega}{10\text{ k}\Omega + 10\text{ k}\Omega} = 1\text{ mAp-p}$

$i_{RC} = \dfrac{i_C R_L}{RC + RL}$

$= \dfrac{2\text{ mAp-p} \times 10\text{ k}\Omega}{10\text{ k}\Omega + 10\text{ k}\Omega} = 1\text{ mAp-p}$

FIGURE 7-95

$Zin = R2\text{ // }(r_e + R1) = 820\text{ }\Omega\text{ // }(20\text{ }\Omega + 180\text{ }\Omega) = 820\text{ }\Omega\text{ // }200\text{ }\Omega = 160\text{ }\Omega$

$i_{in} = Vin/Zin = 400\text{ mVp-p}/160\text{ }\Omega = 2.5\text{ mAp-p}$

$i_E = Vin/(r_e + R2) = 400\text{ mVp-p}/(20\text{ }\Omega + 180\text{ }\Omega) = 400\text{ mVp-p}/200\text{ }\Omega = 2\text{ mAp-p}$

$i_C \approx i_E = 2\text{ mAp-p}$ $i_{RC} = 1\text{ mAp-p}$ $i_o = 1\text{ mAp-p}$

$Vo = I_C r_L = 2\text{ mAp-p} \times 5\text{ k}\Omega = 10\text{ Vp-p}$ or

$Vo = i_o RL = 1\text{ mAp-p} \times 10\text{ k}\Omega = 10\text{ Vp-p}$

$$A_v = V_o/V_{in} = 10 \text{ Vp-p}/400 \text{ mVp-p} = 25$$

$$A_i = A_v \times Z_{in}/R_L = 25 \times 160 \text{ }\Omega/10 \text{ k}\Omega = 0.4 \quad \text{or}$$

$$A_i = i_o/i_{in} = 1 \text{ mAp-p}/2.5 \text{ mAp-p} = 0.4$$

$$A_p = PG = A_v A_i = 25 \times 0.4 = 10 \quad \text{or}$$

$$A_p = P_o/P_{in} = \frac{V_o^2/R_L}{V_{in}^2/Z_{in}} = \frac{10 \text{ V}^2/10 \text{ k}\Omega}{400 \text{ mV}^2\text{p-p}/160 \text{ }\Omega} = \frac{100 \text{ V}/10 \text{ k}\Omega}{0.16 \text{ V}/160 \text{ }\Omega} = \frac{10 \text{ mW}}{1 \text{ mW}} = 10 \quad \text{or}$$

$$A_p = PG = A_v^2 \times Z_{in}/R_L = 25^2 \times 160 \text{ }\Omega/10 \text{ k}\Omega = 10$$

$$PG(dB) = 10 \log 10 = 10$$

THE COMMON COLLECTOR CIRCUIT

The direct current analysis of the common collector circuit is similar to the Figure 7-55 analysis and is repeated here for convenience. Likewise, the alternating current approximation circuit parameters are repeated for convenience in comparing them with results from the equivalent circuit method of analysis.

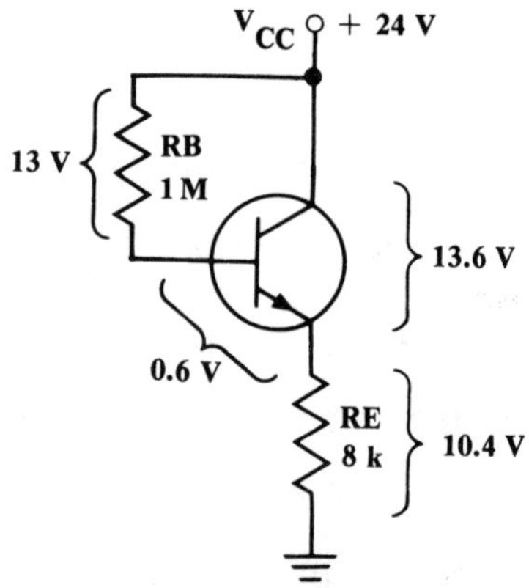

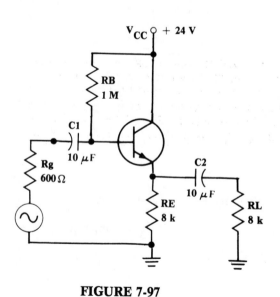

FIGURE 7-96

FIGURE 7-97

$$I_C = \frac{V_{CC} - V_{BE}}{(R_B/\beta) + R_E} = \frac{24 \text{ V} - 0.6 \text{ V}}{(1 \text{ M}\Omega/100) + 8 \text{ k}\Omega} = \frac{23.4 \text{ V}}{10 \text{ k}\Omega + 8 \text{ k}\Omega} = 1.3 \text{ mA}$$

$$I_B = I_C/\beta = 1.3 \text{ mA}/100 = 13 \text{ }\mu A$$

$$V_{RE} = I_E R_E = 1.3 \text{ mA} \times 8 \text{ k}\Omega = 10.4 \text{ V}$$

$$V_{RB} = I_B R_B = 13 \text{ }\mu A \times 1 \text{ M}\Omega = 13 \text{ V}$$

$$V_B = V_{CC} - V_{RB} = 24 \text{ V} - 13 \text{ V} = 11 \text{ V} \quad \text{or}$$

$$V_B = V_{RE} + V_{BE} = 10.4 \text{ V} + 0.6 \text{ V} = 11 \text{ V}$$

$$V_{CE} = V_{CC} - V_{RE} = V_C - V_E = 24 \text{ V} - 10.4 \text{ V} = 13.6 \text{ V}$$

$$r_e = 26 \text{ mV}/I_E = 26 \text{ mV}/1.3 \text{ mA} = 20 \text{ }\Omega$$

$$A_v = \frac{r_L}{r_e + r_L} = \frac{R_E /\!/ R_L}{r_e + (R_E /\!/ R_L)} = \frac{8 \text{ k}\Omega /\!/ 8 \text{ k}\Omega}{20 \text{ }\Omega + (8 \text{ k}\Omega /\!/ 8 \text{ k}\Omega)} = 0.995$$

$$V_o = A_v \times V_{in} = 0.995 \times 4 \text{ V} = 3.98 \text{ Vp-p}$$

$$Z_{in} = R_B /\!/ \beta(r_e + r_L) = R_B /\!/ \beta[r_e + (R_E /\!/ R_L)]$$

$$= 1 \text{ M}\Omega /\!/ 100(20 \text{ }\Omega + 4 \text{ k}\Omega) = 1 \text{ M}\Omega /\!/ 402 \text{ k}\Omega = 286.7 \text{ k}\Omega$$

$$PG = A_v^2 \times Z_{in}/R_L = 0.995^2 \times 286.7 \text{ k}\Omega/8 \text{ k}\Omega = 35.48$$

$$A_i = PG/A_v = 35.48/0.995 = 35.66$$

$$V_{op-p}(max) = 2I_E r_L = 2 \times 1.3 \text{ mA} \times 4 \text{ k}\Omega = 10.4 \text{ Vp-p}$$

$$V_{inp-p}(max) = V_{op-p}(max)/A_v = 10.4 \text{ Vp-p}/0.995 = 10.45 \text{ Vp-p}$$

EQUIVALENT CIRCUIT

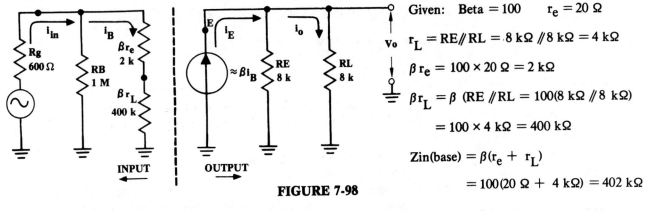

FIGURE 7-98

Given: Beta = 100 $r_e = 20\ \Omega$

$r_L = RE // RL = 8\ k\Omega // 8\ k\Omega = 4\ k\Omega$

$\beta r_e = 100 \times 20\ \Omega = 2\ k\Omega$

$\beta r_L = \beta (RE // RL) = 100(8\ k\Omega // 8\ k\Omega)$

$= 100 \times 4\ k\Omega = 400\ k\Omega$

$Z_{in(base)} = \beta(r_e + r_L)$

$= 100(20\ \Omega + 4\ k\Omega) = 402\ k\Omega$

$Z_{in} = RB // \beta(r_e + r_L) = 1\ M\Omega // 100(20\ \Omega + 4\ k\Omega) = 1\ M\Omega // 400\ k\Omega = 286.7\ k\Omega$

$i_{in} = V_{in}/Z_{in} = 4\ Vp\text{-}p/286.7\ k\Omega = 13.95\ \mu Ap\text{-}p = 14\ \mu Ap\text{-}p$

$i_B = V_{in}/\beta r_L = 4\ Vp\text{-}p/(100 \times 4020\ \Omega) = 4\ Vp\text{-}p/402\ k\Omega = 9.95\ \mu Ap\text{-}p$

$i_C \approx \beta i_B = 100 \times 9.95\ \mu Ap\text{-}p = 995\ \mu Ap\text{-}p$

$V_o = i_B r_L = 9.95\ \mu Ap\text{-}p \times 4\ k\Omega = 3.98\ Vp\text{-}p$

$A_v = V_o/V_{in} = 3.98\ Vp\text{-}p/4\ Vp\text{-}p = 0.995$ or

$A_v = \beta r_L/Z_{in} = (100 \times 4\ k\Omega)/402\ k\Omega = 400\ k\Omega/402\ k\Omega = 0.995$

$A_i = A_v Z_{in}/RL = (0.995 \times 286.7\ k\Omega)/8\ k\Omega = 35.66$

$PG = A_p = A_v A_i = 0.995 \times 35.66 = 35.48$ or

$PG = P_o/P_{in} = \dfrac{V_o^2/RL}{V_{in}^2/Z_{in}} = \dfrac{3.98^2/8\ k\Omega}{4\ V^2/286.7\ k\Omega} = \dfrac{1.98\ MW}{56.8\ \mu W} = 35.48$

$PG(dB) = 10 \log 35.48 \approx 15.5\ dB$

SUMMARY

Approximation and equivalent circuit methods of analysis were used to analyze the common emitter, common collector, and common base circuits. The approximation method uses signal voltage while the equivalent circuit method uses signal current developed across known resistances. Both methods of analysis provide similar results but the approximation techniques, like voltage divider equations, are more direct because they do not have to include signal current flow in the calculations. Essentially, knowing the approximation approach and then learning the equivalent circuit approach stimulates further circuit understanding.

8 | THE UNIVERSAL-BIASED, AMPLIFIER CIRCUIT

GENERAL DISCUSSION

The universal-biased amplifier circuit, like the base-biased amplifier circuit, can be connected into the useful common emitter, common base, and common collector circuit configurations using either NPN or PNP transistors in conventional or upside-down circuits. Additionally, these universal-biased, amplifier-circuit configurations have characteristics similar to those of the base-biased circuits.

As shown in the last chapter, the common emitter circuit has both voltage and current gains and a moderate input impedance; the common collector circuit has current gain, a voltage gain of less than unity, and a high input impedance under normal operating conditions; and the common base circuit has voltage gain, a current gain of less than unity, and a low input impedance. All of these circuit configurations, with their various characteristics, can be used together, successfully, in complex circuits.

As a stage in a complex circuit, the common emitter configurations is most widely used because it has both current and voltage gain capabilities. The common collector circuit configuration generally is used to transform a low impedance into a higher impedance in conjunction with other stages like the common emitter circuit. The common base configuration, because it has a low input impedance, generally is used to match another low impedance, such as the output from a common collector circuit.

In this chapter, all three of the universal-biased, amplifier-circuit configurations will be analyzed for voltage gain, input impedance, power gain, and maximum peak-to-peak output voltage swing.

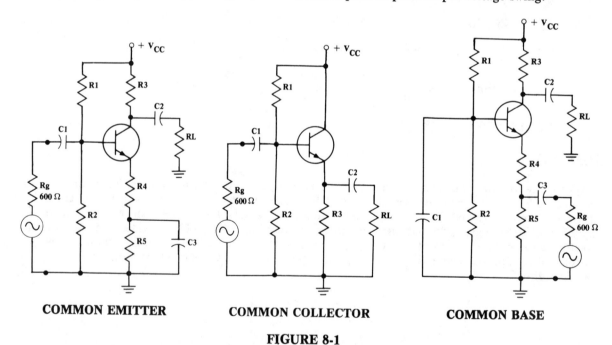

COMMON EMITTER   COMMON COLLECTOR   COMMON BASE

FIGURE 8-1

SECTION I: THE COMMON EMITTER CIRCUIT
GENERAL DISCUSSION

The voltage gain and maximum peak-to-peak output voltage swing of the common-emitter, universal circuits are solved like the base-biased circuits, but the input impedance is generally lower because of the biasing resistors R1 and R2, as shown in Figure 8-1. The R1 and R2 resistors provide improved direct current circuit stability, but in order to do so, successfully, their values are usually much lower than the base resistor of the base-biased circuit. Also, since the signal generator must be connected to the input base in the common emitter configuration, as shown in Figure 8-2, the input impedance has to be lower. Again, notice that the power supply polarities used for the NPN and PNP transistors are opposite, but if the parameters of the devices are identical, then both the DC voltage drops and the AC parameters of the NPN and PNP circuits will be similar.

There is no advantage other than polarity preference in using one device over the other for single stage applications. However, in multistage circuitry, where more than one stage is used to provide a larger voltage gain, higher input impedance, and so forth, the capability of combining NPN and PNP devices will be invaluable.

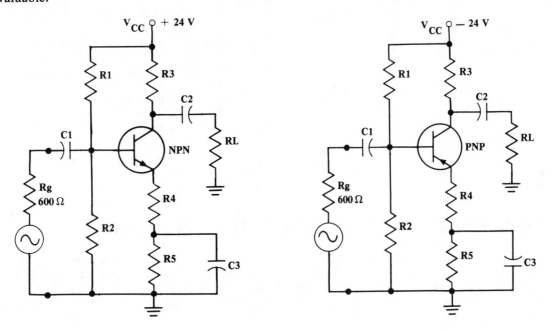

FIGURE 8-2

NOTE: The coupling and bypass capacitors of the NPN and PNP circuits are of opposite polarity, as shown in Figure 8-2.

Part 1: The NPN, Common-Emitter, Universal Circuit

The analysis of the universal circuit, like that of the base-biased circuit, begins with the DC analysis and proceeds to the AC analysis. And, because the NPN and PNP transistor circuits are analyzed in a similar manner with similar results, only the NPN circuit configuration will be fully analyzed.

DIRECT CURRENT ANALYSIS OF THE NPN CIRCUIT

The DC analysis of the NPN circuit was covered in Chapter 6 for both the approximation and exact voltage methods. However, the exact technique used Thevenin's equivalent to solve for I_E, while the approximation method used the much simpler voltage divider approach. Since the difference in the resulting circuit calculations versus the actual measured values in the laboratory, for a well-designed circuit, were too slight to consider one method better than the other, the simpler approximation techniques will be used in all of the remaining universal circuit calculations. However, the equivalent circuit approach will be used later in this chapter to stimulate further circuit understanding and to show that it and the approximation method provide similar results.

The direct current analysis of the common-emitter, universal-biased circuit of Figure 8-3 is as follows:

Given: Beta = 100

$$V_{R1} = \frac{V_{CC} \times R1}{R1 + R2} = \frac{24\text{ V} \times 130\text{ k}\Omega}{130\text{ k}\Omega + 30\text{ k}\Omega} = 19.5\text{ V}$$

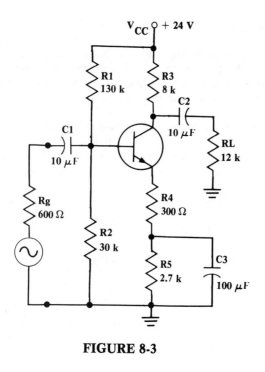

$$V_{R2} = \frac{V_{CC} \times R2}{R1 + R2} = \frac{24\text{ V} \times 30\text{ k}\Omega}{130\text{ k}\Omega + 30\text{ k}\Omega} = 4.5\text{ V}$$

$$V_{RE} = V_{R2} - V_{BE} = 4.5\text{ V} - 0.6\text{ V} = 3.9\text{ V}$$

$$I_E = \frac{V_{RE}}{RE} = \frac{V_{R4} + V_{R5}}{R4 + R5} = \frac{3.9\text{ V}}{300\text{ }\Omega + 2.7\text{ k}\Omega} = 1.3\text{ mA}$$

$$V_{R4} = I_E \times R4 = 1.3\text{ mA} \times 300\text{ }\Omega = 0.39\text{ V}$$

$$V_{R5} = I_E \times R5 = 1.3\text{ mA} \times 2.7\text{ k}\Omega = 3.51\text{ V}$$

$$V_{R3} = I_C \times R3 = I_E \times R3 = 1.3\text{ mA} \times 8\text{ k}\Omega = 10.4\text{ V}$$

$$V_C = V_{CC} - V_{R3} = 24\text{ V} - 10.4\text{ V} = 13.6\text{ V}$$

$$V_B = V_{R2} = 4.5\text{ V}$$

$$V_E = V_{RE} = 3.9\text{ V}$$

$$V_{CE} = V_C - V_E = 13.6\text{ V} - 3.9\text{ V} = 9.7\text{ V}$$

$$V_{RL} = 0\text{ V}$$

FIGURE 8-3

ALTERNATING CURRENT ANALYSIS OF THE NPN CIRCUIT

Once the DC analysis of the circuit is obtained, the AC parameters of voltage gain, input impedance, power gain, and maximum peak-to-peak output voltage swing are solved. However, because of the biasing arrangement of the universal circuit, the input impedance calculation is different from that of the base-biased circuit. All the other parameters are solved in a manner similar to those of the base-biased circuit.

VOLTAGE GAIN

The voltage gain of the common-emitter, universal-biased circuit, like that of the base-biased circuit, is solved by taking the ratio of the effective collector resistance to the effective emitter resistance. If the collector resistance is loaded, as shown in Figure 8-4, then the effective collector resistance is $r_L = R3 \parallel RL$. If the emitter resistance is partially bypassed, the effective emitter resistance is $r_e + R4$. Then, the voltage gain for the AC equivalent circuit of Figure 8-4 is as follows:

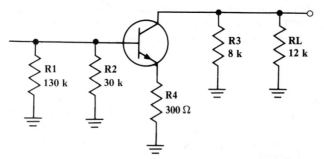

$$A_v = \frac{r_L}{r_e + R4} = \frac{R3 \parallel RL}{r_e + R4}$$

$$= \frac{8\text{ k}\Omega \parallel 12\text{ k}\Omega}{20\text{ }\Omega + 300\text{ }\Omega} = \frac{4.8\text{ k}\Omega}{320\text{ }\Omega} = 15$$

where: $r_e = 26\text{ mV}/I_E = 26\text{ mV}/1.3\text{ mA} = 20\text{ }\Omega$

FIGURE 8-4

NOTE: If the emitter resistance is not bypassed, the gain of the stage would be about 1.6 and, if the emitter resistance is fully bypassed, the gain would be about 240. However, in that latter case, the voltage gain would be too dependent on the unpredictable r_e, which can vary over a 2 to 1 range. Therefore, the partially bypassed resistor, where R4 at 300 Ω "swamps" the unpredictable r_e resistance, provides AC circuit stability and is the configuration most widely used. Also, if the load for the partially bypassed resistor is removed, the gain will be: $A_v = R3/(r_e + R4) = 8\text{ k}\Omega/320\text{ }\Omega = 25$.

INPUT IMPEDANCE

The input impedance to the common-emitter, universal-biased circuit is solved by taking the parallel combination of the biasing resistors R1 and R2 in parallel with the "reflected" emitter resistance. If the effective emitter resistance is $r_e + R4$, then the reflected emitter resistance into the base is $\beta(r_e + R4)$. The input impedance of the circuit of Figure 8-3 would be as follows:

$$Zin = R1 \mathbin{/\mkern-5mu/} R2 \mathbin{/\mkern-5mu/} \beta(r_e + R4) = 130\text{ k}\Omega \mathbin{/\mkern-5mu/} 30\text{ k}\Omega \mathbin{/\mkern-5mu/} 100(20\ \Omega + 300\ \Omega) \approx 13.8\text{ k}\Omega$$

The equivalent circuit for the input impedance is shown in Figure 8-5. Note that V_{CC} is an "AC ground" and R1 and R2 are seen going to ground along with the reflected emitter resistance of 32 kΩ. Again, the reflected emitter resistance is the input impedance looking into the base (Zin[base]), which is the effective emitter resistance $(r_e + R4)$ multiplied by the current gain of the transistor. Therefore:

$$Zin(base) = \beta(r_e + R4) = 100(20\ \Omega + 300\ \Omega) = 100 \times 320\ \Omega = 32\text{ k}\Omega, \text{ also}$$

$$Zin(base) = \beta r_e + \beta R4 = (100 \times 20\ \Omega) + (100 \times 300\ \Omega) = 2\text{ k}\Omega + 30\text{ k}\Omega = 32\text{ k}\Omega$$

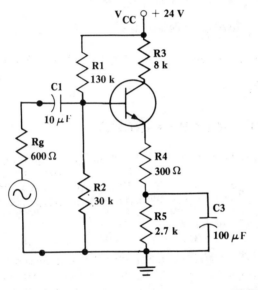

NOTE: Only the unbypassed emitter resistance reflected into the base along with r_e effects the Zin. Again, the high-impedance, collector-base junction isolates collector resistor R3 from the input.

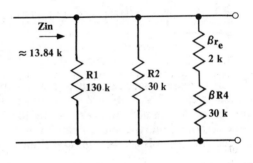

FIGURE 8-5

NOTE: If the emitter resistance is fully bypassed, the emitter resistance would be 20 Ω, the reflected emitter resistance Zin(base) = $100 \times 20\ \Omega = 2\text{ k}\Omega$, and the input impedance Zin = R1 // R2 // βr_e = 130 kΩ // 30 kΩ // 2 kΩ = 1848 Ω. However, if the emitter resistance is not bypassed, then the emitter resistance would be 3020 Ω, the reflected emitter resistance $100 \times 3020\ \Omega = 302\text{ k}\Omega$, and Zin:

$$Zin = R1 \mathbin{/\mkern-5mu/} R2 \mathbin{/\mkern-5mu/} \beta(r_e + R4 + R5) = 130\text{ k}\Omega \mathbin{/\mkern-5mu/} 30\text{ k}\Omega \mathbin{/\mkern-5mu/} 302\text{ k}\Omega \approx 22.5\text{ k}\Omega$$

Therefore, the "trade off" of voltage gain versus input impedance is also evident in the universal-biased circuit, where larger unpassed emitter resistance provides higher Zin and lower voltage gain.

POWER GAIN TO THE LOAD

The power gain calculations are dependent on the voltage gain, the input impedance, and the load resistor RL. Therefore, the power gain to the load is solved from:

$$PG = A_v^2 \times Zin/RL = 15^2 \times 13.84\text{ k}\Omega / 12\text{ k}\Omega = 259.5$$

$$PG(dB) = 10 \log PG = 10 \log 259.5 = 24.14\text{ dB}$$

MAXIMUM PEAK-TO-PEAK OUTPUT VOLTAGE SWING

The maximum peak-to-peak output voltage swing for the universal circuits are solved exactly like those of the base-biased circuits. Therefore, unless the emitter resistor is not bypassed, the maximum peak-to-peak output voltage swing can be calculated from $2V_{CE}$ or $2I_C r_L$. Recall, that for the common emitter circuit, any unbypassed emitter resistance will cause the $2V_{CE}$ portion of the calculations to be slightly less than $2V_{CE}$. For the partially bypassed circuit of Figure 8-6, the peak-to-peak output voltage swing is solved from:

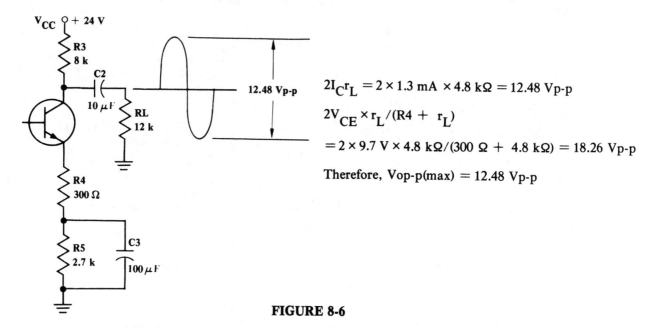

$2I_C r_L = 2 \times 1.3 \text{ mA} \times 4.8 \text{ k}\Omega = 12.48 \text{ Vp-p}$

$2V_{CE} \times r_L/(R4 + r_L)$

$= 2 \times 9.7 \text{ V} \times 4.8 \text{ k}\Omega/(300 \text{ }\Omega + 4.8 \text{ k}\Omega) = 18.26 \text{ Vp-p}$

Therefore, $V_{op\text{-}p(max)} = 12.48 \text{ Vp-p}$

FIGURE 8-6

NOTE: If the slight degeneration of the unbypassed 300 ohms is not used, the $2 V_{CE}$ portion of the calculation is $2 \times 9.7 \text{ V} = 19.4 \text{ Vp-p}$, instead of the precisely calculated 18.26 Vp-p. However, since the $V_{op\text{-}p(max)}$ is determined by $2I_C r_L$ at 12.48 Vp-p, solving for $2V_{CE}$ or $2V_{CE}r_L/(R4 + r_L)$ is purely academic. And, again, the more complex formula: $2V_{CE}r_L/(R4 + r_L)$ should be used only when the unbypassed emitter resistor R4 is greater than 10% of the effective load resistance r_L.

OUTPUT IMPEDANCE

The output impedance of any stage can be measured by monitoring the unloaded and loaded output voltages and knowing the value of the load resistor. The method is much like the input impedance technique of measurement except in the output impedance measurement the load resistor value is known, while in the input impedance method of measurement the series resistor value is known. The calculation for solving the output impedance is:

$$Z_o = R_o = \frac{V_o(\text{no load}) - V_o(\text{loaded})}{V_o(\text{loaded})} \times RL$$

For instance, if the output signal of the unloaded collector resistor is 8 Vp-p, loading the collector with a known 12 kΩ load resistor provides 4.8 Vp-p. Then, the collector resistor can be calculated to be 8 kΩ from:

$$R_o = Z_o = \frac{V_o(\text{no load}) - V_o(\text{loaded})}{V_o(\text{loaded})} \times RL = \frac{8 \text{ Vp-p} - 4.8 \text{ Vp-p}}{4.8 \text{ Vp-p}} \times 12 \text{ k}\Omega = 8 \text{ k}\Omega$$

For verification, assume $I_C = 1 \text{ mAp-p}$ so the 8 Vp-p is developed across the unloaded 8 kΩ resistor RC. Then, if a 12 kΩ load resistor is connected to the circuit, $r_L = RC \parallel RL = 8 \text{ k}\Omega \parallel 12 \text{ k}\Omega = 4.8 \text{ k}\Omega$, and $V_o = 4.8 \text{ Vp-p}$.

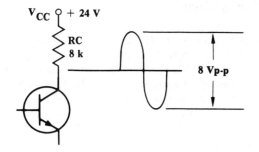

NO LOAD CONDITIONS

$V_{op\text{-}p} = I_C R_C = 1 \text{ mAp-p} \times 8 \text{ k}\Omega = 8 \text{ Vp-p}$

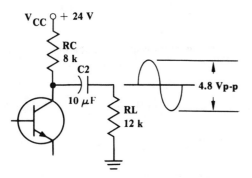

LOADED CONDITIONS

$V_o = I_C r_L = I_C (RC \;/\!/\; RL)$

$= 1 \text{ mAp-p}(8 \text{ k}\Omega \;/\!/\; 12 \text{ k}\Omega) = 4.8 \text{ Vp-p}$

FIGURE 8-7

SOLVING Ro ≈ Zo, USING THEVENIN AND NORTON EQUIVALENTS

The transistor is, essentially, a constant current source that can be conveniently represented by Norton's equivalent circuit. If the previous collector current of $i_C = i_N = 1$ mAp-p is assumed and a loaded condition where $RC = R_N = 8$ kΩ and $RL = 12$ kΩ, then:

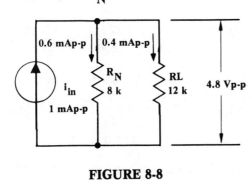

$i_{RL} = \dfrac{i_N \times R_N}{R_N + RL} = \dfrac{1 \text{ mA} \times 8 \text{ k}\Omega}{8 \text{ k}\Omega + 12 \text{ k}\Omega} = 0.4 \text{ mAp-p}$

$V_{RL} = i_{RL} \times RL = 0.4 \text{ mAp-p} \times 12 \text{ k}\Omega = 4.8 \text{ Vp-p}$

$i_{RN} = \dfrac{i_N \times RL}{R_N + RL} = \dfrac{8 \text{ k}\Omega + 12 \text{ k}\Omega}{1 \text{ mAp-p} \times 12 \text{ k}\Omega} = 0.6 \text{ mAp-p}$

$V_{R_N} = i_{R_N} \times R_N = 0.6 \text{ mAp-p} \times 8 \text{ k}\Omega = 4.8 \text{ Vp-p}$

FIGURE 8-8

If the Norton equivalent circuit is converted to a Thevenin equivalent, as shown in Figure 8-9, where $RC = R_{TH} = 8$ kΩ and $V_{TH} = 8$ Vp-p, then:

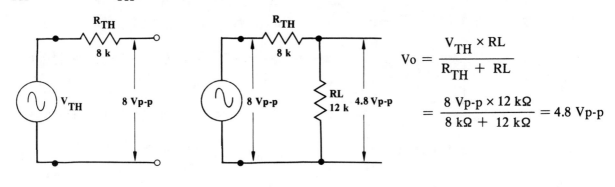

$V_o = \dfrac{V_{TH} \times RL}{R_{TH} + RL}$

$= \dfrac{8 \text{ Vp-p} \times 12 \text{ k}\Omega}{8 \text{ k}\Omega + 12 \text{ k}\Omega} = 4.8 \text{ Vp-p}$

FIGURE 8-9

Ro = Zo DERIVATION

The voltage divider formula used to solve for Vo can be rearranged to solve for Ro, where Ro = R_{TH} and Vo = Vo(loaded).

1. $Vo = \dfrac{V_{TH} \times RL}{R_{TH} + RL}$

2. $Vo(R_{TH} + RL) = V_{TH} \times RL$

3. $VoR_{TH} + VoRL = V_{TH}RL$

4. $VoR_{TH} = V_{TH}RL - VoRL$

5. $R_{TH} = \dfrac{V_{TH}RL - VoRL}{Vo}$

6. $R_{TH} = \dfrac{(V_{TH} - Vo)RL}{Vo}$

Therefore: $Ro = \dfrac{Vo(\text{no load}) - Vo(\text{loaded})}{Vo(\text{loaded})} \times RL$

where: V_{TH} = Vo(no load) and Vo = Vo (loaded)

CIRCUIT SUMMARY AND THE EQUIVALENT CIRCUIT APPROACH

The approximation and the equivalent circuit methods of circuit analysis are now given to show, once more, that either method can be used to analyze circuits. The circuit of Figure 8-10 is the same as that of Figure 8-3, which was analyzed in detail, and it is summarized here for convenience of comparison.

A. APPROXIMATION CALCULATIONS

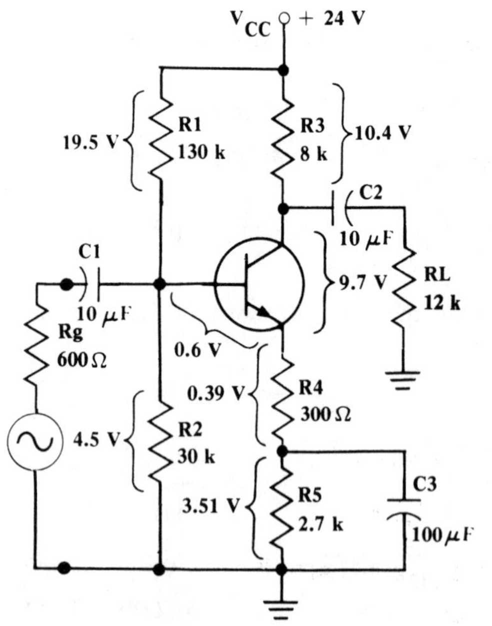

Beta = 100   Vin = 320 mVp-p r_e = 20 Ω

$A_v = \dfrac{R3 \parallel RL}{r_e + R4} = \dfrac{8\ k\Omega \parallel 12\ k\Omega}{20\ \Omega + 300\ \Omega} = \dfrac{4.8\ k\Omega}{320\ \Omega} = 15$

Vo = Vin × A_v = 320 mVp-p × 15 = 4.8 Vp-p

Zin = R1 ∥ R2 ∥ β(r_e + R4) = 130 kΩ ∥ 30 kΩ ∥ 100 × 320 Ω

= 130 kΩ ∥ 30 kΩ ∥ 32 kΩ = 13.84 kΩ

PG = A_v^2 × Zin/RL = 15^2 × 13.84 kΩ /12 kΩ = 259.5

PG(dB) = 10 log PG = 10 log 259.5 = 24.14 dB

A_i = PG/A_v = 259.5/15 = 17.3

Vop-p(max) = 12.48 Vp-p, from:

$2I_C r_L$ = 2 × 1.3 mA × 4.8 kΩ = 12.48 Vp-p

$2V_{CE} r_L /(R4 + r_L)$

= 2 × 9.7 V × 4.8 kΩ/(300 Ω + 4.8 kΩ) = 18.26 Vp-p

$Po(\text{max}) = \dfrac{Vop\text{-}p^2(\text{max})}{8RL} = \dfrac{12.48^2}{8 \times 12\ k\Omega} = 1.62\ mW$

FIGURE 8-10

NOTE: Power is solved from P = V RMS² /RL. However, if Vp-p is used instead of V RMS, dividing by 8 provides a handy conversion. Hence, P = Vop-p² /8RL. See appendix.

B. EQUIVALENT CIRCUIT CALCULATIONS

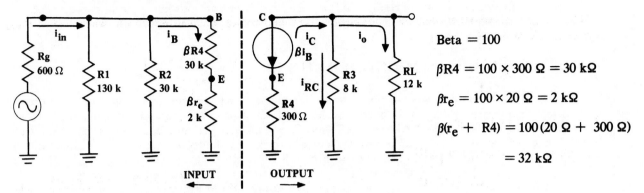

FIGURE 8-11

$r_L = R3 \mathbin{/\mkern-6mu/} RL = 8\,k\Omega \mathbin{/\mkern-6mu/} 12\,k\Omega = 4.8\,k\Omega$

$Z_{in} = R1 \mathbin{/\mkern-6mu/} R2 \mathbin{/\mkern-6mu/} \beta(r_e + R4) = 130\,k\Omega \mathbin{/\mkern-6mu/} 30\,k\Omega \mathbin{/\mkern-6mu/} 100(20\,\Omega + 300\,\Omega) = 13.84\,k\Omega$

$i_{in} = V_{in}/Z_{in} = 320\,mVp\text{-}p/13.84\,k\Omega = 23.12\,\mu Ap\text{-}p$

$i_B = V_{in}/\beta(r_e + R4) = 320\,mVp\text{-}p/100(20\,\Omega + 300\,\Omega) = 320\,mVp\text{-}p/32\,k\Omega = 10\,\mu Ap\text{-}p$

$i_C = \beta i_B = 100 \times 10\,\mu Ap\text{-}p = 1\,mAp\text{-}p$

$i_o = i_C R3/(R3 + RL) = (1\,mA \times 8\,k\Omega)/(8\,k\Omega + 12\,k\Omega) = 0.4\,mAp\text{-}p$

$i_{RC} = i_C R_L/(R3 + RL) = (1\,mAp\text{-}p \times 12\,k\Omega)/(8\,k\Omega + 12\,k\Omega) = 0.6\,mAp\text{-}p$

$V_o = i_C r_L = 1\,mAp\text{-}p \times 4.8\,k\Omega = 4.8\,Vp\text{-}p$, or

$V_o = i_o RL = 0.4\,mAp\text{-}p \times 12\,k\Omega = 4.8\,Vp\text{-}p$, or

$V_o = V_{R3} = i_{R3} R3 = 0.6\,mAp\text{-}p \times 8\,k\Omega = 4.8\,Vp\text{-}p$

$A_v = V_o/V_{in} = 4.8\,Vp\text{-}p/320\,mVp\text{-}p = 15$, or

$A_v = \beta r_L/\beta(r_e + R4) = (100 \times 4.8\,k\Omega)/32\,k\Omega = 15$

$A_i = A_v \times Z_{in}/RL = 15 \times 13.84\,k\Omega/12\,k\Omega = 17.3$

$A_i = i_o/i_{in} = 0.4\,mAp\text{-}p/23.12\,\mu Ap\text{-}p = 17.3$

$A_p = PG = A_v A_i = 15 \times 17.3 = 259.5$

NOTE: Power gain can also be solved from $PG = P_o/P_{in}$, where:

$P_o = V_{opp}^2/8RL = 6\,Vp\text{-}p^2/(8 \times 12\,k\Omega) = 375\,\mu W$

$P_{in} = V_{in\,p\text{-}p}^2/8Z_{in} = 40\,mVp\text{-}p^2/(8 \times 13.84\,k\Omega) = 1.445\,\mu W$

$PG = P_o/P_{in} = 375\,\mu W/1.445\,\mu W = 259.5$

$V_{op\text{-}p(max)}: \quad 2I_C r_L = 2 \times 1.3\,mA \times 4.8\,k\Omega = 12.48\,Vp\text{-}p$

$\qquad\qquad 2V_{CE} r_L/(R4 + r_L) = 2 \times 9.7\,V \times 4.8\,k\Omega/(300\,\Omega + 4.8\,k\Omega) = 18.26\,Vp\text{-}p$

The obvious advantage to solving the same circuit in two different ways is that, providing both methods are correctly done, the second set of results confirms the first. A less obvious but more important reason is that a deeper understanding and appreciation for the signal voltage analysis approach is achieved if both analytical methods are mastered.

FREQUENCY RESPONSE CONSIDERATIONS

The voltage gain of amplifiers is not constant at all frequencies. At low and high frequencies, with reference to mid-frequencies, voltage gain decreases. At mid-frequencies, such as 1 kHz, the circuit coupling and bypass capacitors are considered to be "AC shorts", and all of the circuit capacitances, including stray and device capacitances, are considered to be "opens". In the low frequency range, the "visible" circuit capacitors of C1, C2, and C3 (with reference to Figure 8-12) begin to interact with the associated circuit resistances, and the voltage gain of the amplifier begins to drop. In the high frequency range, the "invisible" capacitances of c_{be}, c_{bc}, c_{miller}, and c_{stray} begin to interact with the associated circuit resistances, and the voltage gain begins to drop. The circuit and a representative frequency response for the circuit are given in Figure 8-12.

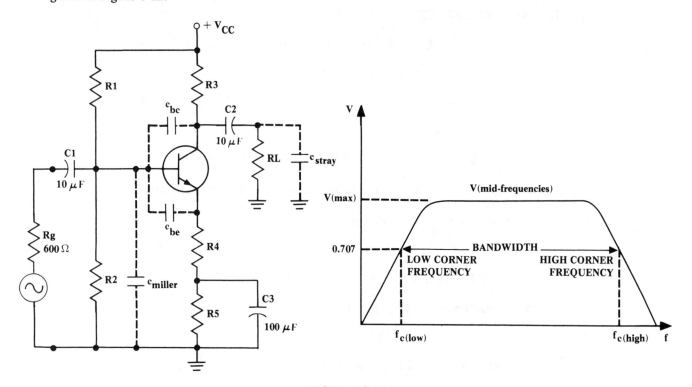

FIGURE 8-12

MEASUREMENT TECHNIQUES

Corner frequencies occur at 0.707 of the mid-frequency amplitude condition. For instance, if the amplitude of the output signal monitored on an oscilloscope is 10 Vp-p at 1 kHz, the signal will eventually drop to 7.07 Vp-p by varying the frequency—either in the low or high frequency directions.

One method of measurement is to set the output amplitude of a mid-frequency 1 kHz signal to a convenient output amplitude such as 5 Vp-p. Then decrease the frequency by multiples of 10 (to 100 Hz, 10 Hz, and so on) until a decrease in amplitude occurs. Then "fine tune" the dial for 0.707 of 5 Vp-p, or approximately 3.5 Vp-p. Repeat the procedure for high frequency roll-off by increasing the frequency by multiples of 10 from the 1 kHz reference (to 10 kHz, 100 kHz, and so on). Again, adjust the dial for approximately 3.5 Vp-p. The low and high frequencies at which Vop-p drops to approximately 3.5 Vp-p are the corner frequencies, and the region between these corner frequencies is the bandwidth for the amplifier under test. See Figure 8-13.

The Bode plot illustrated in Figure 8-14 is a simple way of describing the plotted frequency response. It emphasizes the low and high corner frequencies and the bandwidth, thus eliminating tedious plotting

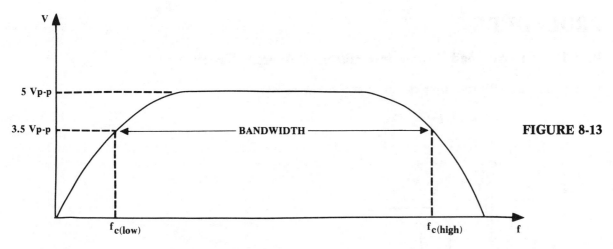

FIGURE 8-13

procedures.

The corner frequency at 0.707 of maximum voltage is also called the half power or $-3\,\mathrm{dB}$ point, because when the voltage drops from 10 V RMS to 7.07 V, for example, the power halves. Thus:

$$P = V^2/R = 10\,V^2/10\,\Omega = 10\,W, \quad \text{but } P = V^2/R = 7.07\,V^2/10\,\Omega = 5\,W$$

Converting a loss of 2 to dB is found from $10 \log A_p = 10 \log 0.5 = -3$ dB, where the minus sign denotes a loss, or the drop from 10 W to 5 W. The ratio of 5 W/10 W $= 0.5$, hence, half power point.

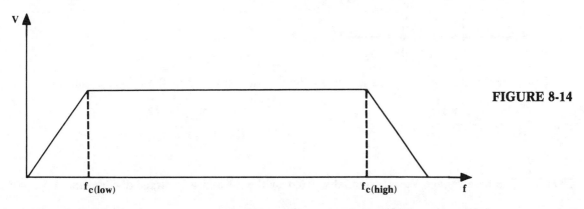

FIGURE 8-14

This brief introduction to frequency response is intended to make the student aware of the concept, and more detailed analysis on frequency response is reserved for more advanced course work. A good reference is Chapter 2 of MacDonald, Lorne: PRACTICAL ANALYSIS OF AMPLIFIER CIRCUITS THROUGH EXPERIMENTATION 3/e, The Technical Education Press, 1981.

SECTION I: LABORATORY EXERCISE (Part 1)

OBJECTIVES

To investigate the DC and AC characteristics of the common-emitter, universal-biased circuit, using an NPN transistor.

LIST OF MATERIALS AND EQUIPMENT

1. Transistors: 2N3904 (one) or equivalent
2. Resistors (one each)   330 Ω   3.3 kΩ   10 kΩ   12 kΩ   15 kΩ   22 kΩ   100 kΩ
3. Capacitors: 10 μF (two)   100 μF (one)
4. Power Supply
5. Voltmeter
6. Signal Generator
7. Oscilloscope

PROCEDURE

Part 1: The No-Load, Common-Emitter, Universal Circuit

1. Connect the NPN transistor circuit shown in Figure 8-15.

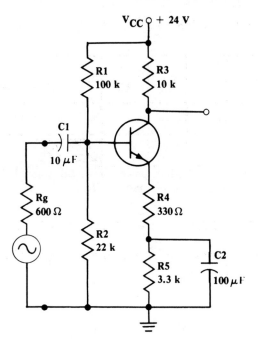

FIGURE 8-15

2. Calculate the DC voltage drops across all the circuit resistors and the collector-emitter of the device from:

a. $V_{R1} = \dfrac{V_{CC} \times R1}{R1 + R2}$
b. $V_{R2} = \dfrac{V_{CC} \times R2}{R1 + R2}$

NOTE: The base voltage V_B is, essentially, fixed by the R1 and R2 voltage divider, and $V_{R2} = V_B$.

c. $V_{RE} = V_{R4} + V_{R5} = V_{(R4 + R5)} = V_{R2} - V_{BE}$

NOTE: V_{BE} is given at approximately 0.6 volts until actual V_{BE} voltage measurements can be made.

d. $I_E = V_{RE}/RE = V_{(R4 + R5)}/(R4 + R5)$

e. $V_{R4} = I_E R4$

$V_{R5} = I_E R5$

f. $V_{R3} = I_C R3$ where: $I_C \approx I_E$

g. Calculate the single point transistor voltages, with respect to ground from:

$V_B = V_{R2}$

$V_E = V_{R4} + V_{R5}$

$V_C = V_{CC} - V_{R3}$

h. $V_{CE} = V_C - V_E$

250

3. Measure the DC Voltage drops around the circuit. Include all resistors and the transistor.

NOTE: All direct current measurements should be made with respect to ground. This helps to avoid ground loops and the resultant incorrect measurements. For instance, a measured value at the collector and then one at the emitter will provide V_{CE} without having to physically measure across the collector-emitter of the device.

4. Insert the calculated and measured values, as indicated, into Table 8-1.

TABLE 8-1	V_{R1}	V_{R2}	V_{R3}	V_{R4}	V_{R5}	V_C	V_E	V_{CE}	$I_C \approx I_E$	V_{BE}
CALCULATED										≈ 0.6 V
MEASURED									/////	

5. VOLTAGE GAIN

   a. Calculate the mid-frequency voltage gain from: $A_v = R3/(r_e + R4)$ where: $r_e \approx 26 \text{ mV}/I_E$

   b. Measure the output voltage using an input signal voltage of 100 mVp-p at a frequency of 1 kHz from:

   $A_v = V_o/V_{in}$ where: $V_{in} = 100$ mVp-p

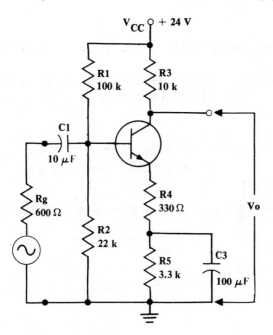

FIGURE 8-16

6. INPUT IMPEDANCE

   a. Calculate the input impedance to the stage from:

   $Z_{in} = R1 \mathbin{/\mkern-5mu/} R2 \mathbin{/\mkern-5mu/} \beta(r_e + R4)$ where: $r_e \approx 26 \text{ mV}/I_E$

NOTE: Use the measured beta in the calculations if known, or use a beta of 100 if not known. The effect of using the exact beta in the universal circuit is not crucial unless it is much lower than 100.

b. Measure the input impedance to the stage. Use a known 15 kΩ resistor and an input signal level of 100 mVp-p at 1 kHz, as shown in Figure 8-17.

$$Zin = \frac{Vin \times RS}{V_{RS}} = \frac{Vin \times RS}{Vg - Vin} \quad \text{where:} \quad Vinp\text{-}p = 100 \text{ mVp-p}$$

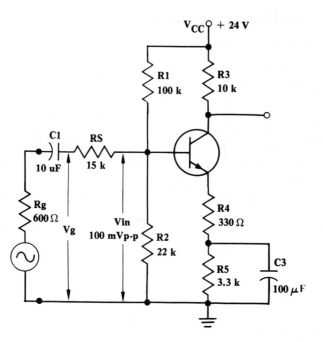

FIGURE 8-17

7. MAXIMUM PEAK-TO-PEAK OUTPUT VOLTAGE SWING

a. Calculate the approximate maximum peak-to-peak output voltage swing from $2I_C R3$ or $2V_{CE}$, whichever is smaller. Again, more accuracy can be achieved by including the effect of the 330 Ω resistor R4 using:

$$2V_{CE} R3/(R3 + R4) \approx 2V_{CE}$$

b. Monitor the collector output and increase the 1 kHz signal level until clipping or distortion, because of alpha crowding, occurs.

NOTE: Remove RS from the circuit when measuring Vop-p(max), because it will alter maximum peak-to-peak output voltage swing conditions.

8. POWER GAIN TO THE COLLECTOR RESISTOR

a. Calculate the power gain to the collector resistor of the circuit using both the measured A_v and Zin values. Calculate from:

$$PG = A_v^2 \times Zin/RC$$

TABLE 8-2	A_v	Vo for Vin = 100 mVp-p	Zin	Vop-p (MAX)	POWER GAIN	PG (dB)	A_i
CALCULATED		/////					
MEASURED					/////	/////	/////

b. Convert to dB from: PG(dB) = 10 log PG.

9. Calculate the circuit current gain from: $A_i = PG/A_v$ or from $A_i = A_v \times Zin/R3$

10. Insert the calculated and measured values, as indicated, into Table 8-2.

Part 2: The Common-Emitter, Universal Circuit with Load

1. Connect the coupling capacitor C3 and the 12 kΩ load resistor to the circuit, as shown in Figure 8-18.

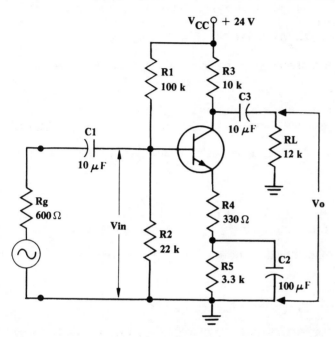

FIGURE 8-18

2. Monitor the DC voltage drops around the circuit to make sure that they have not changed, possibly from a leaky capacitor, from those of the previous circuit. Measure V_C, V_B, V_E, and V_{R5}, with respect to ground, and calculate the remaining voltage drops as indicated in Table 8-3.

3. Insert the calculated and measured voltages, as indicated, into Table 8-3.

TABLE 8-3	V_C	V_E	$V_B = V_{R2}$	V_{R5}	V_{R1}	V_{R3}	V_{R4}	V_{CE}
CALCULATED	/////	/////	/////	/////				
MEASURED					/////	/////	/////	/////

4. VOLTAGE GAIN

   a. Calculate the mid-frequency voltage gain, for the loaded condition of 12 kΩ, from:

   $$A = \frac{R3 \parallel RL}{r_e + R4} = \frac{r_L}{r_e + R4}$$

   b. Measure the mid-frequency voltage gain using an input voltage of 100 mVp-p at a frequency of 1 kHz. Measure Vo across the load resistor and use that value to calculate the measured voltage gain from: $A_v = Vo/Vin$ where: Vin = 100 mVp-p

5. INPUT IMPEDANCE

a. Measure the input impedance in the same way it was measured for the unloaded circuit. Use a known 15 kΩ resistor and a signal level of 100 mVp-p at 1 kHz, and calculate from:

$$Zin = \frac{Vin \times RS}{V_{RS}} = \frac{Vin \times RS}{Vg - Vin} \quad \text{where:} \quad Vin = 100 \text{ mVp-p}$$

NOTE: The loading of the collector resistor by RL has no effect on the input impedance. Comparing the results of the loaded circuit to those of the unloaded circuit is proof enough.

6. MAXIMUM PEAK-TO-PEAK OUTPUT VOLTAGE SWING

a. Calculate the maximum peak-to-peak output voltage swing from $2I_C r_L = 2I_C(R3 \parallel RL)$ or $2V_{CE}$, whichever is smaller.

b. Monitor the output voltage across the 12 kΩ load resistor, and increase the input signal level until clipping or distortion, because of alpha crowding, occurs.

NOTE: Using the more exact $2V_{CE} r_L/(r_L + R4)$ is optional. Also remove RS for this measurement.

7. OUTPUT IMPEDANCE

a. Remove the load resistor and monitor the output signal to the collector, with respect to ground. Adjust the input signal level so the unloaded peak-to-peak signal monitored on the scope is 5 Vp-p.

b. Connect the load resistor, monitor the loaded peak-to-peak output voltage swing, and calculate the output impedance from:

$$Zo = \frac{Vo(\text{no load}) - Vo(\text{loaded})}{Vo(\text{loaded})} \times RL \quad \text{where:} \quad Vo(\text{no load}) = 5 \text{ Vp-p}$$

8. POWER GAIN TO THE LOAD RESISTOR

a. Calculate the power gain to the load resistor using both the A_v and Zin measured values. Calculate from: $PG = A_v^2 \times Zin/RL$ where: $RL = 12$ kΩ

b. Convert to db from: $PG(dB) = 10 \log PG$

NOTE: The power gain is calculated to the load resistor RL, because the power gain to the load is the useful power gain. The power dissipated by the collector resistor is lost power. For instance, the collector resistor and the load resistor contribute to the voltage gain, but only the power developed across the load resistor is, effectively, the power transferred. Using the parallel combination of the collector resistor R3 and the load resistor RL provides the power developed across both resistors, which includes both transferred power and R3 dissipated power used in processing the signal to the load.

9. Calculate the circuit current gain from: $A_i = PG/A_v = A_v \times Zin/RL$.

10. Insert the calculated and measured values, as indicated, into Table 8-4.

TABLE 8-4	A_v	Vo for Vin = 100 mVp-p	Zin	Vop-p (MAX)	POWER GAIN	PG (dB)	A_i	Zo
CALCULATED		/////						
MEASURED					/////	/////	/////	/////

11. Monitor the low and high corner frequencies of the amplifier circuit of Figure 8-18 (loaded). Start by setting the output amplitude of a mid-frequency 1 kHz signal to 5 Vp-p on the oscilloscope, and then proceed as discussed in the text. After measuring the low and high corner frequencies, insert the measured values into the table of Figure 8-19. Also, indicate the available band-pass frequency on the Bode plot.

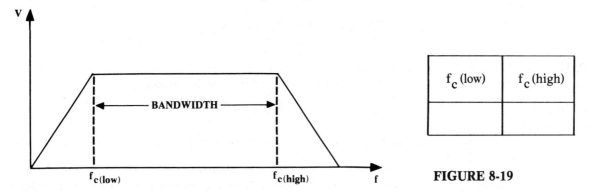

FIGURE 8-19

QUESTIONS

1. Why is the beta change not as significant on the input impedance of the common-emitter, universal-biased circuit as it is on the input impedance of the common-emitter, base-biased circuits?
2. Are the voltage gain and Vop-p(max) calculations different for the universal-biased and base-biased circuits? Why or why not?
3. Why is using the parallel combination of RC and RL not correct when solving for the power gain of the circuit?
4. Is using a nominal beta of 100 valid for the majority of the universal-biased circuits? Why or why not?

Part 2: The PNP, Common-Emitter, Universal Circuit

The PNP universal circuit can be connected in either of two ways — in the standard configuration as shown in Figure 8-20(a), where the polarity of the power supply connections are directly opposite to the NPN device connection, or in the upside-down connection as shown in Figure 8-20(b). Since the analysis of the NPN and PNP device in the standard configuration are identical, except for the polarity differences, the following analysis will be on the upside-down configuration.

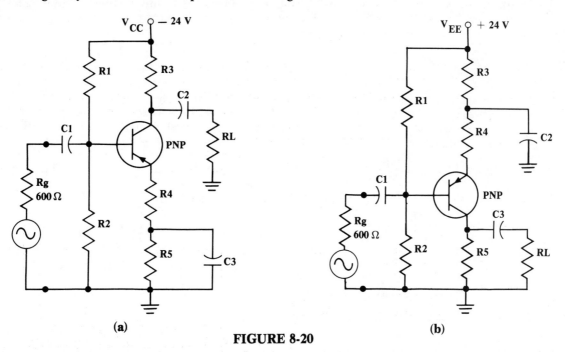

FIGURE 8-20

THE PNP, UPSIDE-DOWN, UNIVERSAL CIRCUIT

The DC analysis for the PNP connection of Figure 8-21 was given in Chapter 6, so the calculations are shown with Figure 8-21 without further explanation.

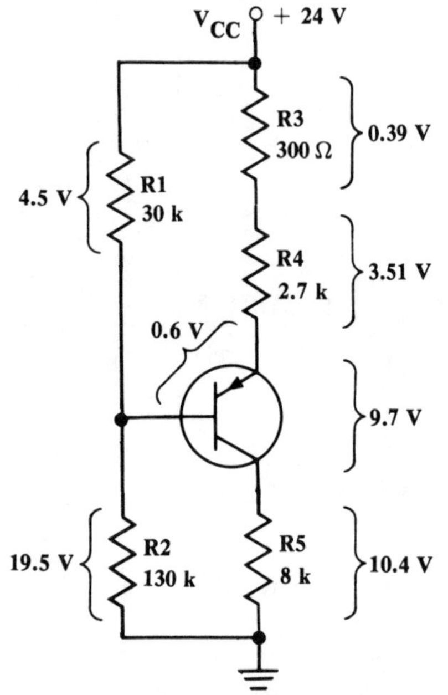

$$V_{R1} = \frac{V_{CC} \times R1}{R1 + R2} = \frac{24 \text{ V} \times 30 \text{ k}\Omega}{30 \text{ k}\Omega + 130 \text{ k}\Omega} = 4.5 \text{ V}$$

$$V_{R2} = \frac{V_{CC} \times R2}{R1 + R2} = \frac{24 \text{ V} \times 30 \text{ k}\Omega}{30 \text{ k}\Omega + 130 \text{ k}\Omega} = 19.5 \text{ V}$$

$$V_{RE} = V_{R1} - V_{BE} = 4.5 \text{ V} - 0.6 \text{ V} = 3.9 \text{ V}$$

$$I_E = V_{RE}/RE = \frac{V_{R3} + V_{R4}}{R3 + R4} = \frac{3.9 \text{ V}}{300 \text{ }\Omega + 2.7 \text{ k}\Omega} = 1.3 \text{ mA}$$

$$V_{R4} = I_E R4 = 1.3 \text{ mA} \times 300 \text{ }\Omega = 0.39 \text{ V}$$

$$V_{R3} = I_E R3 = 1.3 \text{ mA} \times 2.7 \text{ k}\Omega = 3.51 \text{ V}$$

$$V_{R5} = I_C R5 \approx I_E R5 = 1.3 \text{ mA} \times 8 \text{ k}\Omega = 10.4 \text{ V}$$

$$V_C = V_{R5} = 10.4 \text{ V}$$

$$V_E = V_{CC} - V_{RE} = 24 \text{ V} - 3.9 \text{ V} = 20.1 \text{ V}$$

$$V_B = V_{R2} = 19.5 \text{ V}$$

$$V_{CE} = V_{EC} = V_E - V_C = 20.1 \text{ V} - 10.4 \text{ V} = 9.7 \text{ V}$$

FIGURE 8-21

ALTERNATING CURRENT CIRCUIT ANALYSIS

The voltage gain, input impedance, power gain, Vop-p(max), and output impedance are solved in a similar manner to the standard NPN and PNP circuits, but these parameters are detailed again to provide a comparison.

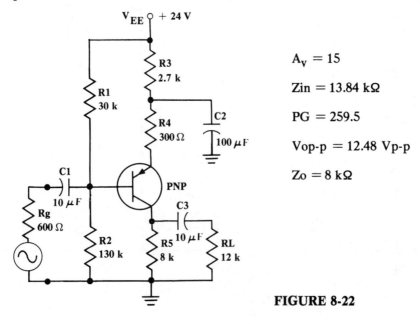

$A_v = 15$

$Z_{in} = 13.84 \text{ k}\Omega$

$PG = 259.5$

$V_{op-p} = 12.48 \text{ Vp-p}$

$Z_o = 8 \text{ k}\Omega$

FIGURE 8-22

VOLTAGE GAIN

The voltage gain of the upside-down PNP circuit is solved by taking the ratio of the effective collector resistance to the effective emitter resistance. The effective collector resistance is r_L = R5 // RL, while the effective emitter resistance is r_e + R4.

NOTE: Capacitor C2 places an effective AC ground at the junction of R3 and R4 causing resistor R3 to be bypassed, since V_{CC} is also an AC ground.

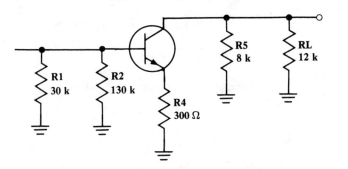

$$A_v = \frac{r_L}{r_e + R4} = \frac{R5 \text{ // } RL}{r_e + R4}$$

$$\frac{8 \text{ k}\Omega \text{ // } 12 \text{ k}\Omega}{20 \text{ }\Omega + 300 \text{ }\Omega} = \frac{4.8 \text{ k}\Omega}{320 \text{ }\Omega} = 15$$

where: $r_e = 26 \text{ mV}/I_E$

$= 26 \text{ mV}/1.3 \text{ mA} = 20 \text{ }\Omega$

FIGURE 8-23

INPUT IMPEDANCE

The input impedance is solved like it was in previous circuits, where Zin equals the effective emitter resistance $\beta(r_e + R4)$ in parallel with R1 and R2.

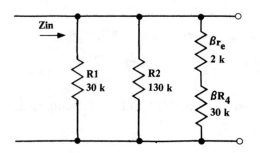

$$Zin = R1 \text{ // } R2 \text{ // } \beta(r_e + R4)$$

$= 30 \text{ k}\Omega \text{ // } 130 \text{ k}\Omega \text{ // } 100(20 \text{ }\Omega + 300 \text{ }\Omega)$

$= 30 \text{ k}\Omega \text{ // } 130 \text{ k}\Omega \text{ // } 32 \text{ k}\Omega = 13.84 \text{ k}\Omega$

FIGURE 8-24

POWER GAIN

Power gain to the load is solved from:

$$PG = A_v^2 \times Zin/RL = 15^2 \times 13.84 \text{ k}\Omega / 12 \text{ k}\Omega = 259.5$$

Converting power gain to decibels:

$$PG(dB) = 10 \log PG = 10 \log 259.5 = 24.14 \text{ dB}$$

MAXIMUM PEAK-TO-PEAK OUTPUT VOLTAGE SWING

The maximum peak-to-peak output voltage swing of the upside-down PNP circuit configuration is, like the previous NPN and PNP circuits, solved from $2V_{CE}$ or $2I_C r_L$, whichever is smaller. Although $2I_C r_L$ provides the maximum voltage swing limit, both the approximate $2V_{CE}$ and more exact $2V_{CE} r_L / (R4 + r_L)$ formulas are given for purpose of comparison.

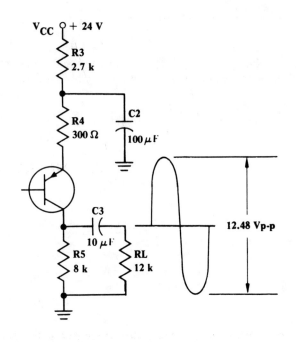

$2I_C r_L = 2 \times 1.3 \text{ mA} \times 4.8 \text{ k}\Omega = 12.48 \text{ Vp-p}$

$2V_{CE} = 2 \times 9.7 \text{ V} = 19.4 \text{ Vp-p, or}$

$2V_{CE} \times r_L/(R4 + r_L)$

$= 2 \times 9.7 \text{ V} \times 4.8 \text{ k}\Omega/(300 \text{ }\Omega + 4.8 \text{ k}\Omega)$

$= 18.26 \text{ Vp-p}$

Therefore, Vop-p(max) = 12.48 Vp-p

FIGURE 8-25

OUTPUT IMPEDANCE

The output impedance of the PNP upside-down circuit, like that of the NPN circuit, is solved from:

$$Ro = \frac{Vo(NL) - Vo(L)}{Vo(L)} \times RL$$

For example, if Vo(NL) equals 8 Vp-p and Vo(L) equals 4.8 Vp-p, where RL is 12 kΩ, then:

$$Ro = \frac{Vo(NL) - Vo(L)}{Vo(L)} \times RL = \frac{8 \text{ Vp-p} - 4.8 \text{ Vp-p}}{4.8 \text{ Vp-p}} \times 12 \text{ k}\Omega = \frac{3.2 \text{ Vp-p} \times 12 \text{ k}\Omega}{4.8 \text{ Vp-p}} = 8 \text{ k}\Omega$$

NOTE: I_C is assumed at 1 mAp-p, like the previously solved Ro of the NPN universal circuit. Hence:

$Vo(NL) = I_C R_C = 1 \text{ mAp-p} \times 8 \text{ k}\Omega = 8 \text{ Vp-p}$

$Vo(L) = I_C r_L = 1 \text{ mAp-p} \times 4.8 \text{ k}\Omega = 4.8 \text{ Vp-p}$

SECTION I: LABORATORY EXERCISE (Part 2)

OBJECTIVES

Investigate the DC and AC characteristics of the upside-down, common-emitter, universal circuit using a PNP transistor.

LIST OF MATERIALS AND EQUIPMENT

1. Transistor: 2N3906 or PNP equivalent
2. Resistors (one each): 330 Ω   3.3 kΩ   10 kΩ   12 kΩ   15 kΩ   27 kΩ   120 kΩ
3. Capacitors: 10 μF (two)   100 μF (one)
4. Power Supply
5. Voltmeter
6. Signal Generator
7. Oscilloscope

PROCEDURE

1. Connect the PNP upside-down transistor circuit as shown in Figure 8-26.

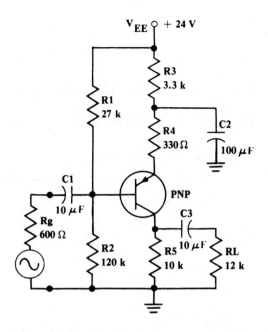

FIGURE 8-26

2. Calculate the DC voltage drop across all the circuit resistors and at the collector emitter of the device. Calculate from:

a. $V_{R1} = \dfrac{V_{CC} \times R1}{R1 + R2}$

b. $V_{R2} = \dfrac{V_{CC} \times R2}{R1 + R2}$

c. $V_{RE} = V_{R3} + V_{R4} = V_{R1} - V_{BE}$ where: $V_{BE} \approx 0.6$ V

d. $I_E = V_{RE}/RE = (V_{R3} + V_{R4})/(R3 + R4)$

e. $V_{R3} = I_E R3$

f. $V_{R4} = I_E R4$

g. $V_{R5} = I_C R5$ where: $I_C \approx I_E$

h. Calculate the single point transistor voltages, with respect to ground.

$V_B = V_{R2}$

$V_E = V_{CC} - V_{RE}$ where: $V_{RE} = V_{R3} + V_{R4}$, also

$V_E = V_B + V_{BE}$

$V_C = V_{R5}$

j. Calculate the voltage drop across the collector emitter terminals of the transistor from:

$V_{CE} = V_{EC} = V_E - V_C$

TABLE 8-5	V_{R1}	V_{R2}	V_{R3}	V_{R4}	V_{R5}	V_C	V_E	V_{CE}	I_E
CALCULATED									
MEASURED									

3. Measure the DC voltage drops around the circuit.

4. Insert the calculated and measured values, as indicated, into Table 8-5.

5. VOLTAGE GAIN

   a. Calculate the mid-frequency voltage gain from: $A_v = \dfrac{RC \,//\, RL}{r_e + R4}$ where: $r_e \approx 26 \text{ mV}/I_E$

   b. Measure the voltage output using an input signal voltage of 100 mVp-p at a frequency of 1 kHz. Calculate A_v from:

   $A_v = Vo/Vin$ where: $Vin = 100 \text{ mVp-p}$

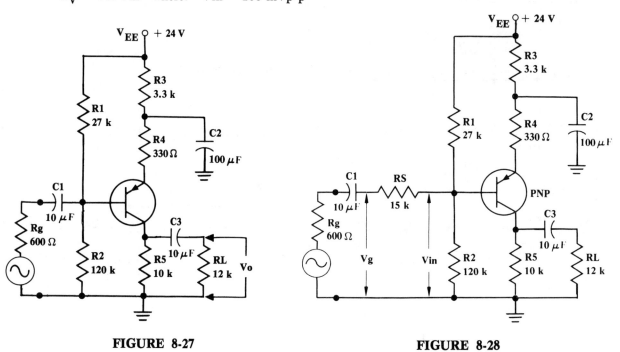

FIGURE 8-27   FIGURE 8-28

6. INPUT IMPEDANCE

   a. Calculate the input impedance to the circuit from: $Zin = R1 \,//\, R2 \,//\, \beta(r_e + R4)$

NOTE: Use the known beta of the device, or substitute a nominal beta of 100 if the device beta is not known.

   b. Measure the input impedance to the circuit using the known series resistor technique. Use an RS resistor value of 15 kΩ and an input signal level of 100 mVp-p at a frequency of 1 kHz, as shown in Figure 8-28.

7. MAXIMUM PEAK-TO-PEAK OUTPUT VOLTAGE SWING

   a. Calculate the maximum peak-to-peak output voltage swing from approximately $2V_{CE}$ or from $2I_C r_L$, whichever is smaller. The more exact formula is optional.
   b. Increase the input signal level and monitor the output signal across the load until clipping or alpha distortion occurs. Remove RS for this measurement.

8. POWER GAIN TO THE LOAD RESISTOR

   a. Calculate the power gain to the load resistor, using both the measured A_v and Zin values. Calculate from:

$$PG = A_v^2 \times Zin/RL \quad \text{where } RL = 12 \text{ k}\Omega$$

b. Convert PG to dB from: $PG(dB) = 10 \log PG$

9. OUTPUT IMPEDANCE

a. Remove the load resistor and monitor the output signal at the collector, with respect to ground. Adjust the input signal level so the unloaded output peak-to-peak signal voltage, monitored on the scope, is 5 Vp-p.

b. Connect the load resistor, monitor the loaded peak-to-peak output voltage swing, and calculate the output impedance from:

$$Zo = \frac{Vo(NL) - Vo(L)}{Vo(L)} \times RL \quad \text{where:} \quad Vo(NL) = 5 \text{ Vp-p}$$

10. Insert the calculated and measured values, as indicated, into Table 8-6.

TABLE 8-6	A_v	Vo for Vin = 100 mVp-p	Zin	Vop-p (MAX)	POWER GAIN	PG(dB)	Zo
CALCULATED		/////					
MEASURED					/////	/////	/////

11. Include the load resistor and monitor the low and high corner frequencies of the amplifier circuit of Figure 8-28. Insert the measured low and high corner frequencies into the table of Figure 8-29 and note the available band-pass frequency on the Bode plot.

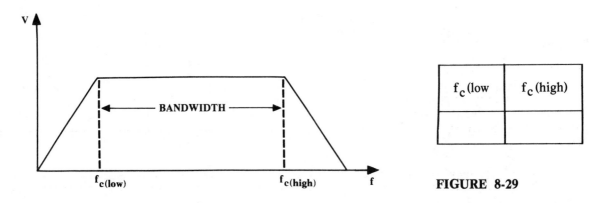

FIGURE 8-29

SECTION QUESTIONS

1. How does the inverted PNP, common-emitter, universal biased circuit differ from the NPN, common-emitter, universal-biased circuit? Are the voltage gain, input impedance, power gain, Vop-p(max), or output impedance parameters solved differently? Explain if they are or are not.

2. Does the 180° phase shift between input and output differ for the NPN or PNP, common-emitter, universal-biased circuits? Explain.

3. Is there any significant AC parameter difference between the NPN and PNP universal connected circuits? Explain.

SECTION PROBLEMS

1. Given: Beta = 100 and $V_{BE} = 0.6$ V
   a. Solve for:
      1. A_v
      2. Zin
      3. Vop-p(max)
   b. With capacitor C3 removed, solve for:
      1. A_v
      2. Zin
   c. With capacitor C3 connected across both R4 and R5, solve for:
      1. A_v
      2. Zin

2. Given: Beta = 100 and $V_{BE} = 0.6$ V
   a. Solve for:
      1. A_v
      2. Zin
      3. Vop-p(max)
   b. With the bypass capacitor C2 connected to the emitter, with respect to ground, solve for:
      1. A_v
      2. Zin

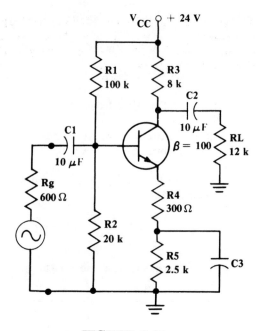

FIGURE 8-30

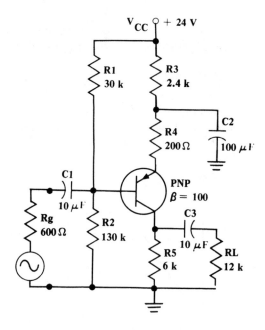

FIGURE 8-31

3. Solve for the input impedance Zin of the circuit of the circuit of Figure 8-32 using a beta of:
   a. 100
   b. 200
   c. 50

   NOTE: $r_e = 26$ mV/I_E

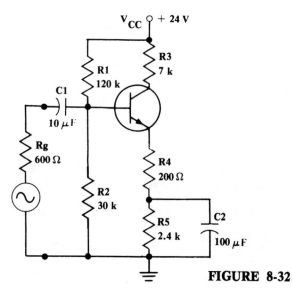

FIGURE 8-32

4. In the circuit of Figure 8-33, when capacitor C3 is removed and resistor R5 is not bypassed, the gain of the amplifier is exactly 2. What is the gain when capacitor C3 bypasses resistor R5?

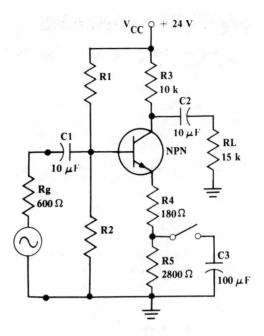

FIGURE 8-33

5. In the circuit of Figure 8-34, with capacitor C2 connected, the gain of the amplifier is 250. What is the value of r_e?

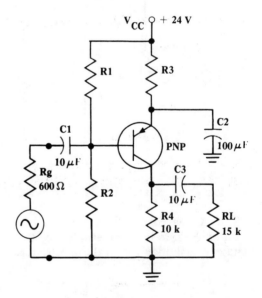

FIGURE 8-34

SECTION II: COMMON-COLLECTOR, UNIVERSAL CIRCUIT

GENERAL DISCUSSION

The DC analysis of the common collector circuit is similar to that of the common emitter circuit except there is no collector resistor and only a single resistor is used from the emitter to ground. The AC analysis is similar to that of the base-biased, common-collector circuit. The circuit to be analyzed is shown in Figure 8-35, where capactior C3 is used to help minimize oscillations that can occur in common collector circuits. Oscillations occur because the common-collector circuit has both in-phase condition and near unity signal condition between the input and output terminals. Therefore, if peaking occurs at the output as some higher frequency, the circuit will break into oscillation. Capacitor C3 helps minimize the possibility by forcing an amplitude roll-off to occur prior to these oscillatory conditions.

FIGURE 8-35

DIRECT CURRENT CIRCUIT ANALYSIS

The DC analysis for the common collector universal circuit is given with Figure 8-36. Also, since a large output voltage swing is determined by the DC conditions of the circuit, R1 and R2 are chosen so that the voltage distribution across the emitter resistor R3 and V_{CE} are about equal.

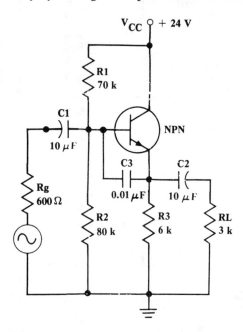

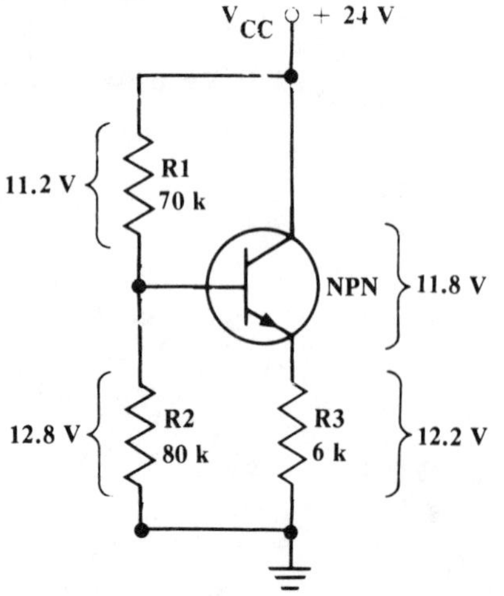

$$V_{R1} = \frac{V_{CC} \times R1}{R1 + R2} = \frac{24\ V \times 70\ k\Omega}{70\ k\Omega + 80\ k\Omega} = 11.2\ V$$

$$V_{R2} = \frac{V_{CC} \times R2}{R1 + R2} = \frac{24\ V \times 80\ k\Omega}{70\ K\Omega + 80\ k\Omega} = 12.8\ V$$

$$V_{R3} = VR2 - VBE = 12.8\ V - 0.6\ V = 12.2\ V$$

$$I_E = V_{R3}/R3 = 12.2\ V/6\ k\Omega = 2.03\ mA$$

$$V_{CE} = V_{CC} = V_{R3} - 24\ V - 12.2\ V = 11.8\ V$$

FIGURE 8-36

VOLTAGE GAIN

The voltage gain of the common collector universal circuit will be slightly less than unity, because a slight amount of signal voltage is dropped across the base-emitter-diode resistance.

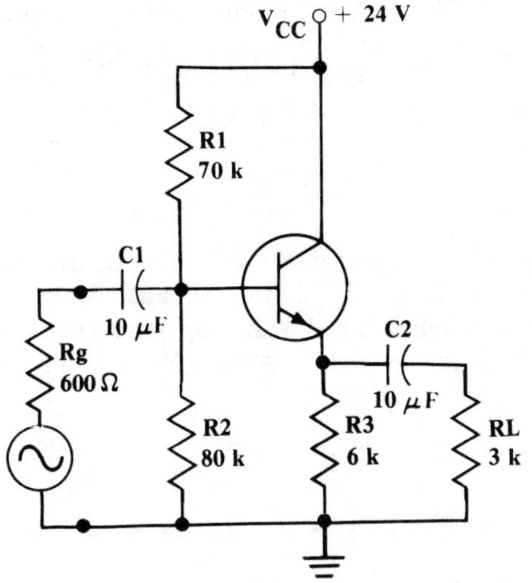

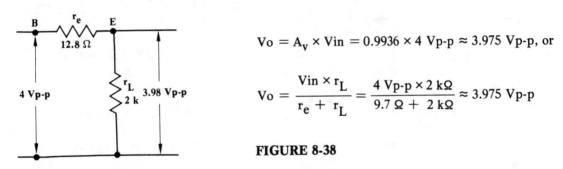

$$A_v = \frac{r_L}{r_e + r_L} = \frac{2\ k\Omega}{12.8\ \Omega + 2\ k\Omega} = 0.9936$$

where: $r_e = 26\ mV/I_E = 26\ mV/2.03\ mA \approx 12.8\ \Omega$

$$r_L = RE \,//\, RL = \frac{R3 \times RL}{R3 + RL}$$

$$= \frac{6\ k\Omega \times 3\ k\Omega}{6\ k\Omega + 3\ k\Omega} = 2\ k\Omega$$

FIGURE 8-37

Therefore, for an input voltage of 4 Vp-p, the output voltage would be:

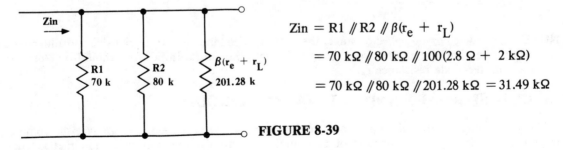

$Vo = A_v \times Vin = 0.9936 \times 4\ Vp\text{-}p \approx 3.975\ Vp\text{-}p$, or

$$Vo = \frac{Vin \times r_L}{r_e + r_L} = \frac{4\ Vp\text{-}p \times 2\ k\Omega}{9.7\ \Omega + 2\ k\Omega} \approx 3.975\ Vp\text{-}p$$

FIGURE 8-38

INPUT IMPEDANCE

The input impedance of the universal circuit is solved by reflecting the effective emitter resistance into the base in parallel with R1 and R2, as shown in Figure 8-39. The effective emitter resistance is r_L in series with r_e. Therefore, Zin(base) = $\beta(r_e + r_L)$ = 201.28 kΩ.

$Zin = R1 \,//\, R2 \,//\, \beta(r_e + r_L)$

$= 70\ k\Omega \,//\, 80\ k\Omega \,//\, 100(2.8\ \Omega + 2\ k\Omega)$

$= 70\ k\Omega \,//\, 80\ k\Omega \,//\, 201.28\ k\Omega = 31.49\ k\Omega$

FIGURE 8-39

POWER GAIN

The power gain to the load resistor is solved from:

$PG = A_v^2 \times Zin/RL = 0.9936^2 \times 31.49\ k\Omega/3\ k\Omega \approx 10.36$

Converting PG to dB:

$$PG(dB) = 10 \log 10.36 = 10.15 \text{ dB}$$

MAXIMUM PEAK-TO-PEAK OUTPUT VOLTAGE SWING

The maximum peak-to-peak output voltage is solved from either $2V_{CE}$ or $2I_E r_L$, whichever is smaller.

$$2V_{CE} = 2 \times 11.8 \text{ V} = 23.6 \text{ Vp-p}$$

$$2I_E r_L = 2 \times 2.03 \text{ mA} \times 2 \text{ k}\Omega = 8.12 \text{ Vp-p}$$

NOTE: A rule of thumb to follow to minimize low Vop-p(max) conditions is to choose RE and RL resistances to be close in value. For instance, lowering the value of RL to 1 kΩ would cause Vop-p(max) to decrease to less than 3 Vp-p. Then, if RE is lowered to 1 kΩ, when RL is 1 kΩ, Vop-p(max) would equal 12.2 Vp-p, because I_E would be about 12.2 mA and 2×12.2 mA $\times$ 500 Ω = 12.2 Vp-p from V_{RE}/RE = 12.2 V/1 kΩ.

OUTPUT IMPEDANCE

The output impedance of the common-collector circuit is solved by removing the load resistor and "looking back" into the emitter. Therefore:

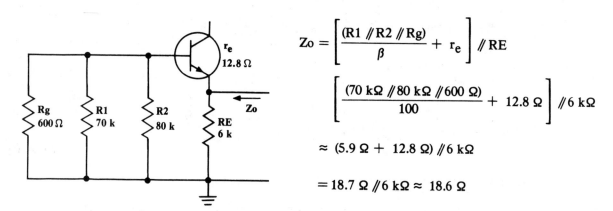

FIGURE 8-40

Therefore, the difference in the output voltage for unloaded or loaded conditions, where RL is 3 kΩ is minimal. Again, Zo is calculated from:

$$Zo = Ro = \frac{Vo(NL) - Vo(L)}{Vo(L)} \times RL$$

NOTE: For the common collector stage, the output impedance is low, obviously, and unless the impedance is reflected into the emitter from the base is high, the output impedance should be close in value to the base-emitter-diode resistance r_e.

CIRCUIT SUMMARY AND EQUIVALENT CIRCUIT

The approximate and equivalent circuit analysis methods are used again to analyze the common collector circuit to provide a comparison of techniques. Essentially, the approximation method uses signal voltage concepts, while the equivalent circuit approach uses signal current concepts. Both methods, however, follow Ohm's law and provide identical results.

A. Approximation Method of Analysis

Beta = 100, Vin = 4 Vp-p, and $r_e \approx 12.8$ Ω

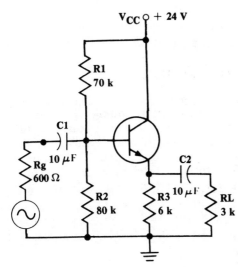

$A_v = r_L/(r_e + r_L) = 2\,k\Omega/(12.8\,\Omega + 2\,k\Omega) = 0.9936$

$V_o = V_{in} \times A_v = 4\,Vp\text{-}p \times 0.9936 = 3.975\,Vp\text{-}p$

$Z_{in} = R1 \,/\!/\, R2 \,/\!/\, \beta(r_e + r_L)$

$\quad = 70\,k\Omega \,/\!/\, 80\,k\Omega \,/\!/\, 201.28\,k\Omega = 31.49\,k\Omega$

$PG = A_v^2 \times Z_{in}/RL = 0.9936^2 \times 31.49\,k\Omega\,/3\,k\Omega = 10.36$

$PG(dB) = 10\,\log\,10.36 = 10.15\,dB$

$A_i = PG/A_v = 10.36/0.9936 = 10.43$

$V_{op\text{-}p}(max) = 2I_C r_L = 2 \times 2.03\,mA \times 2\,k\Omega = 8.12\,Vp\text{-}p$

FIGURE 8-41

B. Equivalent Circuit Method of Analysis

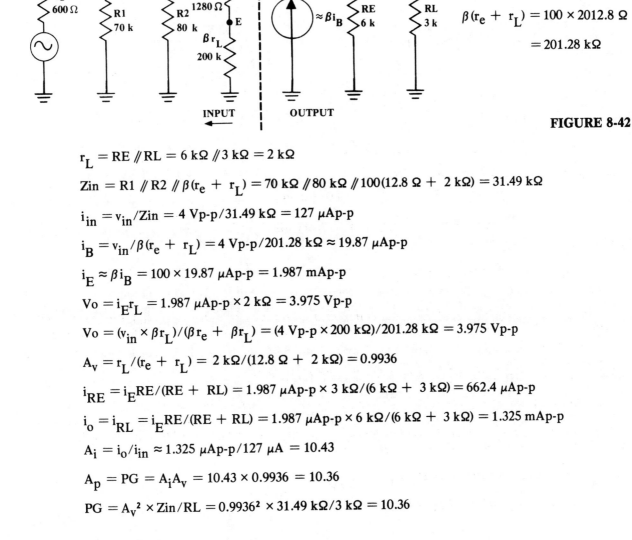

$\beta r_e = 100 \times 12.8\,\Omega = 1280\,\Omega$

$\beta r_L = 100 \times 2\,k\Omega = 200\,k\Omega$

$\beta(r_e + r_L) = 100 \times 2012.8\,\Omega$

$\quad = 201.28\,k\Omega$

FIGURE 8-42

$r_L = RE \,/\!/\, RL = 6\,k\Omega \,/\!/\, 3\,k\Omega = 2\,k\Omega$

$Z_{in} = R1 \,/\!/\, R2 \,/\!/\, \beta(r_e + r_L) = 70\,k\Omega \,/\!/\, 80\,k\Omega \,/\!/\, 100(12.8\,\Omega + 2\,k\Omega) = 31.49\,k\Omega$

$i_{in} = v_{in}/Z_{in} = 4\,Vp\text{-}p/31.49\,k\Omega = 127\,\mu Ap\text{-}p$

$i_B = v_{in}/\beta(r_e + r_L) = 4\,Vp\text{-}p/201.28\,k\Omega \approx 19.87\,\mu Ap\text{-}p$

$i_E \approx \beta i_B = 100 \times 19.87\,\mu Ap\text{-}p = 1.987\,mAp\text{-}p$

$V_o = i_E r_L = 1.987\,\mu Ap\text{-}p \times 2\,k\Omega = 3.975\,Vp\text{-}p$

$V_o = (v_{in} \times \beta r_L)/(\beta r_e + \beta r_L) = (4\,Vp\text{-}p \times 200\,k\Omega)/201.28\,k\Omega = 3.975\,Vp\text{-}p$

$A_v = r_L/(r_e + r_L) = 2\,k\Omega/(12.8\,\Omega + 2\,k\Omega) = 0.9936$

$i_{RE} = i_E RE/(RE + RL) = 1.987\,\mu Ap\text{-}p \times 3\,k\Omega/(6\,k\Omega + 3\,k\Omega) = 662.4\,\mu Ap\text{-}p$

$i_o = i_{RL} = i_E RE/(RE + RL) = 1.987\,\mu Ap\text{-}p \times 6\,k\Omega/(6\,k\Omega + 3\,k\Omega) = 1.325\,mAp\text{-}p$

$A_i = i_o/i_{in} \approx 1.325\,\mu Ap\text{-}p/127\,\mu A = 10.43$

$A_p = PG = A_i A_v = 10.43 \times 0.9936 = 10.36$

$PG = A_v^2 \times Z_{in}/RL = 0.9936^2 \times 31.49\,k\Omega/3\,k\Omega = 10.36$

$$\text{Vop-p(max)} = 2I_E r_L = 2 \times 2.03 \text{ mA} \times 2 \text{ k}\Omega = 8.12 \text{ Vp-p}$$

$$\text{Po(max)} = \text{Vop-p}^2/8RL = 8.12 \text{ Vp-p}^2/(8 \times 3 \text{ k}\Omega) \approx 274.7 \text{ mW}$$

INCREASING THE INPUT IMPEDANCE

The input impedance of the common collector circuit can be increased by adding a base resistor to the circuit, such as R3 in Figure 8-43. The resistor should be large enough in value to increase the input impedance, but it should not be so large in value that the DC voltage distribution is controlled by the beta of the device, as it would be in the base-biased circuit. A good choice is a value of 100 kΩ, because it is large enough to increase the input impedance and it does not effect the DC distribution greatly. The following analysis for the circuit of Figure 8-43 uses the exact Thevenin equivalent theorem method.

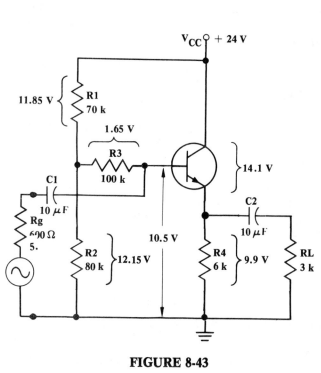

FIGURE 8-43

$$V_{TH} = \frac{V_{CC} \times R2}{R1 + R2} = \frac{24 \text{ V} \times 80 \text{ k}\Omega}{70 \text{ k}\Omega + 80 \text{ k}\Omega} = 12.8 \text{ V}$$

$$R_{TH} = (R1 \parallel R2) + R3 = (70 \text{ k}\Omega \parallel 80 \text{ k}\Omega) + 100 \text{ k}\Omega$$

$$= 37.33 \text{ k}\Omega + 100 \text{ k}\Omega = 137.33 \text{ k}\Omega$$

$$I_E = \frac{V_{TH} - V_{BE}}{(R_{TH}/\beta) + R4} = \frac{12.8 \text{ V} - 0.6 \text{ V}}{(137.33 \text{ k}\Omega/100) + 6 \text{ k}\Omega}$$

$$\frac{12.2 \text{ V}}{1.3733 \text{ k}\Omega + 6 \text{ k}\Omega} = \frac{12.2 \text{ V}}{7.373 \text{ k}\Omega} \approx 1.65 \text{ mA}$$

$$V_{R4} = I_E R_4 = 1.65 \text{ mA} \times 6 \text{ k}\Omega = 9.9 \text{ V}$$

$$V_B = V_{R4} + V_{BE} = 9.9 \text{ V} + 0.6 \text{ V} = 10.5 \text{ V}$$

$$I_B = I_E/(\beta + 1) \approx I_E/\beta = 1.65 \text{ mA}/100 = 16.5 \text{ }\mu\text{A}$$

$$V_{R3} = I_B R3 = 16.5 \text{ }\mu\text{A} \times 100 \text{ k}\Omega = 1.65 \text{ V}$$

$$V_{R2} = V_B + V_{R3} = 10.5 \text{ V} + 1.65 \text{ V} = 12.15 \text{ V}$$

$$V_{R1} = V_{CC} - V_{R2} = 24 \text{ V} - 12.15 \text{ V} = 11.85 \text{ V}$$

$$V_{CE} = V_{CC} - V_{R4} = 24 \text{ V} - 9.9 \text{ V} = 14.10 \text{ V}$$

The DC analysis using Thevenin's equivalent theorem is exact but it involves many steps. The reflected resistance method of DC analysis, which is used in conjunction with Figure 8-44, is not as precise, but it is straight forward and results are well within the tolerance of most resistors used in the construction of the circuit.

In the reflected resistance DC analysis used with Figure 8-44, the 6 kΩ RE resistor is reflected into the base region. Then, the voltage divider equation using R3 (100 kΩ) and RE (600 kΩ) is used to solve for the approximate DC voltage at the base. With this accomplished, V_{RE} is found and I_E and I_B are solved. Next, V_{R3} can be calculated and V_{R2} adjusted to satisfy Ohm's law.

$$VR2 = \frac{V_{CC} \times R2}{R1 + R2} = \frac{24 \text{ V} \times 80 \text{ k}\Omega}{70 \text{ k}\Omega + 80 \text{ k}\Omega} = 12.8 \text{ V}$$

$$VB = \frac{V_{R2} \times \beta R4}{R3 + \beta R4} = \frac{12.8 \text{ V}(100 \times 6 \text{ k}\Omega)}{100 \text{ k}\Omega + (100 \times 6 \text{ k}\Omega)} = 10.97 \text{ V}$$

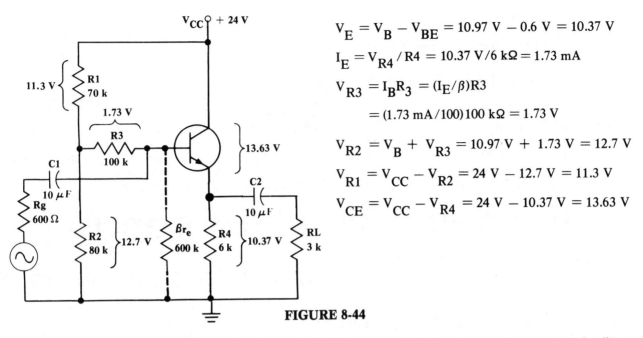

$$V_E = V_B - V_{BE} = 10.97\text{ V} - 0.6\text{ V} = 10.37\text{ V}$$
$$I_E = V_{R4}/R4 = 10.37\text{ V}/6\text{ k}\Omega = 1.73\text{ mA}$$
$$V_{R3} = I_B R3 = (I_E/\beta)R3$$
$$= (1.73\text{ mA}/100)100\text{ k}\Omega = 1.73\text{ V}$$
$$V_{R2} = V_B + V_{R3} = 10.97\text{ V} + 1.73\text{ V} = 12.7\text{ V}$$
$$V_{R1} = V_{CC} - V_{R2} = 24\text{ V} - 12.7\text{ V} = 11.3\text{ V}$$
$$V_{CE} = V_{CC} - V_{R4} = 24\text{ V} - 10.37\text{ V} = 13.63\text{ V}$$

FIGURE 8-44

NOTE: V_{R2} at 12.8 V is modified slightly, from 12.8 V to about 12.7 V, when the base current loading effect is included in the calculations. Also, with regard to the Thevenin equivalent analysis, the reflected resistance method of analysis is not as precise when the base resistance is large.

ALTERNATING CURRENT CIRCUIT ANALYSIS

The input impedance and output impedance differ from the previous common-collector circuit configuration where resistor R3 was not used.

INPUT IMPEDANCE

The input impedance, as seen from the signal generator, is the parallel combination of $\beta(r_e + r_L)$ and $R3 + (R1 \,/\!/\, R2)$. The addition of R3 raises Z_{in} from about 31.49 kΩ to about 81.67 kΩ.

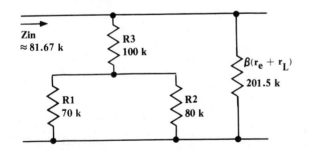

$$Z_{in} = [(R1 \,/\!/\, R2) + R3] \,/\!/\, \beta(r_e + r_L)$$
$$= [(70\text{ k}\Omega \,/\!/\, 80\text{ k}\Omega) + 100\text{ k}\Omega] \,/\!/\, 100(15\text{ }\Omega + 2\text{ k}\Omega)$$
$$= (37.33\text{ k}\Omega + 100\text{ k}\Omega) \,/\!/\, 201.5\text{ k}\Omega$$
$$= 137.33\text{ k}\Omega \,/\!/\, 201.5\text{ k}\Omega \approx 81.67\text{ k}\Omega$$

where: $r_L = R4 \,/\!/\, RL = 6\text{ k}\Omega \,/\!/\, 3\text{ k}\Omega = 2\text{ k}\Omega$

$r_e = 26\text{ mV}/I_E = 26\text{ mV}/1.73\text{ mA} = 15\text{ }\Omega$

FIGURE 8-45

OUTPUT IMPEDANCE

The output impedance of the common collector stage is the impedance "seen" looking into the emitter terminal with respect to ground.

$$Z_o = \left(\frac{[R3 + (R1 \,/\!/\, R2)] \,/\!/\, R_g}{\beta} + r_e\right) \,/\!/\, R4 \approx [(R_g/\beta) + r_e] \,/\!/\, R4$$
$$= [(600\text{ }\Omega/100) + 15\text{ }\Omega] \,/\!/\, 6\text{ k}\Omega = (6\text{ }\Omega + 15\text{ }\Omega) \,/\!/\, 6\text{ k}\Omega = 20.9\text{ }\Omega$$

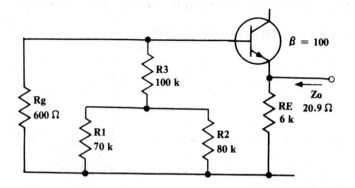

FIGURE 8-46

SECTION II: LABORATORY EXERCISE

OBJECTIVES

Investigate the DC and AC parameters of the common collector (emitter follower), NPN, universal circuit.

LIST OF MATERIALS AND EQUIPMENT

1. Transistors: 2N3904 or equivalent
2. Resistors (one each): 1 kΩ   2.2 kΩ   22 kΩ (two)   47 kΩ
3. Capacitors: 10 μF (two)   0.01 μF (one)
4. Power Supply
5. Voltmeter
6. Signal Generator
7. Oscilloscope

PROCEDURE

1. Connect the NPN transistor in the common-collector circuit as shown in Figure 8-47, where the signal generator and the load resistor are included in the circuit.

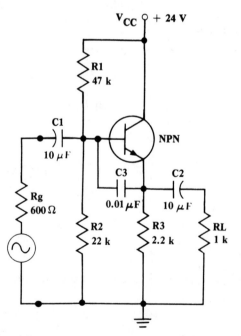

FIGURE 8-47

NOTE: Connect the C3 capacitor (from 0.01 μF to 0.1 μF) across the base-emitter or base-collector terminals to minimize oscillatory conditions. However, if proper breadboarding and measurement techniques are used, such as a 10:1 probe, the C3 capacitor can be eliminated.

2. Calculate and measure the DC voltage drops across all the circuit resistors and the transistor.

3. Insert the calculated and measured values, as indicated, into Table 8-7.

TABLE 8-7	V_{R1}	V_{R2}	V_{R3}	V_C	V_{CE}	I_E	Beta
CALCULATED							
MEASURED							

4. VOLTAGE GAIN

   a. Calculated the mid-frequency voltage gain from:

   $$A_v = \frac{R3 \mathbin{/\mkern-5mu/} RL}{r_e + (R3 \mathbin{/\mkern-5mu/} RL)} \quad \text{where:} \quad r_e \approx 26 \text{ mV}/I_E$$

   b. Measure the mid-frequency voltage gain using an input voltage of 1 Vp-p at a frequency of 1 kHz. Measure the Vo across the load resistor RL. Calculate the measured voltage gain from:

   $$A_v = Vo/Vin \quad \text{where:} \quad Vin = 1 \text{ Vp-p}$$

5. INPUT IMPEDANCE

   a. Calculate the input impedance to the stage from:

   $$Zin = R1 \mathbin{/\mkern-5mu/} R2 \mathbin{/\mkern-5mu/} \beta [(r_e + (R3 \mathbin{/\mkern-5mu/} RL)]$$

   b. Measure the input impedance. Use a known 22 kΩ resistor and an input signal level of 1 Vp-p at 1 kHz, as shown in Figure 8-48. Calculate Zin from:

   $$Zin = \frac{Vin \times RS}{V_{RS}} = \frac{Vin \times RS}{Vg - Vin} \quad \text{where } Vin = 1 \text{ Vp-p}$$

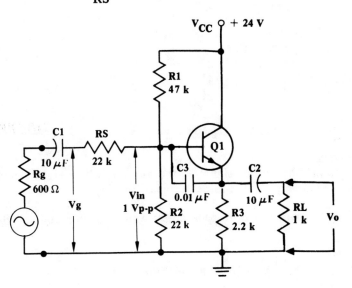

FIGURE 8-48

6. MAXIMUM PEAK-TO-PEAK OUTPUT VOLTAGE SWING

   a. Calculate the maximum peak-to-peak output voltage swing from $2V_{CE}$ or $2I_E r_L$, whichever is smaller. This is where: $r_L = R3 \mathbin{/\mkern-5mu/} RL$; hence, $2I_E(R3 \mathbin{/\mkern-5mu/} RL)$.

b. Monitor the emitter output and increase the 1 kHz input signal level until clipping or distortion, because of alpha crowding, occurs.

NOTE: If the maximum peak-to-peak voltage swing is limited to $2I_E r_L$, then increasing I_E will increase the maximum peak-to-peak output voltage swing. However, increasing the emitter current may cause the device to dissipate more power than the rated power of the device.

7. **POWER DISSIPATED BY THE DEVICE:** The power dissipated by the transistor is calculated from:

$$P(Q1) = V_{CE} \times I_E$$

8. **POWER GAIN TO THE LOAD RESISTOR:** Calculate the power gain to the load resistor using both the measured Av and Zin values. Calculate from:

$$PG = A_v^2 \times Zin/RL \quad \text{where:} \quad RL = 1 \text{ k}\Omega$$

9. **OUTPUT IMPEDANCE**

a. Calculate the output impedance of the circuit from: $Z_o \approx [(Rg/\beta) + r_e] \mathbin{/\mkern-5mu/} R3$

NOTE: Use the beta of the device if it is known. Otherwise, a nominal beta of 100 will provide reasonably accurate results.

b. Measure the output impedance using a 33 Ω load resistor, as shown in Figure 8-49.

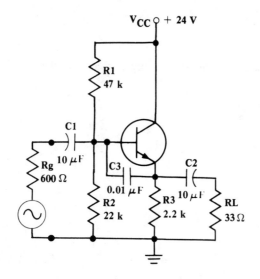

FIGURE 8-49

NOTE: To measure the output impedance:
a. Remove the load resistor and monitor the output signal at the emitter, with respect to ground. Adjust the input signal level so the unloaded output peak-to-peak signal voltage monitored on the scope is about 300 mVp-p.
b. Connect the load resistor, monitor the loaded peak-to-peak output voltage swing, and calculate the output impedance from:

$$Z_o = \frac{Vo(NL) - Vo(L)}{Vo(L)} \times RL, \quad \text{where:} \quad Vo(NL) = 300 \text{ mVp-p}$$

10. Insert the calculated and measured values, as indicated, into Table 8-8.

TABLE 8-8	A_v	Vo for Vin = 1 Vp-p	Vop-p (MAX)	Zin	Zo	POWER GAIN
CALCULATED		/////				
MEASURED						/////

11. Under loaded conditions, monitor the upper and lower frequency 3 dB points. Insert these values on the Bode plot of Figure 8-50.

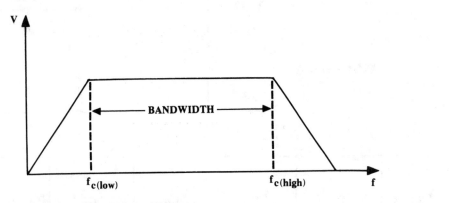

FIGURE 8-50

SECTION QUESTIONS

1. Is the input and output signal voltage of the common collector circuit in phase?
2. Why is the Vop-p(max) for the common collector stage difficult to achieve with most standard laboratory signal generators?
3. Does increasing the emitter current increase the possible Vop-p(max) conditions? Why?
4. What resistor value needs to be lowered to achieve the larger emitter current increase in Figure 8-49? Can raising the value of R2, with regard to R1, also provide a larger emitter current?
5. With reference to Figure 8-52, why is it necessary to include a series base resistor in the common collector universal circuit to increase the input impedance? Why not just increase the proportional resistance value of R1 and R2? Why is this not such a good idea?
6. When adding a base resistor to the common collector universal circuit, why is 100 kΩ about as high a value that should be used, with regard to DC voltage distribution? Does the beta of the device enter into the RB value size? Why?
7. Why is the common collector stage susceptible to oscillatory conditions? What precautionary step can be taken to minimize the possibility of oscillations?

SECTION PROBLEMS

1. With reference to Figure 8-51, solve for:

   a. V_{CE}

   b. Zin

   NOTE: Use the approximation method of analysis.

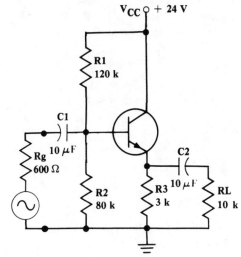

FIGURE 8-51

2. With reference to Figure 8-52, solve for:

a. The DC voltage distribution of the circuit. Use both the Thevenin and the reflected resistance methods of analysis. Insert the calculated results, as indicated, into Table 8-9. The transistor beta is 99.

b. Zin (Use either of the calculated I_E's in solving for r_e.)

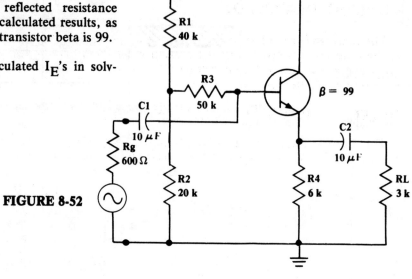

FIGURE 8-52

TABLE 8-9 METHOD	V_{R1}	V_{R2}	V_{R3}	V_{R4}	V_{CE}	I_E
Thevenin						
Reflected Resistance						

SECTION III: THE COMMON-BASE, UNIVERSAL CIRCUIT

GENERAL DISCUSSION

The common base universal amplifier circuit applies the input signal to the emitter and takes the output of the collector. Therefore, like the previously analyzed base-biased, common-base, amplifier circuit, the input impedance is low, the current gain is less than unity, and the voltage gain is about equal to that of the common emitter circuit.

DIRECT CURRENT CIRCUIT ANALYSIS

The direct current analysis of the common base universal circuit is identical to that of the common emitter circuit, and it is repeated here merely for convenience.

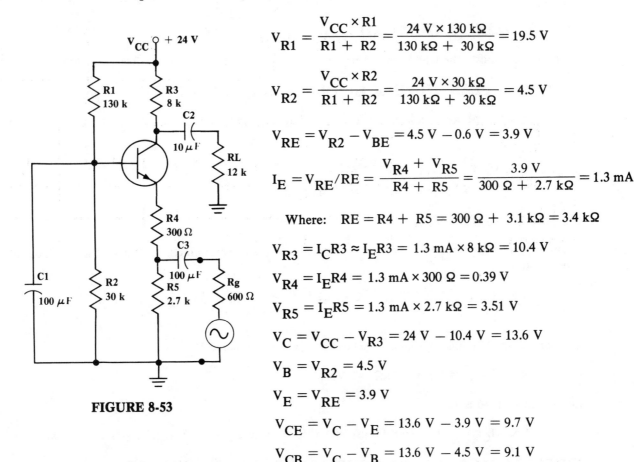

FIGURE 8-53

$$V_{R1} = \frac{V_{CC} \times R1}{R1 + R2} = \frac{24\text{ V} \times 130\text{ k}\Omega}{130\text{ k}\Omega + 30\text{ k}\Omega} = 19.5\text{ V}$$

$$V_{R2} = \frac{V_{CC} \times R2}{R1 + R2} = \frac{24\text{ V} \times 30\text{ k}\Omega}{130\text{ k}\Omega + 30\text{ k}\Omega} = 4.5\text{ V}$$

$$V_{RE} = V_{R2} - V_{BE} = 4.5\text{ V} - 0.6\text{ V} = 3.9\text{ V}$$

$$I_E = V_{RE}/RE = \frac{V_{R4} + V_{R5}}{R4 + R5} = \frac{3.9\text{ V}}{300\ \Omega + 2.7\text{ k}\Omega} = 1.3\text{ mA}$$

Where: $RE = R4 + R5 = 300\ \Omega + 3.1\text{ k}\Omega = 3.4\text{ k}\Omega$

$$V_{R3} = I_C R3 \approx I_E R3 = 1.3\text{ mA} \times 8\text{ k}\Omega = 10.4\text{ V}$$

$$V_{R4} = I_E R4 = 1.3\text{ mA} \times 300\ \Omega = 0.39\text{ V}$$

$$V_{R5} = I_E R5 = 1.3\text{ mA} \times 2.7\text{ k}\Omega = 3.51\text{ V}$$

$$V_C = V_{CC} - V_{R3} = 24\text{ V} - 10.4\text{ V} = 13.6\text{ V}$$

$$V_B = V_{R2} = 4.5\text{ V}$$

$$V_E = V_{RE} = 3.9\text{ V}$$

$$V_{CE} = V_C - V_E = 13.6\text{ V} - 3.9\text{ V} = 9.7\text{ V}$$

$$V_{CB} = V_C - V_B = 13.6\text{ V} - 4.5\text{ V} = 9.1\text{ V}$$

ALTERNATING CURRENT CIRCUIT ANALYSIS

The alternating current analysis of the common base circuit is very similar to that of the base-biased circuit for voltage gain, input impedance, Vop-p(max), power gain, and output impedance.

VOLTAGE GAIN

The voltage gain of the loaded common base circuit is solved by taking the ratio of the effective collector resistance to that of the effective emitter resistance.

$$A_v = \frac{r_L}{r_e + R4} = \frac{RC\ //\ RL}{r_e + R4} = \frac{8\text{ k}\Omega\ //\ 4\text{ k}\Omega}{20\ \Omega + 300\ \Omega} = \frac{4.8\text{ k}\Omega}{320\ \Omega} = 15$$

INPUT IMPEDANCE

Connecting the input to the junction of R4 and R5:

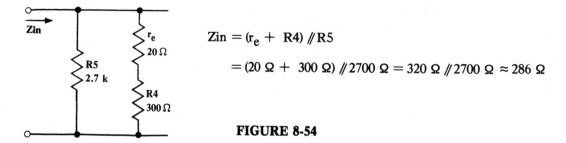

$Zin = (r_e + R4) \parallel R5$

$= (20\ \Omega + 300\ \Omega) \parallel 2700\ \Omega = 320\ \Omega \parallel 2700\ \Omega \approx 286\ \Omega$

FIGURE 8-54

POWER GAIN TO THE LOAD

The power gain to the load is solved from: $PG = A_v^2 \times Zin/RL = 15^2 \times 286\ \Omega / 12\ k\Omega = 5.36$

MAXIMUM PEAK-TO-PEAK OUTPUT VOLTAGE SWING

The maximum peak-to-peak output voltage swing is solved from either $2V_{CB}$ or $2I_C r_L$, whichever is smaller:

$2V_{CB} = 2 \times 9.1\ V = 18.2\ Vp\text{-}p$

$2I_C r_L = 2 \times 1.3\ mA \times 4.8\ k\Omega = 12.48\ Vp\text{-}p$

OUTPUT IMPEDANCE

The output impedance of the common base circuit is solved similarly to that of the common emitter stage from:

$$Zo = Ro = \frac{Vo(NL) - Vo(L)}{Vo(L)} \times RL = \frac{8\ Vp\text{-}p - 4.8\ Vp\text{-}p}{4.8\ Vp\text{-}p} \times 12\ k\Omega = \frac{3.2\ Vp\text{-}p}{4.8\ Vp\text{-}p} \times 12\ k\Omega = 8\ k\Omega$$

NOTE: Vo(NL) assumes an I_C of 1 mAp-p so that the unloaded Vo is:

$Vo = I_C R_C = 1\ mA \times 8\ k\Omega = 8\ Vp\text{-}p$

and the loaded Vo is:

$Vo = I_C r_L = 1\ mAp\text{-}p \times 4.8\ k\Omega = 4.8\ Vp\text{-}p$ where: $r_L = VR \parallel RL = 8\ k\Omega \parallel 12\ k\Omega = 4.8\ k\Omega$

CIRCUIT SUMMARY AND EQUIVALENT CIRCUIT METHOD OF ANALYSIS

Here, again, the approximation analysis and the equivalent circuit analysis of the common base circuit provides a comparison of the signal voltage and the signal current methods of analysis.

A. Approximation Circuit Analysis (See Figure 8-55)

Given: $\beta = 100$, Vin $= 200$ mVp-p, and $r_e = 20\ \Omega$

$$A_v = \frac{r_L}{r_e + R4} = \frac{R3 \parallel RL}{r_e + R4} = \frac{8\ k\Omega \parallel 12\ k\Omega}{20\ \Omega + 300\ \Omega} = 15$$

$Vo = Vin \times A_v = 200\ mVp\text{-}p \times 15 = 3\ Vp\text{-}p$

276

$$Zin = (r_e + R4) \| R5 = (20\ \Omega + 300\ \Omega) \| 2.7\ k\Omega \approx 286\ \Omega$$

$$PG = A_v^2 \times Zin/RL = 15^2 \times 286\ \Omega / 12\ k\Omega = 5.36$$

$$PG\ (dB) = 10 \log 5.36 = 7.29\ dB$$

$$A_i = PG/A_v = 5.36/15 = 0.357$$

$$Vop\text{-}p(max) = 2I_C r_L = 2 \times 1.3\ mA \times 4.8\ k\Omega = 12.48\ Vp\text{-}p$$

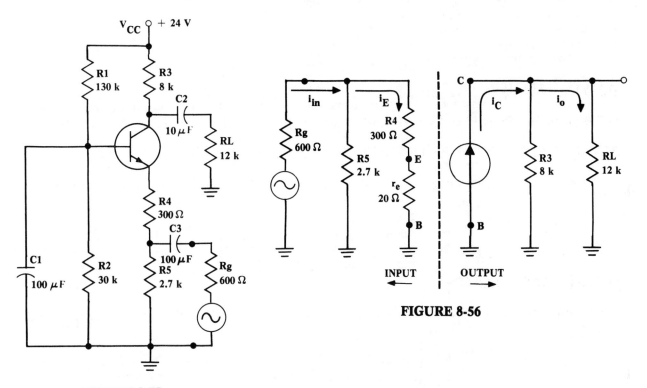

FIGURE 8-55

FIGURE 8-56

B. Equivalent Circuit Analysis (See Figure 8-56)

$$r_L = R3 \| RL = 8\ k\Omega \| 12\ k\Omega = 4.8\ k\Omega$$

$$Zin = (R4 + r_e) \| R5 = (300\ \Omega + 20\ \Omega) \| 2.7\ k\Omega = 286\ \Omega$$

$$i_{in} = v_{in}/Zin = 200\ mVp\text{-}p / 286\ \Omega = 699.3\ \mu Ap\text{-}p$$

$$i_E \approx i_C = v_{in}/(r_e + R4) = 200\ mVp\text{-}p/(20\ \Omega + 300\ \Omega) = 625\ \mu Ap\text{-}p$$

$$Vo = i_C r_L = 625\ \mu Ap\text{-}p \times 4.8\ k\Omega = 3\ Vp\text{-}p$$

$$A_v = Vo/Vin = 3\ Vp\text{-}p / 200\ mVp\text{-}p = 15$$

$$i_o = i_{RL} = (i_C \times RC)/(RC + RL) = (625\ \mu Ap\text{-}p \times 8\ k\Omega)/(8\ k\Omega + 12\ k\Omega) = 250\ \mu Ap\text{-}p$$

$$A_i = i_o/i_{in} = 250\ \mu Ap\text{-}p / 699.3\ \mu Ap\text{-}p = 0.357$$

$$A_p = PG = A_i A_v = 0.357 \times 15 = 5.36$$

$$Vop\text{-}p(max) = 2i_C r_L = 2 \times 1.3\ mA \times 4.8\ k\Omega = 12.48\ Vp\text{-}p$$

SECTION III: LABORATORY EXERCISE

OBJECTIVES
Investigate the DC and AC characteristics of the NPN, common-base, universal circuit.

LIST OF MATERIALS AND EQUIPMENT
1. Transistor: 2N3904
2. Resistors (one each except where indicated): 3.3 kΩ (two)   10 kΩ   12 kΩ   22 kΩ   100 kΩ
3. Capacitors (15-35 V): 10 µF (one)   100 µF (two)
4. Power Supply: 24 Volt
5. Voltmeter
6. Signal Generator
7. Oscilloscope

PROCEDURE
1. Connect the NPN transistor in the common-base, circuit configuration as shown in Figure 8-57, where the signal generator and the load resistor are included in the circuit.

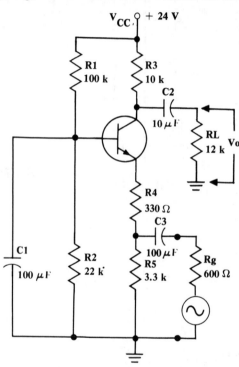

FIGURE 8-57

2. Calculate and measure the DC voltage drops around the circuit. Include all the resistors and the transistor.

NOTE: V_{BE} is given at 0.6 V until the actual V_{BE} voltage measurement can be made. Also, measure the DC voltages with respect to ground to avoid the possibility of ground loops.

3. Insert the calculated and measured values, as indicated, into Table 8-10

TABLE 8-10	V_{R1}	V_{R2}	V_{R3}	V_{R4}	V_{R5}	V_C	V_E	V_{CE}	I_E	Beta
CALCULATED										/////
MEASURED									/////	

5. VOLTAGE GAIN:

   a. Calculate the voltage gain from: $A_v = \dfrac{R3 \parallel RL}{R4 + r_e}$

   b. Measure the mid-frequency voltage gain using an input signal voltage of 100 mVp-p at a frequency of 1 kHz. Measure the Vo and calculate the measured voltage gain from:

   $A_v = Vo/Vin$ where: $Vin = 100$ mVp-p

6. INPUT INPEDANCE:

   a. Calculate the input impedance to the stage from: $Zin = (R4 + r_e) \parallel R5$ where: $r_e \approx 26$ mV/I_E

   b. Measure the input impedance to the circuit. Use a known 330 Ω series resistor and an input signal level of 100 mVp-p at 1 kHz, as shown in Figure 8-58.

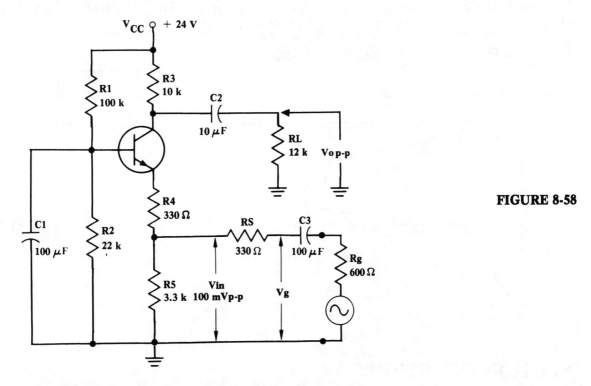

FIGURE 8-58

7. MAXIMUM PEAK-TO-PEAK OUTPUT VOLTAGE SWING:

   a. Calculate the maximum peak-to-peak output voltage swing from $2V_{CB}$ or $2I_C r_L$, whichever is smaller. Again, $r_L = R3 \parallel RL$.

   b. Monitor the output voltage swing across the 12 kΩ load resistor, and increase the input signal level until clipping or distortion, because of alpha crowding, occurs.

NOTE: $2V_{CB}$ exists for common base circuits because of the ground at the base, and because the output voltage across the load is monitored with respect to ground. Also, the AC load resistance is the parallel combination of the collector resistor and the load resistor RL.

8. POWER GAIN TO THE LOAD RESISTOR: Calculate the power gain to the load resistor and use both the measured A_v and Zin values. Calculate from:

   $PG = A_v^2 \times Zin/RL$ where: $RL = 12$ kΩ

9. OUTPUT IMPEDANCE:

a. Remove the load resistor and monitor the output signal at the collector with respect to ground. Adjust the input signal level so the unloaded output peak-to-peak signal monitored on the scope is 5 Vp-p.

b. Connect the load resistor, monitor the loaded peak-to-peak output voltage swing, and calculate the output impedance from:

$$Z_o = \frac{V_o(NL) - V_o(L)}{V_o(L)} \times R_L \quad \text{where:} \quad V_o(NL) = 5 \text{ Vp-p}$$

10. Insert the calculated and measured values, as indicated, into Table 8-11.

TABLE 8-11	A_v	Vo for Vin = 100 mVp-p	Vop-p MAX	Zin	POWER GAIN	Zo
CALCULATED		/////				
MEASURED					/////	/////

11. Under loaded conditions, monitor the upper and lower frequency 3 dB points. Insert these values on the Bode plot of Figure 8-59.

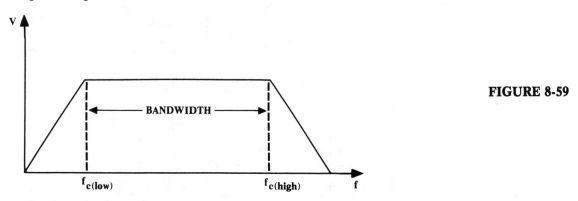

FIGURE 8-59

SECTION QUESTIONS

1. Are the input and output signal voltages of the common base circuit in phase?
2. Why is the Vop-p(max) controlled by $2I_C r_L$ and $2V_{CB}$? Why not $2V_{CE}$?
3. With reference to Figure 8-60, why is R1 and R2 not reflected into the emitter terminal when solving for the input impedance Zin?
4. Why is the gain of the amplifier equal to $(R3 /\!/ RL)/(r_e + R4)$? Does resistor R5 enter into the calculations? Why or why not? (Refer to Figure 8-60.)
5. Is the output impedance of the common base circuit higher than that of the common emitter circuit? Do the equal 10 kΩ collector resistors equalize the Zo for both the common base and common emitter circuits? Why or why not?

SECTION PROBLEMS

1. With reference for Figure 8-60, beta = 100.

   a. Calculate the Zin.

   b. Calculate A_v.

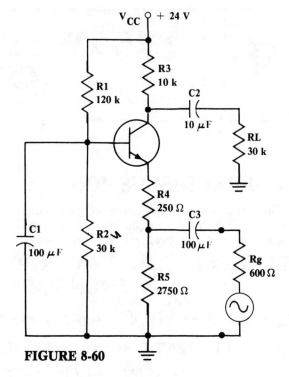

FIGURE 8-60

2. With reference to Figure 8-61, beta = 100. With the base capacitor removed:

   a. Calculate the Zin.

   b. Calculate A_v.

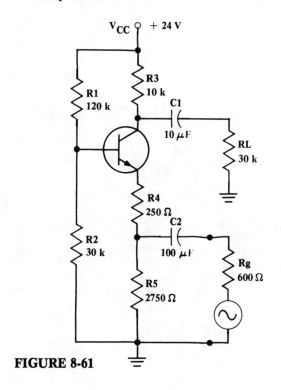

FIGURE 8-61

9 THE TWO-POWER-SUPPLY CIRCUIT

GENERAL DISCUSSION

The two power supply emitter control circuits are solved much like the base-biased and universal circuits. However, there are slight differences in the alternating current analysis, so all three types of circuits will be analyzed and investigated experimentally.

SECTION I: THE TWO-POWER SUPPLY, COMMON-EMITTER CIRCUIT

Part 1: The NPN, Two-Power Supply, Common-Emitter Circuit

DIRECT CURRENT CIRCUIT ANALYSIS

The DC analysis for the two power supply emitter current controlled circuit was developed in detail in Chapter 6. A summary of the analysis is given with Figure 9-1.

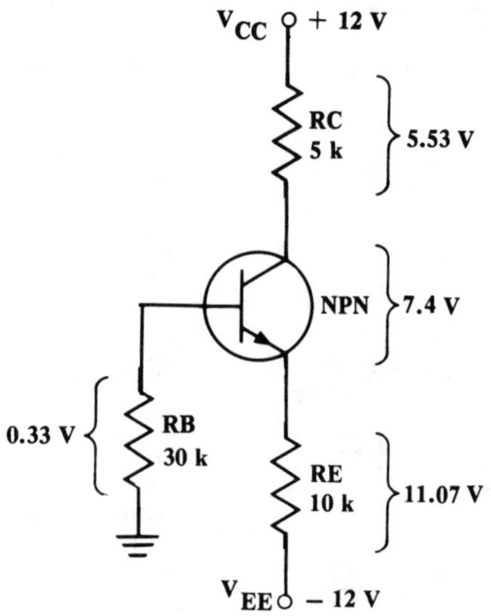

$$I_E = \frac{V_{EE} - V_{BE}}{(R_B/\beta) + R_E} = \frac{|12\ V| - 0.6\ V}{(30\ k\Omega/100) + 10\ k\Omega} = \frac{11.4\ V}{10.3\ k\Omega} \approx 1.107\ mA$$

$$V_{RE} = I_E R_E = 1.107\ mA \times 10\ k\Omega = 11.07\ V$$

$$V_{RB} = I_B R_B \approx (I_E/\beta) R_B = (1.107\ mA/100) \times 30\ k\Omega \approx 0.33\ V$$

$$V_{RC} = I_C R_C \approx I_E R_C = 1.107\ mA \times 5\ k\Omega = 5.53\ V$$

$$V_C = V_{CC} - V_{RC} = 12\ V - 5.53\ V = 6.47\ V$$

$$V_B = -V_{RB} = -0.33\ V$$

$$V_E = V_{EE} - V_{RE} = -12\ V + 11.07\ V = -0.93\ V$$

$$V_E = V_{RB} + V_{BE} = -0.33\ V - 0.6\ V = -0.93\ V$$

$$V_{CE} = V_C - V_E = 6.47\ V - (-0.93\ V) = 7.4\ V$$

$$V_{RC} + V_{CE} + V_{RE} = 5.53\ V + 7.4\ V + 11.07\ V = 24\ V$$

$$V_{CC} - V_{EE} = 12\ V - (-12\ V) = 24\ V$$

$$V_{EE} = V_{RB} + V_{BE} + V_{RE} = 0.33\ V + 0.6\ V + 11.07\ V = |12\ V|$$

FIGURE 9-1

ALTERNATING CURRENT CIRCUIT ANALYSIS

The NPN, two power supply, common emitter circuit and its AC equivalent circuit are shown in Figure 9-2. Note that:

$$R_E = R_3 + R_4 = 300\ \Omega + 9.7\ k\Omega = 10\ k\Omega$$

282

$r_e \approx 26\,mV/I_E = 26\,mV/1.107\,mA \approx 23.5\,\Omega$

$V_{R4} = I_E R4 = 1.107\,mA \times 300\,\Omega = 0.33\,V$

$V_{R5} = I_E R5 = 1.107\,mA \times 9.7\,k\Omega = 10.74\,V$

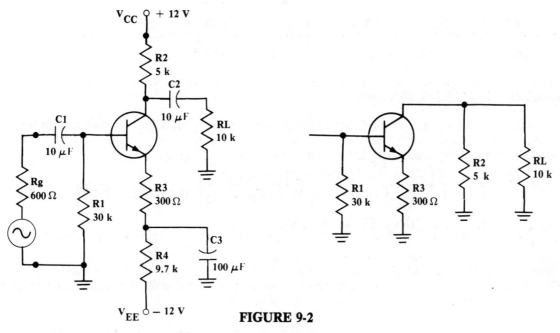

FIGURE 9-2

VOLTAGE GAIN

The voltage gain is solved by taking the ratio of the effective collector resistance the the effective emitter resistance. Therefore:

$$A_v = \frac{r_L}{r_e + R3} = \frac{RC \,//\, RL}{r_e + R3} = \frac{5\,k\Omega \,//\, 10\,k\Omega}{23.5\,\Omega + 300\,\Omega} = \frac{3333.33\,\Omega}{323.5\,\Omega} \approx 10.3$$

INPUT IMPEDANCE

The input impedance is solved by reflecting the effective emitter resistance into the base in parallel with RB. Therefore:

$Z_{in} = RB \,//\, \beta(r_e + R3) = 30\,k\Omega \,//\, 100(23.5\,\Omega + 300\,\Omega)$

$= 30\,k\Omega \,//\, 100 \times 323.5\,\Omega = 30\,k\Omega \,//\, 32.35\,k\Omega = 15.56\,k\Omega$

NOTE: Capacitor C3 places an effective AC ground at the junction of R3 and R4, so only R3 is reflected into the base.

MAXIMUM PEAK-TO-PEAK OUTPUT VOLTAGE SWING

The maximum peak-to-peak output voltage swing is solved from $2I_C r_L$ or approximately $2V_{CE}$, whichever is smaller.

$2I_C r_L = 2 \times 1.107\,mA \times 3333.3 = 7.38\,V_{p\text{-}p}$

$2V_{CE} = 2 \times 7.4\,V = 14.8\,V_{p\text{-}p},$ therefore, $V_{op\text{-}p}(max) = 7.38\,V_{p\text{-}p}$

NOTE: Solving for $2V_{CE} \times RC/(RC + RE)$ to provide further accuracy is academic, since $2I_C r_L$, obviously, limits the voltage swing.

POWER GAIN TO THE LOAD

a. The power gain to the load is solved from:

$$PG = A_v^2 \times Z_{in}/R_L = 10.3^2 \times 15.56 \text{ k}\Omega / 10 \text{ k}\Omega = 165.1$$

b. Converting to dB from: $PG(dB) = 10 \log 165.1 = 22.18$ dB

NOTE: If bypass capacitor C3 is removed, so that both R3 and R4 are left unbypassed, then the voltage gain will drop to less than unity from:

$$A_v = r_L/(r_e + R3 + R4) \approx 3333 \text{ }\Omega/(23.5 \text{ }\Omega + 200 \text{ }\Omega + 10 \text{ k}\Omega) = 3333 \text{ }\Omega/10{,}223.5 \text{ }\Omega \approx 0.326$$

Remember that both V_{CC} and V_{EE} are considered to be at "AC" ground.

Part 2: The PNP, Two-Power Supply, Common-Emitter Circuit

DIRECT CURRENT CIRCUIT ANALYSIS

The DC analysis of the PNP transistor circuit is similar to that of the NPN transistor circuit. Only the direction of the voltage drops across the transistor and resistors of the circuit, and the polarity of the single point voltages of the circuit differ. The DC circuit analysis for the PNP device is summarized in Figure 9-3.

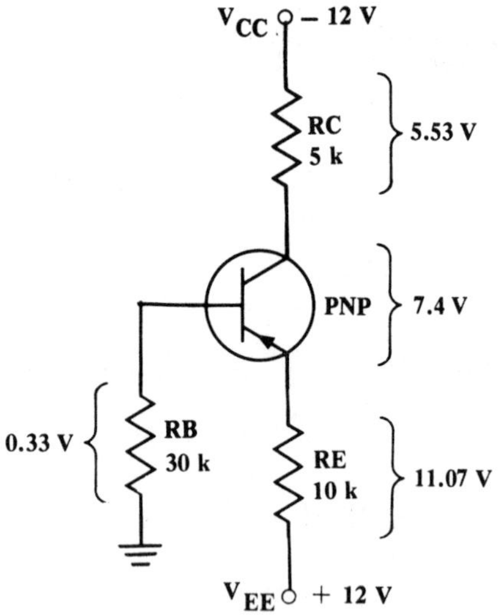

$$I_E = \frac{V_{EE} - V_{BE}}{(R_B/\beta) + R_E} = \frac{12 \text{ V} - 0.6 \text{ V}}{(30 \text{ k}\Omega/100) + 10 \text{ k}\Omega}$$

$$= \frac{11.4 \text{ V}}{300 \text{ }\Omega + 10 \text{ k}\Omega} = \frac{11.4 \text{ V}}{10.3 \text{ k}\Omega} \approx 1.107 \text{ mA}$$

$$V_{RE} = I_E R_E = 1.107 \text{ mA} \times 10 \text{ k}\Omega = 11.07 \text{ V}$$

$$V_{RB} = I_B R_B \approx (I_E/\beta) R_B = 1.107 \text{ mA}/100 \times 30 \text{ k}\Omega \approx 0.33 \text{ V}$$

$$V_{RC} = I_C RC \approx I_E RC = 1.107 \text{ mA} \times 5 \text{ k}\Omega = 5.53 \text{ V}$$

$$V_C = V_{CC} - V_{RC} = -12 \text{ V} + 5.53 \text{ V} = -6.47 \text{ V}$$

$$V_B = V_{RB} = 0.33 \text{ V}$$

$$V_E = V_{EE} - V_{RE} = 12 \text{ V} + -11.07 \text{ V} = 0.93 \text{ V}$$

$$V_{CE} = V_{EC} = V_E - V_C = 0.93 \text{ V} - (-6.47 \text{ V}) = 7.4 \text{ V}$$

$$V_{RE} + V_{CE} + V_{RC} = 11.07 \text{ V} + 7.4 \text{ V} + 5.53 \text{ V} = 24 \text{ V}$$

$$V_{EE} - V_{CC} = 12 \text{ V} - (-12 \text{ V}) = 24 \text{ V}$$

$$V_{EE} = V_{RE} + V_{BE} + V_{RB} = 11.07 \text{ V} + 0.6 \text{ V} + 0.33 \text{ V} = 12 \text{ V}$$

FIGURE 9-3

ALTERNATING CURRENT CIRCUIT ANALYSIS

The AC analysis for the PNP device is summarized in conjunction with the circuit of Figure 9-4. Notice that the AC parameters for the PNP circuit are identical to those for the NPN circuit.

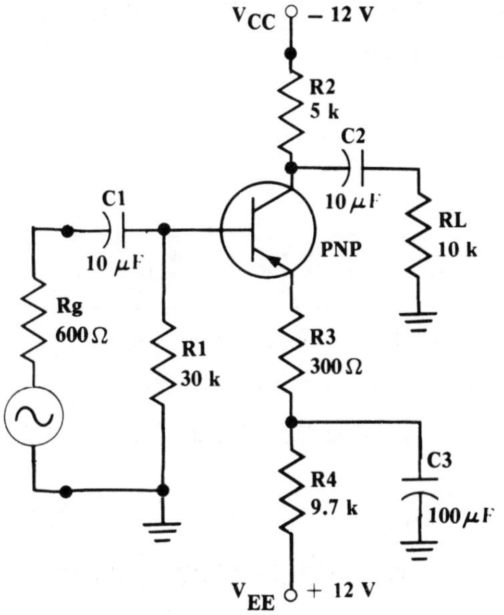

$$A_v = \frac{r_L}{r_e + R3} = \frac{R2 \,\|\, RL}{r_e + R3} = \frac{5\,k\Omega \,\|\, 10\,k\Omega}{23.5\,\Omega + 300\,\Omega} = 10.3$$

where: $r_e = 26\,mV/I_E = 26\,mV/1.107\,mA = 23.5\,\Omega$

$Vo = A_v \times Vin = 10.3 \times 100\,mV = 1.03\,Vp\text{-}p$

$Zin = RB \,\|\, \beta(r_e + R3) = 30\,k\Omega \,\|\, 100(23.5\,\Omega + 300\,\Omega)$
$ = 30\,k\Omega \,\|\, 100(323.5\,\Omega) = 15.56\,k\Omega$

$PG = A_v^2 \times Zin/RL = 10.3^2 \times 15.56\,k\Omega / 10\,k\Omega = 165.1$

$PG(dB) = 10 \log PG = 10 \log 165.1 = 22.18\,dB$

$Vop\text{-}p(max) = 7.38\,Vp\text{-}p$, where:

$2I_C r_L = 2 \times 1.1\,mA \times 3.33\,k\Omega \approx 7.38\,Vp\text{-}p$

$2V_{CE} = 2 \times 7.4\,V = 14.8\,Vp\text{-}p$

FIGURE 9-4

Part 3: The Equivalent Circuit Analysis (NPN or PNP)

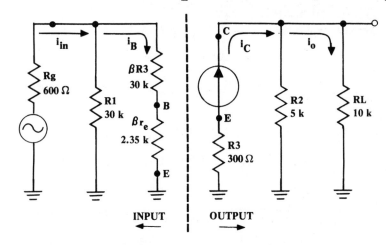

Given: $Vin = 100\,mVp\text{-}p$ and $r_e = 23.5\,\Omega$

$\beta r_e = 100 \times 23.5\,\Omega = 2.35\,k\Omega$

$\beta R3 = 100 \times 300\,\Omega = 30\,k\Omega$

$\beta(r_e + R3) = 100(23.5\,\Omega + 300\,\Omega)$
$ = 100 \times 323.5\,\Omega = 32.35\,k\Omega$

FIGURE 9-5

$r_L = R2 \,\|\, RL = 5\,k\Omega \,\|\, 10\,k\Omega = 3.33\,k\Omega$

$Zin = R1 \,\|\, \beta(R3 + r_e) = 30\,k\Omega \,\|\, 100(300\,\Omega + 23.5\,\Omega) = 30\,k\Omega \,\|\, 32.35\,k\Omega = 15.56\,k\Omega$

$i_{in} = Vin/Zin = 100\,mVp\text{-}p/15.56\,k\Omega = 6.425\,\mu A$

$i_B = Vin/\beta(r_e + R3) = 100\,mVp\text{-}p/32.35\,k\Omega = 3.091\,\mu A$

$i_C = \beta i_B = 100 \times 3.091\,\mu A = 309.1\,\mu A$

$Vo = i_C r_L = 309.1\,\mu A \times 3.33\,k\Omega \approx 1.03\,Vp\text{-}p$

$A_v = Vo/Vin = 1.03\,Vp\text{-}p/100\,mVp\text{-}p = 10.3$

$A_i = A_v \times Zin/RL = 10.3 \times 15.56\,k\Omega/10\,k\Omega = 16.03$

$PG = A_p = A_i A_v = 16.03 \times 10.3 = 165.1$

Further analysis can show that:

$$i_o = (i_C \times R2)/(R2 + RL) \approx (3.09 \text{ mA} \times 5 \text{ k}\Omega)/(5 \text{ k}\Omega + 10 \text{ k}\Omega) = 1.03 \text{ mA}$$

$$A_i = i_o/i_{in} = 1.03 \text{ mA}/64.25 \text{ }\mu\text{A} \approx 16.03$$

$$V_o = i_o RL = 1.03 \text{ mA} \times 10 \text{ k}\Omega = 1.03 \text{ Vp-p}$$

SECTION I: LABORATORY EXERCISE

OBJECTIVES
Investigate the DC and AC characteristics of the NPN, common-emitter, two-power-supply circuit.

LIST OF MATERIALS AND EQUIPMENT
1. Transistors (one each): 2N3904 and 2N3906 or equivalents
2. Resistors (one each except where indicated): 220 Ω   4.7 kΩ   10 kΩ (two)   12 kΩ   22 kΩ
3. Capacitors: 10 μF (two)   100 μF (one)
4. Power Supply ± 12 V
5. Voltmeter
6. Signal Generator
7. Oscilloscope

PROCEDURE
Part 1: The NPN, Two-Power-Supply, Common-Emitter Circuit
1. Connect the NPN transistor in the two power supply circuit shown in Figure 9-6.

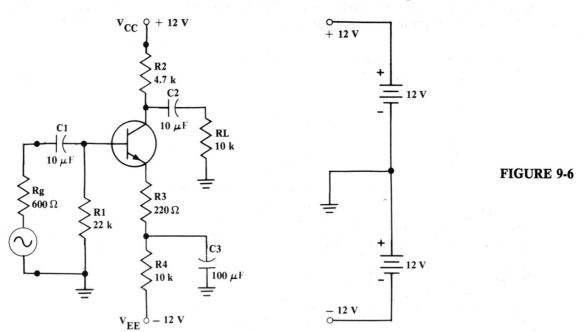

FIGURE 9-6

NOTE: Connecting the two power supplies in series, so that the voltage is series aiding, and referencing ground at the connection of the two supplies as shown in Figure 9-6, provides the plus and minus voltage supply.

2. Calculate and measure the DC voltages around the circuit. Include all resistors and the transistor. Measure V_B, V_E, V_C, and the junction of V_{R3} and V_{R4}, all with respect to ground. Then, calculate the voltage drops across V_{R1}, V_{R2}, V_{R3}, V_{BE}, and V_{CE} instead of measuring directly. This method of measuring, with respect to ground, is a sound approach to making measurements that avoids ground loops.

3. Insert the calculated and measured values, as indicated, into Table 9-1.

TABLE 9-1	I_E	V_{R1}	V_{R2}	V_{R3}	V_{R4}	V_{BE}	V_E	V_C	V_B	V_{CE}	I_B	Beta
CALCULATED						≈ 0.6 V						
MEASURED	/////										/////	

4. **VOLTAGE GAIN**

   a. Calculate the mid-frequency voltage gain from:

   $$Av = (R2 \mathbin{/\mkern-5mu/} RL)/(r_e + R3) \quad \text{where:} \quad r_e \approx 26 \text{ mV}/I_E$$

   b. Measure the mid-frequency voltage gain using an input signal voltage of 100 mVp-p at a frequency of 1 kHz. Measure Vo at the collector, with respect to ground, using an oscilloscope. Calculate the "measured" voltage gain from:

   $$A_v = Vo/Vin \quad \text{where:} \quad Vin = 100 \text{ mVp-p}$$

5. **INPUT IMPEDANCE**

   a. Calculate the input impedance to the circuit from: $Zin = R1 \mathbin{/\mkern-5mu/} \beta(r_e + R3)$

   b. Measure the input impedance, using a known 12 kΩ series resistor and an input signal of 100 mVp-p at 1 kHz, as shown in Figure 9-7. Calculate from:

   $$Zin = \frac{Vin \times RS}{V_{RS}} = \frac{Vin \times RS}{Vg - Vin}$$

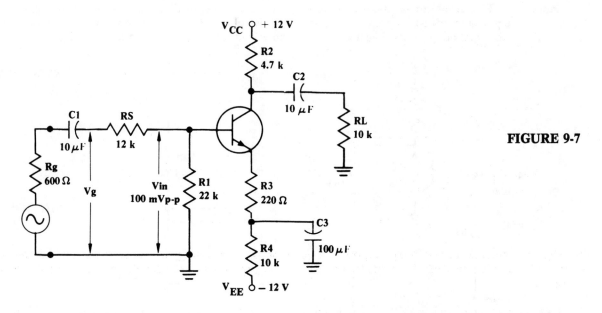

FIGURE 9-7

NOTE: The calculated or measured beta of the device can be used in the calculations to obtain accurate results. However, using a beta of 100 will provide "ball park" versus measured results within the tolerance of most circuit resistors.

6. MAXIMUM PEAK-TO-PEAK OUTPUT VOLTAGE SWING

   a. Calculate the maximum peak-to-peak output voltage swing from $2I_C r_L = 2I_C(R2 \parallel RL)$ or from $2V_{CE}$. Again, more accuracy can be achieved by including the slight effect of the 220 Ω, R3 resistor by solving from:

   $$2(V_{CE} \times r_L)/(r_L + R3)$$

   b. Monitor the output voltage across the load and increase the 1 kHz input signal level until clipping or distortion, because of alpha crowding, occurs. Remove RS before measuring Vop-p(max).

7. POWER GAIN TO THE LOAD RESISTOR: Calculate the power gain to the load resistor of the circuit and use both the measured A_v and Zin values. Calculate from:

   $$PG = A_v^2 \times Zin/RL \quad \text{where:} \quad RL = 10 \text{ k}\Omega$$

8. Insert the calculated and measured values, as indicated, into Table 9-2.

TABLE 9-2	A_v	Vo for Vin = 100 mVp-p	Vop-p MAX	Zin	POWER GAIN	PG(dB)	f_c(low)	f_c(high)
CALCULATED		/////					/////	/////
MEASURED					/////	/////		

9. Under loaded conditions, monitor the upper and lower frequency 3 dB points. Insert these values, as indicated, into Table 9-2.

Part 2: The PNP, Two-Power Supply, Common-Emitter Circuit

1. Connect the PNP transistor in the two-power-supply, common-emitter circuit as shown in Figure 9-8. Notice that the plus and minus voltages are opposite to those of the NPN connection. Also, the C1, C2, and C3 coupling and bypass capacitors, because they are polarized, reflect the DC voltage change.

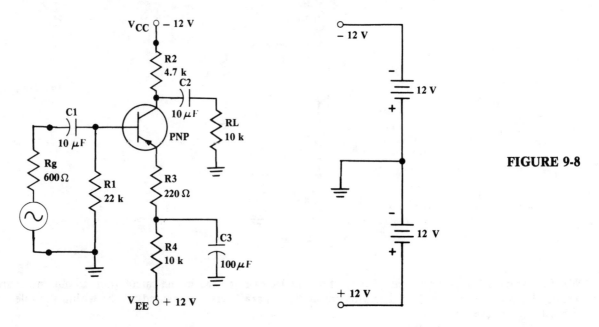

FIGURE 9-8

2. Calculate and measure the DC voltage drops around the circuit. Include all resistors and the transistor. Measure V_B, V_E, V_C, and the junction of V_{R3} and V_{R4}, all with respect to ground. Calculate the voltage drops across V_{R1}, V_{R2}, V_{R3}, V_{BE}, and V_{CE} (instead of measuring) to avoid ground loops.

NOTE: Again voltage drops only require magnitude and the minus sign indicates direction only. Therefore, a $-V_{CE}$ voltage can be written as simply V_{CE} with no sign. However, the minus signs do give the signal point voltage of the circuit, with respect to ground and, hence, allow the remaining voltage drop calculations.

3. Insert the calculated and measured values, as indicated, into Table 9-3.

TABLE 9-3	I_E	V_{R1}	V_{R2}	V_{R3}	V_{R4}	V_C	V_B	V_E	V_{BE}	V_{CE}	I_B	Beta
CALCULATED												/////
MEASURED	/////										/////	

4. **VOLTAGE GAIN**

   a. Calculate the mid-frequency voltage gain from: $A_v = (R2 \parallel RL)/(r_e + R3)$ where: $r_e \approx 26\text{ mV}/I_E$

   b. Measure the mid-frequency voltage gain using an input signal voltage of 100 mVp-p at a frequency of 1 kHz. Measure Vo at the collector, with respect to ground, using an oscilloscope. Calculate the "measured" voltage gain from:

   $$A_v = V_o/V_{in} \quad \text{where:} \quad V_{in} = 100\text{ mVp-p}$$

5. **INPUT IMPEDANCE**

   a. Calculate the input impedance to the circuit from: $Z_{in} = R1 \parallel \beta(R3 + r_e)$

   b. Measure the input impedance using a known 12 kΩ series resistor and an input signal of 100 mVp-p at 1 kHz, as shown in Figue 9-9. Calculate from:

   $$Z_{in} = \frac{V_{in} \times R_S}{V_{R_S}} = \frac{V_{in} \times R_S}{V_g - V_{in}}$$

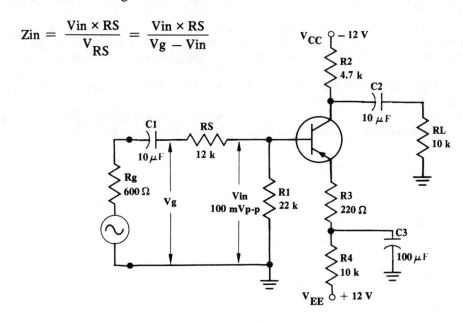

FIGURE 9-9

6. MAXIMUM PEAK-TO-PEAK OUTPUT VOLTAGE SWING

   a. Calculate the maximum peak-to-peak output voltage swing from $2I_C r_L = 2I_C(R2 \;//\; RL)$, or from $\approx 2V_{CE}$. Again, more accuracy can be achieved by including the slight effect of the 220 Ω, R3 resistor by solving from:

   $$2(V_{CE} \times r_L)/(R3 + r_L)$$

   b. Monitor the output signal voltage across the load and increase the 1 kHz input signal level until clipping or distortion, because of alpha crowing, occurs. Remove RS before measuring Vop-p(max).

7. POWER GAIN TO THE LOAD

   a. Calculate the power gain to the load resistor of the circuit and use both the measured A_v and Zin values. Calculate from:

   $$PG = A_v^2 \times Zin/RL \quad \text{where:} \quad RL = 10 \text{ k}\Omega$$

   b. Convert to dB from 10 log PG.

8. Insert the calculated and measured values, as indicated, into Table 9-4.

TABLE 9-4	Av	Vop-p for Vin = 100 mVp-p	Vop-p MAX	Zin	POWER GAIN	PG(dB)	f_c(low)	f_c(high)
CALCULATED		/////					/////	/////
MEASURED					/////			

9. Under loaded conditions, monitor the upper and lower frequency 3 dB points. Insert these values, as indicated, into Table 9-4.

SECTION QUESTIONS

1. Is solving for voltage gain of the NPN and PNP, two power supply, common emitter circuits similar?
2. With reference to Figure 9-2, does the input impedance increase if R4 is increased? Explain.
3. With reference to Figure 9-2, what would happen if the C3 emitter bypass capacitor were removed and why?
4. With reference to Figure 9-2, does increasing the base resistance from 30 kΩ to 60 kΩ double the power gain to the load? Explain why or why not.
5. With reference to Figure 9-2, if beta increases from 100 to 200, the base current I_B will decrease. Does a similar decrease occur for the input signal current i_B? Explain why or why not.

SECTION PROBLEMS

1. Given: Beta = 100 and $V_{BE} = 0.6$ V

   A. Solve for:

   1. V_{R1} 5. V_{CE}
   2. V_{R2} 6. V_C
   3. V_{R3} 7. V_B
   4. V_{R4} 8. V_E

   B. Solve for:

   1. A_v
   2. Zin
   3. PG to the load in dB
   4. Vop-p(max)

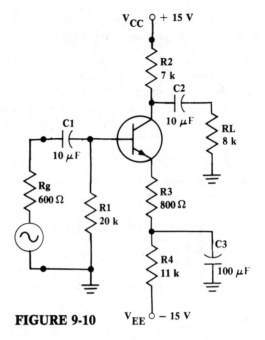

FIGURE 9-10

2. Given: Beta = 100 and $V_{BE} = 0.6$ V

   A. Solve for:

   1. V_{R1} 5. V_{CE}
   2. V_{R2} 6. V_C
   3. V_{R3} 7. V_B
   4. V_{R4} 8. V_E

   B. Solve for:

   1. A_v
   2. Zin
   3. PG to the load in dB
   4. Vop-p(max)

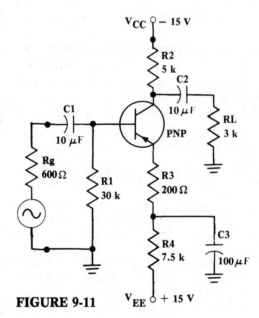

FIGURE 9-11

SECTION II: THE COMMON-COLLECTOR, TWO-POWER-SUPPLY CIRCUIT

DIRECT CURRENT CIRCUIT ANALYSIS

The direct current circuit analysis of the two power supply, common collector circuit is similar to that of the two power supply, common emitter circuit except there is no collector resistor and the V_{CE} and V_{RE} share the V_{CC} voltage. The DC analysis of the circuit is shown in conjunction with Figure 9-12.

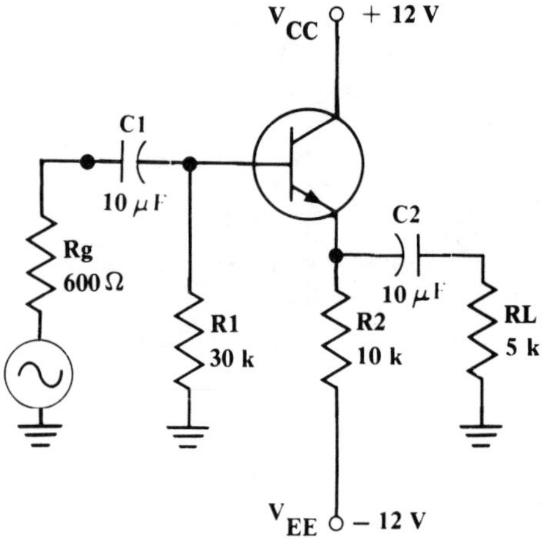

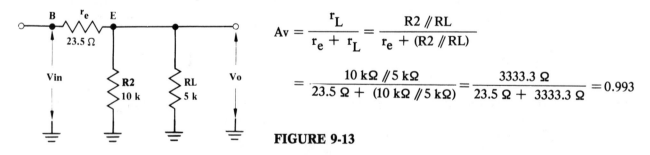

$$I_E = \frac{V_{EE} - V_{BE}}{R2 + R1/\beta} = \frac{12\text{ V} - 0.6\text{ V}}{10\text{ k}\Omega + 30\text{ k}\Omega/100} = \frac{11.4\text{ V}}{10.2\text{ k}\Omega} \approx 1.107\text{ mA}$$

$$V_{R2} = I_E R2 = 1.107\text{ mA} \times 10\text{ k}\Omega = 11.07\text{ V}$$

$$V_{R1} \approx (I_E/\beta) \times R1 = (1.107\text{ mA}/100)30\text{ k}\Omega \approx 0.33\text{ V}$$

$$V_C = V_{CC} = 12\text{ V}$$

$$V_B = V_{R1} = -0.33\text{ V}$$

$$V_E = V_{EE} - V_{R2} = -12\text{ V} + 11.07\text{ V} = -0.93\text{ V} \quad \text{or,}$$

$$V_E = V_{RB} + V_{BE} = -0.33\text{ V} - 0.6\text{ V} = -0.93\text{ V}$$

FIGURE 9-12

$$V_{CE} = V_C - V_E = 12\text{ V} - (-0.93\text{ V}) = 12.93\text{ V}$$

ALTERNATING CURRENT CIRCUIT ANALYSIS

The AC analysis for the two power supply, common collector circuit is similar to the base-biased and universal biased circuits. In fact, the AC equivalent circuits are identical for the base-biased, universal-biased, and two-power-supply-biased circuit configurations.

VOLTAGE GAIN

The voltage gain is solved from:

$$A_v = \frac{r_L}{r_e + r_L} = \frac{R2 \;//\; RL}{r_e + (R2 \;//\; RL)}$$

$$= \frac{10\text{ k}\Omega \;//\; 5\text{ k}\Omega}{23.5\text{ }\Omega + (10\text{ k}\Omega \;//\; 5\text{ k}\Omega)} = \frac{3333.3\text{ }\Omega}{23.5\text{ }\Omega + 3333.3\text{ }\Omega} = 0.993$$

FIGURE 9-13

The voltage out for 4 Vp-p of input is solved from:

$$V_o = A_v \times V_{in} = 0.993 \times 4\text{ Vp-p} \approx 3.97\text{ Vp-p}, \quad \text{or from:}$$

$$V_o = \frac{V_{in} \times r_L}{r_e + r_L} = \frac{4\text{ Vp-p} \times 3333.3\text{ }\Omega}{23.5\text{ }\Omega + 3333.3\text{ }\Omega} \approx 3.97\text{ Vp-p}$$

INPUT IMPEDANCE

The input impedance of the circuit is solved by reflecting the effective emitter resistance into the base in parallel with the base resistor. Therefore:

$$Zin = R1 \;//\; \beta(r_e + r_L) = 30\;k\Omega \;//\; 100(23.5\;\Omega + 3333.3\;\Omega)$$

$$= 30\;k\Omega \;//\; 100(3356.8\;\Omega) = 30\;k\Omega \;//\; 335.68\;k\Omega \approx 27.5\;k\Omega$$

POWER GAIN TO THE LOAD

The power gain to the load is solved from:

$$PG = A_v^2 \times Zin/RL = 0.993^2 \times 27.54\;k\Omega/5\;k\Omega = 5.43$$

Converting PG to dB from: $PG(dB) = 10 \log PG = 10 \log 5.43 \approx 7.35\;dB$

MAXIMUM PEAK-TO-PEAK OUTPUT VOLTAGE SWING

The maximum peak-to-peak output voltage swing is solved from $2I_C r_L$ or $2V_{CE}$, whichever is smaller. Therefore:

$$Vop\text{-}p(max) = 7.38$$

where: $2I_C r_L = 2 \times 1.107\;mA \times 3333.3\;\Omega = 7.38\;Vp\text{-}p$

$2V_{CE} = 2 \times 12.93\;V = 25.86\;Vp\text{-}p$

EQUIVALENT CIRCUIT ANALYSIS

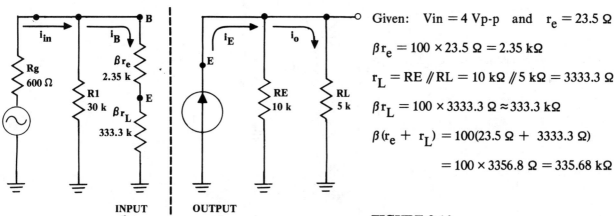

Given: $Vin = 4\;Vp\text{-}p$ and $r_e = 23.5\;\Omega$

$\beta r_e = 100 \times 23.5\;\Omega = 2.35\;k\Omega$

$r_L = RE \;//\; RL = 10\;k\Omega \;//\; 5\;k\Omega = 3333.3\;\Omega$

$\beta r_L = 100 \times 3333.3\;\Omega \approx 333.3\;k\Omega$

$\beta(r_e + r_L) = 100(23.5\;\Omega + 3333.3\;\Omega)$

$= 100 \times 3356.8\;\Omega = 335.68\;k\Omega$

FIGURE 9-14

$Zin = R1 \;//\; \beta(r_e + r_L) = 30\;k\Omega \;//\; 100(23.5\;\Omega + 3333.3\;\Omega) = 30\;k\Omega \;//\; 335.68\;k\Omega = 27.54\;k\Omega$

$i_{in} = Vin/Zin = 4\;Vp\text{-}p/27.54\;k\Omega = 145.24\;\mu A$

$i_B = Vin/\beta(r_e + r_L) = 4\;Vp\text{-}p/335.68\;k\Omega = 11.92\;\mu A$

$i_E = \beta i_B = 100 \times 11.92\;\mu Ap\text{-}p = 1.19\;mAp\text{-}p$

$V_o = i_E r_L = 1.19\;mAp\text{-}p \times 3333.3\;\Omega = 3.972\;mVp\text{-}p$

$A_v = Vo/Vin = 3.972\;Vp\text{-}p/4\;Vp\text{-}p = 0.993$

$A_v = \beta r_L / Zin(base) = (100 \times 3333.3\;\Omega)/335.68\;k\Omega = 0.993$

$A_i = A_v \times Zin/RL = 0.993 \times 27.54\;k\Omega/5\;k\Omega = 5.47$

$$A_p = PG = A_v A_i = 0.993 \times 5.47 = 5.43$$

$$A_p(dB) = PG(dB) = 10 \log PG = 10 \log 5.43 = 7.35 \text{ dB}$$

SECTION II: LABORATORY EXERCISE

OBJECTIVES
Investigate the DC and AC characteristics of the NPN, two power supply, common collector circuit.

LIST OF MATERIALS AND EQUIPMENT
1. Transistor: 2N3904
2. Resistors (one each):   33 Ω   1 kΩ   2.2 kΩ   15 kΩ   33 kΩ
3. Capacitors: 10 μF (two)
4. Power Supply ± 12 V
5. Voltmeter
6. Signal Generator
7. Oscilloscope

PROCEDURE

1. Connect the NPN transistor in the two power supply, common collector circuit as shown in Figure 9-15, where the signal generator and load resistor are included in the circuit.

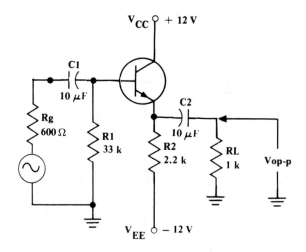

FIGURE 9-15

2. Calculate and measure the DC voltages around the circuit. Include all resistors and the transistor. Measure V_B, V_E, and V_C, with respect to ground, and calculate the voltage drops of V_{R1}, V_{R2}, V_{BE}, and V_{CE}.

NOTE: Use the exact transistor beta value if available. Otherwise, use a nominal beta of 100, but expect slight measured to calculated voltage differences.

3. Insert the calculated and measured values, as indicated, into Table 9-5.

TABLE 9-5	I_E	V_{R1}	V_{R2}	V_C	V_B	V_E	V_{BE}	V_{CE}	I_B	Beta
CALCULATED										/////
MEASURED										

4. **VOLTAGE GAIN**

   a. Calculate the mid-frequency voltage gain from:

   $$A_v = \frac{r_L}{r_e + r_L} = \frac{R2 \; // \; RL}{r_e + (R2 \; // \; RL)} \quad \text{where:} \quad r_L = R2 \; // \; RL$$

   b. Measure the mid-frequency voltage gain using an input signal level of 1 Vp-p at a frequency of 1 kHz. Measure Vo across the load resistor. Calculate the measured voltage gain from:

   $$A_v = Vo/Vin \quad \text{where:} \quad Vin = 1 \; Vp\text{-}p$$

5. **INPUT IMPEDANCE**

   a. Calculate the input impedance to the circuit from:

   $$Zin \approx R1 \; // \; \beta(R2 \; // \; RL) \quad \text{where:} \quad RL = 1 \; k\Omega \; \text{and} \; r_e \; \text{is safely ignored.}$$

   b. Measure the input impedance. Use a known 15 kΩ resistor and an input signal level of 1 Vp-p at 1 kHz as shown in Figure 9-16. Calculate Zin from:

   $$Zin = \frac{Vin \times RS}{V_{RS}} = \frac{Vin \times RS}{Vg - Vin} \quad \text{where:} \quad Vin = 1 \; Vp\text{-}p$$

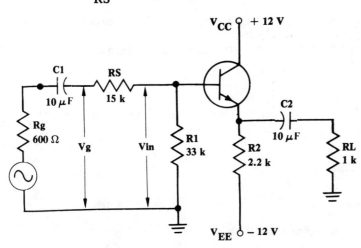

FIGURE 9-16

6. **MAXIMUM PEAK-TO-PEAK OUTPUT VOLTAGE SWING**

   a. Calculate the maximum peak-to-peak output voltage swing from $2I_E r_L = 2I_E(R2 \; // \; RL)$ or $2V_{CE}$, whichever is smaller.

   b. Monitor the output voltage across the load and increase the 1 kHz input signal level until clipping or distortion, because of alpha crowding, occurs. However, because most signal generators are not capable of producing such large peak-to-peak voltage amplitudes, the measurement may not be possible.

7. **POWER GAIN TO THE LOAD:** Calculate the power gain to the load and use both the measured A_v and Zin values. Calculate from:

   $$PG = A_v^2 \times Zin/RL$$

8. **OUTPUT IMPEDANCE**

   a. Calculate the output impedance of the circuit from: $\quad Zo \approx [(Rg/\beta) + r_e] \; // \; R2$

b. Use a 33 ohm load resistor and 400 mVp-p of input signal at 1 kHz.
  1. With the 33 ohm load resistor removed, adjust Vo(NL) at the emitter resistor, with respect to ground, to 400 mVp-p.
  2. Connect the 33 ohm load resistor and again measure the Vo(L) at the emitter, with reference to ground. Make sure the 15 kΩ RS resistor has been removed for this test.
  3. Calculate the measured output impedance of the circuit from:

$$Z_o = R_o = \frac{V_o(NL) - V_o(L)}{V_o(L)} \times R_L \quad \text{where:} \quad V_o(NL) = 400 \text{ mVp-p} \quad \text{and} \quad R_L = 33 \text{ Ω}$$

9. Insert the calculated and measured values, as indicated, into Table 9-6.

TABLE 9-6	A_v	Vo for Vin = 1 Vp-p	Vop-p MAX	Zin	Zo	POWER GAIN	PG(dB)	fc(low)	fc(high)
CALCULATED		/////						/////	/////
MEASURED					/////	/////			

SECTION QUESTIONS

1. With reference to Figure 9-13, does decreasing the load resistor RL from 5 ohms to 330 ohms severely lower the Vo? Explain why or why not.
2. With reference to Figure 9-12, does increasing the beta from 100 to 200 increase the input impedance and why? Is the input impedance doubled by the 100 to 200 beta increase?
3. Why does the signal generator output impedance effect the output impedance of the circuit? Explain.

SECTION PROBLEMS

1. Given: Beta = 100, Vin = 2 Vp-p, and VBE = 0.6 V

   Solve for:
   a. V_{R1} e. Vo
   b. V_{R2} f. Vop-p(max)
   c. V_{CE} g. Zin
   d. r_e h. PG to the load in dB

2. Given: Beta = 100

   Vo(NL) = 3 Vp-p

   Vo(L) = 2.8 Vp-p

   Solve for Zo.

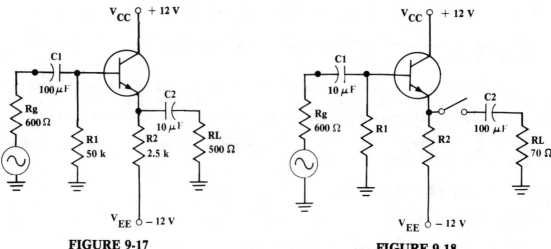

FIGURE 9-17 FIGURE 9-18

SECTION III: THE COMMON-BASE, TWO-POWER SUPPLY CIRCUIT

DIRECT CURRENT CIRCUIT ANALYSIS

The direct current analysis of the two power supply, common base circuit is solved like the common emitter circuit except no base resistor is needed. Hence, the V_{EE} voltage is distributed across V_{R2}, V_{R3}, and V_{BE}.

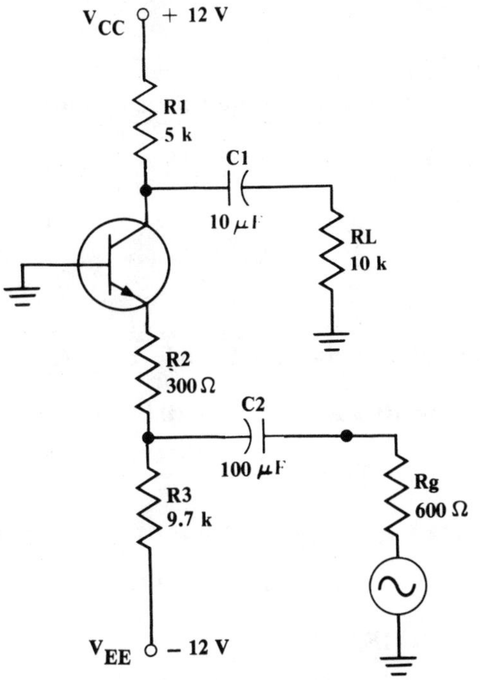

$$I_E = \frac{V_{EE} - V_{BE}}{R2 + R3} = \frac{|12\,V| - 0.6\,V}{300\,\Omega + 9.7\,k\Omega} = \frac{11.4\,V}{10\,k\Omega} = 1.14\,mA$$

$$V_{R1} = I_C R1 \approx I_E R1 = 1.14\,mA \times 5\,k\Omega = 5.7\,V$$

$$V_{R2} = I_E R2 = 1.14\,mA \times 300\,\Omega = 0.342\,V$$

$$V_{R3} = I_E R3 = 1.14\,mA \times 9.7\,k\Omega = 11.058\,V$$

$$V_C = V_{CC} - V_{R1} = 12\,V - 5.7\,V = 6.3\,V$$

$$V_E = V_{EE} - (V_{R2} + V_{R3}) = -12\,V + 11.4\,V = -0.6\,V$$

$$V_B = 0\,V$$

FIGURE 9-19

ALTERNATING CURRENT CIRCUIT ANALYSIS

The alternating current analysis for the two power supply, common base circuit is solved like the base-biased and universal-biased, common base circuits. Therefore, the AC equivalent circuits should be the same.

VOLTAGE GAIN

The voltage gain, like that of the common emitter circuit, is solved from:

$$A_v = \frac{r_L}{r_e + R2} = \frac{R1 \,//\, RL}{r_e + R2} = \frac{5\,k\Omega \,//\, 10\,k\Omega}{22.8\,\Omega + 300\,\Omega} = \frac{3333.3\,\Omega}{322.8\,\Omega} = 10.33$$

where: $r_e = 26\,mV / 1.14\,mA = 22.8\,\Omega$

INPUT IMPEDANCE

The input impedance to the circuit is solved by taking the parallel combination of $r_e + R2$ and R3. Again, as in the previous common base circuit, the base lead is at AC ground. Therefore:

$$Z_{in} = (r_e + R2) \,//\, R3 = (22.8\,\Omega + 300\,\Omega) \,//\, 9.7\,k\Omega = 322.8\,\Omega \,//\, 9.7\,k\Omega \approx 312.4\,\Omega$$

POWER GAIN TO THE LOAD

The power gain to the load is solved from: $PG = A_v^2 \times Z_{in}/RL = 10.33^2 \times 312.4\,\Omega / 10\,k\Omega \approx 3.33$

Convert PG to dB from: $PG(dB) = 10 \log PG = 10 \log 3.33 = 5.22\,dB$

MAXIMUM PEAK-TO-PEAK OUTPUT VOLTAGE SWING

The maximum peak-to-peak output voltage swing is solved from $2I_E r_L$ or approximately $2V_{CB}$, whichever is smaller. Therefore:

$$Vop\text{-}p(max) = 7.6 \text{ Vp-p}$$

where: $2I_E r_L = 2 \times 1.14 \text{ mA} \times 3333 \text{ }\Omega \approx 7.6 \text{ Vp-p}$

$2V_{CB} = 2 \times 6.3 \text{ V} = 12.6 \text{ Vp-p}$

NOTE: The AC ground at the base limits one of the "calculated" maximum peak-to-peak conditions.

EQUIVALENT CIRCUIT ANALYSIS

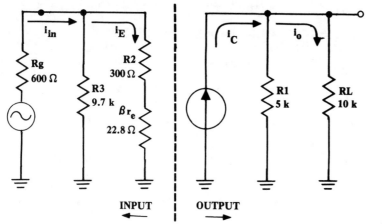

Given: Beta = 100 and r_e = 22.8 Ω

r_L = RC // RL = 5 kΩ // 10 kΩ = 3.33 kΩ

FIGURE 9-20

$Zin = R3 \,//\, (R2 + r_e) = 9.7 \text{ k}\Omega \,//\, (300 \text{ }\Omega + 22.8 \text{ }\Omega) = 9.7 \text{ k}\Omega \,//\, 322.8 \text{ }\Omega = 312.4 \text{ }\Omega$

$i_{in} = Vin/Zin = 100 \text{ mVp-p}/312.4 \text{ }\Omega = 320 \text{ }\mu\text{Ap-p}$

$i_B = Vin/(r_e + R2) = 100 \text{ mVp-p}/322.8 \text{ }\Omega = 309.8 \text{ }\mu\text{Ap-p}$

$i_E \approx i_C = \beta i_B = 100 \times 309.8 \text{ }\mu\text{A} \approx 310 \text{ }\mu\text{Ap-p}$

$Vo = i_C r_L = 310 \text{ }\mu\text{Ap-p} \times 3333.3 \text{ }\Omega = 1.033 \text{ Vp-p}$

$i_o = i_C \times RC/(RC + RL) = 310 \text{ }\mu\text{A} \times 5 \text{ k}\Omega/(5 \text{ k}\Omega + 10 \text{ k}\Omega) = 1.033 \text{ mAp-p}$

$Vo = i_o RL = 103.3 \text{ }\mu\text{Ap-p} \times 10 \text{ k}\Omega = 1.033 \text{ Vp-p}$

$A_v = Vo/Vin = 1.033 \text{ Vp-p}/100 \text{ mVp-p} = 10.33$

$A_i = i_o/i_{in} = 103.3 \text{ }\mu\text{Ap-p}/320 \text{ }\mu\text{Ap-p} = 0.323$

$A_i = A_v \times Zin/RL = 10.33 \times 312.4 \text{ }\Omega/10 \text{ k}\Omega = 0.323$

$PG = A_i \times A_v = 0.323 \times 10.33 = 3.33$

SECTION III: LABORATORY EXERCISE

OBJECTIVES

Investigate the DC and AC characteristics of the NPN, common base, two power supply circuit.

LIST OF MATERIALS AND EQUIPMENT

1. Transistor: 2N3904
2. Resistors (one each except where indicated): 220 Ω (two)   4.7 kΩ   10 kΩ (two)
3. Capacitors: 10 μF   100 μF
4. Power Supply ± 12 V
5. Voltmeter
6. Signal Generator
7. Oscilloscope

PROCEDURE

1. Connect the NPN transistor in the two power supply, common base circuit as shown in Figure 9-21, where the signal generator and the load resistor are included in the circuit.

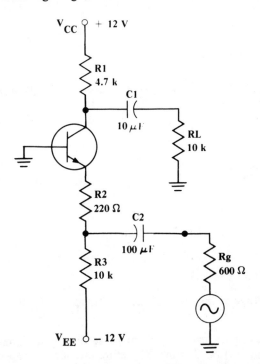

FIGURE 9-21

2. Calculate and measure the DC voltages around the circuit. Include all resistors and the transistor. Measure V_B, V_E, V_C, and the junction of V_{R2} and V_{R3}, all with respect to ground. Calculate the voltage drop across V_{R1}, V_{R2}, V_{R3}, V_{BE}, and V_{CE} instead of measuring directly.

3. Insert the calculated and measured values, as indicated, into Table 9-7.

TABLE 9-7	I_E	V_{R1}	V_{R2}	V_{R3}	V_C	V_B	V_E	V_{CE}	V_{CB}
CALCULATED									
MEASURED	//////								

4. VOLTAGE GAIN

   a. Calculate the voltage gain from: $A_v = (R1 \parallel RL)/(R2 + r_e)$

   b. Measure the mid-frequency voltage gain using an input signal voltage of 100 mVp-p at a frequency of 1 kHz. Measure the Vo and calculate the measured voltage gain from:

$A_v = Vo/Vin$ where: $Vin = 100$ mVp-p

5. **INPUT IMPEDANCE**

   a. Calculate the input impedance to the circuit from: $Zin = (R2 + r_e) // R3$

   b. Measure the input impedance to the circuit. Use a known 220 ohm series resistor and an input signal level of 100 mVp-p at 1 kHz, as shown in Figure 9-22.

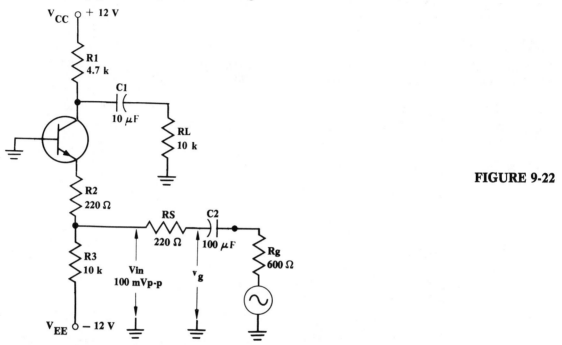

FIGURE 9-22

6. **MAXIMUM PEAK-TO-PEAK OUTPUT VOLTAGE SWING**

   a. Calculate the maximum peak-to-peak output voltage swing from $2V_{CB}$ or $2I_E r_L = 2I_E(R1 // RL)$, whichever is smaller.

   b. Monitor the output voltage swing across the 10 kΩ load resistor, and increase the input signal level until clipping or distortion, because of alpha crowding, occurs. Remove RS before measuring Vop-p(max).

7. **POWER GAIN TO THE LOAD RESISTOR**

   a. Calculate the power gain to the load resistor and use both the measured A_v and Zin values in the calculations. Calculate from:

   $PG = A_v^2 \times Zin/RL$ where: $RL = 10$ kΩ

   b. Convert PG to dB from 10 log PG.

8. Insert the calculated and measured values, as indicated, into Table 9-8.

TABLE 9-8	A_v	Vo for Vin 100 mVp-p	Vop-p MAX	Zin	POWER GAIN	PG(dB)	fc(low)	fc(high)
CALCULATED		/////					/////	/////
MEASURED					/////	/////		

9. Under loaded conditions, monitor the upper and lower frequency 3 dB points. Insert these values, as indicated, into Table 9-8.

SECTION QUESTIONS

1. Does removing the base resistor from the two power supply, common base circuit effect the AC performance? How about the DC voltage distribution? Explain. (Refer to Figure 9-21.)
2. With reference to Figure 9-21, does beta have any effect on the DC voltage distribution or on the AC circuit parameters? Why or why not?

SECTION PROBLEMS

1. Given: beta = 100 and $V_{BE} = 0.6$ V

    A. Solve for:

    1. V_{R1}
    2. V_{R2}
    3. V_{CE}
    4. V_C
    5. V_E

    B. Solve for:

    1. A_v
    2. Zin
    3. PG to the load in dB
    4. Vop-p(max)
    5. A_i

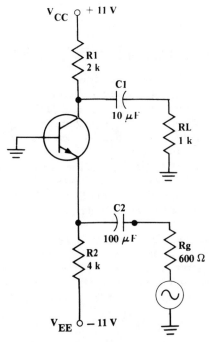

FIGURE 9-24

2. Given: Beta = 100 and $V_{BE} = 0.6$ V

    A. Solve for:

    1. V_{R1}
    2. V_{R2}
    3. V_{R3}
    4. V_C
    5. V_E
    6. V_{CE}

    B. Solve for:

    1. A_v
    2. Zin
    3. PG to the load in dB
    4. Vop-p(max)
    5. A_i

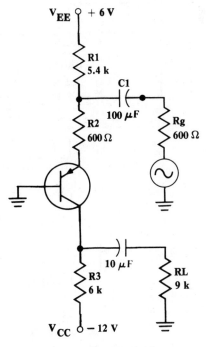

FIGURE 9-23

10 | INTRODUCTION TO JUNCTION FIELD EFFECT TRANSISTORS

GENERAL DISCUSSION

Field effect transistors are high impedance, voltage controlled, solid state devices with characteristics similar to those of vacuum tubes, but with none of the disadvantages of short life span, filament power requirements, and physical size. They can be used as analog switches, voltage controlled variable resistors, and to provide voltage gain and impedance matching. In multistage amplifier circuit, field effect transistors (FETS) are usually used in the front end to take advantage of the high input capabilities of the device.

There are two distinct families of field effect transistors: JFETs and MOSFETs. Junction field effect transistors (JFETs), as the name implies, uses PN junctions in their construction, while metal oxide semiconductor transistors (MOSFETs) do not. At frequencies below 10 MHz, discrete JFETs are more widely used than discrete MOSFETs. In this, and subsequent chapters, only JFETs will be analyzed because they provide the basic foundation for the study of all field effect transistors.

FETs and bipolar devices cannot be directly compared because, for certain applications, each device is superior to the other. FETs have improved and undergone many changes in recent years, both in discrete power applications and monolithic construction, where more than one device exists on a single chip. In discrete form, power FETs, using VMOS and the newer hexfet construction, have obtained breakdown voltages in excess of 300 volts and current capabilities in excess of 100 amps. In low power digital applications, almost all of the newest active devices are MOSFET constructed.

JFET STRUCTURE AND SCHEMATIC SYMBOLS

The discrete JFET transistor is a three terminal device that is constructed by forming two PN junctions on the sides of a bar of semiconductor material, as shown in Figure 10-1. If the bar is N type material, then

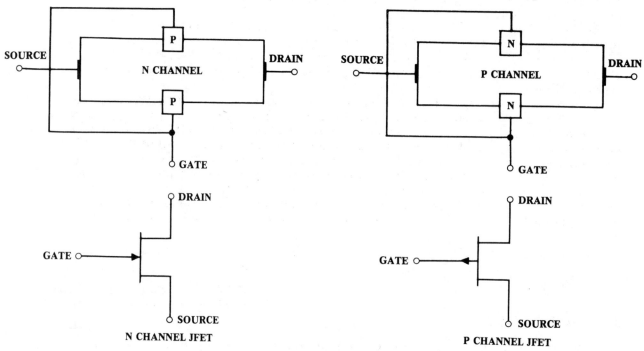

FIGURE 10-1

the side regions are doped P type material. If the bar is P type material, the side regions are N type material. The area between the side regions is called the channel and the complementary devices are aptly called N channel and P channel JFETs.

Properly doping the silicon bar produces three distinct areas: the drain, the source, and the gate regions. The drain is at one end of the semiconductor bar, the source is at the other end of the semiconductor bar, and the gates are the regions on the sides. Metal pads are bonded to each of these regions and the gates, as shown in Figure 10-1, are connected together. Leads are then attached to form the three terminal devices. The cross sectional view of the N channel and P channel devices and their schematic diagrams is shown in Figure 10-1. Also, notice that the PN junction of the N channel device and the NP junction of the P channel device are "PN Junction Diodes" that need to be reverse biased in linear applications.

The power supply connections for the N and P channel devices are given in Figure 10-2. Notice that the gate-to-source junction of both the N and P channel devices are reverse biased. In all other respects, the N channel JFET connection is similar to the NPN bipolar transistor connection and the P channel JFET connection is similar to the PNP bipolar transistor connection. In comparing FETs to bipolar transistors, the drain and collector leads are similar, the source and emitter leads are similar, and the gate and base leads are similar.

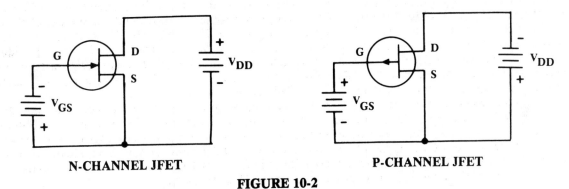

FIGURE 10-2

FET PACKAGING, LEAD INDICATIONS, AND NUMBER SYSTEMS

FETs, like bipolar transistors, are identified by 2N numbers such as 2N5951 and 2N3819. Also used are 3N numbers such as 3N124, and the usual list of proprietary and company numbers, such as the RCA CA40673 and the Siliconix J406.

Also, FETs, like bipolar transistors, come in a variety of plastic and metal packages, and it is difficult to visually identify a FET from a bipolar transistor unless some designated number is given. For instance, JFETs like the 2N5951 or 2N3819 come in plastic packages that are physically similar to the 2N3904 or 2N3392 bipolar transistors. Additionally, TO-3 packaged VMOS power FETs look very much like TO-3 packaged power transistors such as the 2N3055.

Therefore, the best method of determining both the physical and electrical characteristics of a given FET is to look it up in the manufacturer data sheets or data books. In fact, it is always good practice to look up the electrical characteristics and pin configuration for the device, prior to placing it under test, because it is possible that the same numbered devices, from different manufacturers, may be specified differently. Also, the pin configurations for FETs, like transistors, are not necessarily consistent. The bottom views of the 2N5951 and 2N3819 devices are shown in figure 10-3, to illustrate the point. Notice that for the 2N5951 device the drain is in the middle, while for the 2N3819 device it is on one end. Too, most JFETs are symmetrically constructed so that the drain and source terminals can be interchanged with similar electrical results. However, to be sure, use the manufacturer designated lead identification.

FIGURE 10-3

NOTE: As shown, JFET pins are not completely standardized.

JFET OPERATION

JFETs can be connected to operate in two different regions of the output characteristic curves: in the active region where the device is to be used as an amplifier or in the ohmic region where the device is to be used as a voltage controlled variable resistance. In both cases, the gate-to-source junction is reverse biased and the gate-to-source voltage (V_{GS}) is varied in controlling the amount of resistance in the channel, or the amount of current flow through the channel, between the drain and source terminals. Static output characteristic curves are shown in Figure 10-4 to illustrate the ohmic and active regions, where the dotted line called the pinch off voltage line effectively separates the two regions. The ohmic region is between zero volts and the pinch off voltage line, and the active region is between the pinch off voltage line and the voltage breakdown region of the device.

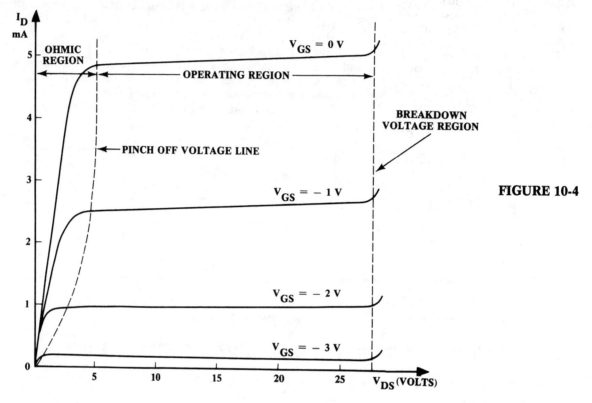

FIGURE 10-4

OHMIC REGION

The ohmic region of the JFET is defined as the area where the drain-to-source voltage of the FET is less than the pinch off voltage line voltage. In practice, the drain-to-source voltage is maintained near zero volts and the channel resistance increases or decreases as the V_{GS} is varied. A representative circuit and a magnified view of the static curve slopes in the low V_{DS} voltage region are given in Figure 10-5.

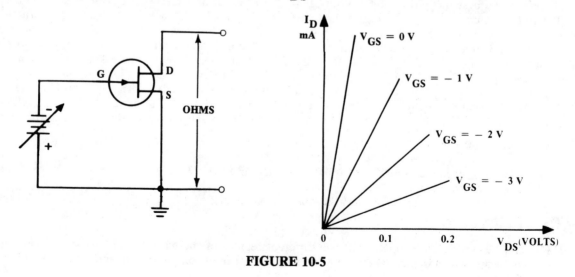

FIGURE 10-5

VARYING THE CHANNEL RESISTANCE

At $V_{GS} = 0$ V, the channel resistance is low, nominally about 10 ohms/cm. However, as the reverse gate-to-source voltage is increased, the depletion region of the reverse biased gate-to-source junction increases, causing fewer current carriers to exist in the channel and the channel resistance to increase. Eventually, the gate-to-source voltage is increased sufficiently so that the depletion regions of the PN regions touch, as shown in Figure 10-6.

The minimum reverse biased gate-to-source voltage that causes the channel of the JFET to be "pinched off" is called the pinch off voltage V_p. The magnitude of the external power supply voltage that causes pinch off to occur is called $V_{GS}(off)$. Therefore, $V_p = V_{GS}(off)$. At pinch off, the channel resistance is nominally about 10 kΩ/cm and, once pinch off is obtained, further increases of V_{GS} do not appreciably increase the channel resistance. Recall from Chapter 2 on diodes, that all reverse-biased diodes exhibit a depletion region that can be controlled as to depth and region by controlling the doping levels of the P and N materials. For JFETs, the depletion region exists mostly in the area of the channel.

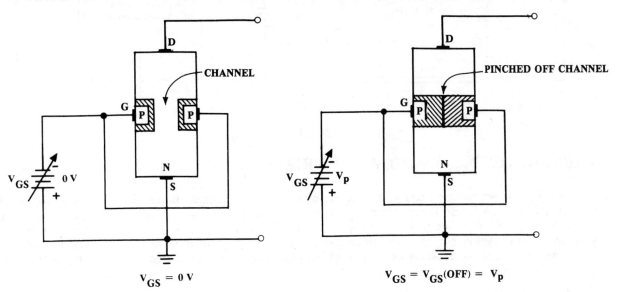

FIGURE 10-6

ACTIVE REGION

The pinch off voltage V_p of the device can be approximated by plotting the static drain characteristic curve for $V_{GS} = 0$ V. For instance, as the drain-to-source voltage V_{DS} is increased from an initial zero volt condition, the characteristic curve of Figure 10-7(a) shows that the drain current increases rapidly at first and then reaches saturation at a theoretical V_p condition. In other words, further increases in V_{DS} beyond V_p, but up to the voltage breakdown region, does not appreciably increase the drain current I_D. For the $V_{GS} = 0$ V condition, $I_D = I_{DSS}$ is in saturation and is the maximum theoretical current flow through the channel. The saturation current I_{DSS}, like the pinch off voltage V_p, is a device parameter.

The reason the drain current does not continue to increase beyond V_p, as V_{DS} is increased, is because as the V_{DS} is increased from an initial zero volt condition, the I_D current flow in the channel increases, causing an increased voltage to be developed across the channel resistance. This increased voltage in the channel, with regard to the zero volts at the gate, causes a reverse biased gate-to-source condition to occur which, in turn, decreases the I_D. Therefore, at pinch off voltage V_p, or beyond, any possibility of increased I_D would cause the V_{GS} to increase and I_D to decrease. Hence, the I_D remains relatively constant beyond the V_p condition, regardless of increased drain-to-source voltage, as long as the maximum drain-to-source breakdown voltage $BV_{DS}(max)$ is not exceeded.

Notice in Figure 10-7(b), however, that the shape of the depletion region differs when a drain-to-source voltage is present. This shape is caused by the voltage differences that exist along the channel with reference to the zero volt gate voltage, with maximum depletion occuring at the drain because of high drain-to-gate voltage and zero depletion occurring at the source because of zero gate-to-source voltage. It is because of the channel-to-gate voltage variance that the resistance at the drain is about 10 kΩ/cm, while the resistance at the source is about 10 Ω/cm.

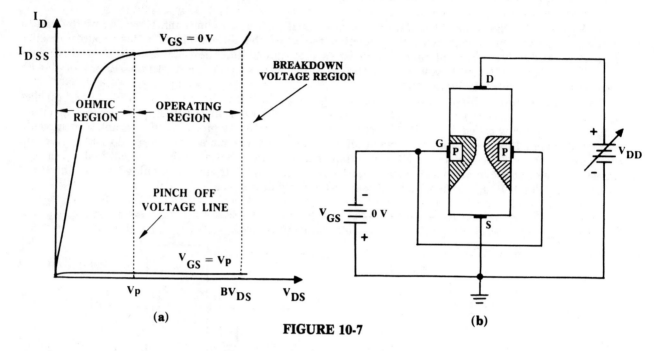

FIGURE 10-7

VARYING THE V_{GS} IN THE ACTIVE REGION

When V_{GS} is set to zero volts and V_{DS} is increased, the channel current I_D reaches a maximum condition at about V_p, and then it remains at a relatively constant current condition as V_{DS} is increased up to the breakdown voltage region of BV_{DS}. However, when an external V_{GS} is used, and the reverse gate-to-source voltage is increased, the amount of current flow in the channel necessary to achieve constant conditions will decrease, creating new pinch off voltages and new characteristic curves, as illustrated in Figure 10-8.

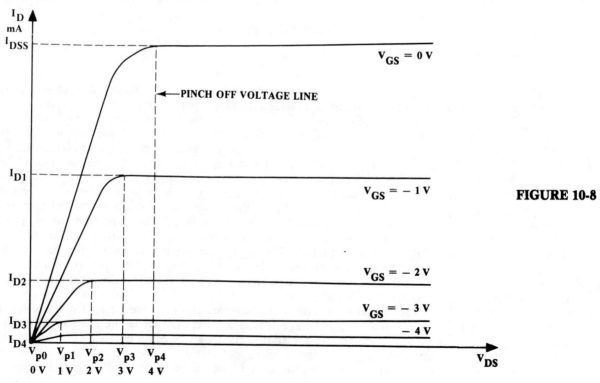

FIGURE 10-8

As shown in Figure 10-8, if V_p of the JFET is 4 V, then at $V_{GS} = 0$ V the pinch off voltage point for that characteristic curve occurs at a V_{DS} of 4 V. With the external V_{GS} supply voltage set to -1 V, the V_{DS} required to obtain pinch off is 3 volts and the magnitude of the constant current is lowered almost half. Then, with a V_{GS} of -2 V, the V_{DS} required to obtain pinch off is 2 V, and with a V_{GS} of -3 V the V_{DS} required to reach pinch off is 1 volt and the constant current is lowered further as the voltage is lowered. Then, with the external gate-to-source supply voltage set at -4 V, the current flow in the channel is reduced to zero, theoretically, and increasing V_{DS} does not appreciably change the approximately 0 mA current flow condition.

Notice, however, that regardless of the external gate-to-source voltage that the pinch off voltage of the device remains constant at 4 volts. For instance, when $V_{GS} = 0$ V, the $V_p = V_{DS}$ occurred at 4 V, or $V_p = V_{p4} - V_{GS} = 4$ V $- 0$ V $= 4$ V. Likewise, when $V_{GS} = -1$ V, the $V_p = V_{DS}$ occurred at 3 V, or $V_p = V_{p3} - V_{GS} = 3$ V $- (-1$ V$) = 4$ V. Similarly, when $V_{GS} = -2$ V, $V_p = V_{DS}$ at 2 V and $V_p = V_{p2} - V_{GS} = 2$ V $- (-2$ V$) = 4$ V. For $V_{GS} = -3$ V, $V_p = V_{DS}$ at 1 V and $V_p = V_{p1} - V_{GS} = 1$ V $-(-3$ V$) = 4$ V. And when $V_{GS} = -4$ V, $V_p = V_{DS}$ at 0 V and $V_p = V_{p4} - V_{GS} = 0$ V $- (-4$ V$) = 4$ V.

For JFETs, the gate-to-channel or gate-to-source junction diode can be reverse biased in two ways: by developing a positive voltage in the cathode (channel), by applying a negative voltage on the anode (gate), or by a combination of both. Figure 10-9 shows both methods of reverse biasing the diode junction.

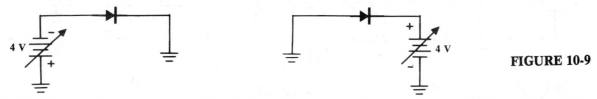

FIGURE 10-9

NOTE: The magnitudes and signs of V_{GS}(off) and V_p are equal but opposite in sign $[V_p = |V_{GS}(\text{off})|]$ and, because of this, they are substituted for each other in the literature. Again, in order to avoid confusion, V_{GS}(off) is the amount of external gate-to-source voltage that limits, theoretically, the drain current I_D to 0 mA, while V_p is the pinch-off voltage of the device, that is defined by the intersection of I_{DSS} and the $V_{GS} = 0$ V curve, as shown in Figure 10-10.

LINEAR OPERATION OF JFETS — MATHEMATICAL APPROACH

FETS are, in theory, square law devices and once the saturation current I_{DSS} and the pinch off voltage of the device are known, then I_D can be solved for any value of V_{GS} between zero volts and V_{GS}(off). The formula for solving for I_D is: $I_D = I_{DSS}(1 - V_{GS}/V_p)^2$. For instance, if the device parameters of $I_{DSS} = 8$ mA and $V_p = 4$ V are given, I_D can then be solved for the V_{GS} conditions between $V_{GS} = 0$ V and $V_{GS} = V_{GS}(\text{off}) = -4$ V.

$V_{GS} = 0$ V

For $V_{GS} = 0$ V, the resistance in the channel is at its lowest and the current flow is maximum. Hence, $I_D = I_{DSS} = 8$ mA, as calculated in conjunction with Figure 10-10.

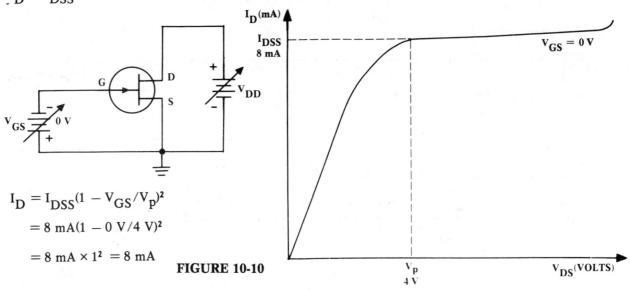

$I_D = I_{DSS}(1 - V_{GS}/V_p)^2$

$= 8$ mA$(1 - 0$ V$/4$ V$)^2$

$= 8$ mA $\times 1^2 = 8$ mA

FIGURE 10-10

NOTE: V_p, like V_{GS}, reverse biases the gate-to-channel junction, similar to the two methods shown in Figure 10-9. However, $V_{GS}(off) = -4$ V, $V_p = 4$ V, and, therefore, $V_p = V_{GS}(off) = |4\ V|$.

$V_{GS} = -1$ V

With V_{GS} set to -1 V, the I_D is lowered to 4.5 mA. Notice, also, that the V_{p1} point on the $V_{GS} = -1$ volt curve, that intersects the 4.5 mA current, is at a V_{DS} of 3 V. This satisfies the $V_p = V_{p1} - V_{GS}$ condition or $V_p = V_{DS} - V_{GS} = 3\ V - (-1\ V) = 4\ V$. See Figure 10-11.

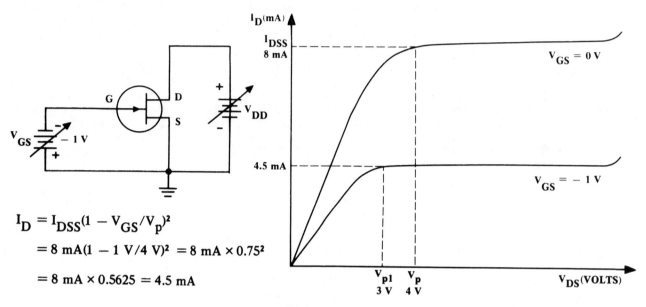

$I_D = I_{DSS}(1 - V_{GS}/V_p)^2$

$= 8\ mA(1 - 1\ V/4\ V)^2 = 8\ mA \times 0.75^2$

$= 8\ mA \times 0.5625 = 4.5\ mA$

FIGURE 10-11

$V_{GS} = -2$ V

When V_{GS} is set to -2 V, the I_D is lowered to 2 mA. Again, notice that the V_{p2} point on the $V_{GS} = -2$ V curve is intersected by $I_D = 2$ mA at a $V_{DS} = V_{p2}$ of 2 V. Hence, $V_p = V_{p2} - V_{GS} = V_{DS} - V_{GS} = 2\ V - (-2\ V) = 4\ V$. See Figure 10-12.

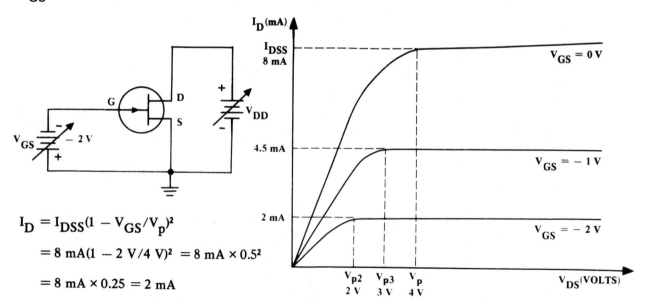

$I_D = I_{DSS}(1 - V_{GS}/V_p)^2$

$= 8\ mA(1 - 2\ V/4\ V)^2 = 8\ mA \times 0.5^2$

$= 8\ mA \times 0.25 = 2\ mA$

FIGURE 10-12

$V_{GS} = -3$ V

When V_{GS} is set to -3 V, the I_D is lowered to 0.5 mA. The V_{p3} point on the $V_{GS} = -3$ V curve is intersected by $I_D = 0.5$ mA at a $V_{DS} = V_{p3}$ of 1 V. Hence $V_p = V_{p3} - V_{GS} = V_{DS} - V_{GS} = 1$ V $- (-3$ V$) = 4$ V. See Figure 10-13.

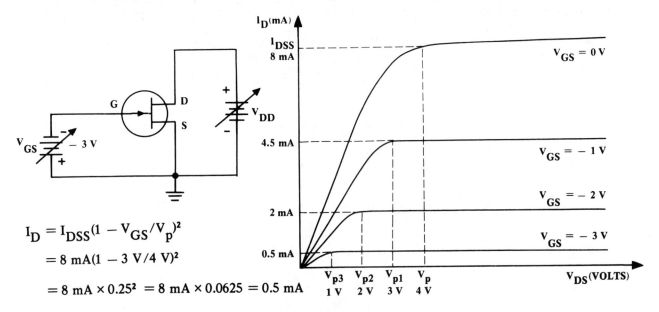

$I_D = I_{DSS}(1 - V_{GS}/V_p)^2$

$= 8$ mA$(1 - 3$ V$/4$ V$)^2$

$= 8$ mA $\times 0.25^2 = 8$ mA $\times 0.0625 = 0.5$ mA

FIGURE 10-13

$V_{GS} = V_{GS}(\text{off}) = -4$ V

When V_{GS} is set to -4 V, the I_D is, theoretically, at 0 mA. Hence, $V_p = V_{p4} - V_{GS} = V_{DS} - V_{GS} = 0$ V $- (-4$ V$) = 4$ V. See Figure 10-14.

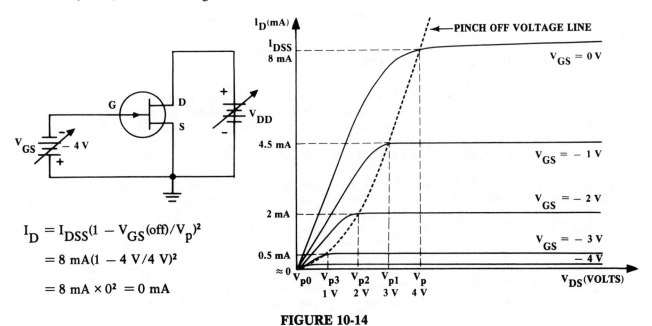

$I_D = I_{DSS}(1 - V_{GS}(\text{off})/V_p)^2$

$= 8$ mA$(1 - 4$ V$/4$ V$)^2$

$= 8$ mA $\times 0^2 = 0$ mA

FIGURE 10-14

NOTE: The line that joins the pinch off voltage points, on each of the characteristic curves, describes the pinch off voltage line. The full set of static output characteristic curves and the pinch off voltage line are shown in Figure 10-14.

TRANSFER CURVE

In theory, I_D follows the $I_D = I_{DSS}(1 - V_{GS}/V_p)^2$ formula, because the device is assumed to be square law. However, in practical applications, the device rarely follows square law principles exactly, and there are usually some discrepancies between calculated and measured values, when the formula is used. Hence, the way to obtain reasonably close calculated to measured circuit values, when using the JFET, is to employ the transfer curve.

Essentially, the transfer curve is a plot of I_D versus V_{GS} plotted from $V_{GS} = 0$ V to $V_{GS} = V_{GS}$(off). It is a carbon copy of the pinch off voltage line, except the I_D for the pinch off voltage line is plotted against V_{DS} and the I_D of the transfer curve is plotted against V_{GS}. The transfer curve of Figure 10-15 is shown referencing the output characteristic curves at each of the V_{GS} and corresponding I_D conditions.

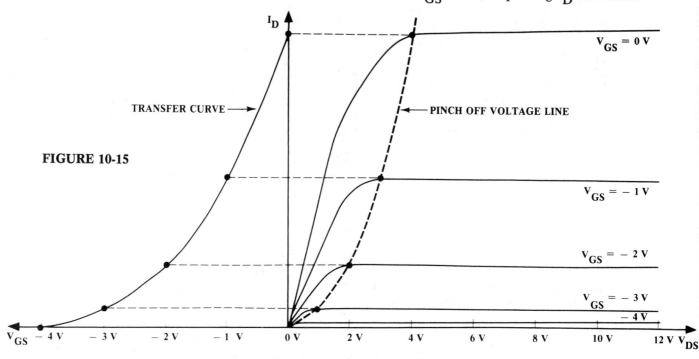

FIGURE 10-15

APPLYING THE TRANSFER CURVE

The transfer curve plots the drain current at every condition of V_{GS} and, once I_D is known, the voltage drops around the circuit are easily solved. Therefore, for design or analysis purposes, knowing the transfer curve of the device and using the inverse slope of the source resistor are all that is necessary in solving for the drain current I_D and the associated gate-to-source voltage V_{GS}. Essentially, the transfer curve and inverse slope of the source resistor are plotted independently, but where they intersect provides the I_D and V_{GS} conditions.

For example, in Figure 10-16(a), a transfer curve similar to that of Figure 10-15 is again shown, but also included on the graph is the inverse slope resistance for a 1 kΩ resistor. Since the voltage-current relationship of resistors is linear, a 1 kΩ resistor will have 1 volt across it at 1 ma through it, 2 volts at 2 mA, 3 volts at 3 mA, and so on. Then, if a straight line is drawn between the 4 V, 4 mA point and the 0 V, 0 mA origin, the inverse slope of the 1 kΩ resistor is obtained. The transfer curve and the 1 kΩ inverse slope line intersect at 2 V and 2 mA, where $I_D = 2$ mA and $V_{GS} = -2$ V on the transfer curve, and 2 volts across the 1kΩ resistor provides 2 mA. The circuit using a 1 kΩ resistor in the source of the JFET device is shown in Figure 10-16(b). The 1 kΩ RS source resistor is chosen last, once V_{GS} and I_D are known.

Developing 2 volts at the source with regard to the zero volts at the gate provides the required 2 volts of reverse bias at the gate-to-source junction of the JFET. This is called self-biasing and it is used extensively in JFET circuits.

In Figure 10-17, the inverse slopes of four more resistors are given, and the inverse slope of each is found by drawing a straight line from the origin of the graph to a voltage-current point that defines each resistor. For the 2.2 kΩ resistor, 4.4 V at 2 mA is used; for the 470 Ω resistor, 2.35 V at 5 mA is used; for the 330 Ω resistor, 1.65 V at 5 mA is used, and for the 150 Ω resistor, 1.2 V at 8 mA is used. All of the respective

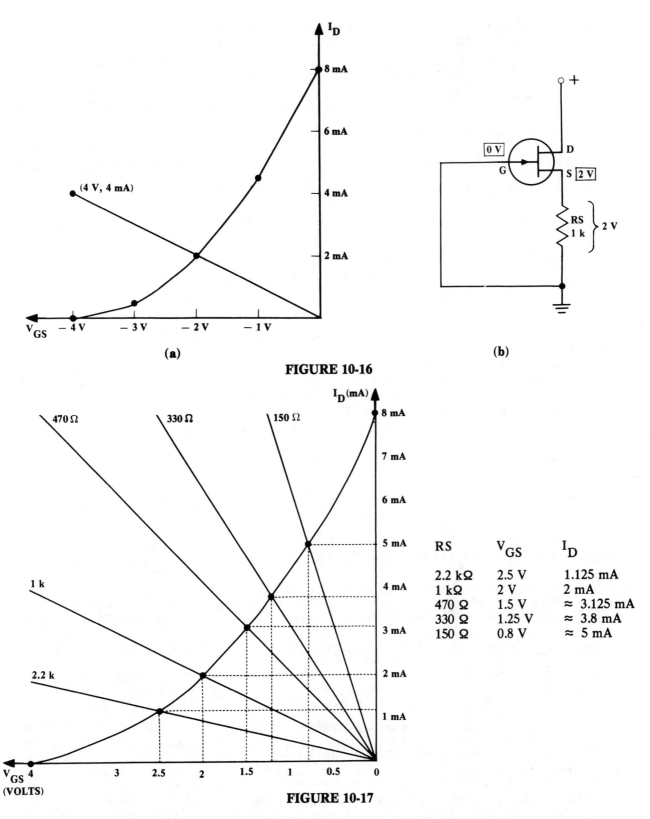

FIGURE 10-16

FIGURE 10-17

RS	V_{GS}	I_D
2.2 kΩ	2.5 V	1.125 mA
1 kΩ	2 V	2 mA
470 Ω	1.5 V	≈ 3.125 mA
330 Ω	1.25 V	≈ 3.8 mA
150 Ω	0.8 V	≈ 5 mA

voltages were selected arbitrarily based on Ohm's Law, but any number of voltage-current values could have been chosen and the slopes of the resistors shown would have remained the same.

From the graph of Figure 10-17, the V_{GS} and the I_D can be read directly off the transfer curve and the intersection of the inverse slope of the resistance of each resistor shown. Mathematically, each V_{GS} and corresponding I_D value can be solved using the standard FET formula, if the transfer curve of the device follows square law principles, which is the case for all of the text examples.

a. $RS = 2.2\ k\Omega$, $V_{GS} = -2.5\ V$

$I_D = 8\ mA(1 - 2.5\ V/4\ V)^2 = 1.125\ mA$ $RS = V_{GS}/I_D = |2.5\ V|/1.125\ mA \approx 2.2\ k\Omega$

b. $RS = 1\ k\Omega$, $V_{GS} = -2\ V$

$I_D = 8\ mA(1 - 2\ V/4\ V)^2 = 2\ mA$, $RS = V_{GS}/I_D = |2\ V|/2\ mA = 1\ k\Omega$

c. $RS \approx 470\ \Omega$, $V_{GS} = -1.5\ V$

$I_D = 8\ mA(1 - 1.5\ V/4\ V)^2 = 3.125\ mA$, $RS = V_{GS}/I_D = |1.5\ V|/3.125\ mA = 480\ \Omega \approx 470\ \Omega$

d. $RS = 330\ \Omega$, $V_{GS} = -1.25\ V$

$I_D = 8\ mA(1 - 1.25\ V/4\ V)^2 = 3.78\ mA$, $RS = V_{GS}/I_D = |1.25\ V|/3.78\ mA \approx 330\ \Omega$

e. $RS = 150\ \Omega$, $V_{GS} = -0.8\ V$

$I_D = 8\ mA(1 - 0.8\ V/4\ V)^2 = 5.12\ mA$, $RS = V_{GS}/I_D = |0.8\ V|/5.12\ mA = 156.25\ \Omega \approx 150\ \Omega$

Therefore, if the JFET is connected into a simple circuit configuration, such as that shown in Figure 10-18, the value of the RS resistor establishes a V_{GS} voltage and the associated I_D for the transfer curve of the device. For instance if a 2.2 kΩ source resistor is connected into the circuit, about 2.5 V will be developed across that source resistor. Likewise, 2 volts will be developed across a 1 kΩ resistor and approximately 1.5 volts across a 470 ohm source resistor, as shown in Figure 10-18.

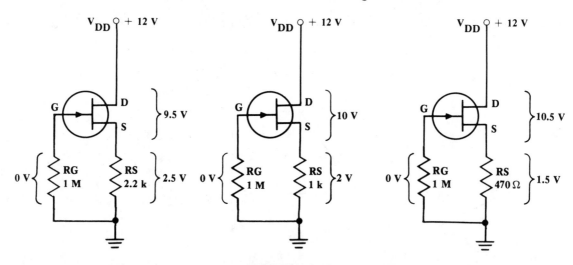

FIGURE 10-18

Essentially, any voltage developed across the source resistor, in the self-biased circuits of Figure 10-18, provides the necessary gate-to-source junction, reverse-bias voltage to make the JFET function. For instance, 2 volts developed across the 1 kΩ RS resistor places 2 volts at the source terminal, with respect to ground. However, because no voltage is developed across the 1 MΩ gate resistor, the voltage at the gate is at zero volts. Hence, the gate-to-source junction is reverse biased at 2 volts. The reason the V_{RG} of 1 MΩ is zero volts is because the reverse biased, gate-to-source diode causes the current flow in the gate to be in the low nano ampere region. For instance, if $I_G = 2$ nano ampere and $R_G = 1\ M\Omega$, then: $V_{RG} = I_G R_G = 10^{-9}\ A \times 1\ M\Omega = 10^{-9} \times 10^6 = 10^{-3}\ V \approx 0\ V$

JFET CURVE TRACER TESTING

The curve tracer provides a convenient method of displaying both the output characteristic curves and the transfer curve of a JFET quickly. In practice, the transfer curve for the JFET can be graphically produced, and any combination of inverse slope lines drawn in minutes. Too, by observing the transfer curve, a convenient point on the curve can be selected and a nominal source resistor value can then be chosen to match the V_{GS} and I_D conditions. As previously stated, the transfer curve is preferred over the output characteristic curves because the I_D and V_{GS} points of the transfer curve are well defined and those of the output characteristic curves are not. In fact, attempting to select the correct I_D and V_{GS} points from the characteristic curves at other than the characteristics shown is pure "guesswork". Only the transfer curve can be considered reliable and it should be used.

Using the curve tracer to test JFETs is similar to using it to test bipolar devices, except the FETs are voltage controlled devices and the bipolars are current contolled devices. Therefore, instead of the base current of bipolars being used, the gate-to-source voltage of the JFET is used. A step-by-step procedure for testing the JFET on the Tektronix 576 and Tektronix 575, the earlier version, using the 2N5951 follows. The device parameters of $I_{DSS} = 8$ mA and $V_p = 4$ V are used to provide continuity with the text discussion.

OUTPUT CHARACTERISTIC CURVES ON THE TEKTRONIX 576 (575)

1. Connect the 2N5951, or equivalent device, into the left or right socket of the curve tracer. Set the toggle switch in the neutral position and set the load resistor to the zero ohm position to observe the full, untruncated, output, response curves. Most JFETs, like the 2N5951, have the drain in the middle and the socket of the Tektronix 576 accommodates that configuration. If the device is tested on the Tektronix 575, the drain and gate leads have to be interchanged to fit the CBE-DGS socket.

2. Set the horizontal selector knob at 1 V/DIV and the vertical knob at 1 mA/DIV. Set the (base) gate-to-source step generator selector knob to 0.5 V/STEP. The Tektronix 575 does not have a 0.5 V/STEP capability. Therefore, when using the 575, connect a 1 kΩ resistor across the base-emitter (gate-source) terminal and set the base current to 0.5 mA/STEP. This procedure develops the necessary 0.5 V/STEP.

3. For the N channel JFET, make sure the polarity switch of the peak voltage (collector) sweep is positive and the invert button for the (base) step generator is inserted. The invert button makes the gate-to-source step voltage negative going. On the Tektronix 575, make the step generator negative (−), and the peak voltage polarity switch positive. Notice that both setup procedures cause the gate-to-source steps to be negative going.

4. Make sure the REP step family button is inserted, along with the step button. It may be necessary to increase the number of steps being used. All three controls are found in the step generator region. On the Tektronix 576, set the display offset control in the normally off mode.

5. Set collector sweep voltage to maximum peak voltage of 15 volts, and increase the horizontal deflection

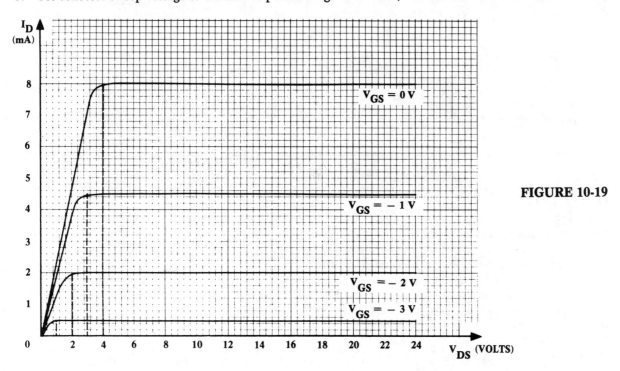

FIGURE 10-19

to 10 full divisions on the CRT. The initial maximum peak voltage on the Tektronix 575 is 20 volts. However, the magnitude of the peak voltage can be increased to 200 volts. On the Tektronix 576, the next voltage level peak is 75 volts, but to obtain this voltage, an inhibit switch has to be activated. The safety interlock is located in the bottom left corner of the socket adapter panel.

6. Following these steps will provide output characteristic curves similar to those shown in Figure 10-19.

TRANSFER CURVE ON THE TEKTRONIX 576 (575)

1. Set the horizontal selector switch at 0.5 V/DIV (base secton) and maintain the vertical knob at one mA/DIV and the (base) gate-to-source selector knob at 0.5 V/STEP. (For the 575, again use the 0.5 mA and the 1 kΩ gate-to-source resistor.)
2. Again, the load resistor setting is zero ohms, the polarity of the peak voltage (collector) sweep is positive, and the inverse button of the step generator is activated. (If the output characteristic curves were obtained, both the step switch and repetitive family are already activated.)
3. Adjust the position of the right most dot to reference the bottom left half corner of the CRT.
4. Throw the toggle switch, increase the variable (collector) supply, and monitor the vertical increase in I_D for each of the V_{GS} steps given. Increase the variable supply to a saturated condition and visually note the transfer curve being plotted on the I_D, V_{GS} dots, as shown in Figure 10-20. The vertical lines to the dots form the retrace lines, and each dot is set at horizontally 0.5 V intervals. Notice that the curve becomes easier to interpret as the number of dots (evaluation points) increase.

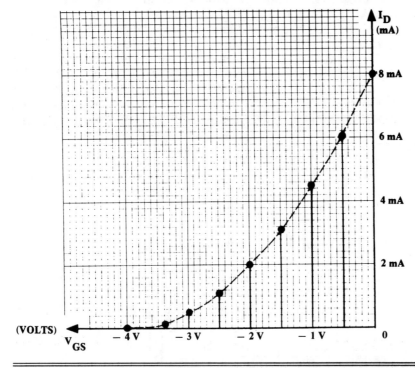

FIGURE 10-20

LABORATORY EXERCISE

OBJECTIVES
To investigate the available techniques for testing JFET transistors — from go and no-go tests to complete parameter measurements.

LIST OF MATERIALS AND EQUIPMENT
1. Transistor: 2N5951 or equivalent (one)
2. Diode: 1N4001 or equivalent (one)
3. Resistors (one each except where indicated): 470 Ω   1 kΩ   2.2 kΩ   100 kΩ   10 kΩ potentiometer (two)
4. Transformer
5. Oscilloscope (horizontal input capability)
6. Power Supply

PROCEDURE

Section I: Go and No-Go Junction Testing of JFETs Using an Ohmmeter

1. Locate the drain, gate, and source leads of the device to be tested. Use the data sheets in the appendix if necessary.
2. Set the ohmmeter at the R × 100 setting and measure the resistance across the gate-to-source junction, first in one direction and then in the reverse direction.
3. Repeat the procedure for the gate-drain leads of the transistor.

NOTE: The R × 100 setting could be changed to a lower resistance setting of the ohmmeter for forward biased junction testing, or it could be set to a higher resistance ohmmeter setting for the reverse biased junction testing. However, since ohmmeter testing is used to indicate generalized go to no-go ratios only, it does not matter. Connect the gate and source leads for this measurement.

4. Measure the drain-source leads of the JFET in both directions with the ohmmeter resistance at the R × 1 kΩ setting.
5. Insert the measured forward and reverse junction resistance for the gate-source, gate-drain, and drain-source juntions into Table 10-1.

TABLE 10-1	DRAIN-GATE		SOURCE-GATE		DRAIN-SOURCE	
	Forward	Reverse	Forward	Reverse	Forward	Reverse
MEASURED						

Section II: Standard Laboratory Equipment Bench Techniques

Part 1A: Obtaining The Transfer Curve

1. Connect the circuit of Figure 10-21. Use a milliameter to measure I_D and a voltmeter to measure V_{GS}.

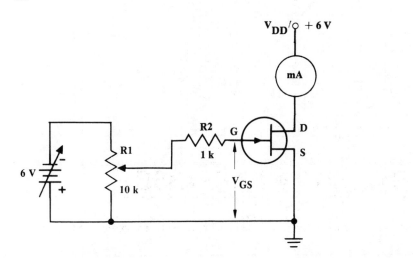

FIGURE 10-21

2. Initially, adjust the variable resistor R1 so the $V_{GS} = 0$ V. Measure $I_D = I_{DSS}$ on the milliammeter. Insert the measured I_{DSS} value into Table 10-2.
3. Increase the reverse biased V_{GS} voltage until I_D decreases to approximately 0.1% of I_{DSS}. Measure the $V_{GS} = V_{GS}(off)$ condition with the voltmeter. Insert the measured pinch off voltage into Table 10-2.

NOTE: In practice, 0.1% of I_{DSS} is usually more than adequate to describe the $V_{GS}(off) = V_p$ condition. Also, a current limiting resistor can be added to protect the device against inadvertent gate-to-source voltage reversals.

4. Measure the I_D at each of the remaining V_{GS} voltage conditions indicated in Table 10-2. Also, insert the measured I_D values for each of the corresponding V_{GS} values into Table 10-2.

NOTE: $V_{GS}(\text{off})$ should occur prior to $V_{GS} = 5$ V. However, measuring V_{GS} beyond the $V_p = V_{GS}(\text{off})$ condition illustrates the reason for $V_{GS}(\text{off})$ being considered at 0.1% of I_{DSS}.

TABLE 10-2											
V_{GS}	0 V		−0.5 V	−1 V	−1.5 V	−2 V	−2.5 V	−3 V	−3.5 V	−4 V	−4.5 V	−5 V
I_D		I_{DSS} 0.1 %										

Part 1B: Plotting The Transfer Curve

1. Use the V_{GS} voltages and corresponding I_D currents of Table 10-2 to plot the transfer curve.
2. Begin the plot with the $V_{GS} = 0$ V, I_{DSS} point. Continue for each remaining point determined by the incremental 0.5 V, V_{GS} voltages and their corresponding currents. Conclude with $V_{GS} = V_{GS}(\text{off})$, where $I_D = 0.001\ I_{DSS} \approx 0$ mA.
3. Join the points together to form the transfer curve. (Use the graph given in Figure 10-22.)

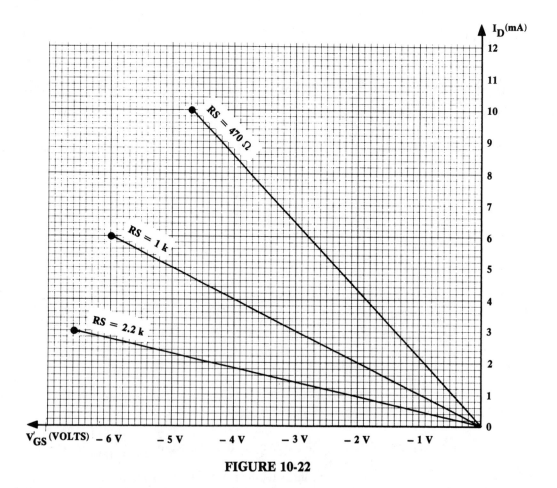

FIGURE 10-22

Part 1C: Applying The Transfer Curve

1. At each of the intersections caused by the transfer curve and each of the predrawn load lines, determine the V_{GS} and corresponding I_D current values. Insert the V_{GS} and I_D values caused by the 2.2 kΩ load line, 1 kΩ load line, and 470 Ω load line into Table 10-3.
2. Using a 2.2 kΩ RS resistor:
   a. Connect the circuit of Figure 10-23, where the RS resistor is 2.2 kΩ and RG = 100 kΩ.

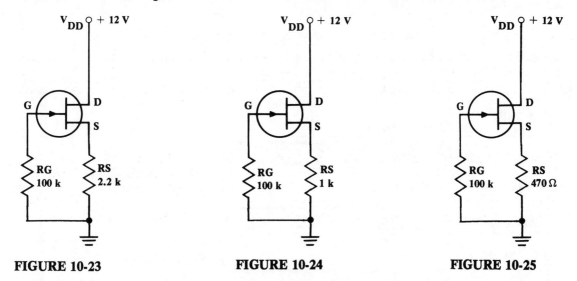

FIGURE 10-23 **FIGURE 10-24** **FIGURE 10-25**

   b. Monitor V_S, V_D, and V_G, all with respect to ground. Insert the measured values into Table 10-3.
   c. Use the measured values to calculate V_{RG}, V_{GS}, V_{DS} and I_D from:

$$V_{RG} = V_G = 0 \text{ V} \qquad\qquad V_{DS} = V_D - V_S$$

$$V_{GS} = V_G - V_S \quad \text{where:} \quad V_{GS} = V_{RS} \qquad I_D = V_{RS}/RS$$

   d. Calculate the theoretical I_D current. Use the I_{DSS} and V_p values from Table 10-2 and the V_{GS} value from Figure 10-23. Calculate I_D from:

$$I_D = I_{DSS}(1 - V_{GS}/V_p)^2 \quad \text{where:} \quad V_p = V_{GS}(\text{off})$$

   e. Insert the calculated values into Table 10-3.

3. Using a 1 kΩ RS resistor:
   a. Connect the circuit of Figure 10-24, where RG remains at 100 kΩ and the RS resistor is 1 kΩ.
   b. Monitor V_S, V_D, and V_G, with respect to ground. Insert the measured values into Table 10-3.
   c. Use the measured values to calculate V_{RG}, V_{GS}, V_{DS}, and I_D from:

$$V_{RG} = V_G = 0 \text{ V} \qquad\qquad V_{DS} = V_D - V_S$$

$$V_{GS} = V_G - V_S \quad \text{where:} \quad V_{GS} = V_{RS} \qquad I_D = V_{RS}/RS$$

   d. Calculate the theoretical I_D current. Use the I_{DSS} and V_p values from Table 10-2 and the V_{GS} value for Figure 10-24. Calculate I_D from:

$$I_D = I_{DSS}(1 - V_{GS}/V_p)^2 \quad \text{where:} \quad V_p = V_{GS}(\text{off})$$

   e. Insert the calculated values into Table 10-3.

4. Using a 470 Ω RS resistor:
   a. Connect the circuit of Figure 10-25, where RG remains at 100 kΩ and the RS resistor is 470 Ω.
   b. Measure V_S, V_D, and V_G with respect to ground. Insert the measured values into Table 10-3.

c. Use the measured values to calculate V_{RG}, V_{GS}, V_{DS}, and I_D from:

$$V_{RG} = V_G = 0\ V \qquad\qquad V_{DS} = V_D - V_S$$
$$V_{GS} = V_G - V_S \qquad\qquad I_D = V_{RS}/RS$$

d. Calculate the theoretical I_D current. Use the V_{GS} voltage of Figure 10-25 and the I_{DSS} and V_p values of Table 10-2. Calculate I_D from:

$$I_D = I_{DSS}(1 - V_{GS}/V_p)^2 \qquad \text{where:}\ V_p = V_{GS}\text{(off)}$$

e. Insert the measured values from the curve, the measured and calculated values of the circuit, and the calculated theoretical I_D, as indicated, into Table 10-3.

TABLE 10-3		CURVE		CIRCUITS							THEORY
		V_{GS}	I_D	V_S	V_D	V_G	V_{GS}	V_{RG}	V_{DS}	I_D	I_D
2.2 k	CALCULATED	///	///	///	///	///					
	MEASURED						///	///	///	///	///
1 k	CALCULATED	///	///	///	///	///					
	MEASURED						///	///	///	///	///
470 Ω	CALCULATED	///	///	///	///	///					
	MEASURED						///	///	///	///	///

Part 2: Plotting The Output Drain Curves

1. Connect the circuit shown in Figure 10-26.

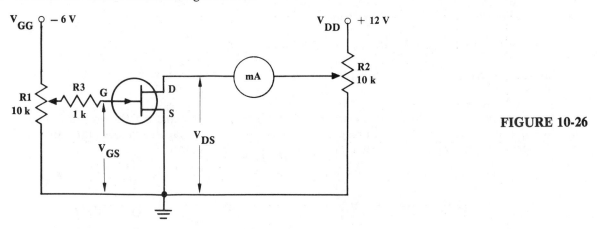

FIGURE 10-26

2. Initially set V_{GS} at 0 V and V_{DS} at 12 V.
   a. Measure I_D at V_{DS} of 12 V and, then, decrease V_{DS} to 6 V, 4 V, 3 V, 2 V, 1 V, 0.5 V, and 0 V by adjusting R2.
   b. Monitor I_D at each of the V_{DS} voltages as indicated in Step a, and insert the measured values into Table 10-4.
3. Set V_{GS} to -1 V and again follow the procedure of Step 2, with regard to V_{DS} voltages and I_D current. Insert the I_D values for the indicated V_{DS} conditions into Table 10-4.
4. Repeat the procedure for V_{GS} values of -2 V and -3 V, and insert the I_D values for the indicated V_{DS} conditions into Table 10-4.

TABLE 10-4	$V_{GS} = 0\ V$							
V_{DS}	12 V	6 V	4 V	3 V	2 V	1 V	0.5 V	0 V
I_D								
	$V_{GS} = -1\ V$							
I_D								
	$V_{GS} = -2\ V$							
I_D								
	$V_{GS} = -3\ V$							
I_D								

5. Use the date obtained in Table 10-4 to graph a set of characteristic curves. Graph each I_D versus V_{DS} curve for the V_{DS} voltages indicated. Use the graph form of Figure 10-27.

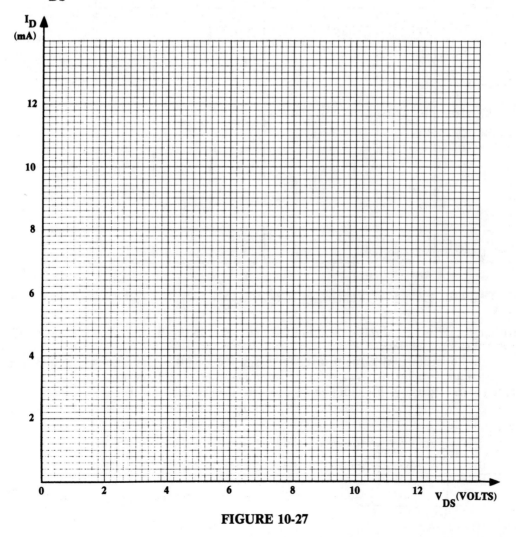

FIGURE 10-27

Section III: Curve Tracer JFET Measurements

Part 1: Output Characteristic Curves — Tektronix 576 (575)

1. Connect a JFET, like the 2N5951 or equivalent device, into the left or right socket of the curve tracer. Set the toggle switch in the neutral position and set the load resistor to the zero ohm position to observe the full untruncated output response curves.
2. Set the horizontal knob at 1 V/DIV and the vertical knob at 1 mA/DIV. Set the step selector knob to 0.5 V/STEP. (For the Tektronix 575, connect a 1 kΩ resistor across the base to emitter terminals and set the step selector knob to 0.5 mA/STEP.)
3. For the N channel JFET, set the polarity of the peak sweep voltage to positive and insert the invert button of the step generator. (For the 575, set the polarity of the step generator to negative.)
4. Set the display switch to repetitive and the maximum peak sweep to 0-15 V. (For the 575, it is in the 0-20 V range.)
5. Increase the horizontal deflection 10 full division, so that V_{DS} will be 10 V. Throw the toggle switch in the direction of the device under test and monitor the output characteristic curves on the screen. It may be necessary to increase the repetitivity or the number of steps being displayed.
6. For each 0.5 V V_{GS} step beginning at $V_{GS} = 0$ V, insert the I_D current at each of the V_{DS} voltages indicated in Table 10-5.

TABLE 10-5 V_{DS}	9 V	6 V	4 V	3 V	2 V	1 V	0.5 V	0V
$V_{GS} = 0$ V								
$V_{GS} = -0.5$ V								
$V_{GS} = -1$ V								
$V_{GS} = -1.5$ V								
$V_{GS} = -2$ V								
$V_{GS} = -2.5$ V								
$V_{GS} = -3$ V								
$V_{GS} = V_{GS}(\text{off})$								

Part 2: Transfer Curve — Tektronix 576 (575)

1. Return the toggle switch to the center position and reset the horizontal knob to 0.5 V/DIV (base). Maintain all other knob settings.
2. Readjust the horizontal position so that the furthermost dot on the right becomes the reference. Throw the toggle switch and increase the peak voltage magnitude to provide the transfer curve.
3. Insert the measured I_{DSS}, $V_{GS}(\text{off})$, and I_D values, as indicated, into Table 10-6.

TABLE 10-6 V_{GS}	0 V	−0.5 V	−1 V	−1.5 V	−2 V	−2.5 V	−3 V	
I_D								≈ 0.0 mA

CHAPTER QUESTIONS

1. Why is a N channel JFET called a unipolar device, while the NPN transistor is called a bipolar device?
2. Why does reverse biasing the gate-to-source junction of a JFET provide a high input impedance? What if the junction were forward biased? What would happen to the input impedance?
3. V_p and V_{GS}(off) are used interchangeably in literature. For clarity, how are V_p and V_{GS}(off) individually defined?
4. Why is it a good idea to make reference to the manufacturer data sheet prior to connecting up an unknown JFET?
5. Why is the drain-to-source voltage of a JFET maintained near zero volts when operated in the ohmic region? Also, why does increasing the reverse biased voltage increase the channel resistance?
6. Based on the data of Table 10-3, which is the most accurate, the transfer curve I_D or the theoretical I_D? Explain.
7. In practice, why is the transfer curve much more reliable than the theoretical square law formula?
8. Since I_{DSS} and V_p are device parameters, what percentage of external V_{GS}, with regard to the device V_p, is required to make the I_D one-fourth of I_{DSS}? (Refer to linear operation of JFET section.)
9. Why is the inverse slope of the source resistor technique so accurate in determining the I_D and V_{GS}, regardless of the source resistor used?

CHAPTER PROBLEMS

1. Given: $V_p = 6$ V, $I_{DSS} = 8$ mA, and $V_{GS} = -3$ V

   find:
   a. I_D
   b. V_{RS}
   c. V_S
   d. V_G
   e. V_{DS}

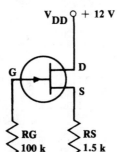

FIGURE 10-28

2. Given: $V_p = 6$ V, $I_{DSS} = 8$ mA, and $V_{GS} = -4.5$ V

   Find:
   a. I_D
   b. V_{RS}
   c. RS

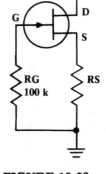

FIGURE 10-29

3. Given: $V_p = 6$ V, $I_{DSS} = 8$ mA, and $V_{GS} = -1.5$ V

   Find:
   a. I_D
   b. V_{RS}
   c. RS

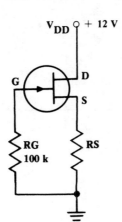

FIGURE 10-30

4. Given: $V_p = 6$ V, $I_{DSS} = 8$ mA, and $V_{GS} = -3$ V

   Find:
   a. I_D
   b. V_{RS}
   c. RS

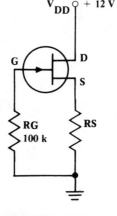

FIGURE 10-31

5. Given: $V_p = 6$ V and $I_{DSS} = 9$ mA

   Solve for:

   a. I_D when $V_{GS} = -1$ V

   b. I_D when $V_{GS} = -3$ V

   c. I_D when $V_{GS} = -5$ V

   Refer to transfer curve of Figure 10-33, and confirm mathematically using the square law formula.

6. With reference to the transfer curve of Figure 10-33:

   a. Locate V_{GS} at -2 V, the associated I_D current, and solve for the value of the required RS source resistor.

   b. Draw the inverse slope line for the known RS resistor and confirm the results using the mathematical square law formula, where: $I_{DSS} = 9$ mA and $V_p = 6$ V.

   c. Repeat Steps a and b for a V_{GS} condition of 4 volts.

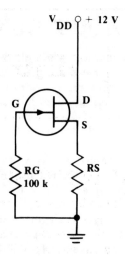

FIGURE 10-32

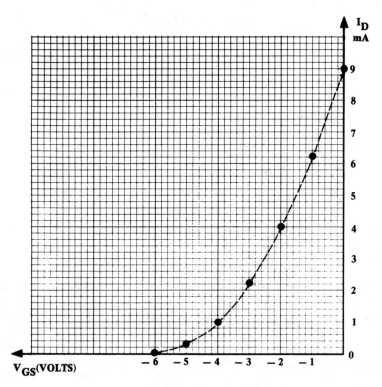

FIGURE 10-33

11 SINGLE STAGE JFET BIASING

GENERAL DISCUSSION

Biasing JFETs can be accomplished through self biasing, universal biasing, or two power supply biasing, as shown in Figure 11-1. Each of the single stage JFET circuits shown is biased in the active region and, regardless of the circuit configuration used, the gate-to-source junction is always reverse biased. Therefore, for the N channel circuits of Figure 11-1, the source will always be at a higher potential, with respect to ground, than the gate.

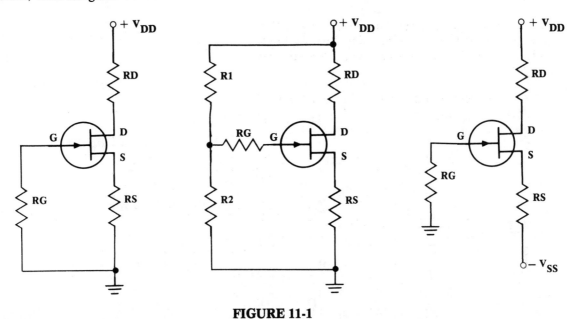

FIGURE 11-1

SECTION I: SELF-BIASED CIRCUITS

GENERAL DISCUSSION

The self-biased, JFET circuit requires a single power supply and a minimum number of components, but the drawback to this form of biasing is that the drain current and voltage drops around the circuit are too dependent on the parameters of the JFET. For instance, if a replacement JFET device with different parameters is used, the DC voltage drops around the circuit will change.

The JFET used in self-biased circuits can be either N channel or P channel. The only difference is the power supply connections, which are connected oppositely, as shown in Figure 11-2.

DIRECT CURRENT CIRCUIT ANALYSIS

The DC analysis of the self-biased circuit begins by solving for I_D. In theory, I_D is solved from the $I_D = I_{DSS}(1 - V_{GS}/V_p)^2$ formula but, in practice, the I_D is found most accurately by using the graphical approach. Both methods of solutions will be used in the example and, in order to provide a comparison, the device will be considered square law. Again, the device parameters of $I_{DSS} = 8$ mA and $V_p = 4$ V are

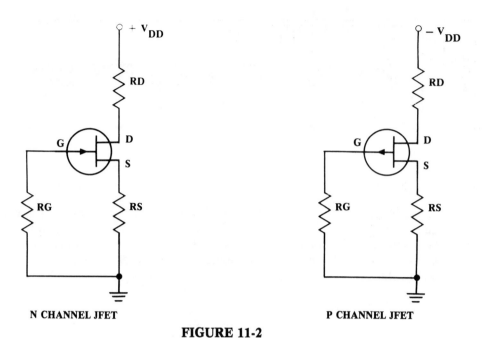

FIGURE 11-2

given. Also, V_{GS} is selected at -2 V (one-half of V_p) so that the I_D will be one-fourth of the I_{DSS}.

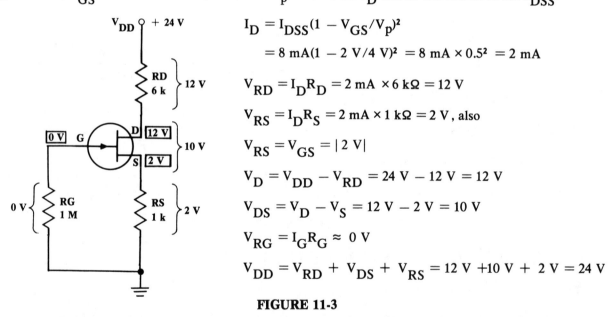

$$I_D = I_{DSS}(1 - V_{GS}/V_p)^2$$
$$= 8\text{ mA}(1 - 2\text{ V}/4\text{ V})^2 = 8\text{ mA} \times 0.5^2 = 2\text{ mA}$$

$V_{RD} = I_D R_D = 2\text{ mA} \times 6\text{ k}\Omega = 12\text{ V}$

$V_{RS} = I_D R_S = 2\text{ mA} \times 1\text{ k}\Omega = 2\text{ V}$, also

$V_{RS} = V_{GS} = |2\text{ V}|$

$V_D = V_{DD} - V_{RD} = 24\text{ V} - 12\text{ V} = 12\text{ V}$

$V_{DS} = V_D - V_S = 12\text{ V} - 2\text{ V} = 10\text{ V}$

$V_{RG} = I_G R_G \approx 0\text{ V}$

$V_{DD} = V_{RD} + V_{DS} + V_{RS} = 12\text{ V} + 10\text{ V} + 2\text{ V} = 24\text{ V}$

FIGURE 11-3

Graphically, the $I_D = 2$ mA and $V_{GS} = -2$ V circuit values are determined by the intersection of the transfer curve and the inverse slope of the 1 kΩ load line, as shown in Figure 11-4. Also, for convenience, a plot of RS is shown, which indicates the 2 volts dropped across the source resistor RS.

SECTION II: TWO-POWER-SUPPLY CIRCUITS

GENERAL DISCUSSION

The main advantage of the two power supply JFET circuit is that I_D is controlled mainly by the circuit and not by the device. Therefore, substituting another JFET device, with different parameters, will cause only slight changes to the DC voltages of the circuit. Hence, the DC stability of two power supply circuit is much superior to that of the self-biased circuit. Again, in the two power supply circuit, either N channel or P channel devices can be used with the only difference being the power supply connections, as

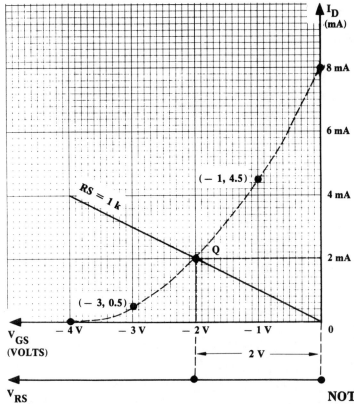

FIGURE 11-4

NOTE: The operating point Q is at the intersection of the transfer curve and the inverse slope line of RS.

shown in Figure 11-5.

DIRECT CURRENT CIRCUIT ANALYSIS

The DC analysis begins by solving for I_D. In practice, I_D is generally found by graphical methods, but in order to provide a comparison, the device will be considered square law and the mathematical approach will also be used. As with the self-biasing circuit, the gate-to-source of the two power supply circuit has to be reverse biased, so V_S must be at a higher DC voltage than V_G. Therefore, as shown in Figure 11-6, the positive 2 volts at the source terminal causes 10 volts to be dropped across RS, when V_{SS} is at -8 volts.

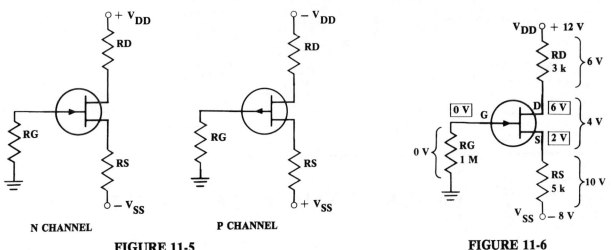

FIGURE 11-5 **FIGURE 11-6**

$$I_D = I_{DSS}(1 - V_{GS}/V_p)^2 = 8\text{ mA}(1 - 2\text{ V}/4\text{ V})^2 = 2\text{ mA}$$

$$V_{RD} = I_D R_D = 2\text{ mA} \times 3\text{ k}\Omega = 6\text{ V}$$

$$V_{RS} = I_D R_S = 2 \text{ mA} \times 5 \text{ k}\Omega = 10 \text{ V}$$

$$V_{RG} = I_G R_G = 10 \text{ nA} \times 1 \text{ M}\Omega = 0.01 \text{ V} \approx 0 \text{ V}$$

$$V_D = V_{DD} - V_{RD} = 12 \text{ V} - 6 \text{ V} = 6 \text{ V}$$

$$V_S = V_{SS} + V_{RS} = -8 \text{ V} + 10 \text{ V} = 2 \text{ V}$$

$$V_{DS} = V_D - V_S = 6 \text{ V} - 2 \text{ V} = 4 \text{ V}$$

$$V_{DD} - V_{SS} = V_{RD} + V_{DS} + V_{RS}$$

$$12 \text{ V} - (-8 \text{ V}) = 6 \text{ V} + 4 \text{ V} + 10 \text{ V} = 20 \text{ V}$$

Graphically, the $I_D = 2$ mA and $V_{GS} = -2$ V circuit values can be determined at the intersection of the transfer curve and the 5 kΩ inverse slope line, as shown in Figure 11-7. Note, however, that V_S is at 2 V and V_{SS} at -8 V, and 10 V are dropped across RS, since RS = 5 kΩ, $I_D = V_{RS}/RS = 10$ kΩ/5 kΩ = 2 mA. Also, for convenience, a plot of the V_{RS} voltage of 10 V is shown in Figure 11-7.

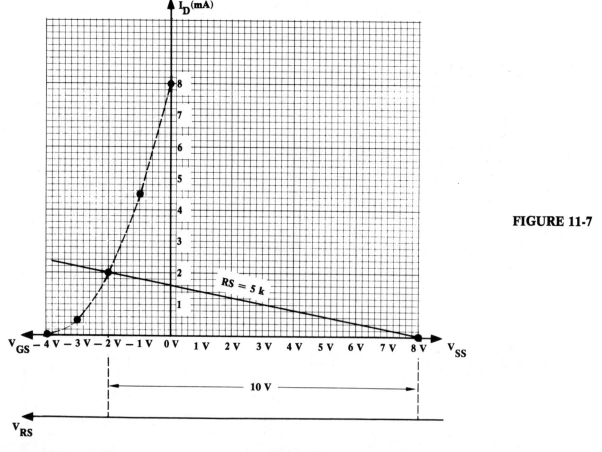

FIGURE 11-7

SECTION III: UNIVERSAL-BIASED JFET CIRCUITS
GENERAL DISCUSSION

The advantage of the universal biased JFET circuit is that it requires only a single power supply, but it provides control of the I_D current by establishing a fixed voltage at the gate using the voltage divider network. Also, like the two power supply circuit, the additional gate voltage increases the V_{RS} voltage and this is a determining factor in establishing the I_D current. Therefore, when JFET devices with different parameters are used, the DC voltages of the circuit will vary slightly more than those of the two power supply circuit, but the universal biased circuit has much better DC stability than the self-biased circuit. Again, either N or P channel JFET devices can be used, which simply requires changing power supply connections, as shown in Figure 11-8.

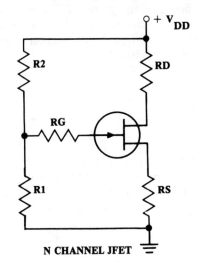
N CHANNEL JFET

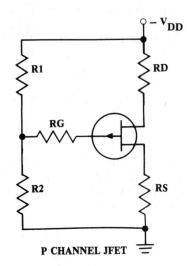
P CHANNEL JFET

FIGURE 11-8

DIRECT CURRENT CIRCUIT ANALYSIS

The DC analysis of the circuit of Figure 11-9 begins by solving for V_{R1} and V_{R2} and then I_D. In practice I_D is generally found by graphical techniques but, again, the device will be considered square law, so the mathematical approach can be used. As with the previous circuits, the gate-to-source junctions must remain reverse biased, so the source resistor must be V_{GS} volts higher than the gate. For instance, if V_G equals 3 volts and V_{GS} equals 2 volts, the source has to be 5 volts. Also, because I_G is extremely low, the voltage drop across R_G is approximately zero volts and $V_G \approx V_{R2}$.

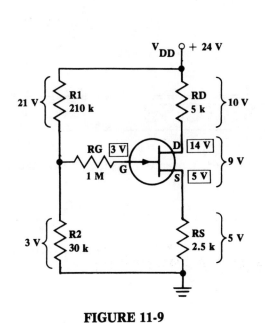

FIGURE 11-9

$$V_{R1} = \frac{V_{DD} \times R1}{R1 + R2} = \frac{24\,V \times 210\,k\Omega}{210\,k\Omega + 30\,k\Omega} = 21\,V$$

$$V_{R2} = \frac{V_{DD} \times R2}{R1 + R2} = \frac{24\,V \times 30\,k\Omega}{210\,k\Omega + 30\,k\Omega} = 3\,V$$

$$I_D = I_{DSS}(1 - V_{GS}/V_p)^2$$
$$= 8\,mA(1 - 2\,V/4\,V)^2 = 2\,mA$$

$$V_{RD} = I_D R_D = 2\,mA \times 5\,k\Omega = 10\,V$$

$$V_{RS} = I_D R_S = 2\,mA \times 2.5\,k\Omega = 5\,V$$

$$V_{RG} = I_G R_G = 10 \times 10^{-9} \times 1\,M\Omega = 0.01\,V \approx 0\,V$$

$$V_G = V_{R2} + V_{RG} = 3\,V + 0\,V = 3\,V$$

$$V_{GS} = V_G - V_S = 3\,V - 5\,V = -2\,V$$

$$V_D = V_{DD} - V_{RD} = 24\,V - 10\,V = 14\,V$$

$$V_{DS} = V_D - V_S = 14\,V - 5\,V = 9\,V$$

$$V_{DD} = V_{RD} + V_{DS} + V_{RS} = 24\,V$$

Graphically, the $I_D = 2$ mA and $V_{GS} = -2$ V circuit values are determined at the intersection of the transfer curve and the 2.5 kΩ inverse slope line. Like the two power supply circuit, the universal circuit provides a comparatively large voltage drop across RS. As shown in Figure 11-9, 3 volts at the gate increased V_{RS} by 3 volts over the V_{GS} value. Therefore, the inverse slope line, like that of the two power supply circuit, originates in the positive polarity region as shown on the graph of Figure 11-10. Hence, a positive voltage at the gate or a negative voltage connected to the source resistor provide the same effect of increasing the voltage drop across RS which, in turn, minimizes the effect V_{GS} change has on I_D. Also, as shown in Figure 11-10, the V_{RS} plot provides a voltage drop of 5 volts.

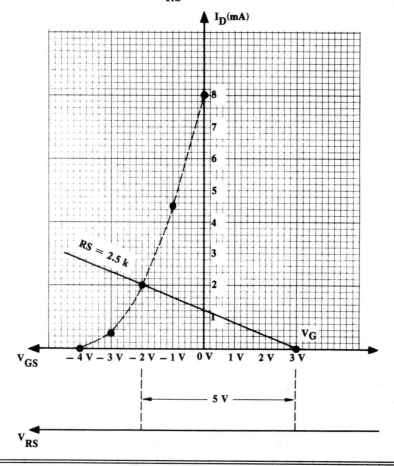

FIGURE 11-10

LABORATORY EXERCISE

OBJECTIVES

To investigate the JFET, DC biasing techniques using self-biased, universal biased, and two power supply biased JFET circuits.

LIST OF MATERIALS AND EQUIPMENT

1. Transistor: 2N5951 or equivalent
2. Resistors (one each): 1 kΩ   2.2 kΩ   4.7 kΩ   10 kΩ   47 kΩ   100 kΩ   330 kΩ
3. Power Supply
4. Voltmeter

PROCEDURE
Section I: Self-Biased JFET Circuits
Part 1: Measurement Techniques

1. Connect the self-biased circuit of Figure 11-11.

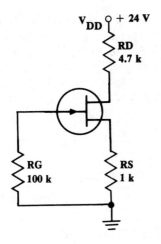

FIGURE 11-11

2. Measure V_D, V_S, and V_G, all with respect to ground. Insert the measured values into Table 11-1.
3. Use the measured values to calculate V_{RD}, V_{DS}, V_{RG}, $V_{RS} = V_{GS}$, and I_D from:

   a. $V_{RD} = V_{DD} - V_D$
   b. $V_{RS} = V_S$
   c. $V_{DS} = V_D - V_S$
   d. $V_{RG} = V_G \approx 0\ V$
   e. $V_{GS} = V_G - V_S$ where: $V_{GS} = V_{RS}$
   f. $I_D = V_{RS}/RS$

NOTE: In the self-biased circuits, the voltage developed across the source resistor provides the gate-to-source biasing voltage, because $V_G = V_{RG} \approx 0\ V$.

4. Insert the calculated values, as indicated, into Table 11-1.

TABLE 11-1	V_D	V_S	V_G	V_{RD}	V_{RS}	V_{DS}	V_{RG}	V_{GS}	I_D
CALCULATED	///	///	///						
MEASURED				///	///	///	///	///	///

Part 2: Transfer Curve Analysis Techniques

1. Using the graph of Figure 11-12, plot the transfer curve of the JFET device, where the inverse slope line of the 1 kΩ resistor RS is provided. The intersection determines the V_{GS} and I_D values. Insert these values into Table 11-2. Refer back to Table 10-2 if the same device is used.

NOTE: The inverse slope line is drawn between the (0,0) origin and the 6 V, 6 mA point. Since the slope line is linear, it intersects 1 V, 1 mA; 2 V, 2 mA; and so forth, until 6 V, 6 mA. Also, use the most convenient method learned in Chapter 10 to plot the transfer curve. In fact, if the same device is used, as that used in Chapter 10, then make use of the same data established for the device in that chapter.

TABLE 11-2	I_D	V_{GS}	V_{RD}	V_{RS}	V_{RG}	V_{DS}
CALCULATED	///	///				
MEASURED						

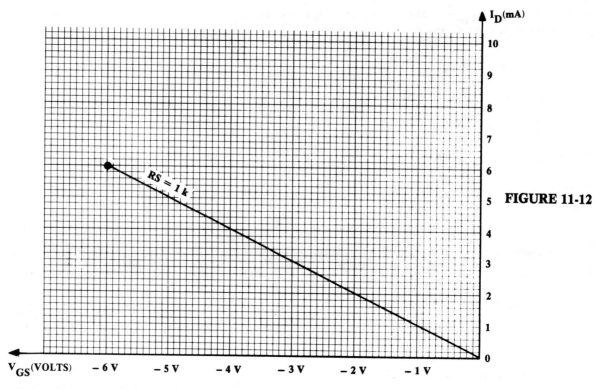

FIGURE 11-12

2. Using the I_D and V_{GS} values obtained from the graph of Figure 11-12, calculate:

   a. $V_{RD} = I_D R_D$

   b. $V_{RS} = I_D R_S$

   c. $V_{RG} = I_G R_G = 0\ V$

   d. $V_{DS} = V_{DD} - (V_{RD} + V_{RS})$

3. Insert the calculated values, as indicated, into Table 11-2.

Section II: The Universal-Biased Circuit
Part 1: Measurement Techniques

1. Connect the universal circuit shown in Figure 11-13.

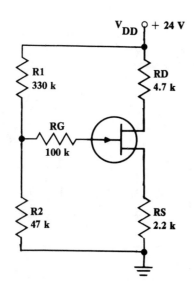

FIGURE 11-13

2. Measure V_{R2}, V_G, V_S, and V_D, all with respect to ground.
3. Use the measured values to calculated V_{RD}, V_{RS}, V_{R1}, V_{RG}, V_{GS}, and I_D from:

   a. $V_{RD} = V_{DD} - V_D$
   b. $V_{RS} = V_S$
   c. $V_{R1} = V_{DD} - V_{R2}$
   d. $V_{RG} = V_{R2} - V_G$
   e. $V_{GS} = V_G - V_S$
   f. $I_D = V_{RS}/RS$

4. Insert the calculated values, as indicated, into Table 11-3.

NOTE: In the universal biased circuit, $V_{RS} = V_{R2} + V_{GS}$ and V_{R2} value is confirmed through calculations.

TABLE 11-3	V_D	V_S	V_G	V_{R2}	V_{R1}	V_{RD}	V_{RS}	V_{GS}	V_{DS}	V_{RG}	I_D
CALCULATED	/////	/////	/////	/////							
MEASURED					/////	/////	/////	/////	/////	/////	/////

Part 2: Transfer Curve Analysis Techniques

1. Use the graph of Figure 11-4 and plot the transfer curve of the JFET device. The intersection determines the V_{GS} and I_D values. Insert these values into Table 11-4.

NOTE: The inverse slope line is drawn between the (3 V, 0 mA) origin and a convenient 1.4 V, 2 mA point. This provides a V_{RS} of 4.4 volts and an I_D of $I_D = V_{RS}/RS = 4.4 \text{ V}/2.2 \text{ k}\Omega = 2 \text{ mA}$. Too, the transfer curve should be plotted identically to that of the previous self-biased circuit. If the same device is being used, use the same data as before.

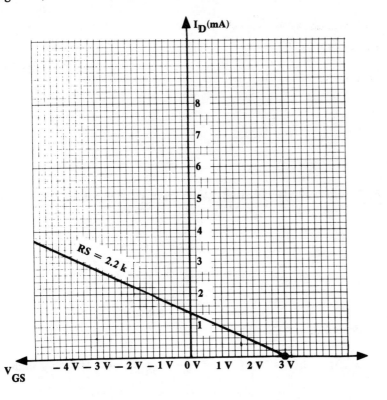

FIGURE 11-14

2. Using the plotted I_D and V_{GS} values from the graph of Figure 11-14, calculate:

$$V_{RD} = I_D R_D \qquad\qquad V_G = V_{RS} - V_{GS}$$
$$V_{RS} = I_D R_S, \text{ or } V_{RS} = V_G + V_{GS} \qquad\qquad V_{DS} = V_{DD} - (V_{RD} + V_{RS})$$

3. Calculate:

$$VR1 = \frac{V_{DD} \times R1}{R1 + R2} \quad \text{and} \quad VR2 = \frac{V_{DD} \times R2}{R1 + R2}$$

$$V_{RG} = I_G R_G \approx 0 \text{ V}$$

4. Insert the calculated values, as indicated, into Table 11-4.

TABLE 11-4	I_D	V_{GS}	V_{RD}	V_{RS}	V_G	V_{DS}	V_{R1}	V_{R2}	V_{RG}
CALCULATED	////	////							
MEASURED			////	////	////	////	////	////	////

Section III: The Two-Power Supply Circuit
Part 1: Measurement Techniques

1. Connect the two power supply circuit shown in Figure 11-15.

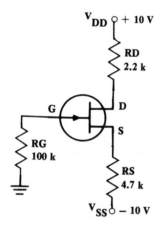

FIGURE 11-15

2. Measure V_D, V_S and V_G, all with respect to ground. Insert the measured values into Table 11-5.
3. Use the measured values to calculate V_{RD}, V_{DS}, V_{RS}, V_{RG}, and I_D from:

   a. $V_{RD} = V_{DD} - V_D$ \qquad d. $V_{RG} = V_G$
   b. $V_{RS} = V_S - V_{SS}$ \qquad e. $V_{GS} = V_G - V_S$
   c. $V_{DS} = V_D - V_S$ \qquad f. $I_D = V_{RS} / RS$

4. Insert the calculated values, as indicated, into Table 11.5.

NOTE: In the two power supply circuit, I_D is primarily controlled by V_{SS} and RS and only slightly by

V_{GS}. However, $V_G \approx 0$ V and V_S will be a positive voltage with respect to ground, ensuring the reverse biased gate-to-source voltage condition.

TABLE 11-5	V_D	V_S	V_G	V_{RD}	V_{RS}	V_{DS}	V_{RG}	V_{GS}	I_D
CALCULATED	//////	//////	//////						
MEASURED				//////	//////	//////	//////	//////	//////

Part 2: Transfer Curve Analysis Techniques

1. Use the graph of Figure 11-16 to plot the transfer curve of the device and the inverse slope of the 4.7 kΩ RS resistor. The intersection determines the V_{GS} and I_D values. Insert these values, as indicated, into Table 11-6. Again, use the transfer curve of the prevous device.

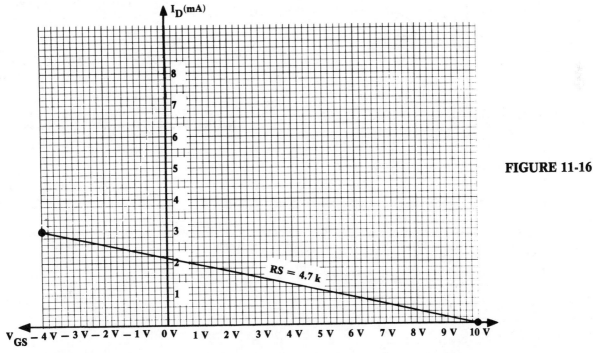

FIGURE 11-16

2. Using the I_D and V_{GS} values determined from the graph of Figure 11-16, calculate:

$$V_{RD} = I_D R_D \qquad V_{RG} = I_G R_G \qquad V_{DS} = V_D - V_S$$
$$V_{RS} = I_D R_S \qquad V_S = V_{SS} + V_{RS}$$

3. Insert the calculated values, as indicated, into Table 11-6.

TABLE 11-6	I_D	V_{GS}	V_{RD}	V_{RS}	V_{RG}	V_S	V_{DS}
CALCULATED	//////	//////					
MEASURED			//////	//////	//////	//////	//////

CHAPTER QUESTIONS

1. What is an advantage of the self-biased circuit?
2. What major advantage does the two power supply circuit have over the self-biased circuit?
3. Why would the universal circuit be used instead of the two power supply circuit? What is the tradeoff?
4. Why is the two power supply circuit more stable than the universal circuit and how is it accomplished?
5. Why is it that the source terminal of the N channel device is always more positive than the gate, regardless of the circuit configuration used?

CHAPTER PROBLEMS

1. Given: $I_{DSS} = 6$ mA, $V_p = 6$ V, and $V_{GS} = -3$ V

   Solve for:
   a. I_D
   b. RS
   c. V_{RD}
   d. V_{DS}
   e. V_{RG}

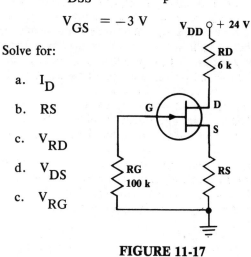

FIGURE 11-17

NOTE: Use the mathematical approach.

2. Using the transfer curve of Figure 11-4 and the inverse slope technique, solve for:
   a. I_D
   b. V_{RS}
   c. V_{RD}
   d. V_{DS}
   e. V_{RG}

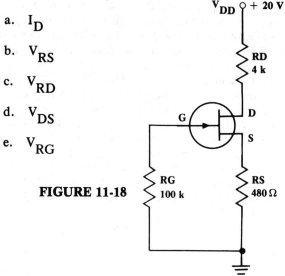

FIGURE 11-18

3. Using the transfer curve of Figure 11-10 and a V_{GS} of 1.2 V, solve for:
   a. I_D
   b. VRS
   c. RS
   d. V_{RD}
   e. V_{DS}
   f. V_{R1}
   g. V_{R2}

4. Using the transfer curve of Figure 11-7 and a V_{GS} of 1.6 V, solve for:
   a. I_D
   b. V_{RS}
   c. RS
   d. V_{RD}
   e. V_{DS}
   f. V_{RG}

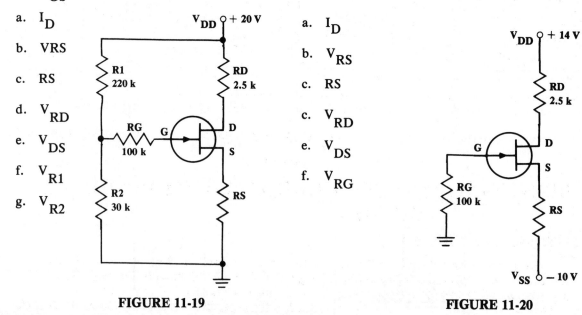

FIGURE 11-19

NOTE: Confirm with the inverse slope technique.

FIGURE 11-20

NOTE: Confirm with the inverse slope technique.

12 | THE COMMON SOURCE JFET CIRCUITS

GENERAL DISCUSSION

Field effect transistors are normally used in the front end of multistage amplifiers to take advantage of the high input impedance that they can provide. They are also used for large impedance transfer ratios and, in some instances, impedance matching. Like bipolar transistors, field effect transistors can be connected in the three useful configurations shown in Figure 12-1.

The first circuit, the common-source configuration of Figure 12-1(a), provides both voltage gain and high input impedance and it is similar to the bipolar common-emitter connection. The input signal for the common-source circuit is applied to the gate and the output is taken off the drain.

The second circuit, the common-drain configuration of Figure 12-1(b), is noted for its very high input impedance to moderately low output impedance ratio. The common-drain circuit, like the bipolar transistor common-collector connection, has a voltage gain of less than unity. The input signal is applied to the gate and the output is taken off the source.

The third circuit, the common-gate circuit of Figure 12-1(c), has a moderately low input impedance and voltage gain, but it is not widely used because it offers no advantages over the similarly connected bipolar common-base transistor circuit. In this chapter, only the common-source configuration will be analyzed, but in Chapters 13 and 14, analysis will be provided for the common-drain and common-gate circuits.

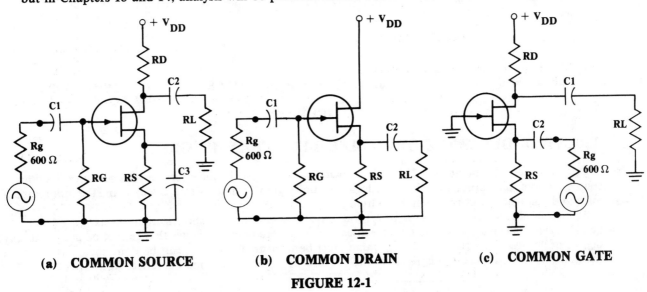

(a) COMMON SOURCE (b) COMMON DRAIN (c) COMMON GATE

FIGURE 12-1

SECTION I: INTRODUCTION TO COMMON SOURCE CIRCUITS

GENERAL DISCUSSION

The common-source circuit can be connected in the self-biased circuit, the universal-biased circuit, or the two power supply circuit, as shown in Figure 12-2. However, because the self-biased circuit is the simplest, it will be used to introduce the AC circuit parameters of JFET amplifiers.

The analysis of field effect transistor circuits, like the analysis of bipolar transistor circuits, begins with DC biasing and then proceeds to the AC analysis. And, like bipolar circuits, the circuit parameters normally solved include voltage gain, input impedance, output impedance, and power gain. The Vop-p(max) parameter of JFETs is not normally solved for, because JFET characteristics are non-linear, and large peak-to-peak voltage swings are difficult to accurately predict through calculations. In fact, the accurate, non-distorted, Vop-p(max) can only be obtained through bench measurements.

There are two methods of solving for the voltage gain of common source JFET amplifiers. The first method uses the transconductance parameter g_m, while the second method uses the inverse of the transconductance parameter $(1/g_m)$ or r_s'.

NOTE: The r_s' method, a method developed by the author, allows FET circuits to be solved like bipolar transistor circuits, and it is the preferred technique in this text, because it is straight-forward. However, because most texts continue to use the transconductance formulas, both methods will be used to show that either provides identical results.

The transconductance term is expressed in literature by both mhos and Siemens. For example, the quantity of 500 μmhos can also be expressed by the IEEE standard of 500 μS. Therefore, whenever convenient, both the mho and Siemen terms will be used.

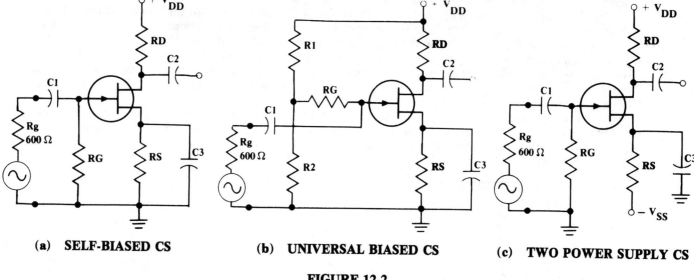

(a) SELF-BIASED CS (b) UNIVERSAL BIASED CS (c) TWO POWER SUPPLY CS

FIGURE 12-2

TRANSCONDUCTANCE OF FIELD EFFECT TRANSISTORS

Once the transconductance of JFETs is known, it can be used to solve for the voltage gain of common-source circuits, the output impedance of less than unity gain common-drain circuits, and the input impedance and voltage gain of common-gate circuits.

Transconductance can be shown in several ways in manufacturer specification sheets: as g_m, y_{f_s}, or g_{f_s}. However, the values given in most specification sheets are nominal values that cannot be used with any degree of accuracy. Therefore, the g_m value must be measured and that can be done through graphical methods or, if the device is square law, which it rarely is, through the mathematical formula:

$$g_m = 2/V_p \sqrt{I_D \times I_{DSS}}$$

USING THE TRANSFER CURVE

In theory, JFETs are considered to be square law and, when this is assumed to be so, then the transconductance of JFETs can be solved both mathematically and graphically. However, in practical analysis, only the more accurate graphical approach is used.

For instance, if the transfer curve developed in the previous chapter is used, where $I_{DSS} = 8$ mA and $V_p = 4$ V, then the V_{GS} points at 1 volt intervals can be selected and the associated I_D currents calculated. In fact, the V_{GS} and associated I_D currents were solved in the previous chapter from:

$$I_D = I_{DSS}(1 - V_{GS}/V_p)^2$$

These values are shown in the Table given in association with Figure 12-3.

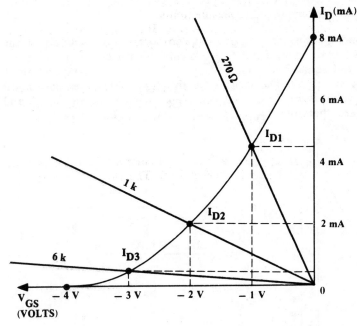

Given: $I_{DSS} = 8$ mA and
$V_p = V_{GS}(\text{off}) = |4\ V|$

$I_D = 8$ mA	$V_{GS} = 0$ V
$I_D = 4.5$ mA	$V_{GS} = -1$ V
$I_D = 2$ mA	$V_{GS} = -2$ V
$I_D = 0.5$ mA	$V_{GS} = -3$ V
$I_D = 0$ mA	$V_{GS} = -4$ V

FIGURE 12-3

DERIVING THE TRANSCONDUCTANCE

Graphically, the transconductace is found from $g_m = \Delta I_D / \Delta V_{GS}$, where ΔV_{GS} is a change in voltage about the operating point and ΔI_D is the corresponding change in current. Mathematically, g_m is solved from:

$$g_m = (2/V_p)\sqrt{I_D \times I_{DSS}}$$

which is derived from the $I_D = I_{DSS}(1 - (V_{GS}/V_p)^2$ using a calculus technique called partial derivatives.

The highest transconductance (g_{m0}) occurs when $I_D = I_{DSS}$, and it is the transconductance value given in data sheets. Since $I_D = I_{DSS}$ occurs when $V_{GS} = 0$ V, the transconductance is given with the subscript 0 or g_{m0}. Knowing g_{m0} and $V_p[V_{GS}(\text{off})]$, and assuming the device follows square law principles, the g_m can be solved for any V_{GS} condition from:

$$g_m = g_{m0}(1 - V_{GS}/V_p)$$

SOLVING FOR TRANSCONDUCTANCE

Continuing to use the $I_{DSS} = 8$ mA and $V_p = 4$ V values, the g_m can be solved at each of the DC operating points indicated in Figure 12-3 — beginning with $V_{GS} = 0$ V and ending with $V_{GS}(\text{off}) = -4$ V.

A. $g_{m0} = g_m = 4000\ \mu\text{mho} = 4$ mS

The highest transconductance for a JFET device with parameters of $I_{DSS} = 8$ mA and $V_p = 4$ V occurs when $V_{GS} = 0$ V. At $V_{GS} = 0$ V and $I_D = I_{DSS} = 8$ mA, g_m is solved from:

$$g_m = (2/V_p)\sqrt{I_D I_{DSS}} = 2/V_p \times I_{DSS} = 2/4V \times 8\ \text{mA} = 4000\ \mu\text{mho} = 4\ \text{mS}$$

B. $g_m = 3000\ \mu\text{mho} = 3$ mS

Mathematically, $g_m = 3000\ \mu\text{mho}$ when V_{GS} is set at -1 V and $I_D = 4.5$ mA from:

$$g_m = 2/V_p \sqrt{I_D I_{DSS}} = 2/4 \text{ V} \sqrt{4.5 \text{ mA} \times 8 \text{ mA}} = 2/4 \text{ V} \sqrt{36 \times 10^{-6} \text{ A}}$$
$$= 2/4 \text{ V} \times 6 \times 10^{-3} \text{ A} = 3 \times 10^{-3} \text{ mho} = 3000 \text{ } \mu\text{mho} = 3 \text{ mS}$$

Graphically, the change in voltage ΔV and the corresponding change in current ΔI occur about the operating point of $V_{GS} = -1$ V and $I_D = 4.5$ mA. Then, if the ΔV_{GS} is conveniently chosen to be ± 1 V about the operation point of -1 V, ΔV_{GS} will equal 2 V (from 0 V and -2 V). The corresponding ΔI will be 6 mA (from 8 mA at 0 V and 2 mA at -2 V). The subscript 0 through 4 references the known V_{GS} voltage. For instance, the extreme V_{GS} voltages are $V_{GS} = 0$ V and $V_{GS} = -2$ V, hence, V_{GS0} and V_{GS2}. The corresponding current changes are, therefore, I_{D0} and I_{D2}, as shown in Figure 12-4.

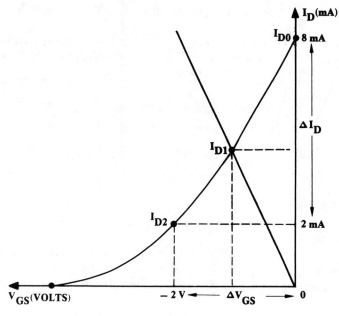

FIGURE 12-4

C. $g_m = 2000 \text{ } \mu\text{mho} = 2 \text{ mS}$

Mathematically, $g_m = 2000 \text{ } \mu\text{mho}$ when V_{GS} is set at -2 V and $I_D = 2$ mA from:

$$g_m = 2/V_p \sqrt{I_D I_{DSS}} = 2/4 \text{ V} \sqrt{2 \text{ mA} \times 8 \text{ mA}} = 2/4 \text{ V} \sqrt{16 \times 10^{-6} \text{ A}}$$
$$= 2/4 \text{ V} \times 4 \times 10^{-3} \text{ A} = 2 \times 10^{-3} \text{ mho} = 2000 \text{ } \mu\text{mho} = 2 \text{ mS}$$

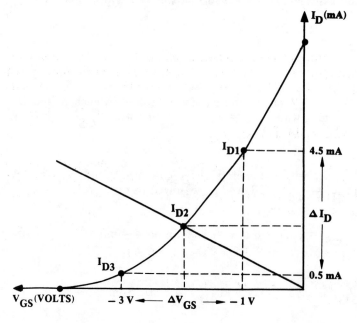

FIGURE 12-5

Graphically, the change in voltage of ±1 V occurs between −1 V and −3 V and the corresponding currents change between 4.5 mA and 0.5 mA. Hence, the operating voltage is V_{GS2} and the change in voltage between V_{GS1} and V_{GS3} produces associated I_{D1} and I_{D3} current changes, as shown in Figure 12-5.

D. $g_m = 1000\ \mu\text{mho} = 1\ \text{mS}$

Mathematically $g_m = 1000\ \mu\text{mho}$, when V_{GS} is set at −3 V and $I_D = 0.5$ mA, from:

$$g_m = 2/V_p \sqrt{I_D I_{DSS}} = 2/4\ \text{V} \sqrt{0.5\ \text{mA} \times 8\ \text{mA}} = 2/4\ \text{V} \sqrt{4 \times 10^{-6}\ \text{A}}$$

$$= 2/4\ \text{V} \times 2 \times 10^{-3}\ \text{A} = 1 \times 10^{-3}\ \text{mho} = 1000\ \mu\text{mho} = 1\ \text{mS}$$

Graphically, the change in voltage of ±1 V occurs between −2 V and −4 V and the corresponding current changes between 2 mA and 0 mA. Hence, the operating voltage is V_{GS3}, and the change in voltage between V_{GS2} and V_{GS4} produces associated I_{D2} and I_{D4} current changes.

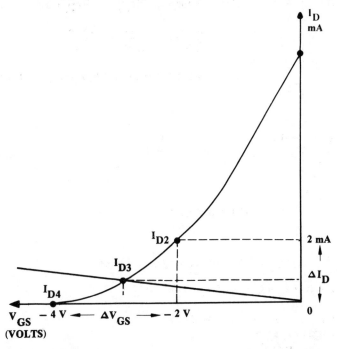

$$g_m = \frac{\Delta I_D}{\Delta V_{GS}} = \frac{I_{D2} - I_{D4}}{V_{GS2} - V_{GS4}}$$

$$= \frac{2\ \text{mA} - 0\ \text{mA}}{-2\ \text{V} - (-4\ \text{V})} = \frac{2\ \text{mA}}{2\ \text{V}}$$

$$= 1 \times 10^{-3}\ \text{mho} = 1000\ \mu\text{mho} = 1\ \text{mS}$$

FIGURE 12-6

E. $g_m = 0\ \mu\text{mho}$

When $V_{GS} = -4$ V and $I_D = 0$ mA, the transconductance equals 0 μmhos, because the channel is pinched off. Hence 0 μmho and 4000 μmho represent the extreme transconductance conditions of the JFET and, for the JFET to operate as an amplifier, it must be biased somewhere between these two extreme transconductance conditions.

USING g_{m0} TO SOLVE FOR THE THEORETICAL g_m

Data sheets normally provide g_{m0}, the parameter when $I_D = I_{DSS}$ and $V_{GS} = 0$ V. However, the theoretical transconductance is easily solved from $g_m = g_{m0}(1 - V_{GS}/V_p)$ when $V_p = V_{GS}(\text{off})$ is also known. Therefore, using the previous parameters of $g_m = 4000\ \mu\text{mho}$ and $V_p = V_{GS}(\text{off}) = -4$ V, g_m can be solved for V_{GS} conditions of −4 V through 0 V from:

$$g_m = g_{m0}(1 - V_{GS}/V_p) = 4000\ \mu\text{mho}(1 - 4\ \text{V}/4\ \text{V}) = 0\ \mu\text{mho} = 0\ \text{mS}$$

$$g_m = g_{m0}(1 - V_{GS}/V_p) = 4000\ \mu\text{mho}(1 - 3\ \text{V}/4\ \text{V}) = 1000\ \mu\text{mho} = 1\ \text{mS}$$

$$g_m = g_{m0}(1 - V_{GS}/V_p) = 4000 \text{ } \mu\text{mho}(1 - 2\text{ V}/4\text{ V}) = 2000 \text{ } \mu\text{mho} = 2 \text{ mS}$$

$$g_m = g_{m0}(1 - V_{GS}/V_p) = 4000 \text{ } \mu\text{mho}(1 - 1\text{ V}/4\text{ V}) = 3000 \text{ } \mu\text{mho} = 3 \text{ mS}$$

$$g_m = g_{m0}(1 - V_{GS}/V_p) = 4000 \text{ } \mu\text{mho}(1 - 0\text{ V}/4\text{ V}) = 4000 \text{ } \mu\text{mho} = 4 \text{ mS}$$

CONVERTING g_m TO r_s'

Mathematically $r_s' = 1/g_m$, which can be solved directly from $r_s' = V_p/2\sqrt{I_D I_{DSS}}$. Graphically, r_s' can be solved from $r_s' = \Delta V_{GS}/\Delta I_D$. The r_s' of JFETs is low, nominally around 1000 Ω, and it is the impedance seen looking into the source terminal of the device, with respect to ground. The drain-to-source resistance of the device is in parallel with the source resistance r_s' but, because it is assumed large, it is not included in the calculations. Also, as can be seen in Figure 10-7(b) of Chapter 10, the channel resistance near the source is at its lowest because in this area the depletion region is non-existent, while the channel resistance near the drain is at its highest.

The r_s' of JFETs and the r_e of bipolar devices are similar, where the r_e of bipolars is solved from $r_e = 26 \text{ mV}/I_E$ and the r_s' of JFETs is solved from $1/g_m$. A relative comparison of the r_s' of JFETs and the r_e of bipolars is shown in Figure 12-7.

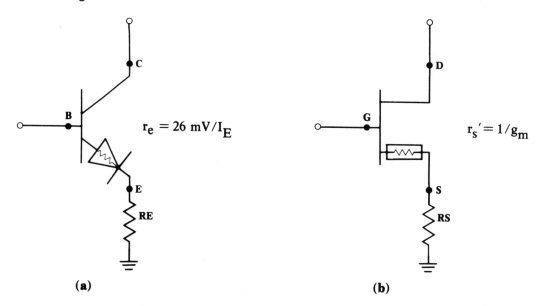

FIGURE 12-7

The r_s' for a JFET with a transconductance of 1000 μmho is 1000 ohms, for a JFET with a g_m of 2000 μmhos it is 500 ohms, and for a JFET with a g_m of 3000 μmho the r_s' is 333 ohms. These three conditions are solved from:

$$r_s' = 1/g_m = 1/1000 \text{ } \mu\text{mho} = 1/10^{-3} = 1000 \text{ ohms}$$

$$r_s' = 1/g_m = 1/2000 \text{ } \mu\text{mho} = 1/2 \times 10^{-3} = 1000 \text{ } \Omega/2 = 500 \text{ ohms}$$

$$r_s' = 1/g_m = 1/3000 \text{ } \mu\text{mho} = 1/3 \times 10^{-3} = 1000 \text{ } \Omega/3 = 333 \text{ ohms}$$

OPERATING POINT CONSIDERATIONS

Voltage gain, transconductance, drift because of temperature change, and undistorted maximum peak-to-peak output voltage swing of any JFET circuit are primarily controlled by the operating point of the transfer curve. For instance, the highest transconductance occurs when $I_D = I_{DSS}$. Therefore, the as stated, setting V_{GS} to 0.5 of V_p will provide an I_D that is, theoretically, 0.25 of I_{DSS}.

The lowest drift because of temperature change occurs when the drain current of the JFET is low with regard to I_{DSS}. However, if the drain current is too low, the I_D is in the non-linear region of the transfer curve and this will cause unequal peak-to-peak voltage swing distortion. In fact, the minimum peak-to-peak signal distortion, because of the device characteristics, occurs when the I_D is greater than 0.5 of I_{DSS}, as shown in Figure 12-8(a).

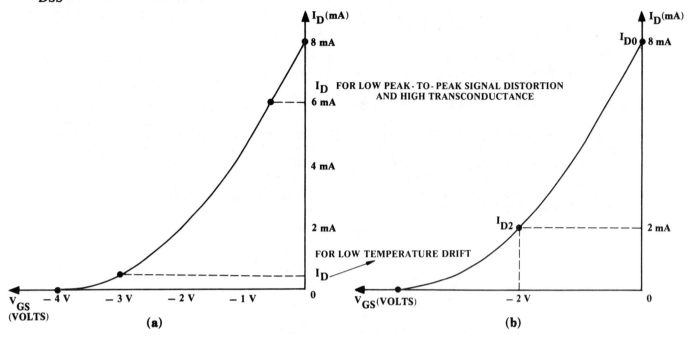

FIGURE 12-8

With reference to Figure 12-8(b), a good starting point in the design of any single stage JFET circuit is to set the V_{GS} to one-half of the $V_{GS}(\text{off}) = V_p$, which causes the theoretical I_D to be one-fourth of I_{DSS}. For instance, if $V_{GS} = -2$ V for a V_p of 4 V, the $I_D = 2$ mA, when $I_{DSS} = 8$ mA, as illustrated. Therefore, as stated, setting V_{GS} to 0.5 of V_p will provide an I_D to be, theoretically, 0.25 of I_{DSS}.

LOAD LINE CONSIDERATIONS

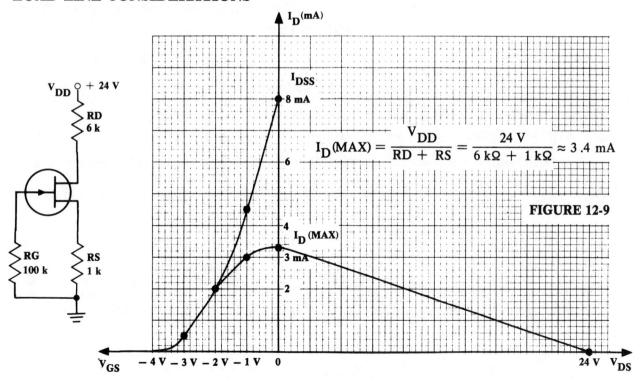

$$I_D(\text{MAX}) = \frac{V_{DD}}{RD + RS} = \frac{24 \text{ V}}{6 \text{ k}\Omega + 1 \text{ k}\Omega} \approx 3.4 \text{ mA}$$

FIGURE 12-9

The one condition that will cause graphical analysis to be inaccurate is when the $I_D(SAT)$ of the circuit is much lower than the I_{DSS} of the device. For example, if $I_{DSS} = 8$ mA and the $I_D(SAT) = 3$ mA, then the I_D has to be lower than 3 mA. The V_{GS} at -1 V, in Figure 12-9, shows the problem exactly. The operating I_D current is too close to the $I_D(SAT)$ condition of 3 mA. Therefore, the transconductance is effected and there is the possibility of distortion.

NOTE: The ideal transfer curve and the effect on the curve by a low $I_D(SAT)$, as illustrated in Figure 12-9, can be duplicated on a curve tracer simply by dialing in a relatively high drain resistance.

VOLTAGE GAIN OF JFETs USING g_m AND r_s'

As shown, g_m can be solved graphically, by using the transfer curve, or theoretically, by using the mathematical approach. For instance, when the device parameters are $I_{DSS} = 8$ mA and $V_p = 4$ V, biasing the device at a V_{GS} of -2 V will produce an I_D of 2 mA and a g_m of 2000 µmho. Then, by using standard FET formulas, the gain of the bypassed source resistor circuit is calculated from $A_v = g_m RD$. However, the voltage gain can also be solved from $A_v = RD/r_s'$, where $r_s' = 1/g_m$. Both methods of solution are given in conjunction with the circuit of Figure 12-10.

A. COMMON SOURCE, BYPASSED SOURCE RESISTOR CIRCUIT

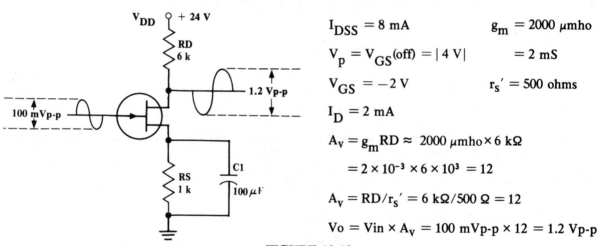

$I_{DSS} = 8$ mA $g_m = 2000$ µmho

$V_p = V_{GS}(off) = |4\text{ V}|$ $= 2$ mS

$V_{GS} = -2$ V $r_s' = 500$ ohms

$I_D = 2$ mA

$A_v = g_m RD \approx 2000$ µmho $\times 6$ kΩ

$\quad = 2 \times 10^{-3} \times 6 \times 10^3 = 12$

$A_v = RD/r_s' = 6$ kΩ/500 $\Omega = 12$

$V_o = V_{in} \times A_v = 100$ mVp-p $\times 12 = 1.2$ Vp-p

FIGURE 12-10

NOTE: Mid-frequency conditions of 1 kHz are used and capacitor C1 equals 1.59 Ω from:

$$X_C = 1/2\pi fC = 0.159/(1 \text{ kHz} \times 100 \text{ µF}) = 0.159/(10^{-3} \times 10^{-4}) = 1.59 \text{ }\Omega$$

Also, a comparison and derivation of the formulas follow in the next section of this chapter.

B. COMMON SOURCE, UNBYPASSED SOURCE RESISTOR CIRCUIT

For the unbypassed source resistor circuit, the voltage gain, using the standard FET formula, is solved from $A_v = g_m RD/[(g_m RS) + 1]$. It can also be solved from $A_v = RD/(r_s' + RS)$. Therefore, once RS is unbypassed, the RS resistor is, effectively, in series with r_s'. Both methods of solution are given in conjunction with the circuit of Figure 12-11.

JFET VOLTAGE GAIN FORMULA DERIVATIONS

The voltage gain formulas used for JFETS can be similarly applied to bipolar devices, because the formulas are interchangeable. For single stage JFET or bipolar transistor circuits, using r_e or r_s' to solve for the voltage gain is the most direct method, and the easiest method to understand, because the formula is reduced to resistance ratios. For instance, $A_v = g_m RD = RD/r_s'$ is fairly straight-forward, for the fully bypassed source resistor circuit, because it uses a simple $g_m = 1/r_s'$ substitution since:

$$A_v = g_m RD = 1/r_s' \times RD = RD/r_s'$$

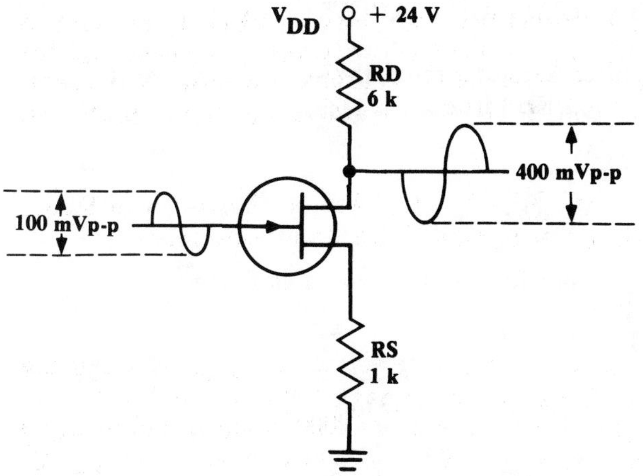

$$A_v = \frac{g_m RD}{1 + (g_m RS)} = \frac{2000\ \mu\text{mho} \times 6\ \text{k}\Omega}{1 + (2000\ \mu\text{mho} \times 1\ \text{k}\Omega)}$$

$$\frac{2 \times 10^{-3} \times 6 \times 10^3}{1 + (2 \times 10^{-3} \times 10^3)} = 12/3 = 4$$

$$A_v = \frac{RD}{r_s' + RS} = \frac{6\ \text{k}\Omega}{500\ \Omega + 1\ \text{k}\Omega} = \frac{6\ \text{k}\Omega}{1500\ \Omega} = 4$$

$$Vo = Vin A_v = 100\ \text{mVp-p} \times 4 = 400\ \text{mVp-p}$$

FIGURE 12-11

For the unbypassed source resistor circuit, the voltage gain FET formula is more complex, but by using similar substitution methods it can be reduced to "bipolar form" where:

$$A_v = \frac{g_m RD}{(g_m RS) + 1} = \frac{RD/r_s'}{RS/r_s' + 1} = \frac{RD}{RS + r_s'}$$

Essentially, the unbypassed source resistor formula is solved where g_m is substituted by $1/r_s'$, which provides RD/r_s' and RS/r_s'. Next, both the numerator and denominator are multiplied by r_s', leaving RD in the numerator and $RS + r_s'$ in the denominator, as shown in the previous equation.

A_v OF BIPOLAR TRANSISTORS USING BOTH r_e AND g_m — A COMPARISON

Since JFETs can be solved using r_s' or g_m, then bipolar transistors can be solved, similarly, using r_e or g_m, because the formulas are interchangeable. Both methods of solution are shown in the following examples.

A. COMMON-EMITTER, BYPASSED EMITTER RESISTOR CIRCUIT

For the bypassed resistor, common-emitter circuit, the voltage gain can be solved from either $A_v = RC/r_e$ or $A_v = g_m RC$ because:

$$A_v = RC/r_e = RC/(1/g_m) = g_m RC \quad \text{where:} \quad r_e = 1/g_m$$

For the circuit of Figure 12-12, using both formulas, the voltage gain is solved as follows:

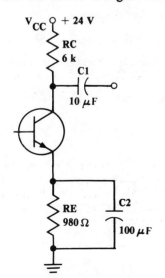

$V_{CC} = 24\ \text{V} \qquad I_C = 1.3\ \text{mA}$

$RC = 6\ \text{k}\Omega \qquad RE = 980\ \Omega$

$r_e \approx 26\ \text{mV}/I_E = 26\ \text{mV}/1.3\ \text{mA} = 20\ \Omega$

$g_m = 1/r_e = 1/20\ \Omega = 0.05\ \text{mho} = 50\ \text{mmho} = 50{,}000\ \mu\text{mho} = 50\ \text{mS}$

1. $A_v = RC/r_e = 6\ \text{k}\Omega/20\ \Omega = 300$

2. $A_v = g_m RC = 50\ \text{mmho} \times 6\ \text{k}\Omega = 300$

FIGURE 12-12

B. COMMON-EMITTER, UNBYPASSED EMITTER RESISTOR CIRCUIT

For the unbypassed common-emitter, unbypassed emitter resistor circuit, the voltage gain can be solved from either $A_v = RC/(r_e + RE)$ or $A_v = g_m RC/[1 + (g_m RE)]$ from:

$$A_v = \frac{RC}{r_e + RE} = \frac{RC/r_e}{r_e/r_e + RE/r_e} = \frac{g_m RC}{1 + (g_m RE)}$$

For the circuit of Figure 12-13, the voltage gain, using both formulas, is solved as follows:

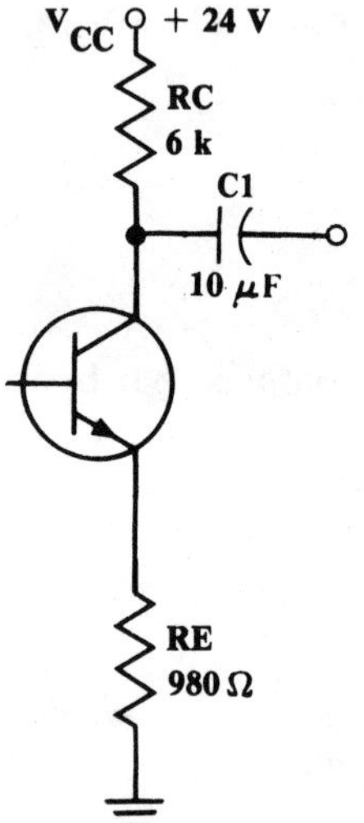

$RC = 6\ k\Omega \qquad RE = 980\ \Omega$

$r_e = 20\ \Omega \qquad g_m = 50\ mmho = 50\ mS$

1. $A_v = \dfrac{RC}{r_e + RE} = \dfrac{6\ k\Omega}{20\ \Omega + 980\ \Omega} = \dfrac{6\ k\Omega}{1\ k\Omega} = 6$

2. $A_v = \dfrac{g_m \times RC}{1 + (g_m \times RE)} = \dfrac{50\ mmho \times 6\ k\Omega}{1 + (50\ mmho \times 980\ \Omega)}$

$\dfrac{50 \times 10^{-3} \times 6 \times 10^3}{1 + (49 \times 10^{-3} \times 10^3)} = \dfrac{300}{1 + 49} \approx \dfrac{300}{50} = 6$

FIGURE 12-13

SECTION II: THE SELF-BIASED JFET AMPLIFIER
DIRECT CURRENT CIRCUIT ANALYSIS

The DC voltages of the self-biased, JFET amplifier circuit of Figure 12-14 must be known before the AC parameters can be solved. Therefore, using the information given for the circuit, the DC analysis calculations and voltage drops for the circuit are as follows:

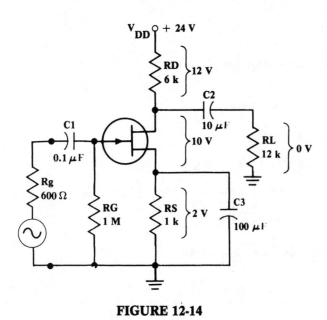

FIGURE 12-14

$V_p = V_{GS}(\text{off}) = |4\ V| \qquad I_{DSS} = 8\ mA$

$V_{GS} = -2\ V$

$I_D = I_{DSS}(1 - V_{GS}/V_p)^2$

$\quad = 8\ mA(1 - 2\ V/4\ V)^2 = 8\ mA \times 0.5^2 = 2\ mA$

$R_S = V_{GS}/I_D = 2\ V/2\ mA = 1\ k\Omega$

$V_{RS} = I_D R_S = 2\ mA \times 1\ k\Omega = 2\ V$

$V_{RD} = I_D R_D = 2\ mA \times 6\ k\Omega = 12\ V$

$V_{RG} = V_G = I_G R_G = 0\ V$

$V_D = V_{DD} - V_{RD} = 24\ V - 12\ V = 12\ V$

$V_{DS} = V_D - V_S = 12\ V - 2\ V = 10\ V$

$V_{GS} = V_G - V_S = 0\ V - 2\ V = -2\ V$

BYPASSED SOURCE RESISTOR — ALTERNATING CURRENT ANALYSIS

Once the DC voltages are known, the AC circuit parameters of voltage gain, input impedance, output impedance, power gain, and maximum peak-to-peak output voltage swing are determined. The I_D current, the device parameters, and the solved AC circuit values to be used in the AC circuit analysis are given in conjunction with the circuit of Figure 12-15.

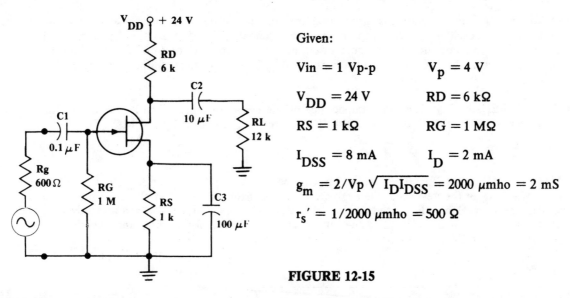

Given:

Vin = 1 Vp-p	V_p = 4 V
V_{DD} = 24 V	RD = 6 kΩ
RS = 1 kΩ	RG = 1 MΩ
I_{DSS} = 8 mA	I_D = 2 mA
$g_m = 2/V_p \sqrt{I_D I_{DSS}}$ = 2000 μmho = 2 mS	
r_s' = 1/2000 μmho = 500 Ω	

FIGURE 12-15

VOLTAGE GAIN: The voltage gain of the self-biased JFET amplifier circuit with a fully bypassed source resistor can be solved by multiplying the transconductance of the device by the effective load resistance RD ∥ RL, or by taking the ratio of the effective load resistance RD ∥ RL to the effective source resistance r_s' as follows:

$$A_v = g_m(RD \parallel RL) = 2000 \, \mu mho \, (6 \, k\Omega \parallel 12 \, k\Omega) = 2 \times 10^{-3} \times 4 \times 10^3 = 8$$

$$A_v = (RD \parallel RL)/r_s' = (6 \, k\Omega \parallel 12 \, k\Omega)/500 \, \Omega = 4 \, k\Omega/500 \, \Omega = 8$$

VOLTAGE OUTPUT SWING: The output voltage across the load resistor RL equals the input voltage times the voltage gain of 8. Therefore, if Vin is 1 Vp-p, then Vo = 8 Vp-p, as shown in Figure 12-16. Note that the output signal is 180° out of phase with the input signal for the common source circuit. This is similar to the 180° phase shift capabilities of the bipolar transistor, common-emitter circuit.

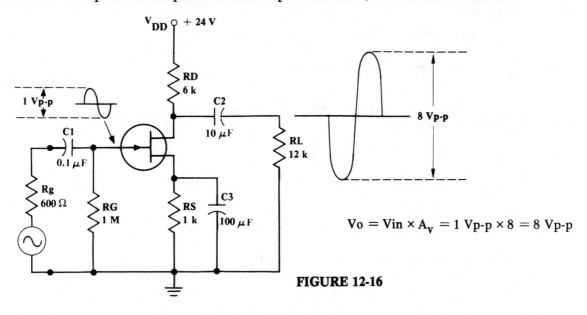

$Vo = Vin \times A_v = 1 \, Vp\text{-}p \times 8 = 8 \, Vp\text{-}p$

FIGURE 12-16

MAXIMUM PEAK-TO-PEAK OUTPUT VOLTAGE SWING: The Vop-p(max) for the JFET amplifier is difficult to predict accurately because of the distortion that can occur because of the pinch off voltage line, the load line's effect on the input voltage swing, and the non-linearity of the device's characteristic curve. Therefore, because of all these limitation, the Vop-p(max) of JFETs can rarely be as large as it can be with bipolar transistor devices. Also, because of these limiting factors, the Vop-p(max) formula for single stage JFET circuit will not be developed.

NOTE: Vop-p(max) is most easily attained through measurement, where excessive distortion can be monitored on an oscilloscope and, at the same time, detected by using a DC voltmeter. The DC voltmeter technique is to monitor the DC voltage at the drain, with respect to ground, and continue to increase the input signal until the DC voltage begins to shift. The DC voltage shift indicates a distorted Vop-p condition.

INPUT IMPEDANCE: The input impedance of the common-source circuit is equal to the RG resistor alone, because the input impedance of the JFET is very high, nominally 100 MΩ. Therefore:

$$Zin \approx RG = 1\ M\Omega$$

OUTPUT IMPEDANCE: The output impedance of the JFET is comparable to bipolar transistors. Again, the R_{DS} of the device can be evaluated directly from the slope of the characteristic curves at the operating current. Hence, R_{DS} is evaluated from $R_{DS} = \Delta V_{DS}/\Delta I_D$, as shown and solved in conjunction with Figure 12-17. In this example, one output curve is shown, for clarity, and the slope is taken from nominal values of ΔV_{DS} and ΔI_D.

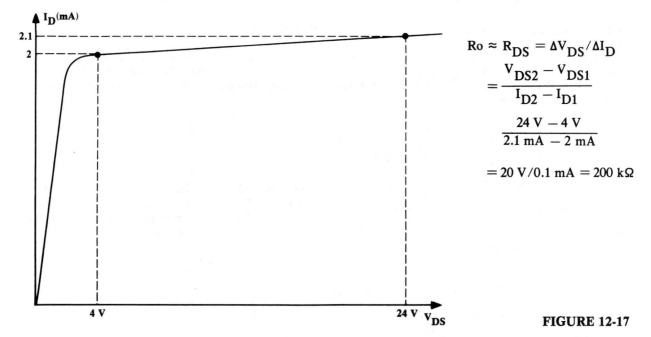

$$Ro \approx R_{DS} = \Delta V_{DS}/\Delta I_D$$
$$= \frac{V_{DS2} - V_{DS1}}{I_{D2} - I_{D1}}$$
$$\frac{24\ V - 4\ V}{2.1\ mA - 2\ mA}$$
$$= 20\ V/0.1\ mA = 200\ k\Omega$$

FIGURE 12-17

Therefore, if Ro = 200 kΩ and RD = 6 kΩ, then the output impedance of the circuit is approximately RD, alone, which equals 6 kΩ from:

$$Zo = Ro\ //\ RD \approx 200\ k\Omega\ //\ 6\ k\Omega = 5825\ \Omega \approx 6\ k\Omega$$

POWER GAIN TO THE LOAD RESISTOR: The power gain of JFETs is solved similarly to bipolars. However, JFETs have low voltage gains and high input impedances.

$$PG = A_v^2 \times Zin/RL = 8^2 \times 1\ M\Omega/12\ k\Omega = 5333$$

$$PG(dB) = 10 \log A_p = 10 \log 5333 = 37.27\ dB$$

EQUIVALENT CIRCUIT OF THE CS BYPASSED SOURCE RESISTOR CIRCUIT

The equivalent circuits of JFETs are solved much like those of bipolar devices, so only the common source, bypassed source resistor circuit will be analyzed to show the similarity. The previously analyzed circuit is summarized to allow an easy comparison of the two methods of analysis.

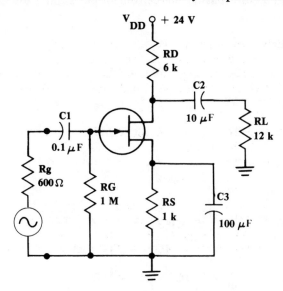

$A_v = (RD \mathbin{/\mkern-5mu/} RL)/r_s' = (6\ k\Omega \mathbin{/\mkern-5mu/} 12\ k\Omega)/500\ \Omega$

$= 4\ k\Omega/500\ \Omega = 8$

$Vop\text{-}p = Vin A_v = 1\ Vp\text{-}p \times 8 = 8\ Vp\text{-}p$

$Zin \approx RG = 1\ M\Omega$

$PG = A_v^2 \times Zin/RL = 8^2 \times 1\ M\Omega/12\ k\Omega = 5333$

$PG(dB) = 10 \log 5333 = 37.27\ dB$

FIGURE 12-18

The equivalent circuit is shown in Figure 12-19 and, like bipolars, developed signal currents are used to provide the method of analysis.

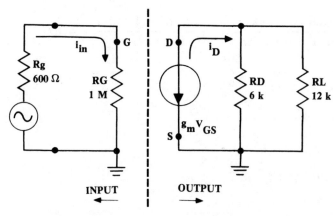

$Vin = 1\ Vp\text{-}p$

$r_L = RD \mathbin{/\mkern-5mu/} RL = 6\ k\Omega \mathbin{/\mkern-5mu/} 12\ k\Omega = 4\ k\Omega$

$g_m = 2000\ \mu mho = 2\ mS$

FIGURE 12-19

$i_{in} = Vin/Zin = 1\ Vp\text{-}p/1\ M\Omega = 1\ \mu Ap\text{-}p$

$A_v = g_m r_L = 2\ mS \times 4\ k\Omega = 8$

$Vo = A_v Vin = 8 \times 1\ Vp\text{-}p = 8\ Vp\text{-}p$

$i_D = Vop\text{-}p/r_L = 8\ Vp\text{-}p/4\ k\Omega = 2\ mAp\text{-}p$, or

$i_D = g_m v_{GS} = gm Vin = Vin/r_s' = 1\ Vp\text{-}p/500\ \Omega = 2\ mAp\text{-}p$

$i_o = \dfrac{i_D \times RD}{RD + RL} = \dfrac{2\ mAp\text{-}p \times 6\ k\Omega}{6\ k\Omega + 12\ k\Omega} = 666\ \mu Ap\text{-}p$

$A_i = i_o/i_{in} = 666\ \mu Ap\text{-}p/1\ \mu Ap\text{-}p = 666$

$PG = A_v \times A_i = 8 \times 666 = 5333$ $\qquad PG(dB) = 10 \log 5333 = 37.27\ dB$

UNBYPASSED SOURCE RESISTOR — ALTERNATING CURRENT ANALYSIS

Bypassing the source resistor with a 100 μF capacitor essentially places an AC ground at the source terminal at the mid-band frequency of 1 kHz, but it does not change the DC voltage distribution of the circuit. Therefore, only the AC analysis is required. The unbypassed source resistor circuit and the device parameters are given in Figure 12-20.

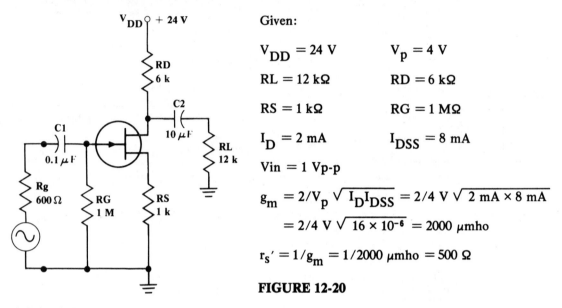

Given:

$V_{DD} = 24$ V $\qquad V_p = 4$ V

$R_L = 12$ kΩ $\qquad R_D = 6$ kΩ

$R_S = 1$ kΩ $\qquad R_G = 1$ MΩ

$I_D = 2$ mA $\qquad I_{DSS} = 8$ mA

$V_{in} = 1$ Vp-p

$g_m = 2/V_p \sqrt{I_D I_{DSS}} = 2/4\text{ V} \sqrt{2\text{ mA} \times 8\text{ mA}}$

$\qquad = 2/4\text{ V} \sqrt{16 \times 10^{-6}} = 2000$ μmho

$r_s' = 1/g_m = 1/2000$ μmho $= 500$ Ω

FIGURE 12-20

VOLTAGE GAIN:

$$A_v = \frac{g_m(R_D /\!/ R_L)}{1 + g_m R_S} = \frac{2000\text{ μmho}(6\text{ kΩ} /\!/ 12\text{ kΩ})}{1 + (2000\text{ μmho} \times 1\text{ kΩ})} = \frac{(2 \times 10^{-3})(4 \times 10^3)}{1 + (2 \times 10^{-3})(1 \times 10^3)} = 8/3 = 2.67$$

$$A_v = \frac{R_D /\!/ R_L}{R_S + r_s'} = \frac{6\text{ kΩ} /\!/ 12\text{ kΩ}}{1\text{ kΩ} + 500\text{ Ω}} = \frac{4\text{ kΩ}}{1500\text{ Ω}} = 2.67$$

VOLTAGE OUT:

$V_{op\text{-}p} = V_{inp\text{-}p} \times A_v = 1$ Vp-p $\times 2.67 = 2.67$ Vp-p

INPUT IMPEDANCE AND OUTPUT IMPEDANCE

$Z_{in} \approx R_G = 1$ MΩ

$Z_o \approx R_D = 6$ kΩ

POWER GAIN TO THE LOAD RESISTOR:

$PG = A_v^2 \times Z_{in}/R_L = 2.67^2 \times 1$ MΩ$/12$ kΩ ≈ 592.6

$PG(dB) = 10 \log A_p = 10 \log 592.6 = 27.73$ dB

FREQUENCY RESPONSE MEASUREMENTS

The JFET, like bipolars, have coupling and bypass capacitors that contribute to the low frequency response, and device and stray capacitance that contribute to the high frequency response. In Figure 12-21, the circuit showing the capacitors is given, along with a representative Bode plot. Again, the corner frequencies, both high and low, are measured at −3 dB or 0.707 of V(max), where V(max) is measured at a mid-frequency of around 1 kHz.

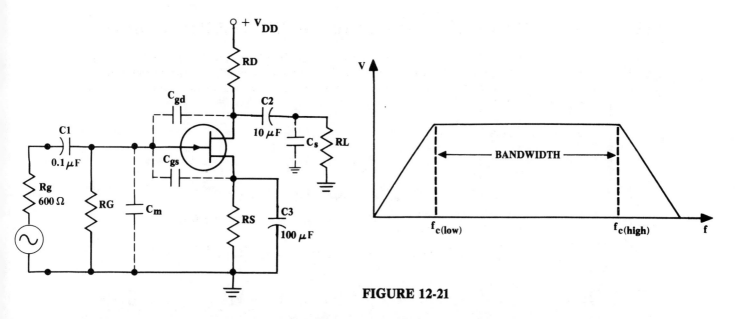

FIGURE 12-21

SECTION III: THE UNIVERSAL JFET AMPLIFIER CIRCUIT
DIRECT CURRENT CIRCUIT ANALYSIS

The connections for the universal JFET amplifier are shown in Figure 12-22. Included with the illustration are the basic equations for the direct current analysis — repeated here for convenience.

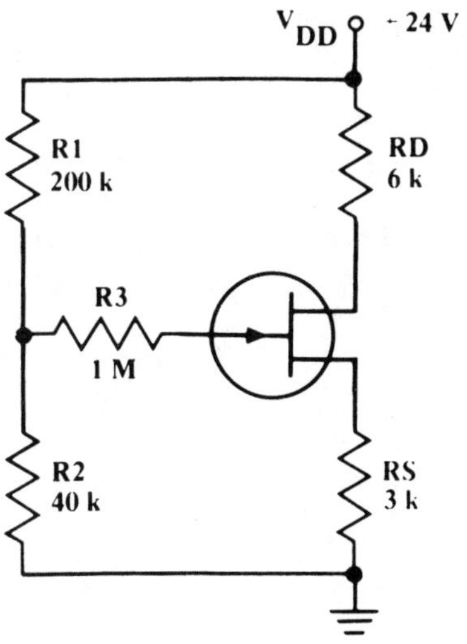

$$V_{R1} = \frac{V_{DD} \times R1}{R1 + R2} = \frac{24\text{ V} \times 200\text{ k}\Omega}{200\text{ k}\Omega + 40\text{ k}\Omega} = 20\text{ V}$$

$$V_{R2} = \frac{V_{DD} \times R2}{R1 + R2} = \frac{24\text{ V} \times 40\text{ k}\Omega}{200\text{ k}\Omega + 40\text{ k}\Omega} = 4\text{ V}$$

$$I_D = I_{DSS}\left(1 - \frac{V_{GS}}{V_p}\right)^2 = 8\text{ mA}\left(1 - \frac{2}{4}\right)^2 = 8 \times 0.5^2 = 2\text{ mA}$$

$$V_{RS} = I_D RS = 2\text{ mA} \times 3\text{ k}\Omega = 6\text{ V}$$

$$V_{RD} = I_D RD = 2\text{ mA} \times 6\text{ k}\Omega = 12\text{ V}$$

$$V_D = V_{DD} - V_{RD} = 24\text{ V} - 12\text{ V} = 12\text{ V}$$

$$V_{DS} = V_D - V_{RS} = 12\text{ V} - 6\text{ V} = 6\text{ V}$$

$$V_{GS} = V_{R2} - V_{RS} = 4\text{ V} - 6\text{ V} = -2\text{ V}$$

$$V_{R3} = I_G R3 = 10\text{ nA} \times 1\text{ M}\Omega = 10^{-8} \times 10^6 = 10^{-2} = 0.01\text{ V}$$

FIGURE 12-22

NOTE: The gate current I_G of 10 nanoamps is estimated to demonstrate the minimal voltage drop to be expected across R3.

Figure 12-23 shows all the DC voltages of the circuit. Remember that all the calculated and measured voltage drops between V_{DD} and ground must equal the applied V_{DD}.

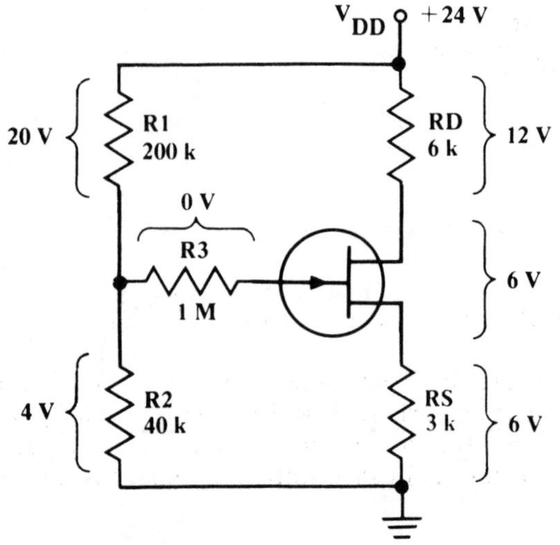

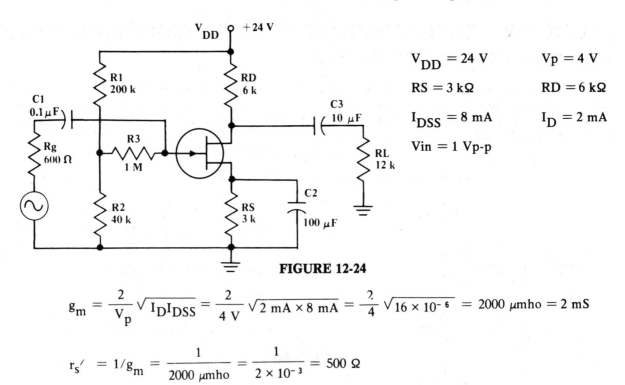

FIGURE 12-23

$$V_{DD} = V_{RD} + V_{DS} + V_{RS}$$
$$= 12\text{ V} + 6\text{ V} + 6\text{ V} = 24\text{ V}$$
$$V_{DD} = V_{R1} + V_{R2} = 20\text{ V} + 4\text{ V} = 24\text{ V}$$
$$V_{GS} = V_{R2} - V_{RS} = 4\text{ V} - 6\text{ V} = -2\text{ V}$$

SECTION IV: THE COMMON SOURCE AMPLIFIER CIRCUIT
ALTERNATING CURRENT CIRCUIT ANALYSIS

Once the DC voltages are known, then the alternating current parameters are found. Voltage gain, input impedance, output impedance, power gain, and peak-to-peak output voltage swing for the input signal used are solved.

The connections for the common source amplifier circuit configuration are shown in Figure 12-24. From the information included with the illustration, the alternating current parameters will be solved.

FIGURE 12-24

$V_{DD} = 24\text{ V}$ $\qquad V_p = 4\text{ V}$
$RS = 3\text{ k}\Omega$ $\qquad RD = 6\text{ k}\Omega$
$I_{DSS} = 8\text{ mA}$ $\qquad I_D = 2\text{ mA}$
$V_{in} = 1\text{ Vp-p}$

$$g_m = \frac{2}{V_p}\sqrt{I_D I_{DSS}} = \frac{2}{4\text{ V}}\sqrt{2\text{ mA} \times 8\text{ mA}} = \frac{2}{4}\sqrt{16 \times 10^{-6}} = 2000\text{ }\mu\text{mho} = 2\text{ mS}$$

$$r_s' = 1/g_m = \frac{1}{2000\text{ }\mu\text{mho}} = \frac{1}{2 \times 10^{-3}} = 500\text{ }\Omega$$

VOLTAGE GAIN:

$$A_v = g_m r_L = g_m(RD \mathbin{/\mkern-5mu/} RL) = 2000\text{ }\mu\text{mho}(6\text{ k}\Omega \mathbin{/\mkern-5mu/} 12\text{ k}\Omega) = 2 \times 10^{-3} \times 4 \times 10^3 = 8$$

$$A_v = \frac{r_L}{r_S'} = \frac{RD \parallel RL}{r_S'} = \frac{6\ k\Omega \parallel 12\ k\Omega}{500\ \Omega} = \frac{4\ k\Omega}{500\ \Omega} = 8$$

VOLTAGE OUT:

$$Vop\text{-}p = Vinp\text{-}p \times A_v = 1\ Vp\text{-}p \times 8 = 8\ Vp\text{-}p$$

INPUT IMPEDANCE:

$$Zin = R3 + (R1 \parallel R2) = 1\ M\Omega + (200\ k\Omega \parallel 40\ k\Omega) = 1\ M\Omega + 33.3\ k\Omega = 1033\ k\Omega$$

NOTE: R3 is used specifically in this universal circuit to increase the input impedance. Without R3, the input impedance would be 16.6 kΩ. Also, since little gate current exists, little DC voltage is dropped across R3. However, increasing input impedance and increasing R3 to too large a value (much greater than 10 MΩ) would cause a larger voltage drop and provide increased circuit instability. For this reason, extremely large values of R3 are not used.

OUTPUT IMPEDANCE:

$$Zo \approx RD = 6\ k\Omega$$

POWER GAIN TO THE LOAD RESISTOR:

$$PG = A_v^2 \times \frac{Zin}{RL} = 8^2 \times \frac{1033\ k\Omega}{12\ k\Omega} = 5509.3$$

$$PG(dB) = 10\ \log 5509.3 = 37.41\ dB$$

SECTION V: TWO-POWER SUPPLY JFET AMPLIFIER CIRCUIT
DIRECT CURRENT CIRCUIT ANALYSIS

The DC analysis for the two power supply circuit is only slightly different from that of the self-biased JFET circuit, and the AC analysis is the same. For convenience, the DC analysis is shown in Figure 12-25. It is solved in terms of drain current using the device characteristics of pinch-off voltage Vp and saturation current I_{DSS}.

Given: $V_{DD} = 24\ V$ $V_p = 4\ V$ $V_{SS} = -6\ V$

$I_{DSS} = 8\ mA$ $V_{GS} = -2\ V$

$$I_D = I_{DSS}\left(1 - \frac{V_{GS}}{V_P}\right)^2 = 12\ mA\left(1 - \frac{2\ V}{4\ V}\right)^2$$

$$= 8\ mA \times 0.5^2 = 8\ mA \times 0.25 = 2\ mA$$

$V_{RS} = I_D RS = 2\ mA \times 4\ k\Omega = 8\ V$

$V_{RD} = I_D RD = 2\ mA \times 6\ k\Omega = 12\ V$

$V_D = V_{DD} - V_{RD} = 24\ V - 12\ V = 12\ V$

$V_S = -V_{SS} + V_{RS} = -6\ V + 8\ V = 2\ V$

$V_G = 0\ V$

$V_{DS} = V_D - V_S = 12\ V - 2\ V = 10\ V$

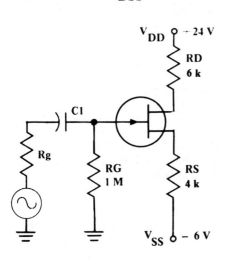

FIGURE 12-25

Figure 12-26 shows the DC distribution of the circuit. It is important to remember that all calculated and measured voltage drops between V_{DD} and $-V_{SS}$ must equal be equal to the applied voltage.

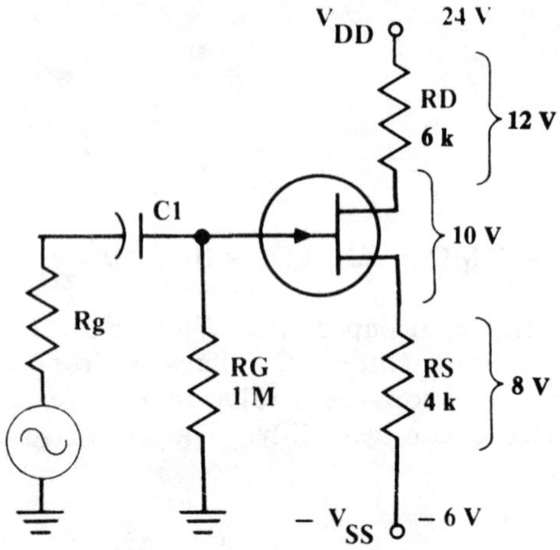

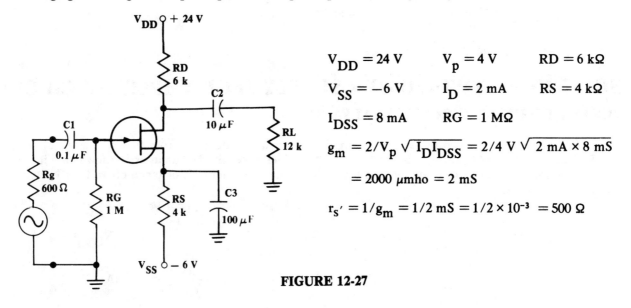

$V_{DD} = V_{RD} + V_{DS} + V_{RS} - V_{SS}$

$24\,V = 12\,V + 10\,V + 8\,V - 6\,V$ or

$V_{DD} + V_{SS} = V_{RD} + V_{DS} + V_{RS}$

$24\,V + 6\,V = 12\,V + 10\,V + 8\,V = 30\,V$

$V_{GS} = V_G - V_S = 0\,V - 2\,V = -2\,V$

FIGURE 12-26

ALTERNATING CURRENT CIRCUIT ANALYSIS

The connections for the two power supply circuit are given in **Figure 12-27**, and the AC parameters of voltage gain, voltage out, input impedance, output impedance, and power gain are solved.

$V_{DD} = 24\,V \qquad V_p = 4\,V \qquad RD = 6\,k\Omega$

$V_{SS} = -6\,V \qquad I_D = 2\,mA \qquad RS = 4\,k\Omega$

$I_{DSS} = 8\,mA \qquad RG = 1\,M\Omega$

$g_m = 2/V_p \sqrt{I_D I_{DSS}} = 2/4\,V \sqrt{2\,mA \times 8\,mS}$

$= 2000\,\mu mho = 2\,mS$

$r_s' = 1/g_m = 1/2\,mS = 1/2 \times 10^{-3} = 500\,\Omega$

FIGURE 12-27

VOLTAGE GAIN:

$A_v = g_m r_L = g_m(RD \parallel RL) = 2000\,\mu mho\,(6\,k\Omega \parallel 12\,k\Omega) = 2 \times 10^{-3} \times 4 \times 10^3 = 8$

$A_v = r_L/r_s' = (RD \parallel RL)/r_s' = (6\,k\Omega \parallel 12\,k\Omega)/500\,\Omega = 4\,k\Omega/500\,\Omega = 8$

VOLTAGE OUT:

$V_o = V_{in} A_v = 1\,V_{p-p} \times 8 = 8\,V_{p-p}$

INPUT IMPEDANCE:

$Z_{in} \approx RG = 1\,M\Omega$

OUTPUT IMPEDANCE:

$$Z_o \approx R_D = 6 \text{ k}\Omega$$

POWER GAIN TO THE LOAD RESISTOR:

$$PG = A_v^2 \times Z_{in}/R_L = 8^2 \times 1 \text{ M}\Omega/12 \text{ k}\Omega = 5333$$

$$PG(dB) = 10 \log A_p = 10 \log 5333 = 37.27 \text{ dB}$$

LABORATORY EXERCISE

OBJECTIVES
To investigate the alternating current characteristics of self-biased, common-source, JFET amplifiers.

LIST OF MATERIALS AND EQUIPMENT
1. Transistor: JFET 2N5951 or equivalent
2. Resistors: 1 kΩ (one)   4.7 kΩ (one)   10 kΩ (one)   100 kΩ (two)
3. Capacitors: 10 μF (two)   100 μF (one)
4. Signal Generator
5. Voltmeter
6. Oscilloscope

PROCEDURE
Section I: Self-Biased, Common-Source Amplifier Circuit with Bypassed Source Resistor
Part 1: Mathematical Analysis
1. Connect the self-biased amplifier circuit of Figure 12-28.

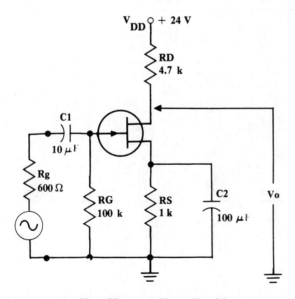

FIGURE 12-28

2. Measure the V_D, V_S, and V_G, all with respect to ground. Insert the measured values, as indicated, into Table 12-1.
3. Use the measured values to calculate V_{RD}, V_{DS}, V_{RG}, $V_{RS} = V_{GS}$, and I_D, from:

$$V_{RD} = V_{DD} - V_D \qquad V_{DS} = V_D - V_S \qquad V_{GS} = V_G - V_S$$

$$V_{RS} = V_S \qquad V_{RG} = V_G = 0 \text{ V} \qquad I_D = V_{RS}/R_S$$

353

4. Insert the calculated values, as indicated, into Table 12-1.

TABLE 12-1	V_D	V_S	V_G	V_{RD}	V_{RS}	V_{DS}	V_{RG}	V_{GS}	I_D
CALCULATED	/////	/////	/////						
MEASURED				/////	/////	/////	/////	/////	/////

5. Measure the voltage out, using an input signal voltage of 100 mVp-p at a frequency of 1 kHz.

6. Calculate the voltage gain from:

   $A_v = V_o/V_{in}$ where: $V_{in} = 100$ mVp-p

7. Calculate the source resistance r_s' from:

   $r_s' = R_D/A_v$

8. Calculate the transconductance from:

   $g_m = 1/r_s'$ or $g_m = A_v/R_D$

9. Insert the measured and calculated values, as indicated, into Table 12-2(a).

Part 2: Graphical Analysis

1. Plot the transfer curve for the device and the inverse slope line for the 470 ohm RS resistor. Use the graph of Figure 12-29 to plot the curve. Reference the V_{GS} on the transfer curve, at the interesection of the inverse slope line, and determine, from the graph, both the V_{GS} and associated I_D currents. Insert the DC values into Table 12-2(b).

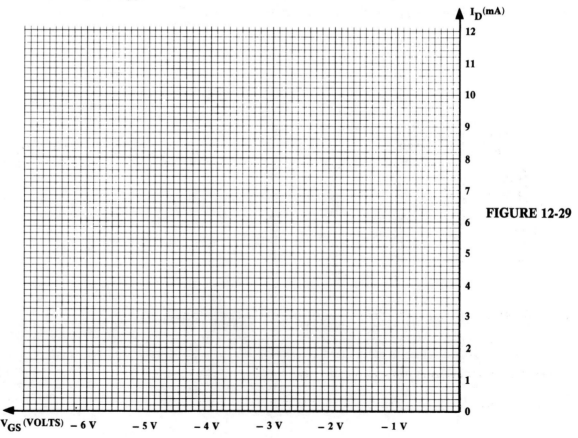

FIGURE 12-29

2. Determine the two points on the transfer curve that are approximately ±0.5 V of the intersected V_{GS} point. Label the V_{GS1} ($V_{GS} + 0.5$ V) and the V_{GS2} ($V_{GS} - 0.5$ V), and determine, from the graph, the associated I_D currents. Calculate:

$$g_m = \frac{\Delta I_D}{\Delta V_{GS}} = \frac{I_{D1} - I_{D2}}{V_{GS1} - V_{GS2}} \approx \frac{I_{D1} - I_{D2}}{1\text{ V}}$$

$$r_s' = \frac{\Delta V_{GS}}{\Delta I_D} = \frac{V_{GS1} - V_{GS2}}{I_{D1} - I_{D2}} \approx \frac{1\text{ V}}{I_{D1} - I_{D2}}$$

3. Insert the calculated g_m and r_s', as indicated, into Table 12-2(b).

TABLE 12-2	(a) Circuit Analysis Techniques				(b) Transfer Curve Techniques					
	V_o	A_v	r_s'	g_m	I_D	V_{GS}	ΔV_{GS}	ΔI_D	g_m	r_s'
CALCULATED	/////				/////					
MEASURED		/////	/////				≈ 1 V	/////	/////	/////

Section II: The Self-Biased, Common-Source Amplifier Circuit with Unbypassed Source Resistor

Mathematical Analysis

1. Connect the circuit shown in Figure 12-30, where the source resistor is not bypassed.

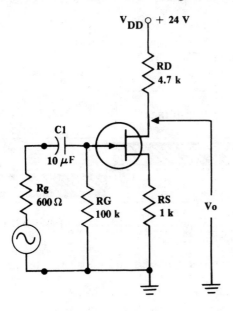

FIGURE 12-30

2. VOLTAGE GAIN: Calculate the voltage gain of the circuit from:

$A_v = RD/(r_s' + RS)$ (Use the r_s' from Table 12-2[a])

Measure the voltage gain using an input signal voltage of 100 mVp-p at 1 kHz.

3. INPUT IMPEDANCE: Calculate the input impedance from: $Z_{in} \approx RG$.

Measure the input impedance using the known series resistor technique, previously shown for bipolars. Use an R1 of 100 kΩ, as shown in Figure 12-31.

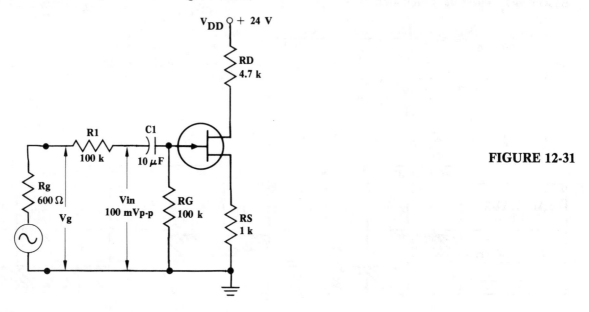

FIGURE 12-31

NOTE: Oscilloscopes with an input impedance of 1 MΩ will have some effect on the measured input impedance of approximately 100 kΩ.

4. MAXIMUM PEAK-TO-PEAK OUTPUT VOLTAGE SWING: Monitor the output voltage and increase the voltage level at 1 kHz until clipping or distortion occurs.

5. POWER GAIN TO THE DRAIN RESISTOR: Calculate the power gain to the drain resistor from:

$$PG = A_v^2 \times Zin/RD$$

6. Insert the calculated and measured values, as indicated, into Table 12-3.

TABLE 12-3	A_v	Zin	Vop-p(max)	POWER GAIN
CALCULATED			/////////	
MEASURED				/////////

Section III: The Loaded Amplifier Circuit

1. Connect the circuit of Figure 12-32, where the source resistor is bypassed and a load resistor has been added.

2. VOLTAGE GAIN: Calculate the voltage gain from:

$$A_v = r_L/r_s' = (RD \mathbin{/\mkern-5mu/} RL)/r_s'$$

Measure the voltage gain using an input signal voltage of 100 mVp-p at 1 kHz.

3. INPUT IMPEDANCE: Calculate the input impedance from: Zin ≈ RG.

Measure the input impedance using the known series resistor technique. Use a series resistor of 100 kΩ, similar to the previous input impedance measurements.

4. **MAXIMUM PEAK-TO-PEAK OUTPUT VOLTAGE SWING:** Monitor the output across the load resistor and increase the signal voltage level at 1 kHz until clipping or distortion occurs. Distortion can also be observed by monitoring the DC voltage at the drain, with respect to ground.

NOTE: JFETs are primarily used in small signal applications. They can tolerate only a few tenths of a volt of positive voltage swing that can be caused by a Vin that is slightly greater than $2V_{GS}$.

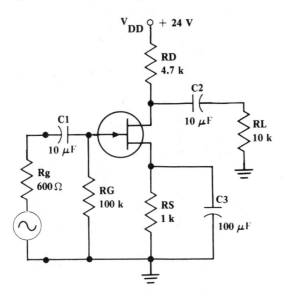

FIGURE 12-32

5. **POWER GAIN TO THE LOAD:** Calculate the power gain to the load resistor from:

$$PG = A_v^2 \times Zin/RL \quad \text{where:} \quad RL = 10\ k\Omega$$

Convert the power gain to dB from: $PG(dB) = 10 \log dB$.

6. Insert the calculated and measured values, as indicated, into Table 12-4.

7. Measure the low and high corner frequencies and provide a representative Bode plot for the circuit of Figure 12-32.

TABLE 12-4	A_v	Zin	Vop-p(max)	POWER GAIN / dB
CALCULATED				
MEASURED				

CHAPTER QUESTIONS

1. Why should the transconductance of JFETs be measured, instead of using manufacturer data sheet values?
2. Compare the r_s' from the tables of 12-2(a) and 12-2(b). If the values are not close, refer to the I_{DSS} and $I_D(SAT)$ conditions and comment on the possible reasons for the discrepancy.
3. Graphical technique is a reasonably accurate method of obtaining r_s' or g_m. However, under what conditions does this method become grossly inaccurate?
4. Name two possible reasons why the Vop-p(max) of JFETs is smaller than that of bipolar devices.
5. Why is the input impedance of self-biased JFET circuits approximately equal to the value of the gate resistor alone?

CHAPTER PROBLEMS

1. Given: $I_{DSS} = 4$ mA, $V_{GS} = -2$ V, and $V_p = 4$ V

   Find:
   a. V_{RG}
   b. V_{RD}
   c. V_{DS}
   d. r_s'
   e. g_m
   f. A_v
   g. PG

2. When switch S1 is closed and the C3 capacitor is across the RS resistor, $A_v = 9$. What is the A_v when the switch is open and the RS resistor is not bypassed?

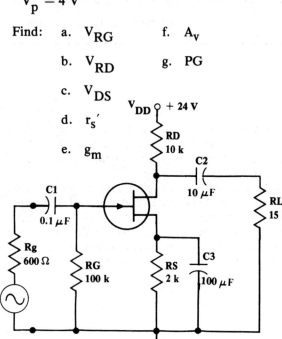

FIGURE 12-33

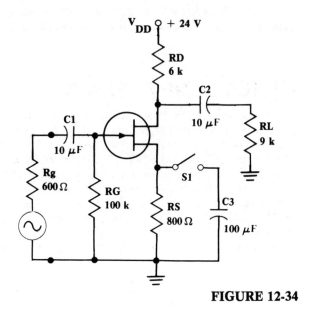

FIGURE 12-34

3. Given: $I_{DSS} = 4$ mA, $V_{GS} = -2$ V, and $V_p = 4$ V

   Find:
   a. V_{R1}
   b. V_{R2}
   c. V_{R3}
   d. V_{R4}
   e. V_{R5}
   f. V_{DS}
   g. A_v
   h. Zin
   i. PG

4. Given: $I_{DSS} = 4$ mA, $V_{GS} = -2$ V, and $V_p = 4$ V

   Find:
   a. V_{R1}
   b. V_{R2}
   c. V_{R3}
   d. V_{DS}
   e. A_v
   f. Zin
   g. PG

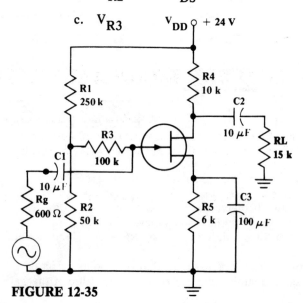

FIGURE 12-35

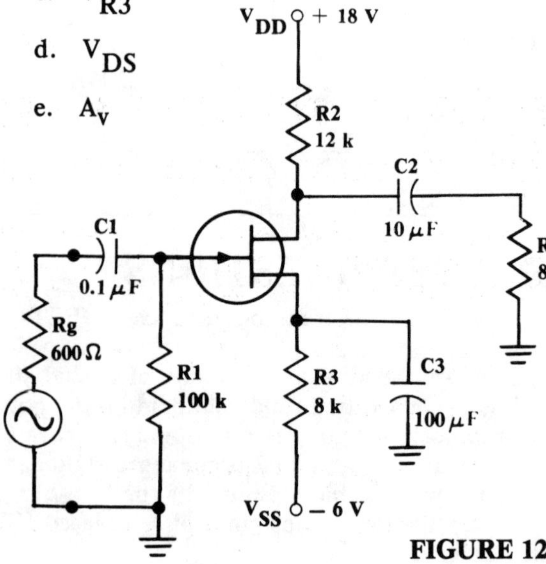

FIGURE 12-36

13 | COMMON DRAIN JFET CIRCUITS

GENERAL DISCUSSION

The common drain circuit is used primarily for the impedance transfer of very high input impedances to moderately low output impedances. The voltage gain of the common drain circuit, like that of the bipolar common collector circuit, is less than unity, but the advantage of the JFET over bipolars is that the value of the gate resistor, alone, usually determines the input impedance of the stage.

In other words, the gate resistor can be as high as 1 MΩ and there need be no concern about the leakage current of the device causing a gate-to-source DC offset, or about the input impedance being effected by the input impedance of the device alone, which is normally greater than 10 MΩ. Another advantage to JFETs is that the source resistance is not reflected into the gate like bipolars, which reflect by beta the emitter resistance into the base. Hence, the value of the source resistance can vary and not effect the input impedance. The self-biased, universal, and two power supply biased circuits are shown in Figure 13-1.

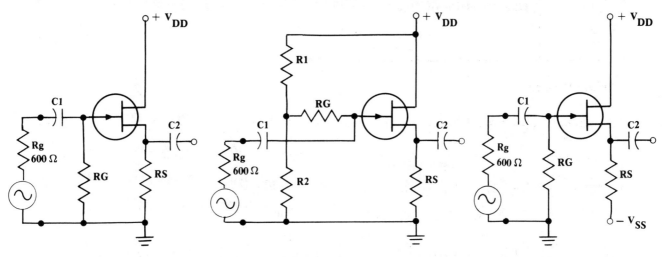

FIGURE 13-1

SECTION I: THE SELF-BIASED COMMON DRAIN AMPLIFIER

DIRECT CURRENT CIRCUIT ANALYSIS

The DC voltage of the self-biased common drain amplifier circuit, in theory, can be mathematically derived if I_{DSS} and $V_p = V_{GS}(off)$ are known. The DC voltage drops of the circuit are given in conjunction with Figure 13-2, where an I_{DSS} of 8 mA and a V_p or 4 volts are used.

$$I_D = I_{DSS}(1 - V_{GS}/V_p)^2 = 8 \text{ mA}(1 - 2\text{ V}/4\text{ V})^2 = 8 \text{ mA} \times 0.5 \text{ V}^2 = 8 \text{ mA} \times 0.25 \text{ V} = 2 \text{ mA}$$

$$R_S = V_{GS}/I_D = |2\text{ V}|/2\text{ mA} = 1 \text{ k}\Omega$$

$$V_{RS} = I_D R_S = 2 \text{ mA} \times 1 \text{ k}\Omega = 2 \text{ V}$$

$$V_{DS} = V_{DD} - V_{RS} = 12 \text{ V} - 2 \text{ V} = 10 \text{ V}$$

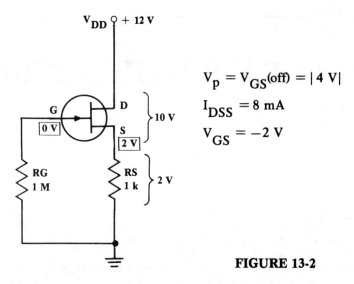

FIGURE 13-2

ALTERNATING CURRENT CIRCUIT ANALYSIS

The theoretical alternating current circuit analysis of voltage gain, input impedance, power gain, and peak-to-peak voltage swing at the output are determined once the DC biasing conditions are established. The AC parameters are given with reference to Figure 13-3.

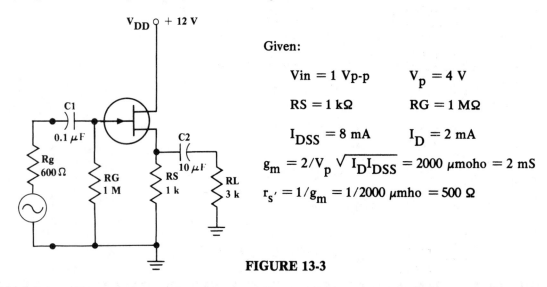

Given:

$V_{in} = 1$ Vp-p $\quad V_p = 4$ V

$RS = 1$ kΩ $\quad RG = 1$ MΩ

$I_{DSS} = 8$ mA $\quad I_D = 2$ mA

$g_m = 2/V_p \sqrt{I_D I_{DSS}} = 2000$ μmoho $= 2$ mS

$r_s' = 1/g_m = 1/2000$ μmho $= 500$ Ω

FIGURE 13-3

VOLTAGE GAIN: The voltage gain of the common drain circuit of Figure 13-3 can be solved in two ways: by the standard FET formula or by using the r_s' method. Both methods follow:

$$A_v = \frac{g_m r_L}{1 + g_m r_L} = \frac{g_m(RS \mathbin{/\mkern-6mu/} RL)}{1 + [g_m(RS \mathbin{/\mkern-6mu/} RL)]} = \frac{2 \text{ mS}(1 \text{ k}\Omega \mathbin{/\mkern-6mu/} 3 \text{ k}\Omega)}{1 + [2 \text{ mS}(1 \text{ k}\Omega \mathbin{/\mkern-6mu/} 3 \text{ k}\Omega)]} = \frac{1.5}{2.5} = 0.6$$

$$A_v = \frac{r_L}{r_s' + r_L} = \frac{1 \text{ k}\Omega \mathbin{/\mkern-6mu/} 3 \text{ k}\Omega}{500 \text{ }\Omega + (1 \text{ k}\Omega \mathbin{/\mkern-6mu/} 3 \text{ k}\Omega)} = \frac{750 \text{ }\Omega}{500 \text{ }\Omega + 750 \text{ }\Omega} = \frac{750 \text{ }\Omega}{1250 \text{ }\Omega} = 0.6$$

THE PEAK-TO-PEAK OUTPUT VOLTAGE SWING: The output voltage swing can be solved in two ways. Also, notice that in JFET circuits the Vo is much smaller than the Vin. In the common collector bipolar circuits, r_e is normally so small, with reference to r_L, that Vo $\approx$ Vin. This is not so with JFETs, where $r_s' = 1/g_m$ is much larger. Therefore, solving for the output voltage when Vin $= 2$ Vp-p:

$V_o = A_v \times V_{in} = 0.6 \times 2$ Vp-p $= 1.2$ Vp-p, or

$$V_o = \frac{V_{in} \times r_L}{r_s' + r_L} = \frac{2\text{ Vp-p} \times 750\text{ }\Omega}{750\text{ }\Omega + 500\text{ }\Omega} = 1.2\text{ Vp-p}$$

NOTE: The r_s' can be calculated, if Vin, Vo, and r_L are known, from:

$$r_s' = \frac{r_L(V_{in} - V_o)}{V_o} = \frac{750\text{ }\Omega(2\text{ Vp-p} - 1\text{ Vp-p})}{1.2\text{ Vp-p}} = 500\text{ }\Omega$$

INPUT IMPEDANCE: The input impedance Zin is approximately equal to RG. Therefore:

$$Z_{in} \approx R_G = 1\text{ M}\Omega$$

POWER GAIN TO THE LOAD: The power gain is solved from:

$$PG = A_v^2 \times Z_{in}/R_L = 0.6^2 \times 1\text{ M}\Omega / 3\text{ k}\Omega = 0.36 \times 333.3 = 120$$

$$PG(dB) = 10 \log 120 = 20.79\text{ dB}$$

SECTION II: THE UNIVERSAL CIRCUIT

DIRECT CURRENT CIRCUIT ANALYSIS

One advantage of the common drain universal circuit is that the voltage drop across the source resistor can be set to any DC value by the ratio of the R1 and R2 resistors. Hence, the Vop-p(max) of this circuit is not controlled entirely by the device, as it was in the self-biased circuit. In fact, because the voltage drop across RS is normally larger than that for self-biased circuits, the Vop-p can be larger. The direct current circuit analysis of the universal circuit is given in conjunction with Figure 13-4, where, again, $V_p = 4$ V and $I_{DSS} = 8$ mA.

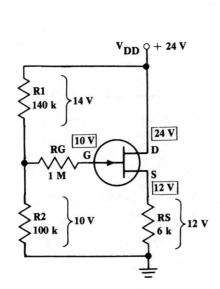

$$V_p = V_{GS}(\text{off}) = |4\text{ V}| \qquad I_{DSS} = 8\text{ mA}$$

$$V_{GS} = -2\text{ V}$$

$$I_D = I_{DSS}(1 - V_{GS}/V_p)^2 = 8\text{mA}(1 - 2\text{ V}/4\text{ V})^2$$
$$= 8\text{ mA} \times 0.5^2 = 8\text{ mA} \times 0.25 = 2\text{ mA}$$

$$V_{R1} = \frac{V_{DD} \times R1}{R1 + R2} = \frac{24\text{ V} \times 140\text{ k}\Omega}{140\text{ k}\Omega + 100\text{ k}\Omega} = 14\text{ V}$$

$$V_{R2} = \frac{V_{DD} \times R2}{R1 + R2} = \frac{24\text{ V} \times 100\text{ k}\Omega}{140\text{ k}\Omega + 100\text{ k}\Omega} = 10\text{ V}$$

$$V_{RS} = V_{R2} + V_{GS} = 10\text{ V} + |2\text{ V}| = 12\text{ V}$$

$$RS = V_{RS}/I_D = 12\text{ V}/2\text{ mA} = 6\text{ k}\Omega$$

FIGURE 13-4

ALTERNATING CURRENT CIRCUIT ANALYSIS

The alternating current circuit analysis of the universal-biased, common drain circuit is solved in a manner similar to that of the self-biased circuit. Therefore, the calculations given in conjunction with the circuit of Figure 13-5, are made without comment. Two analysis methods are given for both A_v and Vo.

VOLTAGE GAIN:

$$A_v = \frac{g_m r_L}{1 + g_m r_L} = \frac{g_m(RS \,/\!/\, RL)}{1 + g_m(RS \,/\!/\, RL)} = \frac{2\,mS(6\,k\Omega \,/\!/\, 3\,k\Omega)}{1 + 2\,mS(6\,k\Omega \,/\!/\, 3\,k\Omega)}$$

$$= \frac{2 \times 10^{-3} \times 2\,k\Omega}{1 + 2 \times 10^{-3} \times 2\,k\Omega} = \frac{4}{1+4} = 4/5 = 0.8$$

$$A_v = \frac{r_L}{r_s' + r_L} = \frac{6\,k\Omega \,/\!/\, 3\,k\Omega}{500\,\Omega + (6\,k\Omega \,/\!/\, 3\,k\Omega)} = \frac{2\,k\Omega}{500\,\Omega + 2\,k\Omega} = 0.8$$

PEAK-TO-PEAK VOLTAGE OUTPUT SWING:

$$Vo = A_v \times Vin = 0.8 \times 2\,Vp\text{-}p = 1.6\,Vp\text{-}p$$

$$Vo = \frac{Vin \times r_L}{r_s' + r_L} = \frac{2\,Vp\text{-}p \times 2\,k\Omega}{500\,\Omega + 2\,k\Omega} = 1.6\,Vp\text{-}p$$

INPUT IMPEDANCE:

$$Zin \approx RG + (R1 \,/\!/\, R2) = 1\,M\Omega + (140\,k\Omega \,/\!/\, 100\,k\Omega) = 1\,M\Omega + 58.33\,k\Omega \approx 1.058\,M\Omega$$

NOTE: As shown in The AC equivalent circuit of Figure 13-6, Zin is approximately equal to RG plus the parallel combination of R1 and R2, because of the location of the input signal connection. Again, the high input impedance of the JFET is ignored in the calculations without effect, because it is usually higher than 10 MΩ. Also, X_{C1} and X_{C2} are considered as AC shorts at mid-frequency conditions of about 1 kHz.

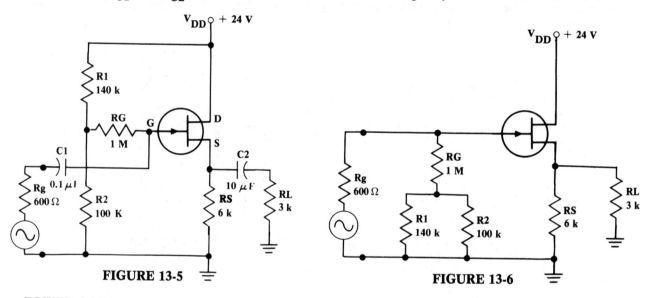

FIGURE 13-5

FIGURE 13-6

POWER GAIN:

$$PG = A_v^2 \times Zin/RL = 0.8^2 \times 1.058\,M\Omega/3\,k\Omega = 0.64 \times 352 = 225.7$$

$$PG(dB) = 10\,\log 225.7 = 23.5\,dB$$

SECTION III: THE TWO-POWER SUPPLY CIRCUIT

The two power supply common drain configuration is similar to the universal circuit but it has the advantage of requiring fewer components. Additionally, if equal plus and minus power supplies are used, the voltage dropped across the source resistor will be larger than V_{SS} or V_{DD}, because of the positive voltage at the source. The DC analysis is shown in conjunction with the two power supply circuit of Figure 13-7. Again, $I_{DSS} = 8$ mA and $V_p = 4$ V.

DIRECT CURRENT CIRCUIT ANALYSIS

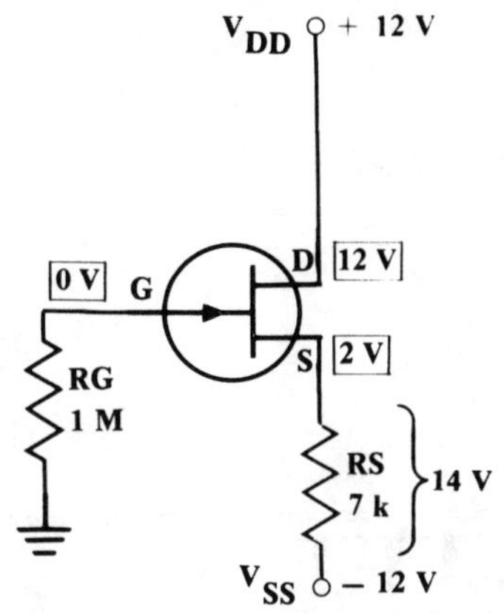

$I_D = I_{DSS}(1 - V_{GS}/V_p)^2 = 8$ mA$(1 - 2$ V$/4$ V$)^2 = 2$ mA

$V_{RS} = I_D R_S = 2$ mA $\times 7$ k$\Omega = 14$ V

$V_{RG} = 0$ V

$V_D = V_{DD} = 12$ V

$V_S = V_{GS} = |2$ V$|$

$V_{DS} = V_D - V_S = 12$ V $- 2$ V $= 10$ V

FIGURE 13-7

ALTERNATING CURRENT CIRCUIT ANALYSIS

The AC analysis of the circuit of Figure 13-8 is similar to that of the other common drain circuits. However, as discussed in conjunction with two power supply bipolar circuits, all power supply voltages are effective AC grounds. Therefore, RS is seen connected to AC ground and in parallel with RL, as shown in the equivalent AC circuit of Figure 13-9.

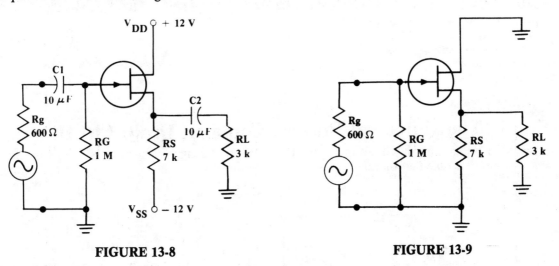

FIGURE 13-8 **FIGURE 13-9**

VOLTAGE GAIN:

$$A_v = \frac{g_m r_L}{1 + g_m r_L} = \frac{g_m(RS /\!/ RL)}{1 + g_m(RS /\!/ RL)} = \frac{2 \text{ mS}(7 \text{ k}\Omega /\!/ 3 \text{ k}\Omega)}{1 + 2 \text{ mS} \times (7 \text{ k}\Omega /\!/ 3 \text{ k}\Omega)}$$

$$= \frac{2 \times 10^{-3} \times 2.1 \text{ k}\Omega}{1 + 2 \times 10^{-3} \times 2.1 \text{ k}\Omega} = \frac{4.2}{1 + 4.2} = \frac{4.2}{5.2} = 0.808$$

$$A_v = \frac{r_L}{r_s' + r_L} = \frac{RS /\!/ RL}{r_s' + (RS /\!/ RL)} = \frac{7 \text{ k}\Omega /\!/ 3 \text{ k}\Omega}{500 \text{ }\Omega + (7 \text{ k}\Omega /\!/ 3 \text{ k}\Omega)} = \frac{2.1 \text{ k}\Omega}{500 \text{ }\Omega + 2.1 \text{ k}\Omega} = 0.808$$

VOLTAGE OUT:

$$Vo = A_v \times Vin = 0.808 \times 2 \text{ Vp-p} \approx 1.61 \text{ Vp-p, or}$$

$$Vo = \frac{Vin \times r_L}{r_s' + r_L} = \frac{2 \text{ Vp-p} \times 2.1 \text{ k}\Omega}{500 \text{ }\Omega + 2.1 \text{ k}\Omega} \approx 1.61 \text{ Vp-p}$$

INPUT IMPEDANCE:

$$Zin \approx RG = 1 \text{ M}\Omega$$

OUTPUT IMPEDANCE:

$$Zo = r_s' \parallel RS = 500 \text{ }\Omega \parallel 7 \text{ k}\Omega \approx 467 \text{ }\Omega$$

POWER GAIN TO THE LOAD:

$$PG = A_v^2 \times Zin/RL = 0.808^2 \times 1 \text{ M}\Omega/3 \text{ k}\Omega = 217.6$$

$$PG(dB) = 10 \log PG = 10 \log 217.6 = 23.38 \text{ dB}$$

LABORATORY EXERCISE

OBJECTIVES

Investigate the DC and AC characteristics of the universal-biased, N channel, common drain circuit using measurements techniques.

LIST OF MATERIALS AND EQUIPMENT

1. Transistor: 2N5951 JFET or equivalent
2. Resistors (one each except where indicated):   2.2 kΩ   4.7 kΩ   100 kΩ (three)   120 kΩ
3. Capacitors (one each):   10 μF (two)
4. Power Supply
5. Voltmeter
6. Signal Generator
7. Oscilloscope

PROCEDURE
Section I: The Universal-Biased, Common Drain Circuit

1. Connect the common drain circuit of Figure 13-10, where the signal generator is connected into the gate and the output is taken off the source.

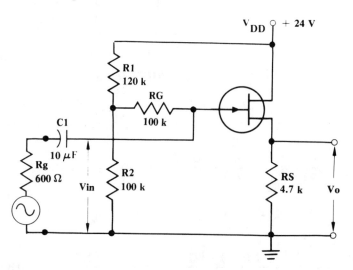

FIGURE 13-10

2. Measure V_D, V_S, V_G, and V_{R2}, with respect to ground. Insert the measured values into Table 13-1.

3. Use the measured values to calculate V_{RS}, V_{DS}, V_{GS}, V_{RG}, V_{R1}, and I_D from:

   a. $V_{RS} = V_S$
   b. $V_{DS} = V_D - V_S$
   c. $V_{GS} = V_G - V_S$
   d. $I_D = V_{RS}/RS$
   e. $V_{RG} = V_{R2} - V_G$
   f. $V_{R1} = V_{DD} - V_{R2}$

4. Insert the calculated values, as indicated, into Table 13-1.

TABLE 13-1	V_D	V_S	V_G	V_{R2}	V_{RS}	V_{DS}	V_{GS}	V_{RG}	V_{R1}	I_D
CALCULATED	/////	/////	/////	/////						
MEASURED					/////	/////	/////	/////	/////	/////

5. ALTERNATING CURRENT CIRCUIT ANALYSIS (Figure 13-10)

   a. Measure the voltage output swing using an input voltage of 1 Vp-p at a frequency of 1 kHz.

   b. Calculate the voltage gain A_v, from:

   $$A_v = V_o/V_{in} \quad \text{where:} \quad V_{in} = 1 \text{ Vp-p}$$

   c. Calculate the source resistance r_s' from:

   $$r_s' = \frac{RS(V_{in} - V_o)}{V_o} \quad \text{where:} \quad r_s' \text{ was derived from } V_o = \frac{V_{in} \times RS}{r_s' + RS}$$

   NOTE: The r_s' can also be obtained from the transfer curve.

   d. Calculate the transconductance from:

   $$g_m = 1/r_s'$$

6. Insert the measured and calculated values, as indicated, into Table 13-2.

TABLE 13-2	V_o	A_v	r_s'	g_m
CALCULATED	/////			
MEASURED		/////	/////	/////

Section II: The Common Drain Circuit with Load Resistor

1. Connect the circuit of Figure 13-11, where the load resistor has been added.

2. VOLTAGE GAIN: Calculate the voltage gain from:

   $$A_v = r_L/(r_s' + r_L) \quad \text{where:} \quad r_L = RS \parallel RL$$

   Measure the voltage gain using an input signal voltage of 1 Vp-p at 1 kHz.

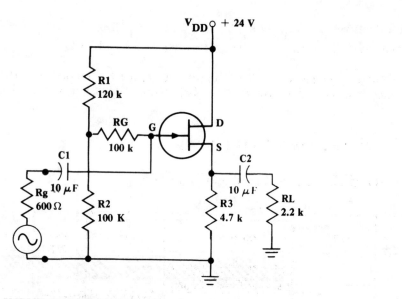

FIGURE 13-11

3. **INPUT IMPEDANCE:** Calculate the input impedance from:

   $Zin \approx RG + (R1 \mathbin{/\mkern-6mu/} R2)$

Measure the input impedance using the known series resistor method, as shown in Figure 13-12. The series resistor R3 equals 100 kΩ.

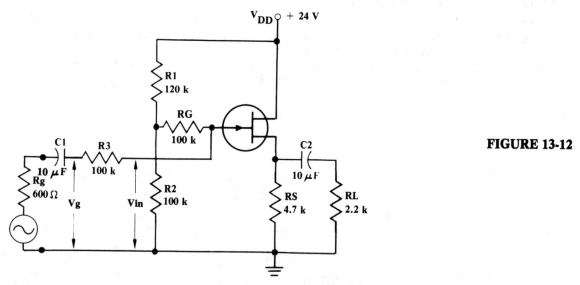

FIGURE 13-12

4. **MAXIMUM PEAK-TO-PEAK OUTPUT VOLTAGE SWING:** Monitor the output across the load resistor and increase the signal voltage level at 1 kHz, until clipping or distortion occurs.

NOTE: JFETs are primarily used in small signal applications because the non-linearity of their output characteristic curves will distort large peak-to-peak output waves. Additionally, the gate-to-source junction of JFETs must remain reverse biased and can only tolerate a few tenths of a volt of positive voltage swing.

5. **POWER GAIN TO THE LOAD RESISTOR:** Calculate the power gain to the load resistor from:

   $PG = A_v^2 \times Zin/RL$ where: $RL = 2.2$ kΩ

Calculate the power gain in dB from:

   $PG(dB) = 10 \log PG$

6. OUTPUT IMPEDANCE:

   a. Calculate the output impedance from: $Z_o \approx r_s' \parallel R_S$ where: $R_S = 4.7\ k\Omega$

   b. Measure the output impedance by removing the 2.2 kΩ load and setting the unloaded Vo(NL) to 1 Vp-p. Replace the 2.2 kΩ load resistor and measure the loaded Vo(L). Calculate the measured Zo from:

   $$Z_o = \frac{V_o(NL) - V_o(L)}{V_o(L)} \times R_L \quad \text{where: } V_o(NL) = 1\ Vp\text{-}p$$

7. Insert the calculated and measured values, as indicated, into Table 13-3.

TABLE 13-3	A_v	Zin	Vop-p(max)	PG	PG (dB)	Zo
CALCULATED			/////			
MEASURED				/////	/////	

8. Measure the low and high corner frequencies. Use the Bode plot representation.

CHAPTER QUESTIONS

1. What prevents the source resistance from effecting the input impedance of the common drain circuit?
2. Why is the percentage of input signal reaching the load so much smaller in common drain JFET circuits than it is in bipolar common collector circuits?
3. Why is the common drain, self-biased circuit not so widely used as either the two power supply or universal biased circuits?

CHAPTER PROBLEMS

1. Given: $V_p = 6$ V $I_{DSS} = 8$ mA
   $V_{GS} = -3$ V   Vin = 1.5 Vp-p

   Solve for:

   a. r_s' from $r_s' = 2/V_p \sqrt{I_D I_{DSS}}$
   b. Vo across RS
   c. r_s' from $r_s' = RS(Vin - Vo)/Vo$

2. Given: $V_p = 6$ V $I_{DSS} = 8$ mA
   $V_{GS} = -3$ V   Vin = 3.5 Vp-p

   Solve for:

   a. RS
   b. r_s'
   c. Vo across RL

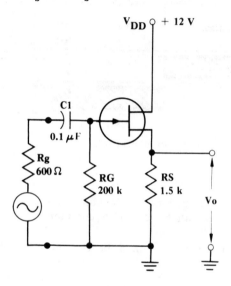

FIGURE 13-13

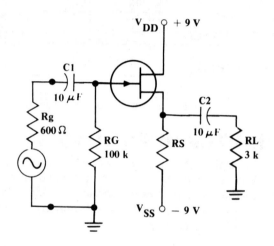

FIGURE 13-14

3. Given: $V_p = 6$ V $I_{DSS} = 8$ mA
   $V_{GS} = -3$ V   Vin = 2.5 Vp-p

   Solve for:

   a. I_D
   b. Zin
   c. Vo

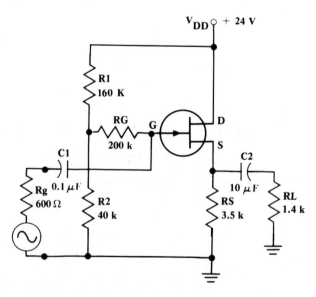

FIGURE 13-15

14 | COMMON GATE JFET CIRCUITS

GENERAL DISCUSSION

The common gate circuit is rarely used in single stage applications because it has no advantage over bipolar common base circuits connected in similar configurations. In fact, the voltage gain of the common gate circuit is considerably lower than that of the common base circuit. In common gate circuits, the signal is applied to the source and taken off the drain and, because r_s' of JFETs is larger than r_e of bipolar devices, the input impedance of JFETs is normally higher. For example, in the sample problems used, representative r_s' values used have been about 500 ohms, while the r_e values of bipolar devices used have been about 20 ohms. In Figure 14-1, the self-biased, universal-biased, and two power supply biased common gate circuits are shown.

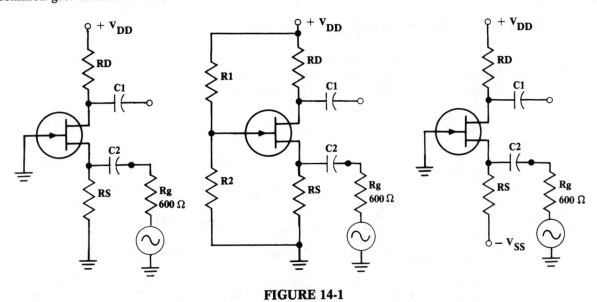

FIGURE 14-1

SECTION I: SELF-BIASED, COMMON GATE AMPLIFIER

DIRECT CURRENT CIRCUIT ANALYSIS

The DC voltages of the self-biased, common gate circuit are solved using the mathematical approach, and assuming the device is square law. The DC analysis was solved previously and the equations are shown in conjunction with the circuit of Figure 14-2. Again, an I_{DSS} of 8 mA, a V_p of 4 V, and a V_{GS} of -2 V are used.

$$I_D = I_{DSS}(1 - V_{GS}/V_p)^2 = 8 \text{ mA}(1 - 2V/4 \text{ V})^2 = 8 \text{ mA} \times 0.5 \text{ V}^2 = 8 \text{ mA} \times 0.25 = 2 \text{ mA}$$

$$R_S = V_{GS}/I_D = |2 \text{ V}|/2 \text{ mA} = 1 \text{ k}\Omega$$

$$V_{RS} = I_D R_S = 2 \text{ mA} \times 1 \text{ k}\Omega = 2 \text{ V}$$

$$V_D = V_{DD} - V_{RD} = 24 \text{ V} - 12 \text{ V} = 12 \text{ V}$$

$$V_{DS} = V_D - V_S = 12\text{ V} - 2\text{ V} = 10\text{ V}$$

$$V_{DD} = V_{RD} + V_{DS} + V_{RS}$$

$$24\text{ V} = 12\text{ V} + 10\text{ V} + 2\text{ V}$$

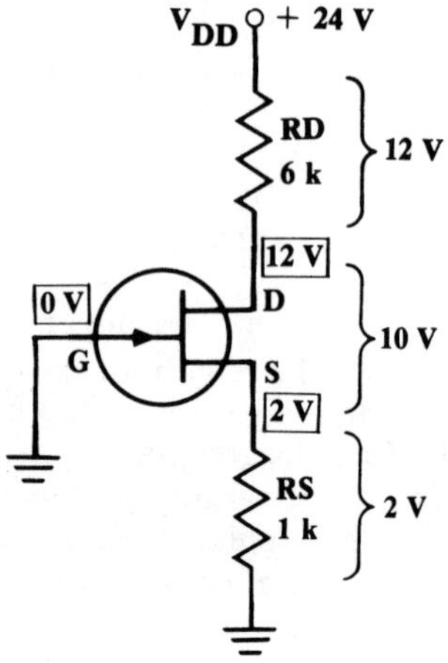

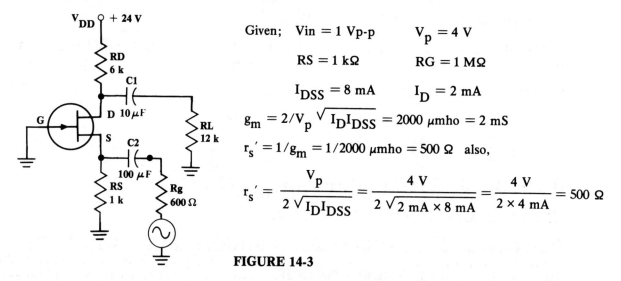

FIGURE 14-2

ALTERNATING CURRENT CIRCUIT ANALYSIS

The theoretical alternating current circuit analysis of voltage gain, input impedance, power gain, and peak-to-peak voltage swing at the output are determined once the direct current biasing conditions are established. These parameters are given in conjunction with Figure 14-3.

Given; $V_{in} = 1$ Vp-p $V_p = 4$ V

$RS = 1$ kΩ $RG = 1$ MΩ

$I_{DSS} = 8$ mA $I_D = 2$ mA

$g_m = 2/V_p \sqrt{I_D I_{DSS}} = 2000\ \mu$mho $= 2$ mS

$r_s' = 1/g_m = 1/2000\ \mu$mho $= 500\ \Omega$ also,

$$r_s' = \frac{V_p}{2\sqrt{I_D I_{DSS}}} = \frac{4\text{ V}}{2\sqrt{2\text{ mA} \times 8\text{ mA}}} = \frac{4\text{ V}}{2 \times 4\text{ mA}} = 500\ \Omega$$

FIGURE 14-3

VOLTAGE GAIN: The voltage gain of the common gate circuit is solved like that of the common source circuit. The two methods of solution are:

$$A_v = g_m r_L = g_m(RD \parallel RL) = 2\text{ mS}(6\text{ k}\Omega \parallel 12\text{ k}\Omega) = 2 \times 10^{-3} \times 4 \times 10^3 = 8$$

$$A_v = r_L/r_s' = (RD \parallel RL)/r_s' = (6\text{ k}\Omega \parallel 12\text{ k}\Omega)/500\ \Omega = 4\text{ k}\Omega/500\ \Omega = 8$$

THE OUTPUT VOLTAGE: The output voltage is amplified by 8 and, like that of the bipolar, common base circuit, it is in phase with the input signal. With reference to Figure 14-4, this in-phase condition is explained by assuming a positive-going wave at the source, which causes the voltage at the source to increase with respect to ground which, in turn, causes the I_D to decrease. Hence, the voltage across RD decreases, causing the voltage at the drain, with respect to ground, to increase. The negative-going wave provides just the opposite, in-phase, condition. Therefore, with a Vin of 1 Vp-p, the output is in phase and equals 8 Vp-p from:

$$V_O = V_{in} \times A_V = 1 \text{ Vp-p} \times 8 = 8 \text{ Vp-p}$$

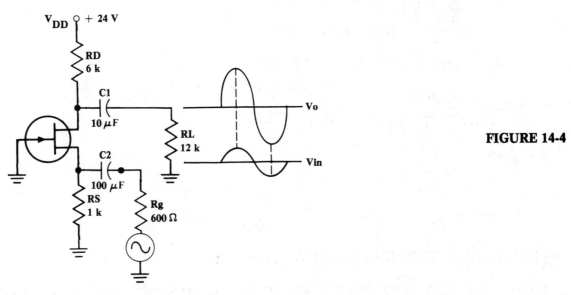

FIGURE 14-4

INPUT IMPEDANCE: The input impedance to the self-biased circuit, looking into the source terminal, "sees" r_s' in parallel with RS. Therefore:

$$Z_{in} = r_s' \mathbin{/\mkern-6mu/} RS = 500 \text{ }\Omega \mathbin{/\mkern-6mu/} 1 \text{ k}\Omega = 333.3 \text{ }\Omega$$

NOTE: Standard formulas show Zin being solved from $Z_{in} = (1/g_m) \mathbin{/\mkern-6mu/} RS$ but, since $g_m = 1/r_s'$, the formulas are identical.

POWER GAIN TO THE LOAD: The power gain to the load is solved from:

$$PG = A_V^2 \times Z_{in}/RL = 8^2 \times 333.3 \text{ }\Omega / 12 \text{ k}\Omega = 1.78$$

$$PG(dB) = 10 \log 1.78 = 2.5$$

SECTION II: THE UNIVERSAL CIRCUIT
DIRECT CURRENT CIRCUIT ANALYSIS

The direct current analysis of the universal-biased, common gate circuit shown in Figure 14-5 is similar to that of the common source circuits, and the alternating current analysis is similar to that of the self-biased, common source circuit. For convenience, the DC analysis is given in conjunction with the circuit of Figure 14-5 where, again, $I_{DSS} = 8$ mA, $V_p = 4$ V, and $V_{GS} = -2$ V.

$$I_D = I_{DSS}(1 - V_{GS}/V_p)^2 = 8 \text{ mA}(1 - 2\text{ V}/4\text{ V})^2 = 8 \text{ mA} \times 0.5 \text{ V}^2 = 8 \text{ mA} \times 0.25 = 2 \text{ mA}$$

$$V_{R1} = \frac{V_{DD} \times R1}{R1 + R2} = \frac{24 \text{ V} \times 180 \text{ k}\Omega}{180 \text{ k}\Omega + 20 \text{ k}\Omega} = 21.6 \text{ V}$$

$$V_{R2} = \frac{V_{DD} \times R2}{R1 + R2} = \frac{24 \text{ V} \times 20 \text{ k}\Omega}{180 \text{ k}\Omega + 20 \text{ k}\Omega} = 2.4 \text{ V}$$

$$V_{RS} = V_{R2} + V_{GS} = 2.4 \text{ V} + 2 \text{ V} = 4.4 \text{ V}$$

$$RS = V_{RS}/I_D = 4.4\,V/2\,mA = 2.2\,k\Omega$$

$$V_{RD} = I_D R_D = 2\,mA \times 6\,k\Omega = 12\,V$$

$$V_D = V_{DD} - V_{RD} = 24\,V - 12\,V = 12\,V$$

$$V_{DS} = V_D - V_S = 12\,V - 4.4\,V = 7.6\,V$$

$$V_{DD} = V_{RD} + V_{DS} + V_{RS} = 24\,V = 12\,V + 7.6\,V + 4.4\,V$$

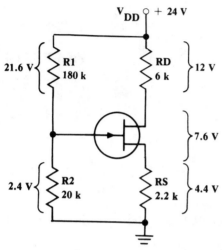

$$V_p = V_{GS}(\text{off}) = |4\,V| \qquad I_{DSS} = 8\,mA$$

$$V_{GS} = -2\,V$$

FIGURE 14-5

ALTERNATING CURRENT CIRCUIT ANALYSIS

The alternating current circuit analysis begins by knowing g_m or r_s', and both can be solved, theoretically, from the DC parameters of the circuit as shown previously. Therefore:

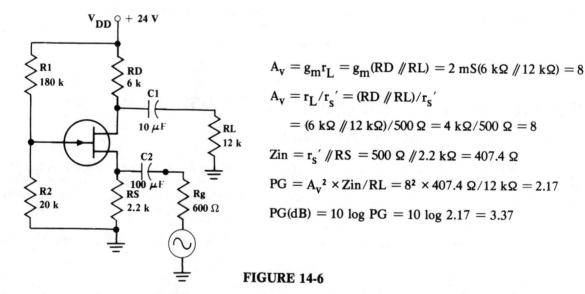

$$A_v = g_m r_L = g_m(RD \,//\, RL) = 2\,mS(6\,k\Omega \,//\, 12\,k\Omega) = 8$$

$$A_v = r_L/r_s' = (RD \,//\, RL)/r_s'$$

$$= (6\,k\Omega \,//\, 12\,k\Omega)/500\,\Omega = 4\,k\Omega/500\,\Omega = 8$$

$$Z_{in} = r_s' \,//\, RS = 500\,\Omega \,//\, 2.2\,k\Omega = 407.4\,\Omega$$

$$PG = A_v^2 \times Z_{in}/RL = 8^2 \times 407.4\,\Omega/12\,k\Omega = 2.17$$

$$PG(dB) = 10 \log PG = 10 \log 2.17 = 3.37$$

FIGURE 14-6

SECTION III: THE TWO-POWER SUPPLY CIRCUIT

DIRECT CURRENT CIRCUIT ANALYSIS

The two power supply common gate connection is similar to the universal circuit, but it has the advantage of requiring fewer components. Reviewing the DC analysis, which was solved previously, where $V_p =$

4 V, $I_{DSS} = 8$ mA, and $V_{GS} = -2$ V:

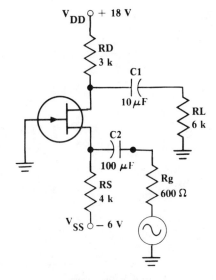

FIGURE 14-7

$I_D = I_{DSS}(1 - V_{GS}/V_p)^2 = 8$ mA$(1 - 2$ V$/4$ V$)^2 = 2$ mA

$V_{RS} = I_D R_S = 2$ mA $\times 4$ k$\Omega = 8$ V

$V_{RG} = 0$ V

$V_{RD} = I_D R_D = 2$ mA $\times 3$ k$\Omega = 6$ V

$V_S = V_{GS} = |2$ V$|$

$V_D = V_{DD} - V_{RD} = 18$ V $- 6$ V $= 12$ V

$V_{DS} = V_D - V_S = 12$ V $- 2$ V $= 10$ V

$V_{DD} - V_{SS} = V_{RD} + V_{DS} + V_{RS}$

18 V $- (-6$ V$) = 6$ V $+ 10$ V $+ 8$ V

24 V $= 6$ V $+ 10$ V $+ 8$ V

ALTERNATING CURRENT CIRCUIT ANALYSIS

The alternating current circuit analysis of the two power supply circuit of Figure 14-8 begins by knowing g_m or r_s' which are solved, theoretically, from the DC parameters of the circuit. Therefore:

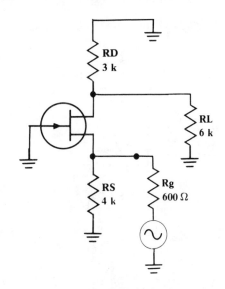

FIGURE 14-8   **FIGURE 14-9**

$g_m = 2000$ μmho $= 2$ mS

$r_s' = 1/2$ mS $= 1/(2 \times 10^{-3}) = 1000/2 = 500$ Ω

$A_v = g_m r_L = g_m$(RD // RL) $= 2$ mS$(3$ kΩ // 6 k$\Omega) = 4$ or

$A_v = r_L/r_s' = $ (RD // RL)$/r_s' = (3$ kΩ // 6 k$\Omega)/500$ $\Omega = 2$ k$\Omega/500$ $\Omega = 4$

$Z_{in} = $ RS // $r_s' = 4$ kΩ // 500 $\Omega = 444.4$ Ω

$PG = A_v^2 \times Z_{in}/RL = 4^2 \times 444.4$ $\Omega/6$ k$\Omega \approx 1.19$ PG(dB)$ = 10 \log 1.19 \approx 0.74$

NOTE: Zin is approximately equal to RS $\parallel r_s'$, because the input signal is connected to the source and the V_{SS} supply is an effective AC ground. The equivalent circuit is shown in Figure 14-9. Again, the reactances of C1 and C2 are shown "shorted" to simplify the illustration.

LABORATORY EXERCISE

OBJECTIVES

Investigate the two power supply biased, N channel, common gate circuit configuration.

LIST OF MATERIALS AND EQUIPMENT

1. Transistor: 2N5951 JFET or equivalent
2. Resistors (one each):   470 Ω   3.3 kΩ   4.7 kΩ   10 kΩ (two)
3. Capacitors (one each):   10 μF   100 μF
4. Power Supply
5. Voltmeter
6. Signal generator
7. Oscilloscope

PROCEDURE
Section I: The Two-Power-Supply, Common-Gate Circuit

1. Connect the two-power-supply, common-gate circuit of Figure 14-10, where the signal generator is connected into the source lead. Note, that for common gate circuits, the gate can be grounded, as shown. Unlike bipolar transistor devices, there is no reflected resistance between the gate and source leads.

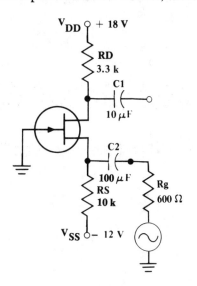

FIGURE 14-10

2. Measure V_D, V_S, and V_G, all with respect to ground. Insert the measured values into Table 14-1.

3. Use the measured values to calculate V_{RD}, V_{DS}, V_{GS}, V_{RS}, and I_D from:

   a. $V_{RD} = V_{DD} - V_D$

   b. $V_{RS} = V_S - V_{SS}$

   c. $V_{DS} = V_D - V_S$

   d. $V_{GS} = V_G - V_S$

   e. $I_D = V_{RS}/RS$

4. Insert the calculated values into Table 14-1.

TABLE 14-1	V_D	V_S	V_G	V_{RD}	V_{RS}	V_{DS}	V_{GS}	I_D
CALCULATED	//////	//////	//////					
MEASURED				//////	//////	//////	//////	//////

5. ALTERNATING CURRENT CIRCUIT ANALYSIS

   a. Measure the voltage out using an input signal voltage of 100 mVp-p at a frequency of 1 kHz.

   b. Calculate the voltage gain from: $A_v = V_o/V_{in}$ where: $V_{in} = 100$ mVp-p

   c. Calculate the source resistance from: $r_s' = R_D/A_v$

   d. Calculate the transconductance from: $g_m = 1/r_s'$ or $g_m = A_v/R_D$

NOTE: The r_s' or the g_m can be calculated, also, directly off the transfer curve.

6. Insert the measured and calculated values, as indicated, into Table 14-2.

TABLE 14-2	Vo	A_v	r_s'	g_m
CALCULATED	//////			
MEASURED		//////	//////	//////

Section II: The Common Gate Circuit with Load Resistor Added

1. Connect the circuit of Figure 14-11, where the load resistor of 10 kΩ has been added.

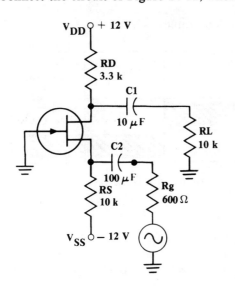

FIGURE 14-11

2. **VOLTAGE GAIN:** Calculate the voltage gain from: $A_v = r_L/r_s' = (R_D \parallel R_L)/r_s'$

Measure the voltage gain using an input signal voltage of 100 mVp-p at 1 kHz.

3. INPUT IMPEDANCE: Calculate the input impedance from: $Zin = RS \parallel r_s'$.

Measure the input impedance using the known series resistor technique, as shown in Figure 14-12 where the series resistor has a 470 ohm value.

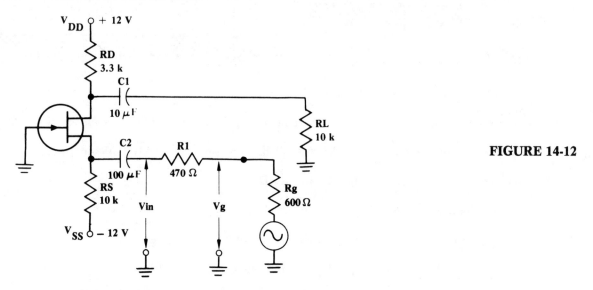

FIGURE 14-12

4. MAXIMUM PEAK-TO-PEAK OUTPUT VOLTAGE SWING: Monitor the output across the load resistor and increase the 1 kHz signal voltage level until clipping or distortion occurs.

NOTE: As previously noted, JFETs are used primarily in small signal applications because the non-linearity of their output characteristic curves will distort large peak-to-peak output waves. Additionally, the gate-to-source junction of JFETs must remain reverse biased because the device can tolerate only a few tenths of a volt of positive voltage swing, which can be caused by a Vin that is slightly greater than $2V_{GS}$.

5. POWER GAIN TO THE LOAD RESISTOR: Calculate the power gain to the load resistor from:

$PG = A_v^2 \times Zin/RL$ where: $RL = 10\ k\Omega$

Calculate PG in dB from: $PG(dB) = 10 \log PG$

6. Insert the calculated and measured values, as indicated, into Table 14-3.

TABLE 14-3	A_v	Zin	Vop-p(max)	PG	PG(dB)
CALCULATED			/////		
MEASURED				/////	/////

7. Measure the low and high corner frequencies and represent them on a Bode plot.

CHAPTER QUESTIONS

1. Why is the input impedance of the common gate circuit relatively low?
2. Does connecting the gate directly to ground lower the input impedance? Why or why not?
3. Does r_s' remain constant as the I_D changes or does it also change? Explain what happens to r_s' when I_D is increased.
4. Why is the power gain of the common gate circuits so low?

CHAPTER PROBLEMS

1. Given: $V_p = 6$ V $I_{DSS} = 8$ mA
   $V_{GS} = -3$ V   Vin = 1 Vp-p

   Solve for:

   a. Zin

   b. A_v

   c. Vo across RL

2. Given: $V_p = 4$ V $I_{DSS} = 8$ mA
   $V_{GS} = -2$ V   Vin = 1 Vp-p

   Solve for:

   a. Zin

   b. A_v

   c. PG to the load

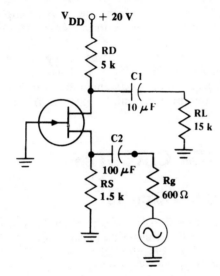

FIGURE 14-13

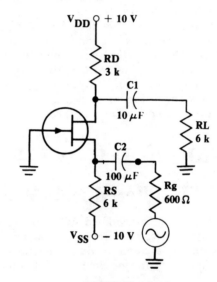

FIGURE 14-14

15 | DC POWER SUPPLIES — Unregulated AC to DC Converters

GENERAL DISCUSSION

Most of today's electronic equipment is powered by DC voltages that are obtained from batteries, DC to DC converters, or AC to DC converters. Both batteries and DC to DC converters imply portability, and the degree of portability is determined by the amount of power consumed. AC to DC converters, on the other hand, are not normally portable since they convert AC line power from the power company to DC power. However, this latter technique for obtaining power for electronic equipment is efficient, convenient, and the method most commonly used.

The AC to DC converter consists of a "bulk" supply and, in most instances, regulating circuitry. This is where the "bulk" supply provides the initial conversion of AC to DC voltage, and where the regulating circuitry then provides an output voltage or current that will remain constant under varying degrees of loading. The regulating circuitry also minimizes the ripple content, that usually exists at the output of the "bulk" supply, from getting to the load.

SECTION I: HALF-WAVE RECTIFIER CIRCUITS

Part 1: Unfiltered Half-Wave Rectifier Circuits

The bulk supply unregulated AC to DC converter consists of a transformer, rectifiers or a bridge rectifier, load resistance, and filtering capacitors. The transformer fills the needs of stepping up or stepping down the line voltage and providing isolation. The diodes, or rectifiers, provide the AC to DC conversion, and the capacitors provide the necessary filtering so that most of the AC voltage can be converted to DC voltage.

The means by which the transformer steps up or steps down the line voltage is simply a ratio of turns on the primary to those of the secondary. For instance, a transformer with 240 turns on the primary and 120 turns on the secondary will step down the line voltage of 120 volts RMS across the primary to 60 volts RMS across the secondary. This is shown in **Figure 15-1**.

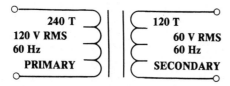

FIGURE 15-1

However, to insure adequate flux cutting, necessary in providing power transformation between primary and secondary transformer windings, iron core material must be used. This is because power transformers operating at line frequencies of 60 Hz can only produce a magnetic field, necessary in the transfer of electrical energy, if close coupling and adequate iron core material are used. Therefore, in power transformers, the secondary is wound directly on top of the primary and the E's and I's of the core material are interleaved to minimize eddy current losses.

An example of a power transformer capable of providing 60 V RMS at 4 amps with a high degree of efficiency is one with 240 turns on the primary, with 120 turns on the secondary, and with three square inches of core material. The bobbin used to wind the wire on the transformer should be 1½ by 2 inches.

Transformer isolation is used to prevent the possibilities of lethal voltages existing between chassis ground and earth ground. The isolation is obtained by physically separating the primary and secondary windings from each other with layers of insulation. High potential voltage tests are usually performed on transformers to insure this isolation.

AC to DC converters that do not use transformers have no isolation. They must rely on keyed power plugs to guard against a "hot" chassis condition. This is because the chassis is usually used at the common ground for the circuit, and when power companies provide 120 volts AC, one wire is "hot" at 120 volts AC and the other is at ground condition. Therefore, an inadvertently reversed plug in the power outlet can cause the chassis to be at 120 volts AC with respect to the power company and earth ground. Some portable TVs, for instance, are designed without transformers to cut down on weight, cost, and size; and these TVs are good examples of "hot" chassis possibilities.

NOTE: If work must be done on the portable (possible) hot chassis TV, where the chassis is exposed, then a simple VOM reading should be made, prior to connecting any grounded test equipment (scope) leads to the chassis. That is, if severe and spectacular ground loops are to be avoided.

One grounded test lead is common practice, with most quality test equipment, where the three prong plug for maximum safety is used. (This is where the power company provides two of the three wires, 120 volts AC and ground, and the user provides an additional safety ground, green in color and connected to a water pipe or ground rod.) The safety ground is then connected to the chassis, references the power company ground, and if at any time the chassis attempts to go "hot" (inadvertently of course), the line fuse will simply blow, until the problem is corrected. Both isolation transformers or keyed plugs can circumvent this problem, but isolation transformers provide the ultimate safety provision.

For rectification purposes, the diode (rectifier) is a device that is an effective switch — a "short" when forward biased and an "open" when reverse biased. **See Figure 15-2.** Therefore, when a sinusoidal

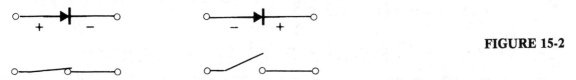

FIGURE 15-2

peak-to-peak wave is applied to the diode, only the positive-going pulses are processed to the load. This is because the positive-going wave turns the diode "on" and the negative-going wave turns the diode "off", as shown in Figure 15-3. However, if the diode were to be reversed, then only the negative-going pulses would be processed to the load.

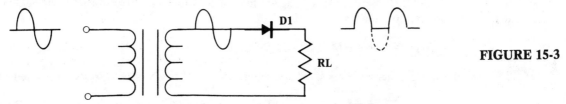

FIGURE 15-3

The reason for processing only one half of the peak-to-peak positive-going and negative-going wave is because the average of the peak-to-peak wave is zero. However, for the positive only or negative only pulse, a zero reference is established and the average voltage is solved by dividing Vpeak by pi (π).

The power company line voltage is normally stated in RMS voltage and can be read off of a standard voltmeter at approximately 120 volts RMS. However, when an oscilloscope is used, the monitored voltage will be in peak-to-peak voltage. **Therefore**, the relationship between peak-to-peak, peak, and RMS voltage is shown in Figure 15-4, where $Vp = \sqrt{2}$ V RMS and Vp-p = 2 Vp or, with regard to peak-to-peak voltage, Vp = Vp-p/2 and V RMS = $1/\sqrt{2}$ Vp or 0.707 Vp. Hence, 120 volts RMS equals 169.7 Vp and 339.4 Vp-p.

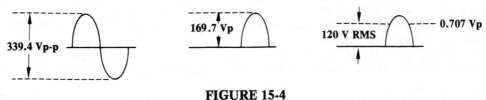

FIGURE 15-4

Part 2: Unfiltered AC to DC "Bulk" Power Supplies
THE HALFWAVE RECTIFIER — CIRCUIT ANALYSIS

The halfwave rectifier circuit, shown in Figure 15-5, uses the transformer to step down the 120 V RMS line voltage to 30 V RMS and the diodes to process only the positive-going pulse to the load. At the load, the pulse contains both DC and AC components, where the DC voltage is the average of the pulse and the AC ripple content is the amplitude of the pulse. The analysis for Figure 15-5 is shown solved, where the 30 V RMS of the secondary voltage is rectified to provide 13.5 V DC and 16.36 V AC at the load. The 16.36 V AC is the effective AC value of the positive-going pulse and represents that portion of the 30 V RMS secondary that is not converted to DC.

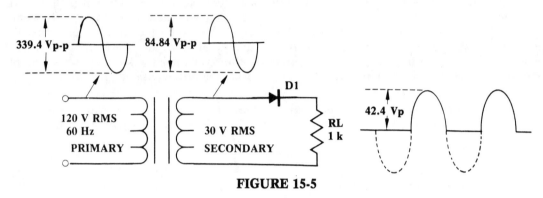

FIGURE 15-5

1. a) Solving for the peak-to-peak voltage across the primary for 120 V RMS of line voltage, where the 120 V line voltage is multiplied by $2\sqrt{2}$:

$$V_{p\text{-}p} = 2\sqrt{2} \times V\text{ RMS} = 2\sqrt{2} \times 120\text{ V} = 339.4\text{ Vp-p}$$

b) Solving for the V RMS across the secondary for a 4 : 1 stepdown transformer, where N represents the primary to secondary stepdown turns ratio:

$$V_{sec} = V_{pri}/N = 120\text{ V RMS}/4 = 30\text{ V RMS}$$

c) Solving for the Vp-p across the secondary, where the 30 V RMS secondary is multiplied by $2\sqrt{2}$:

$$V_{p\text{-}p}(sec) = 2\sqrt{2} \times V\text{ RMS}(sec) = 2\sqrt{2} \times 30\text{ V RMS} = 84.84\text{ Vp-p}$$

2. a) Solving for the peak voltage at the load, where the peak-to-peak secondary voltage is "halved" by the rectifier D1.

$$V_p = V_{p\text{-}p}/2 = 84.84\text{ Vp-p}/2 = 42.42\text{ V}$$

NOTE: The peak-to-peak voltage at the secondary is "chopped" in half by the diode and only the positive-going pulse reaches the load. Also, the forward biased diode voltage of about 0.6 V is small in comparison with the V DC across the load and can be safely ignored in the calculations.

b) Solving for the peak current through the load and, hence, through the series diode by dividing the peak voltage by the 1 kΩ load resistor.

$$I_{peak} = V_{peak}/R_L = 42.42\text{ Vp}/1\text{ k}\Omega = 42.42\text{ mA peak}$$

3. a) Solving for the DC voltage content at the load, which is solved by dividing the peak voltage by π:

$$V\text{ DC} = V_{peak}/\pi = 42.42\text{ Vp}/3.1415 = 13.5\text{ V DC}$$

NOTE: The DC voltage at the load is the average voltage for the positive-going pulse when taken over one complete cycle. Therefore, the V DC is approximately 0.3183 of peak voltage, as shown in Figure 15-6, where $0.3183 = 1/\pi$.

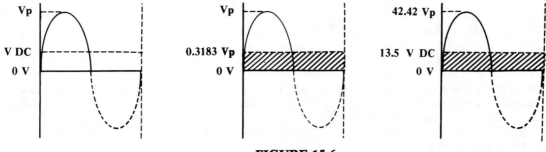

FIGURE 15-6

b) The V DC = V_p/π formula is dervied by using the classical approach of integrating the area under the positive-going pulse, averaged over one complete (2π) cycle.

$$V\ DC = V_{av} = \frac{1}{T} \int V\ dt = \frac{1}{2\pi} \int_0^\pi V_p \sin \omega t\ dt = \frac{V_p}{2\pi} \left[-\cos \omega t \right]_0^\pi$$

$$= \frac{V_p}{2\pi} [-(\cos \pi - \cos 0)] = \frac{V_p}{2\pi} (-\cos \pi + \cos 0) = \frac{V_p}{2\pi} [-(-1) + 1] = \frac{V_p}{\pi}$$

NOTE: Another technique used in solving for the DC voltage content at the load, for the halfwave rectifier circuit, is to multiply the RMS voltage value by 0.45. This is where the ratio of the V DC to RMS equals 0.45 from V DC/ V RMS = 0.3183/0.707.

4. Solving for the DC Current through the load, by dividing the V DC by the 1 kΩ load resistor:

   I DC = V DC/R_L = 13.5 V DC/1 kΩ = 13.5 mA

5. The total RMS voltage at the output load can be solved in two ways.

   a) From the standard theoretical formula:

      V RMS = $V_{peak}/2$ = 42.42 $V_p/2$ = 21.21 V

   b) From the algebraic summation of the V DC and V AC:

      $$V\ RMS = \sqrt{V\ DC^2 + V\ AC^2} = \sqrt{13.5^2 + 16.36^2}$$
      $$= \sqrt{182.25 + 267.65} = \sqrt{449.9} = 21.21\ V$$

NOTE: The total V RMS at the load is comprised of the DC voltage and the effective AC value of the peak-to-peak ripple voltage. Therefore, the V RMS can be solved from the calculus resulting in V RMS = $V_{peak}/2$, or it can be solved from the algebraic summation of V DC and V AC. Essentially, V DC is the average load voltage, V AC is the effective AC value of the ripple voltage at the load, and V RMS is the algebraic summation of the two. Both V DC and V AC are measured using standard voltmeters.

6. Solving for the effective AC value of the ripple voltage at the load:

   VAC = 0.3856 × V_p = 0.3856 × 42.42 V = 16.36 V

NOTE: The V AC (effective value of the peak-to-peak ripple voltage) is solved knowing that the total V RMS is the algebraic summation of the DC voltage content and the effective AC value of the ripple voltage at the load. Therefore, manipulating the equations and solving:

$$V\ RMS = \sqrt{V\ DC^2 + V\ AC^2}\ ,\ \text{therefore,}\ V\ RMS^2 = V\ DC^2 + V\ AC^2$$

$$VAC = \sqrt{V\ RMS^2 - V\ DC^2} = \sqrt{(Vp/2)^2 - (Vp/\pi)^2} = Vp\sqrt{(1/2)^2 - (1/\pi)^2}$$
$$= Vp\sqrt{0.5^2 - 0.3183^2} = Vp\sqrt{0.25 - 0.10132} = 0.3856\ Vp$$

7. a) Solving for the peak inverse voltage of the diode:

   $PIV = Vp = 42.42\ V$

NOTE: The PIV is the parameter of the diode, which denotes the maximum reverse voltage that can be applied across the diode before the device breakdown occurs. The least expensive diodes have a PIV of about 50 volts and diodes rated at 200 PIV to 400 PIV can be obtained for a few pennies more. The 1N4001 is rated at 50 V, but the actual breakdown could exceed 200 V. The circuit PIV of 42.42 V is well within the rated PIV of the 1N4001.

   b) The peak inverse voltage (PIV) for the diode of the halfwave rectifier circuit of Figure 15-7 takes place when the applied negative-going wave turns diode D1 off causing zero volts to exist at the cathode at the same time as the applied negative-going wave reaches its maximum of −42.42 volts at the anode. At this point in time, 42.42 volts is applied across the diode in a reverse direction, and the PIV parameter of the diode must be higher than 42.42 volts to avoid having the device destroyed. For the halfwave circuit, the PIV occurs 60 times per second.

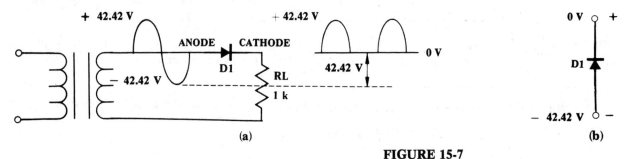

FIGURE 15-7

8. Percent ripple is a comparison of the DC voltage content to the effective value of the AC content at the load. For the ideal power supply, the percent ripple is zero, but for the halfwave rectifier circuit it is 121%. Therefore, solving for the percent ripple for the unfiltered halfwave rectifier circuit in two ways:

   a) By taking the ratio of V AC to V DC and multiplying by 100:

   % ripple = V AC/V DC × 100 = 16.36 V/13.5 V × 100 = 121%

   b) By using the classical form factor approach, where $F = V\ RMS/V\ DC$:

$$\%\ ripple = \sqrt{F^2 - 1} = \sqrt{\left(\frac{V\ RMS}{V\ DC}\right)^2 - 1} = \sqrt{\left(\frac{Vp}{2} \Big/ \frac{Vp}{\pi}\right)^2 - 1}$$

$$= \sqrt{\left(\frac{Vp}{2} \times \frac{\pi}{Vp}\right)^2 - 1} = \sqrt{\left(\frac{\pi}{2}\right)^2 - 1} = \sqrt{\left(\frac{3.14}{2}\right)^2 - 1}$$

$$= \sqrt{1.57^2 - 1} = \sqrt{2.465 - 1} = \sqrt{1.467} = 1.21$$

and, in terms of percent, 1.21 × 100 = 121 %

Part 3: The Filtered Half-Wave Rectifier Circuit

A large capacitor placed across the load, as shown in **Figure 15-8**, reduces the high ripple content associated with the output of the unfiltered half wave rectifier circuit. The capacitor repetitively charges up to the peak value of the positive-going pulses, 60 times per second, and then discharges on the negative-going pulses. Because the capacitor cannot discharge in 1/60 of a second under light loading, the amount of ripple will be relatively small. Since the ripple voltage ΔV rides on the DC voltage, the DC voltage will equal the peak voltage minus one half of the ripple voltage. If ΔV is small, the peak voltage approximately equals the DC voltage.

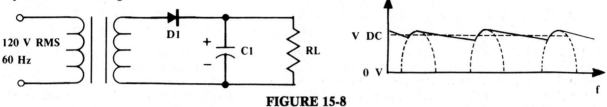

FIGURE 15-8

The ripple voltage has a sawtooth shape because the positive-going pulses (60 per second) charge the capacitor to peak value, and then the capacitor discharges into the load with the time constant $R_L C$ — a relatively long discharge time. The magnitude of the ripple voltage is a function of frequency, capacitance, and load current. The ripple voltage across the filter capacitor is solved from: $\Delta V_C \approx T/C \times V_p/R_L$. The inclusion of the 3 ohm RS resistor serves two purposes: it limits the surge current and it allows peak recurrent measurements.

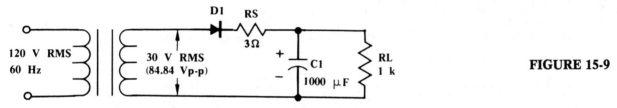

FIGURE 15-9

1. Solving for the peak-to-peak voltage at the secondary:

    Vp-p = V RMS × 2√2 = 30 V × 2.828 = 84.84 Vp-p

2. Solving for the peak voltage at the load:

    Vpeak = Vp-p/2 = 84.84 Vp-p/ 2 = 42.42 Vpeak

3. Solving for the peak current through the load resistor:

    Ipeak = Vpeak/RL = 42.42 Vp/1 kΩ = 42.42 mA

4. Solving for the initial DC voltage at the load:

    V DC ≈ Vpeak ≈ 42.4 V DC

NOTE: While it is impossible to calculate, initially, the exact DC voltage across RL, an estimation of V DC ≈ Vp ≈ 42.4 V is acceptable, because the ripply voltage is rarely designed much above 1 Vp-p across the filter capacitor. The more precise V DC is solved only after the (ΔV) is know.

5. Solving for the ripple voltage magnitude across the filter capacitor from:

$$\Delta V_{C1} = \frac{T}{C} \times I_{RL} = \frac{T}{C} \times \frac{V_p}{R_L} = \frac{1/60 \times 42.4 \text{ V}}{1000 \text{ }\mu F \times 1 \text{ k}\Omega} = \frac{16.6 \times 10^{-3} \times 42.4 \text{ V}}{10^{-3} \times 10^3} = 706 \text{ mVp-p}$$

$$\Delta V_{C1} = 1/fC_1 \times V_p/R_L = V_p/fC_1 R_L = 42.4 \text{ V}/(60 \text{ Hz} \times 1000 \text{ }\mu F \times 1 \text{ k}\Omega) = 706 \text{ mVp-p}$$

6. The more precise V DC is slightly smaller than the Vp voltage because the ripple voltage is 706 mVp-p and rides on the DC voltage. The DC voltage is approximately 42.07 V DC and the ripple voltage is ± 0.35 volts. This is shown in Figure 15-10, where the ripple voltage limit is the peak voltage and the actual DC is solved by subtracting one-half the ΔV_{C1} ripple voltage from Vp. Therefore:

    V DC = Vp − ΔV_{C1}/2 = 42.42 V − 0.7 V/2 = 42.07 V DC

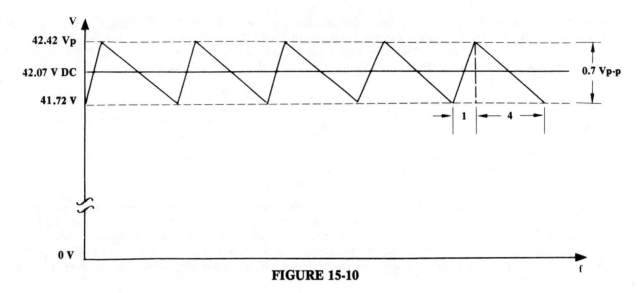

FIGURE 15-10

NOTE: ΔV across the capacitor carries the subscript ΔV_{C1}.

7. Solving for the DC current through the load by dividing the more precise 42.07 V by the 1 kΩ load resistor:

$$I\ DC = V\ DC/R_L = 42.07\ V\ DC/1\ k\Omega \approx 42\ V\ DC/1\ k\Omega = 42\ mA$$

8. The recurring or repetitive DC charging current through the diode, based on a sample 4 : 1 discharging-charging ratio, is measured across the load, as shown in Figure 15-11. It is solved by multiplying the DC load current by 4.

$$I\ DC(recurring) = 4 \times I\ DC(load) = 4 \times 42\ mA = 168\ mA$$

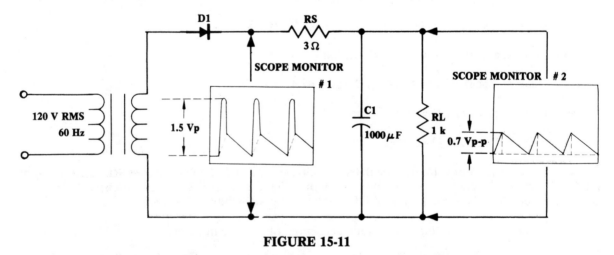

FIGURE 15-11

NOTE: The advantages of filtering are reduced ripple and higher DC content to the load, which approaches Vpeak in magnitude. However, the disadvantage is increased current flow into the capacitor during the charging time and, hence, increased current flow through the series diode D1. This increased current occurs because the discharge time has a longer duration than the charge time Therefore, current flow drawn out of the capacitor during discharge time must be replenished in the shorter span of charge time. For instance, 42 mA delivered during discharge will charge to approximately 168 mA if the charge time is one-fourth the discharge time. The charge-discharge sawtooth wave shape can be monitored across the load with an oscilloscope, and it is similar to the sample 4 : 1 discharge-charge wave shape shown in Figure 15-10 and Figure 15-11. Also, because the charge time occurs with each positive-going pulse, the 168 mA of DC charging current is recurring - that is, it occurs 60 times a second for the halfwave rectifier.

The DC charging current should not exceed the DC current rating of the diode, and it is one ampere for the 1N4001 device.

A second form of recurring current is peak recurring current. It too can be measured, as shown in Figure 15-11. For example, if 1.5 Vp is developed across the 3 ohm RS resistor, then peak current charges through the diode 60 times per second. It is solved from:

$$Ip(recurring) = V\ RS(peak)/RS = 1.5\ V/3\ \Omega = 500\ mA$$

NOTE: The rated peak recurring current for the 1N4001 diode is 10 amperes.

9. **SURGE CURRENT.** The maximum peak, worst-case, surge current is approximated by dividing Vp by the 3 ohm current limiting resistor. For the worst-case condition, the DC resistance of the transformer and the diode are ignored.

$$I(surge) = Vp/RS = 42.42\ V/3\ \Omega \approx 14\ A$$

NOTE: Surge current is a one time, momentary current surge that occurs at the instant power is turned on for the circuit. Essentially, the momentary high current condition occurs because, with reference to Figure 15-11, the uncharged C1 filter capacitor looks like an effective short circuit the instant power is applied and only the RS resistor, along with the transformer and diode resistances, limit the instantaneous peak current flow. Using only RS in solving for the peak surge current provides a worst-case current condition, which aids in selecting a diode that is superior to actual surge current conditions. A 1 ampere diode like the 1N4001 can withstand surge current levels up to 30 amperes.

10. Solving for the effective value of the AC ripple voltage at the load:

$$V\ AC(load) = \Delta V_{C1}/2\sqrt{3} = 706\ mVp\text{-}p/3.464 \approx 204\ mV$$

NOTE: The effective AC value of a sawtooth ripple voltage wave, as shown in Figure 15-11, is approximated by dividing the peak-to-peak, sawtooth-ripple voltage by $2\sqrt{3}$.

11. Solving for the percent ripple:

$$\%\ ripple = (V\ AC/V\ DC)100 = (204\ mV/42.07\ V)100 \approx 0.485\%$$

NOTE: Percent ripple is solved by taking the ratio of the effective value of the AC ripple voltage to the DC voltage and multiplying by 100 to attain the percentage.

12. Solving the the peak inverse voltage of the Diode D1:

$$PIV = 2\ Vp \approx 2 \times 42.42\ V = 84.84\ V$$

For well filtered circuits, the charged filter capacitor remains at, or about, the Vp value of 42 volts. However, when the applied negative-going peak voltage reaches 42.42 volts, the reverse voltage across the diode is at approximately 84 volts. This results in a PIV of 2 Vp, as shown in Figure 15-12.

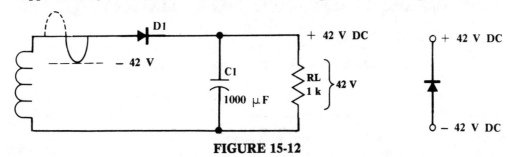

FIGURE 15-12

NOTE: The PIV of the 1N4001 is rated at 50 V, but in practice the PIV can exceed 200 V. If in doubt, use the 1N4002 rated at 100 V, the 1N4003 rated at 200 V, or the 1N4004 rated at 400 V. In all other parameter respects, these diodes are alike.

13. Solving for the percent regulation:

$$\% \text{ regulation} = \frac{V(\text{no load}) - V(\text{load})}{V(\text{no load})} \times 100 = \frac{42.42 \text{ V} - 42.07 \text{ V}}{42.42 \text{ V}} \times 100 = \frac{0.35 \text{ V}}{42.42 \text{ V}} \times 100 \approx 82.5\%$$

$$\% \text{ regulation} = \frac{V(\text{no load}) - V(\text{load})}{V(\text{load})} \times 100 = \frac{42.42 \text{ V} - 42.07 \text{ V}}{42.07 \text{ V}} \times 100 = \frac{0.35 \text{ V}}{42.07 \text{ V}} \times 100 \approx 83.2\%$$

NOTE: Two methods are used to show percent regulation because the Federal Communication Commission (FCC) uses the latter method and most of the "real world" uses the former method. Most engineering standards are based on the normal 0 to 100% scale, while the FCC, in this instance, operates on a zero to infinity scale. However, as shown in the solved examples, the choice of using one method over the other is purely academic.

DIODE SELECTION — BASED ON PIV AND CIRCUIT CURRENTS

Selecting the diode for the circuit is based on the current carrying capabilities and the peak inverse voltage (PIV) of the device. The actual PIV, of most diodes, is higher than the rated values, but the surge current, repetitive current (recurring), or DC current through the device should not exceed the rated values.

Therefore, to review the current ratings of diodes, the surge current happens when the power is turned on. It is one time and instantaneous. The rated surge current of diodes is high and, for the 1N4001, it is 30 amperes. The repetitive, or recurring, peak current happens each time the capacitor is being charged - 60 times a second for the halfwave filtered circuits, and the diode rating for the 1N4001 is 10 amperes. The DC, or DC charging, current rating of diodes is low and for the 1N4001 it is 1 ampere. The DC charging current for the halfwave filtered circuit also occurs 60 times per second. Also, because the DC current rating of the diode is the lowest, it is the current most likely to destroy the device.

EFFECTS OF LOAD RESISTOR OR FILTER CAPACITOR VARIATIONS

Using a load resistor of 1 kΩ and a C1 capacitor of 1000 µF provides a ripple voltage ΔV_{C1} of about 700 mVp-p. However, if the load resistor is decreased to 100 ohms across the 1000 µF capacitor, the ripple voltage will increase to about 7 Vp-p. The large ripple voltage of 7 Vp-p will also occur if the load is maintained at 1 kΩ, but the filter capacitor lowered to 100 µF. Too, because the ripple voltage "rides" on the DC voltage and the maximum positive-going excursion of the ripple voltage only goes to about Vp, then the V DC voltage across the capacitor will drop to about 39.9 V, as shown in Figure 15-13(b). Both circuits of Figure 15-13 are shown solved, but in large ripple voltage conditions the measured to calculated values are not as accurate as with smaller ripple voltage levels of about 1 Vp-p or less.

A. $R_L = 1$ kΩ and $C1 = 100$ µF

$$\Delta V_{C1} \approx V_p / fCR_L = 42.4 \text{ V} / 60 \times 100 \text{ µF} \times 1 \text{ k}\Omega = 42.4 \text{ V}/6 \approx 7 \text{ Vp-p}$$

$$V \text{ DC} \approx V_p - \Delta V_{C1}/2 = 42.4 \text{ V} - 7 \text{ V}/2 = 42.4 \text{ V} - 3.5 \text{ V} = 39.9 \text{ V}$$

B. $R_L = 100$ Ω and $C1 = 1000$ µF

$$\Delta V_{C1} \approx V_p / fCR_L = 42.4 \text{ V} / 60 \times 1000 \text{ µF} \times 100 \text{ }\Omega = 42.4 \text{ V}/6 \approx 7 \text{ Vp-p}$$

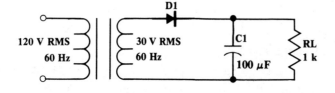

FIGURE 15-13(a)

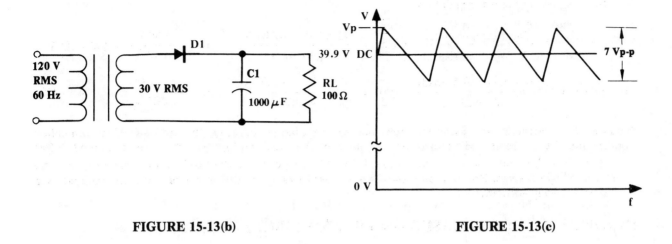

FIGURE 15-13(b)   **FIGURE 15-13(c)**

The % ripple for a V DC of 39.9 V and an effective ripple voltage of 7 Vp-p is solved from:

$$\% \text{ ripple} = \frac{V\ AC \times 100}{V\ DC} = \frac{7\ Vp\text{-}p/2\sqrt{3} \times 100}{39.9\ V} = 5.06\ \%$$

NOTE: Filtering of 5.06% is considered intolerable. Reasonable accuracy in the calculations can be achieved for light loading, only when the ripple voltage magnitude is kept below 1 Vp-p. For instance:

$$\Delta V_{C1} \approx T/C \times Vp/RL = 1/fC \times I = I/fC$$

and, for the halfwave circuit, a 1000 μF capacitor and 60 mA will provide 1 Vp-p of ripple voltage. Further combinations of C1 and I(Vp/RL) can be used, such as 200 Ω × 5000 μF or 10 kΩ × 100 μF, which will not exceed 1 Vp-p, if 60 mA is maintained. Therefore, 1 Vp-p can be used as an initial "benchmark" in selecting filter capacitors for the load current needed.

SECTION I: LABORATORY EXERCISE

OBJECTIVES
To analyze and investigate the halfwave, unfiltered and filtered, rectifier circuits.

LIST OF MATERIALS
1. Transformer: 120 V : 24 V RMS or equivalent (not exceeding 30 volts)
2. Diode: 1N4001 or equivalent
3. Resistors (one each):   10 Ω   1 kΩ
4. Capacitor: 1000 μF
5. Oscilloscope (dual trace if available)
6. Voltmeter
7. Ammeter

NOTE: 24 volts is a standard transformer secondary voltage, but any transformer secondary voltage between 12 V RMS and 30 V RMS can be used, successfully, to work the experiments.

PROCEDURE
Part 1: The Halfwave Rectifier Circuit
1. Connect the circuit as shown in Figure 15-14.

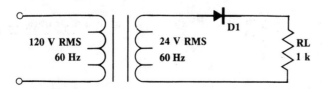

FIGURE 15-14

2. Across the secondary of the transformer:
   a) Measure the V RMS with a voltmeter.
   b) Measure the Vp-p with an oscilloscope.
   c) Convert the Vp-p scope reading to V RMS from:

$$\text{V RMS(sec)} = \frac{\text{Vp-p(sec)}}{2\sqrt{2}} \approx \frac{\text{Vp-p(sec)}}{2.828}$$

NOTE: Use the measured oscilloscope values in all of the calculations. Also, calibrate the equipment as good measuring techniques will insure proper calculated to measured results.

3. a) Calculate Vp at the load from Vp-p(sec)/2.
   b) Measure the peak voltage Vp at the load.
   c) Calculate the peak current (Ip) from: $I_p = V_p/R_L$.
4. a) Calculate the DC voltage content at the load from: $V\ DC = V_p/\pi$.
   b) Measure the DC voltage across the load.

NOTE: Use the DC, AC switch on the oscilloscope to see the effect of the ripple voltage riding on the DC.

   c) Calculate the DC current from: $I\ DC = V\ DC/R_L$.
   d) Measure the DC current. Use an ammeter or, if the load resistance is known accurately, use the calculation of $I\ DC = V_{R_L}/R_L$.
5. a) Calculate the value of the ripple voltage at the load from: $V\ AC = 0.3856 \times V_p$.
   b) Measure the effective AC value of the ripple voltage at the load using a standard AC voltmeter.
6. Calculate the total RMS voltage at the load from:

   a) $V\ RMS = V_p/2$

   b) $V\ RMS = \sqrt{V\ DC^2 + V\ AC^2}$

NOTE: $V\ RMS = V_p/2$ should be reasonably close in value to $V\ RMS = \sqrt{V\ DC^2 + V\ AC^2}$, but enter only the latter result, Step b, into Table 15-1.

7. a) Calculate the peak inverse voltage PIV from: $PIV = V_p$.
   b) Measure the PIV of D1 with a dual trace oscilloscope. Monitor the output load with one oscilloscope probe and monitor the anode of D1 with the other oscilloscope probe.
8. Calculate the percent ripple at the load from: % ripple = (V AC/V DC)100

NOTE: Use the measured V DC and V AC values in the calculations.

9. Insert the calculated and measured values, as indicated, into Table 15-1.

TABLE 15-1	Vp-p Sec	V RMS Sec	Vp Load	Ip Load	V DC Load	I DC Load	V AC Load	PIV	V RMS Load	%Ripple
CALCULATED	//////									
MEASURED			//////						//////	//////

Part 2: Filtering the Halfwave Rectifier Circuit

1. Connect the circuit as shown in Figure 15-15, where a 1000 µF filter capacitor and a 10 ohm RS resistor have been added.
2. V DC to the load:
   a) Calculate the approximate V DC from: $V\ DC \approx V_p - \Delta V_{C1}/2 \approx V_p$
   b) Measure the DC voltage across the load.

NOTE: The "exact" value of V DC, solved by subtracting $\Delta V_{C1}/2$ from Vp, is academic. It becomes significant only when ΔV_{C1} exceeds 2 Vp-p.

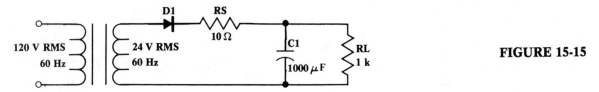

FIGURE 15-15

3. I DC in the load:

   Calculate the DC current in the load from: I DC = V DC/RL

NOTE: IDC can be verified by placing an ammeter in series with RL.

4. Ripple voltage across C1:

   a) Calculate the ripple voltage from:

   $$\Delta V_{C1} \approx T/C1 \times Vp/RL = Vp/fC_1R_L \quad \text{where:} \quad f = 60 \text{ Hz}$$

   b) Measure the ripple voltage with an oscilloscope.

NOTE: To measure ΔV_{C1}, set the oscilloscope in the 100 mV to 1 V per division range and in the AC mode.

5. a) Calculate the effective value of the ΔV_{C1} from: V AC = $\Delta V_{C1}/2 \sqrt{3}$.

   b) Measure the V AC at the load with an AC voltmeter.

5. RECURRING CURRENT:

   a) Monitor the wave slope at the load (SCOPE #1), as shown in Figure 15-16, and calculate the DC charging current through the diode from:

   I DC(recurring) = I DC × discharging time/charge time

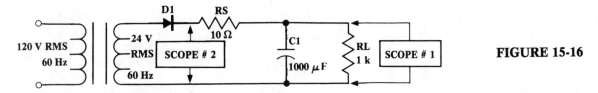

FIGURE 15-16

   b) Monitor the wave slope across RS (SCOPE #2), as shown in Figure 15-16, and calculate the peak charging current through the diode from:

   Ip(recurring) = V_{RS}(peak)/RS

7. PEAK INVERSE VOLTAGE:

   a) Calculate the PIV of the D1 diode from: PIV(D1) = 2 Vp.

   b) Measure the PIV using a dual scope.

NOTE: Using a dual-trace oscilloscope will show the positive DC voltage at the cathode and the negative-going peak pulse of the peak-to-peak input wave at the anode. On the oscilloscope, use the 10 V to 20 V per division range and set the scope in the DC mode.

8. Calculate the % ripple from: % ripple = V AC/ V DC × 100

9. Insert the calculated and measured values, as indicated, into Table 15-2.

TABLE 15-2	LOAD V DC	LOAD I DC	I(recurring) DC	I(recurring) peak	ΔV_{C1}	V AC	PIV	% ripple
CALCULATED								
MEASURED			///////	///////				///////

SECTION II: FULL WAVE RECTIFIER CIRCUITS

The advantages of full wave over halfwave rectification are that twice the number of pulses (frequency) exist at the output load, the DC content is doubled, the AC content in decreased, and the percent ripple is lowered. Two methods used to provide full wave rectification are the full wave bridge and the full wave center-tapped rectifier.

Part 1: The Full-Wave, Center-Tapped, Rectifier Circuit

The full-wave, center-tapped, rectifier circuit of Figure 15-17 uses two diodes and a "grounded" center-tapped transformer to produce a full-wave condition at the load. The effect of this center-tapped connection is to produce two back-to-back half wave rectifier circuits 180° out of phase with each other. Therefore, the amplitude of the waveshape remains the same, but double the amount of pulses exist at the load. Hence, the average, or DC voltage, is doubled.

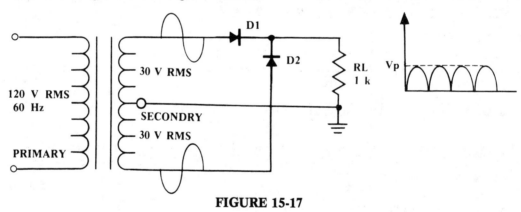

FIGURE 15-17

1. a) Solving for the p-p voltage at the anode of D1 with respect to the center-tap ground:

   Vp-p = $2\sqrt{2}$ V RMS ≈ 2.828 × 30 V RMS = 84.85 Vp-p

   b) Solving for the p-p voltage at the anode of D2 with respect to the center-tap ground:

   Vp-p = $2\sqrt{2}$ V RMS ≈ 2.828 × 30 V RMS = 84.85 Vp-p

NOTE: As shown in Figure 15-17, a 180° phase shift difference exists between the peak-to-peak voltage at the anode of D1, with respect to the anode of D2. Also, since the center tap of the transformer is at ground, each halfwave rectifier connects to only 30 V RMS, or half of the full 60 V center-tapped secondary.

2. a) Solving for the peak voltage at the load resulting from D1 and the upper portion (30 V RMS) of the center-tapped transformer:

   Vp(D1) = Vp-p/2 = 84.85 Vp-p/2 ≈ 42.42 Vp

b) Solving for the peak voltage at the load resulting from D2 and the lower portion (30 V RMS) of the center-tapped transformer:

$$V_p(D2) = V_{p-p}/2 = 84.85\ V_{p-p}/2 \approx 42.42\ V_p$$

NOTE: The peak voltage to the load that is alternately processed by D1 and then by D2 is of the same magnitude as the half wave rectifier circuit, but there are twice as many pulses. Therefore, the frequency increase of the full wave. over the halfwave, doubles — from 60 Hz to 120 Hz.

3. a) Solving for the peak current through the load resistor:

$$I_p = V_p/R_L = 42.42\ V_p/1\ k\Omega = 42.42\ mA$$

b) Solving for the peak current through each diode:

$$I_p(D1) = V_p/R_L = 42.42\ V_p/1\ k\Omega = 42.42\ mA$$

$$I_p(D2) = V_p/R_L = 42.42\ V_p/1\ k\Omega = 42.42\ mA$$

NOTE: This is similar to the halfwave rectifier circuit where the peak current that flows through the load flows through the diode. For the full wave center-tapped rectifier, $I_p(D1) = I_p(D2)$.

4. a) Solving for the DC voltage content of the peak voltage at the load:

$$V\ DC = 2 \times V_{peak}/\pi \approx 2 \times 42.42\ V_p/3.14 = 2 \times 13.5\ V = 27\ V$$

NOTE: The DC voltage is the average of the pulsating positive-going peak voltage and, since twice the number of pulses exist that existed with respect to the half wave, twice as much DC content also exists. The V DC is 0.6366 of the peak voltage for the full-wave, unregulated, rectifier circuit. Therefore, V DC = $0.6366 \times 42.42\ V_p = 27\ V\ DC$, as shown in Figure 15-18.

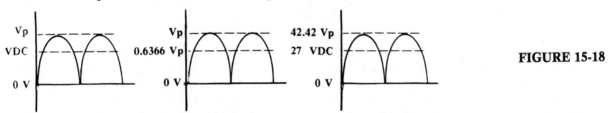

FIGURE 15-18

b) Another technique used in solving for the DC voltage content at the load for a full-wave rectifier circuit is to multiply the RMS voltage value by 0.9. This is the V DC to RMS ratio where V DC/RMS = 0.6366/0.707 = 0.9.

$$V\ DC = 0.9\ V\ RMS = 0.9 \times 30\ V\ RMS = 27\ V\ DC$$

5. a) Solving for the DC current at the load:

$$I\ DC = V\ DC/R_L = 27\ V/1\ k\Omega = 27\ mA$$

b) Solving for the DC current through each diode:

$$I\ DC(D1) = I\ DC(load)/2 = 27\ mA/2 = 13.5\ mA$$

$$I\ DC(D2) = I\ DC(load)/2 = 27\ mA/2 = 13.5\ mA$$

NOTE: Since the full wave center-tapped rectifier circuit is effectively two half wave rectifier circuits 180° removed, each circuit delivers to the load the equivalent of what a halfwave circuit does. The effect is that each diode delivers one half of the DC load current. For instance, if one of the diodes is temporarily disconnected, the circuit would again be halfwave and the frequency would drop from 120 Hz to 60 Hz, the average DC voltage from 27 V DC to 13.5 V DC, and the DC current from 27 mA to 13.5 mA.

6. Solving for the effective AC value of the ripple voltage at the load:

$$V\ AC = 0.3077 \times Vp = 0.3077 \times 42.42\ V = 13.05\ V$$

NOTE: Rearranging the total V RMS formula and solving for the V AC in terms of the full-wave V DC and V RMS:

$$V\ RMS = \sqrt{V\ DC^2 + V\ AC^2} \quad \text{therefore,} \quad V\ RMS^2 = V\ DC^2 + V\ AC^2$$

$$V\ AC = \sqrt{V\ RMS^2 - V\ DC^2} = \sqrt{(Vp/\sqrt{2})^2 - (2Vp/\pi)^2}$$

$$= Vp\sqrt{(1/\sqrt{2})^2 - (2/\pi)^2} = Vp\sqrt{0.7071^2 - 0.6366^2}$$

$$= Vp\sqrt{0.5 - 0.4053} = 0.3077\ Vp$$

7. a) Solving for the total RMS at the output load from $Vp/\sqrt{2}$:

$$V\ RMS = Vp/\sqrt{2} \approx 42.42\ V / 1.414 = 29.995\ V \approx 30\ V$$

b) The total RMS at the load can also be solved from:

$$V\ RMS = \sqrt{V\ DC^2 + V\ AC^2} = \sqrt{27\ V^2 + 13.05\ V^2}$$

$$= \sqrt{729 + 170.3} = \sqrt{899.3} = 29.98\ V \approx 30\ V$$

NOTE: Since both pulses are processed to the load during full wave rectification, the applied V RMS equals the V RMS at the load.

8. Solving for the peak inverse voltage (PIV) of diodes D1 and D2:

$$PIV(D1) = 2\ Vp = 2 \times 42.42\ V \approx 84.84\ V$$

$$PIV(D2) = 2\ Vp = 2 \times 42.42\ V \approx 84.84\ V$$

NOTE: For the full-wave, center-tapped, rectifier circuit, the positive-going wave, processed by diode D1, coincides with the negative-going wave at the anode of diode D2. Therefore, at that point in time, a peak 42.42 V exists at the output load and a −42.42 V peak exists at the anode of D2, for 84.84 V of reverse biased voltage across D2. Likewise, a positive-going 42.42 V processed by D2 coincides with the negative-going −42.42 V at the anode of D1, causing a PIV for D1 of 84.84 V. This 180° phase shift difference, the output voltage, and the effective PIV are shown in **Figure 15-19**.

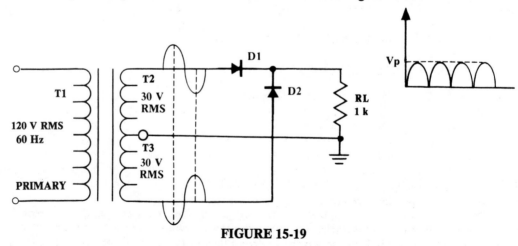

FIGURE 15-19

9. Percent ripple is a comparison of the DC voltage content to the effective value of the AC ripple voltage

at the load. For the full-wave rectifier circuit it is 48.3%. The percent ripple for the unfiltered full-wave rectifier circuit can be solved in two ways:

a) By taking the ratio of the V AC to the V DC and multiplying by 100:

% ripple = V AC/V DC × 100 = 13.05 V/27 V × 100 = 48.3%

b) By using the classical form factor approach, where $F = $ V RMS/V DC:

% ripple = $\sqrt{F^2 - 1}$

$$= \sqrt{\left(\frac{V\ RMS}{V\ DC}\right)^2 - 1} = \sqrt{\left(\frac{Vp}{\sqrt{2}} \bigg/ \frac{2\ Vp}{\pi}\right)^2 - 1} = \sqrt{\left(\frac{Vp}{\sqrt{2}} \times \frac{\pi}{2\ Vp}\right)^2 - 1}$$

$$= \sqrt{\left(\frac{\pi}{2\sqrt{2}}\right)^2 - 1} = \sqrt{\left(\frac{3.14}{2.828}\right)^2 - 1} = \sqrt{1.11^2 - 1} \approx \sqrt{1.234 - 1} \approx 0.483$$

and, in terms of percent, 0.483 × 100 = 48.3%. Also,

$$\% \text{ ripple} = \sqrt{F^2 - 1} = \sqrt{\left(\frac{V\ RMS}{V\ DC}\right)^2 - 1} = \sqrt{\left(\frac{0.707}{0.6366}\right)^2 - 1} = \sqrt{1.11^2 - 1} = 0.483$$

and, in terms of percent, 0.483 × 100 = 48.3%

CAPACATIVE FILTERING OF THE FULL WAVE RECTIFIER CIRCUIT

The full-wave rectifier circuit, such as that of Figure 15-20, is easier to filter than the halfwave rectifier circuit because twice as many pulses (120 Hz) exist across the load. Therefore, the magnitude of the ripple voltage ΔV_{C1} is cut in half and, in comparison to the halfwave rectifier, the full wave circuit needs only one half the capacitance, or one half the value of load resistance, in providing equal ripple voltage magnitude.

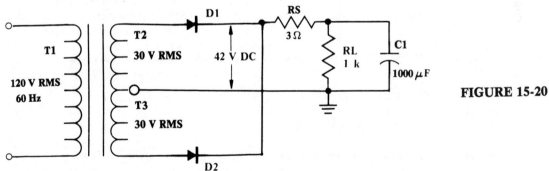

FIGURE 15-20

1. Solving for the peak-to-peak voltage at the anodes of D1 and D2:

Vp-p = VRMS × 2$\sqrt{2}$ = 30 V × 2.828 = 84.84 Vp-p

NOTE: The center tapped full wave rectifier circuit is effectively two separate half wave circuits, operating 180° out of phase with each other. Therefore, each diode "sees" only one half of the transformer—30 V RMS or 84.84 Vp-p. They are each analyzed in the same way.

2. Solving for the peak voltage at, and across, capacitor C1, where either half of the secondary (30 V RMS) can be used in calculating the peak voltage at the load.

Vpeak = Vp-p/2 = 84.84 Vp-p/2 = 42.42 Vpeak

3. Solving for the peak current through the load resistor:

$$I_{peak} = V_{peak}/R_L = 42.42 \text{ V}/1 \text{ k}\Omega = 42.42 \text{ mA}$$

4. Solving for the DC voltage at the load:

$V_{DC} \approx V_{peak}$, or more precisely:

$$V_{DC} = V_p - (\Delta V_{C1}/2) \approx 42.42 \text{ V} - 0.35 \text{ V}/2 = 42.42 \text{ V} - 0.175 = 42.245 \text{ V} \approx 42.25 \text{ V}$$

5. Solving for the magnitude of the ripple voltage:

$$\Delta V_{C1} = T/C \times I_{RL} = T/C \times V_{DC}/R_L = (1/120 \times 42.25 \text{ V})/(1000 \text{ }\mu\text{F} \times 1 \text{ k}\Omega) \approx 0.35 \text{ Vp-p}$$

$$\Delta V_{C1} = V_p/fC_1R_L = 42.4 \text{ V}/(120 \text{ Hz} \times 1000 \text{ }\mu\text{F} \times 1 \text{ k}\Omega) = 0.35 \text{ Vp-p}$$

NOTE: For the full wave rectifier circuit, the pulse repetition rate is 120 Hertz per second.

6. The DC voltage, the peak voltage, and the ΔV_{C1} ripple voltage for the full wave rectifier circuit of Figure 15-20 is shown in Figure 15-21. The DC voltage is ≈ 42.25 volts and the ripple voltage of ≈ 0.35 Vp-p rides on it. The peak excursion of the ripple voltage is 42.42 volts, the minimum excursion is 42.07 volts, and the V DC is 42.25 volts or, more precisely, 42.245 volts. Again, notice that the difference voltage between V_p and the more precise V DC is slight, so in practical applications it is ignored.

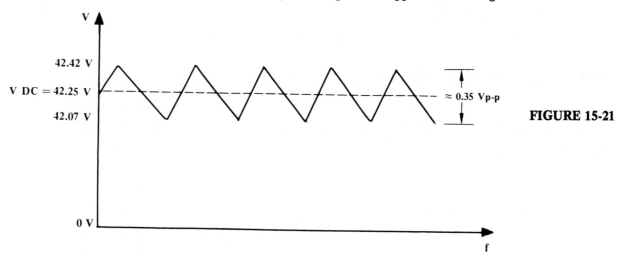

FIGURE 15-21

7. Solving for the DC current through the load:

$$I_{DC} = V_{DC}/R_L = 42.25 \text{ V}/1 \text{ k}\Omega = 42.25 \text{ mA}$$

8. Solving for the DC charging current through the diodes where, as an example, the ratio of discharge time to charge time is given at 4 : 1 — as shown in Figure 15-22. In this example, the capacitance is discharged into (by) the load over a period of time and it is recharged to peak value in one-fourth the discharge time. Therefore, the charge current into the capacitor must be four times larger than the discharge current.

$$I_{DC(recurring)} = \frac{I_{DC} \times \text{discharge time}}{\text{charge time}} = \frac{42.25 \text{ mA} \times 4}{1} = 170 \text{ mA}$$

NOTE: If the discharge-charge ratio is selected at 4 : 1 again, the recurring current flow through the diodes will be four times greater than the DC current into the load. Hence, $I_{D1} = I_{D2} = 170$ mA, since the DC charging current through each diode occurs 60 times per second, or every alternate pulse.

9. Solving for the peak recurring charging current through each diode where, again, 1.5 volts is shown developed across a 3 ohm RS resistor. Refer to Figure 15-22.

Ip(recurring) = 1.5 Vp/3 Ω = 500 mA

NOTE: The 3 ohm RS resistor, as it did in the halfwave rectifier circuit, limits the surge current to about 14 amperes.

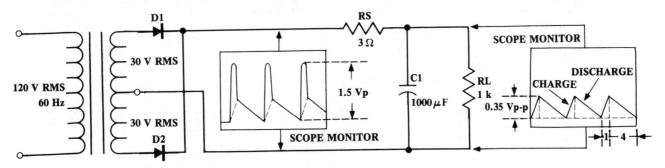

FIGURE 15-22

10. Solving for the effective AC value of the ripple voltage by dividing the peak-to-peak ripple voltage by $2\sqrt{3}$.

$$V\,AC = \Delta V_{C1}/2\sqrt{3} = 0.35\,\text{Vp-p}/4.464 = 0.101\,V$$

11. Solving for the percent ripple:

$$\%\,\text{ripple} = (V\,AC/V\,DC)100 = (0.101\,V/42.25\,V)100 \approx 0.24\%$$

12. Solving for the PIV of diodes D1 and D2:

$$PIV = 2\,V\text{peak} = 2 \times 42.42\,V = 84.84\,V$$

NOTE: The PIV for the diodes in the full wave center-tapped rectifier circuit is the same, with or without a filter capacitor.

13. Solving for the percent regulation:

$$\%\,\text{regulation} = (V[NL]-V[L]/\,V[NL])100 = ([42.42\,V - 42.25\,V]/42.42)100 \approx 0.40\%$$

$$\%\,\text{regulation} = (V[NL]-V[L]/\,V[L])100 = ([42.42\,V - 42.25\,V]/42.25\,V)100 \approx 0.414\%$$

NOTE: Obviously, either method of solution for percent regulation is valid (mathematically close) for low ripple voltage conditions. Again, approximate V DC values alter calculations slightly.

Part 2: The Full Wave Bridge Rectifier

The full wave bridge rectifier circuit, such as that shown in Figure 15-23, uses four diodes and a non-center-tapped transformer secondary to provide full wave rectification to the load. Therefore, it has the

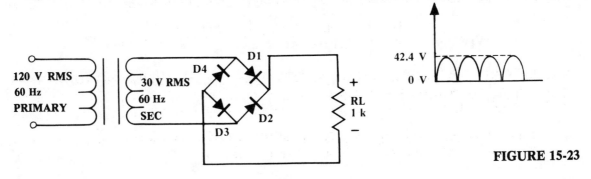

FIGURE 15-23

same output as that provided by the full-wave, center-tapped rectifier, but it does not require a center-tapped transformer and the PIV rating of the diodes need be only half that of those used in the latter circuit. The only minor drawback is that two extra diodes are required.

The full wave operation is best described by applying DC input voltage to show diode switching. See Figure 15-24(a). The positive battery voltage turns on diodes D1 and D3 and turns off diodes D2 and D4 to provide a series of diodes D1, D3, and the load resistor RL, as shown in Figure 15-24 (b). Then a negative

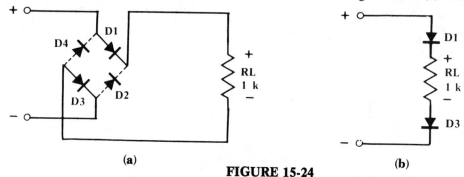

FIGURE 15-24

battery voltage turns on diodes D2 and D4 and turns off diodes D1 and D3. See Figure 15-25(a). This provides a series circuit of diodes D2, D4, and the load resistor RL, as shown in Figure 15-25(b).

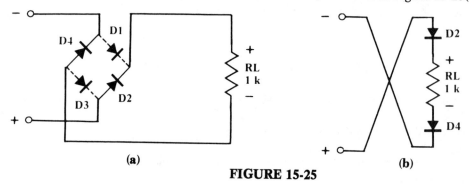

FIGURE 15-25

In circuit operation, series diodes D1 and D3, and series diodes D2 and D4, alternately switch. Therefore, no matter what polarity exists at the input to the bridge, the direction of the voltage drop across RL will remain the same. As shown in Figure 15-26, the full wave bridge rectifies the positive-going and negative-going excursions of the sine wave which provides the load with positive-going pulses only.

1. Solving for the peak-to-peak voltage at the secondary:

    $$V_{p\text{-}p} = V\,RMS \times 2\sqrt{2} = 30\,V \times 2.828 = 84.84\,V_{p\text{-}p}$$

2. Solving for the peak voltage at the load:

    $$V_{peak} = V_{p\text{-}p}/2 = 84.84\,V_{p\text{-}p}/2 = 42.42\,V_{peak}$$

NOTE: The frequency of the full wave waveshape, at the load, is 120 Hz.

3. Solving for the peak current through the load resistor:

    $$I_p = V_{peak}/RL = 42.42\,V/1\,k\Omega = 42.42\,mA$$

4. Solving for the peak current through the diode series D1 and D3 and the diode series D2 and D4:

    $$I_p(D1, D3) = V_{peak}/RL = 42.42\,V/1\,k\Omega = 42.42\,mA$$

    $$I_p(D2, D4) = V_{peak}/RL = 42.42\,V/1\,k\Omega = 42.42\,mA$$

NOTE: With reference to the circuit of Figure 15-26, peak current is a function of peak voltage and the load resistor. However, peak voltage amplitude remains the same, regardless of whether a halfwave or full wave waveshape exists at the load. Therefore, peak current flows through each diode string every other pulse (alternately), and that peak current is 42.42 mA.

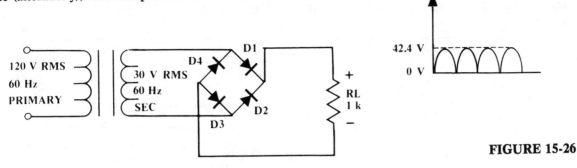

FIGURE 15-26

5. Solving for the DC voltage content at the load:

$$V\ DC = 2 \times V_p/\pi = 2 \times 42.42\ V/3.14 = 2 \times 13.5\ V = 27\ V\ DC$$

Another technique used to solve for the DC content at the load, for the full-wave rectifier circuit, is to multiply the RMS voltage values by 0.9. Therefore, $V\ DC = 0.9 \times V\ RMS = 0.9 \times 30\ V\ RMS = 27\ V\ DC$. (The 0.9 constant was derived in the previous discussion on the center-tapped, full-wave, rectifier circuit.)

6. Solving for the DC current at the load:

$$I\ DC = V\ DC/R_L = 27\ V/1\ k\Omega = 27\ mA$$

7. Solving for the DC current through each diode of the series diodes D1, D3 and D2, D4:

$$I\ DC(D1, D2) = I\ DC(load)/2 = 27\ mA/2 = 13.5\ mA$$

$$I\ DC(D3, D4) = I\ DC(load)/2 = 27\ mA/2 = 13.5\ mA$$

NOTE: If one or both diodes in a diode string were opened, halfwave rectification would occur at the load. However, the peak voltage and peak current remain the same (halfwave or full wave) but the DC current to the load is halved. Therefore, the DC current is supplied (shared) by both diode strings.

8. Solving for the effective AC value of the ripple voltage at the load:

$$V\ AC = 0.3077\ V_p = 42.42\ V \times 0.3077 \approx 13.05\ V$$

9. Solving for the total V RMS at the output load:

$$V\ RMS = V_p/\sqrt{2} = 42.42\ V/1.414 \approx 30\ V$$

NOTE: V RMS can also be solved from:

$$V\ RMS = \sqrt{V\ DC^2 + V\ AC^2} = \sqrt{27^2 + 13.05^2} \approx 30\ V$$

This equation was derived in the previous discussion on the center-tapped full wave circuit.

10. Solving for the PIV of the Diodes D1, D3 and D2, D4:

$$PIV = V_p = 42.42\ V$$

NOTE: The PIV of the diodes in the full wave bridge circuit is equal to the peak voltage. See Figure 15-27(a) and 15-27(b), where the forward biased diodes have been removed to help explain the circuit operation. In Figure 15-27(a), a postive-going pulse turns on diodes D1 and D3 to provide a reverse bias voltage across diodes D2 and D4 — equal to the voltage developed across the load resistor. Since the voltage at the load is Vp, then PIV of diodes D2 and D4 is Vp or 42.42 V. For the negative-going pulse, as shown in Figure 15-27(b), just the opposite occurs. The PIV of diodes D1 and D3 also equals the peak voltage Vp. (The turned on diodes D1 and D3 of Figure 15-27(a) and diodes D2 and D4 of Figure 15-27(b) are represented by a "shorted" condition to show PIV conditions.)

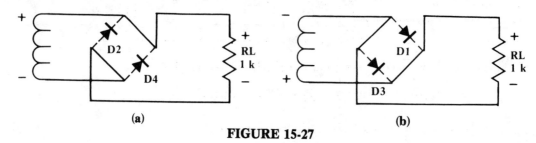

FIGURE 15-27

11. The percent ripple of the full wave bridge rectifier circuit is the same as that of the full wave center-tapped rectifier circuit. Therefore:

% ripple = V AC/V DC × 100 = 13.05 V/27 V × 100 = 48.3%

NOTE: The waveshapes most commonly associated with both unfiltered and filtered "bulk" power supplies are the sine wave, halfwave, full wave, and sawtooth wave. Each of these waves are normally converted to effective AC values and, since they have different shapes and areas, the converting (dividing) factor is slightly different for each wave. The peak-to-peak wave shapes and their effective AC converting factors are shown in Figure 15-28, where all ripple voltage amplitudes are considered peak-to-peak. The oscilloscope is used to measure the peak-to-peak wave and the V AC is either converted, as shown in Figure 15-28, or measured directly with an AC voltmeter. The reason V AC, the effective value of the peak-to-peak ripple voltage, is measured directly is because standard AC voltmeters and the AC mode of the oscilloscope measure only the AC portion of the wave and the DC is excluded.

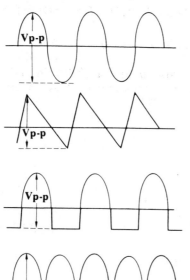

SINE WAVE

$$V\ AC = \frac{V_{p\text{-}p}}{2\sqrt{2}} \approx 0.3536\ V_{p\text{-}p}$$

SAWTOOTH WAVE

$$V\ AC = \frac{V_{p\text{-}p}}{2\sqrt{3}} \approx 0.2887\ V_{p\text{-}p}$$

HALFWAVE (RECTIFIED)

$$V\ AC \approx 0.3856\ V_{p\text{-}p}$$

FULL WAVE (RECTIFIED)

$$V\ AC \approx 0.3077\ V_{p\text{-}p}$$

FIGURE 15-28

CAPACITIVE FILTERING OF THE FULL WAVE BRIDGE RECTIFIER CIRCUIT

The filtered full wave bridge rectifier circuit of Figure 15-29 has all the advantages of full wave rectification without the need for a center-tapped transformer secondary. Therefore, the transformer secondary can be 30 V RMS, instead of 60 V(CT) as required in the full wave center-tapped circuit configuration.

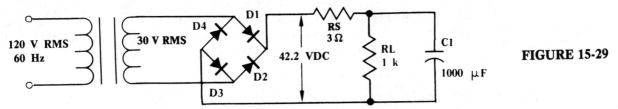

FIGURE 15-29

1. Solving for the peak-to-peak voltage at the anodes of D1 and D2:

    $V_{p-p} = V\ RMS \times 2\sqrt{2} = 30 \times 2.828 = 84.84\ V_{p-p}$

2. Solving for the peak voltage at, and across, capacitor C1:

    $V_{peak} = V_{p-p}/2 = 84.84\ V_{p-p}/2 = 42.42\ V_{peak}$

3. Solving for the peak current through the load resistor:

    $I_{peak} = V_{peak}/R_L = 42.42\ V/1\ k\Omega = 42.42\ mA$

4. Solving for the DC voltage at the load:

    $V\ DC \approx V_{peak} \approx 42.4\ V$ or, more precisely,

    $V\ DC = V_p - (\Delta V_{C1}/2) = 42.42\ V - (0.35\ V_{p-p}/2) \approx 42.25\ V$

5. Solving for the magnitude of the ripple voltage:

    $\Delta V_{C1} = T/C \times I_{RL} \approx T/C \times V\ DC/R_L = (1/120 \times 42.42\ V)/(1000\ \mu F \times 1\ k\Omega) = 0.35\ V_{p-p}$

    $\Delta V_{C1} = I_{RL}/fC_1 = V_p/fC_1R_L = 42.42\ V/(120\ Hz \times 1000\ \mu F \times 1\ k\Omega) = 0.35\ V_{p-p}$

6. The DC voltage, the peak voltage, and the ripple voltage are shown in Figure 15-30. The DC voltage is approximately 42.25 volts and the ripple voltage of 0.35 Vp-p rides on it. These values are the same as those for the full wave rectifier circuit.

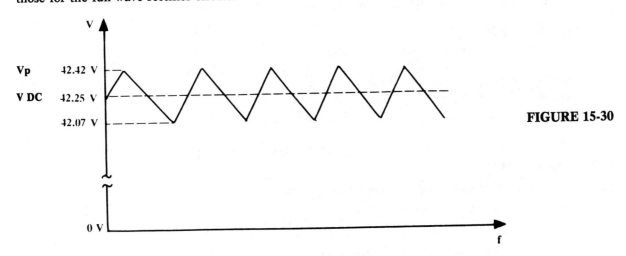

FIGURE 15-30

7. Solving for the effective AC of the peak-to-peak ripple voltage:

    $V\ AC = \Delta V_{C1}/2\sqrt{3} = 350\ mV/3.464 \approx 101\ mV$

8. Solving for the DC current through the load:

   I DC = V DC/RL ≈ 42.25 V/1 kΩ = 42.25 mA

9. Solving for the recurring DC charging current from the diode bridge when the ratio of charge time to discharge time is 4 : 1:

   I DC(recurring) = $\dfrac{\text{discharge time}}{\text{charge time}} \times$ I DC = $\dfrac{1}{4} \times$ 42.5 mA = 170 mA

10. Solving for the recurring peak charging current through series diodes D1, D3 and D2, D4:

    Ip(recurring) ≈ Vp(RS)/RS = 1.5 V/3 Ω = 500 mA

NOTE: Each diode series string (D1, D3 and D2, D4) alternately contributes 500 mA of recurring peak charging current. Therefore, approximately the same amount of peak charging current (approximately 500 mA) flows through each diode. Also, as in previous circuits, I(surge) = 14 amperes.

11. Solving for the PIV of diodes D1, D2, D3, and D4:

    PIV = Vp = 42.42 V

NOTE: The PIV across the diodes of the unfiltered and filtered full wave bridge rectifier circuits is the same. The PIV for the bridge circuit was illustrated in Figure 15-27.

12. Solving for the percent ripple:

    % ripple = V AC/V DC × 100 = ($\Delta V_{C1}/2\sqrt{3}$)/V DC × 100

    = (0.35 Vp-p/3.464)/42.25 V × 100 = (101 mV/42.25 V) × 100 ≈ 0.24%

SECTION II: LABORATORY EXERCISE

OBJECTIVES
To investigate the unfiltered full wave bridge rectifier circuit.

LIST OF MATERIALS
1. Transformer: 120 V : 24 V RMS or equivalent
2. Diodes: 1N4001 (four) or equivalents
3. Resistors (one each):   1 kΩ 10 Ω
4. Capacitor: 1000 μF (one)
5. Oscilloscope
6. Voltmeter
7. Ammeter

PROCEDURE
Part 1: The Full-Wave-Bridge, Rectifier Circuit
1. Connect the circuit shown in Figure 15-31.

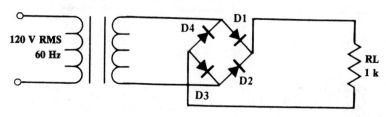

FIGURE 15-31

2. Across the secondary of the transformer:
   a) Measure the peak-to-peak voltage with a scope.
   b) Measure the RMS voltage with a voltmeter.
3. a) Calculate the Vp at the load from: $Vp = Vp\text{-}p/2$
   b) Measure the Vp at the load
4. Calculate Ip from: $Ip = Vp/RL$
5. a) Calculate the V DC at the load from: $V\,DC = 2 \times Vp/\pi$.
   b) Measure V DC.
6. a) Calculate the DC current flow through the load resistor from: $I\,DC = V\,DC/RL$.
   b) Measure the DC current flow through the load. Measure with an ammeter or by monitoring the voltage drop across the known load resistor.
7. a) Calculate the DC current flow through the D1, D3 and D2, D4 diode strings. Calculate from:

   $$I(D1, D3) = I(D2, D4) = I_{RL}/2$$

   b) Measure the DC current flow through each diode string, as shown in Figure 15-32, by placing an ammeter in series with diodes D1 or D3 and then with D2 or D4.

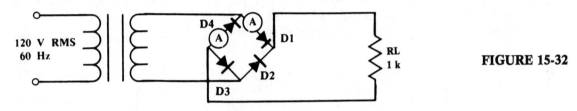

FIGURE 15-32

8. a) Calculate the ripple voltage VAC, at the load, from: $V\,AC = 0.3077 \times Vp$.
   b) Measure the AC ripple voltage, at the load, using a standard AC voltmeter.
9. Calculate the total RMS voltage at the load from:

   $$V\,RMS = Vp/\sqrt{2} \quad \text{or} \quad V\,RMS = \sqrt{V\,DC^2 + V\,AC^2}$$

10. a) Calculate the PIV of (all) the diodes in the bridge circuit from:

    $$PIV(D1) = PIV(D2) = PIV(D3) = PIV(D4) = Vp$$

NOTE: The PIV of the diodes can be measured with a dual trace oscilloscope. Monitor the output with one scope probe and, alternately, monitor the anodes of D1, D2, D3, and D4 with the other probe.

11. Calculate the percent ripple from: % ripple $= V\,AC/V\,DC \times 100$
12. Insert the calculate and measured values, as indicated, into Table 15-3.

TABLE 15-3	Vp-p SEC	V RMS SEC	Vp LOAD	Ip LOAD	V DC LOAD
CALCULATED	/////	/////			
MEASURED				/////	

	I DC			V AC	V RMS TOTAL	PIV				% RIPPLE
	LOAD	D1, D3	D2, D4			D1	D3	D2	D4	
CALCULATED										
MEASURED		/////	/////		/////	/////	/////	/////	/////	/////

Part 2: Filtering the Full-Wave-Bridge, Rectifier Circuit

1. Connect the circuit as shown in Figure 15-33.

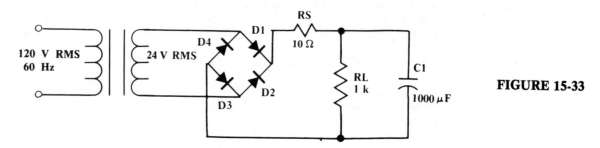

FIGURE 15-33

2. a) Calculate the V DC from: $V\ DC \approx V_p$ or, more precisely, $V\ DC = V_p - \Delta V_{C1}/2$.
   b) Measure the DC voltage at the load.
3. a) Calculate the DC current through the load from: $I\ DC = V\ DC/R_L$.
   b) Measure the DC current through the load with an ammeter.
4. a) Calculate the AC ripple voltage from: $\Delta V_{C1} = T/C \times I_{RL} = T/C \times V_p/R_L = V_p/fC_1R_L$.
   b) Measure the ripple voltage at the load with an oscilloscope.
5. Calculate the DC and peak recurring charging current out of the bridge (refer to Fig. 15-34):

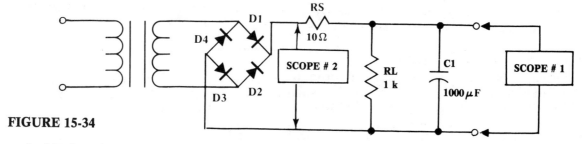

FIGURE 15-34

a) Monitor the sawtooth ΔV_{C1} wave shape at the load. Note the discharge-charge ratio and calculate from:

$I\ DC(\text{recurring}) = I\ DC \times \text{discharge time}/\text{charge time}$ (Scope #1)

b) Monitor the wave shape on the bridge side of the rectifier with respect to ground. Measure the peak recurring charging voltage and calculate the peak recurring charging current from:

$I_p(\text{recurring}) = V_p(\text{recurring})/R_S$ (Scope #2)

6. Calculate the peak inverse voltage of the bridge diodes from: $PIV = V_p$.

7. a) Calculate the effective AC value of the peak-to-peak ripple voltage from: $V\ AC = \Delta V_{C1}/2\sqrt{3}$

   b) Use a standard AC voltmeter to measure the V AC at the load.

8. Calculate the percent ripple from: $\%\ \text{ripple} = V\ AC/V\ DC \times 100$.
9. Insert the calculated and measured values, as indicated, into Table 15-4.

TABLE 15-4	LOAD		I(recurring)		ΔV_{C1}	V AC	PIV D1-D3 D2-D4	% ripple
	V DC	I DC	DC	peak				
CALCULATED								
MEASURED			/////	/////			/////	

402

CHAPTER QUESTION

1. List two reasons for using a transformer in power supply circuits.
2. Why does the VDC of unfiltered full wave circuits double as compared to unfiltered halfwave circuits?
3. The VDC of filtered halfwave and full wave bridge circuits are about the same for like secondary voltages. Therefore, what advantage is there to using the full wave bridge over the halfwave rectifier circuit?
4. What is the major disadvantage to using center-tapped, full-wave circuits in comparison to bridge, full-wave circuits?
5. Does the PIV of the diodes in the full-wave, center-tapped circuits change from unfiltered to filtered circuits? If so, why or why not?
6. Does the PIV of the diode in the halfwave circuit change if the circuit is changed from unfiltered to filtered? If so, by how much?
7. What is the PIV of the four bridge diodes before and after filtering for the full wave bridge rectifier circuit?
8. In calculating the ripple voltage, the measured ripple voltage is converted to V AC and then compared to the V DC. Why?
9. What is I(recurring), and why is it a factor in the diode selection for filtered circuits?
10. Why is the peak repetitive current not as important a circuit parameter, with regard to diode selection, as is the DC charging current?
11. Diode surge current can be greater than 20 times higher that the maximum DC current flow through the diode. How can this be?
12. In the unfiltered circuits, the V DC and V AC can be measured directly on a standard voltmeter. Is it possible, therefore, to measure directly the total V RMS with a standard voltmeter? Why or why not?

CHAPTER PROBLEMS

1. Using the circuit and values of Figure 15-35, calculate:

   a. Vpeak   d. PIV (D1)
   b. V DC   e. V AC
   c. I DC   f. % ripple

2. Using the circuit and values of Figure 15-36, calculate:

   a. ΔV_{C1} d. PIV (D1)
   b. V DC   e. V AC
   c. I DC   f. % ripple

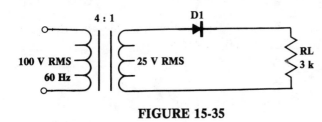

FIGURE 15-35

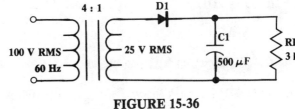

FIGURE 15-36

3. Using the circuit and values of Figure 15-37, calculate:

   a. Vpeak
   b. V DC
   c. I DC
   d. PIV (D1)
   e. V AC
   f. % ripple

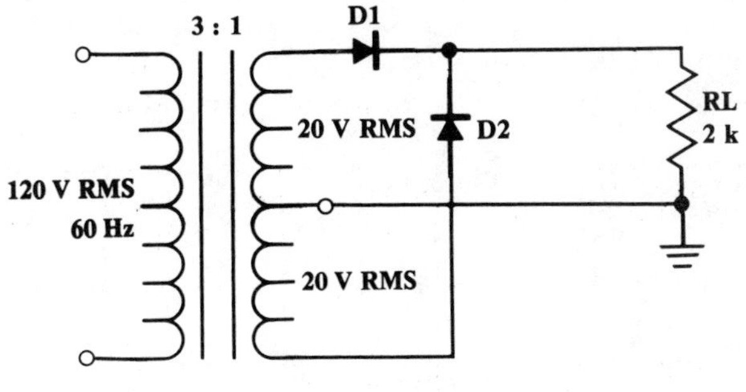

FIGURE 15-37

4. Using the circuit and values of Figure 15-38, calculate:

   a. ΔV_{C1}

   b. V DC

   c. I DC

   d. PIV (D1)

   e. V AC

   f. % ripple

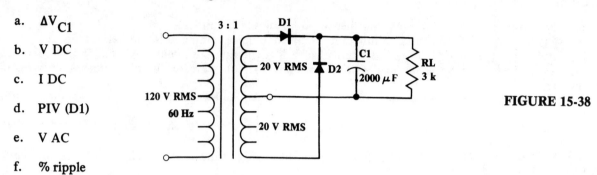

FIGURE 15-38

5. Using the circuit and values of Figure 15-39, calculate:

   a. Vpeak

   b. V DC

   c. I DC

   d. PIV (D1)

   e. V AC

   f. % ripple

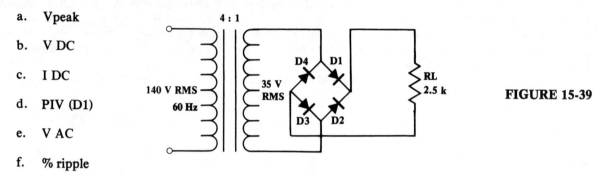

FIGURE 15-39

6. Using the circuit and values of Figure 15-40, calculate:

   a. ΔV_{C1}

   b. V DC

   c. I DC

   d. PIV (D1)

   e. V AC

   f. % ripple

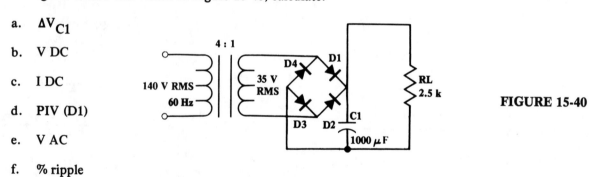

FIGURE 15-40

16 | INTRODUCTION TO REGULATED POWER SUPPLIES

GENERAL DISCUSSION

Bulk power supplies alone do not make very good regulators because, with increased loading, the DC voltage decreases and ripple voltage increases. Therefore, additional circuitry, along with the bulk supply, is needed to insure against voltage output changes for varying degrees of load current changes. In solid state circuitry, the device most widely used as a voltage regulator is the zener diode.

REVIEW OF ZENER DIODE PRINCIPLES

All diodes exhibit a linear breakdown region when the device is subject to reverse biased conditions. For regular diodes this region is the PIV voltage, where excessive power caused by the combination of high voltage and current can destroy the device. However, zener diodes are doped to take advantage of this zener or avalanche characteristics, where lower breakdown voltages at higher currents are used to advantage in low to medium power devices. Typical breakdown characteristics, for both the regular diode and the zener diode, are shown in Figure 16-1.

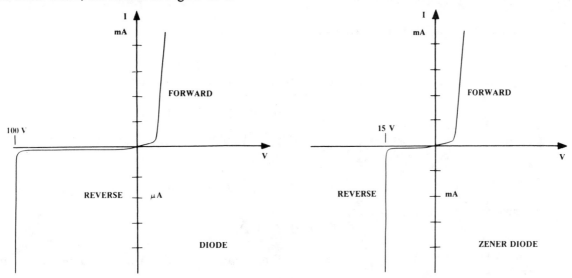

FIGURE 16-1

Zener or avalanche breakdown occurs when the zener diode is connected in a reverse biased condition. This is where the positive voltage is on the cathode and the negative voltage is on the anode, as shown in Figure 16-2(a). In this circuit configuration, various degrees of input voltages can be applied but the voltage across the zener diode will remain constant, with the excess voltage being dropped across the series resistor R1. However, the current flow through the zener diode will increase with increased voltage across R1 — the maximum current being determined by the power dissipating capabilities of the device.

For instance, too low a current flow through the zener will cause the device to come out of regulation or become increasingly noisy as it approaches the knee of the curve. Therefore, the current flow through the zener diode is maintained at between 1 mA and 5 mA. When the zener diode is in a forward biased condition, as shown in Figure 16-2 (b), it acts like any normally forward biased diode. For the silicon zener diode, the forward bias voltage drop is approximately 0.6 volts — which is the same as that for the normal silicon (rectifier) diode.

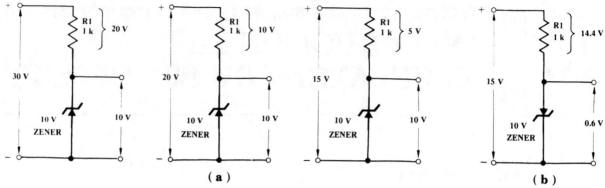

FIGURE 16-2

ZENER DIODE LOADING

Placing a load resistor across the zener diode does not change the breakdown voltage of the zener if enough current is available to supply both the zener diode and the load resistor. This is because the zener needs only enough current through it to remain below the knee of the curve and in zener condition. Therefore, from 1 mA to 5 mA of current must always flow through the zener or else it will come out of regulation and revert to a normally reverse biased diode, exhibiting a high (greater than 100 kΩ) resistance.

For instance, if 30 V DC is applied to a series 1 kΩ resistor and a 10 V zener diode, as shown in Figure 16-3(a), 20 volts are dropped across the 1 kΩ resistor and 10 volts are dropped across the zener diode. However, when a load resistor is placed across the zener, a portion of available R1 current (20 mA) flows into the load resistor and the remaining current flows through the zener. This is shown in Figure 16-3, for varying degrees of no-load and loaded conditions.

    a. For the no load condition, $I_Z = I_{R1} = 20$ mA.

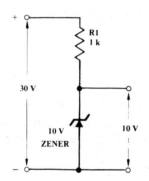

$V_Z = 10$ V

$V_{R1} = V_{in} - V_Z = 30\text{ V} - 10\text{ V} = 20$ V

$I_{R1} = V_{R1}/R1 = 20\text{ V}/1\text{ k}\Omega = 20$ mA

$I_Z = I_{R1} = 20$ mA

FIGURE 16-3(a)

    b. For the 2 kΩ load condition, I_{R1} remains at 20 mA, but $I_{RL} = 5$ mA and the zener diode current is at 15 mA.

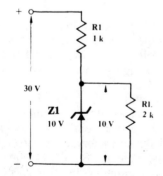

$V_Z = 10$ V

$V_{R1} = V_{in} - V_Z = 30\text{ V} - 10\text{ V} = 20$ V

$I_{R1} = V_{R1}/R1 = 20\text{ V}/1\text{ k}\Omega = 20$ mA

$I_{RL} = V_{RL}/RL = V_Z/RL = 10\text{ V}/2\text{ k}\Omega = 5$ mA

$I_Z = I_{R1} - I_{RL} = 20\text{ mA} - 5\text{ mA} = 15$ mA

FIGURE 16-3(b)

c. For the 1 kΩ load condition, I_{R1} remains at 20 mA, but I_{RL} = 10 mA and the zener diode current is at 10 mA.

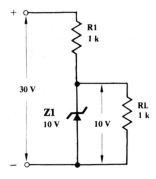

$V_Z = 10$ V

$V_{R1} = V_{in} - V_Z = 30$ V $- 10$ V $= 20$ V

$I_{R1} = V_{R1}/R1 = 20$ V$/1$ kΩ $= 20$ mA

$I_{RL} = V_{RL}/RL = 10$ V $/1$ kΩ $= 10$ mA

$I_Z = I_{R1} - I_{RL} = 20$ mA $- 10$ mA $= 10$ mA

FIGURE 16-3(c)

d. For the 500 Ω load condition, I_{R1} remains at 20 mA, but I_{RL} = 20 mA and the zener current is at 0.0 mA. At I_Z = 0.0 mA, the zener diode is out of regulation.

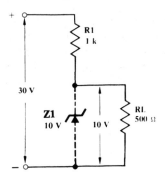

$V_Z = 10$ V

$V_{R1} = V_{in} - V_Z = 30$ V $- 10$ V $= 20$ V

$I_{R1} = V_{R1}/R1 = 20$ V$/1$ kΩ $= 20$ mA

$I_{RL} = V_{RL}/RL = V_Z/RL = 10$ V$/500$ Ω $= 20$ mA

$I_Z = I_{R1} - I_{RL} = 20$ mA $- 20$ mA $= 0.0$ mA

FIGURE 16-3(d)

NOTE: At I_Z = 0.0 mA, the zener diode is out of regulation and acts like a reverse biased diode at a high resistance of approximately 100 kΩ.

e. Reducing RL to 200 ohms maintains I_Z at 0.0 mA and the 30 volts are distributed across R1 and RL. Hence, V_{RL} = 5 V and the 10 V zener diode is simply a reverse biased diode with 5 V across it.

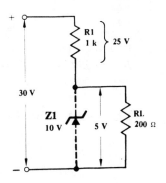

$V_{RL} = \dfrac{V_{in} \times RL}{R1 + RL} = \dfrac{30 \text{ V} \times 200 \text{ Ω}}{1 \text{ kΩ} + 200 \text{ Ω}} = 5$ V

$V_{R1} = \dfrac{V_{in} \times R1}{R1 + RL} = \dfrac{30 \text{ V} \times 1000 \text{ Ω}}{1 \text{ kΩ} + 200 \text{ Ω}} = 25$ V

FIGURE 16-3(e)

THE ZENER DIODE REGULATOR

Connecting a 15 volt zener diode and a 2.2 kΩ current limiting resistor across a bulk power supply will provide a regulated 15 V DC output voltage, having a ripple voltage of less than 1 mV. As shown in the regulated zener diode circuit of Figure 16-4, the transformer steps down the line voltage to 30 V RMS and the rectifier bridge and filter capacitor combine to establish a VDC of about 42.4 V across filter capacitor

C1. Therefore, about 42.4 volts are developed across the series R1 and Z1 components and, since $V_{Z1} = 15$ V, then $V_{R1} = 27.4$ V. Also, because $V_{R1} = 27.4$ V, $I_{R1} = V_{R1}/R1 = 27.4$ V$/2.2$ k$\Omega = 12.5$ mA, which is the amount of current that flows through the zener diode Z1 for the no load connection.

The ripple voltage across the filter capacitor C1 is solved by knowing f, T, C, and (for this circuit) I_{R1}. Since $f = 120$ Hz, $T = 1/f = 1/120$ Hz, $C = 1000$ μF, and $I_{R1} = 12.5$ mA, then $\Delta V_{C1} = 104.2$ mVp-p. The ripple voltage across the zener diode is solved, once the dynamic resistance of the zener diode is known, by using the voltage divider equation. Since $R1 = 2.2$ kΩ and R_{Z1} is assumed at 15 Ω, ΔV_{Z1} is approximately equal to 0.7 mVp-p.

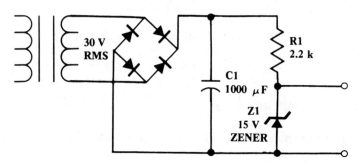

FIGURE 16-4

1. Solving for the peak-to-peak voltage at the secondary:

   Vp-p = (V RMS) $2\sqrt{2}$ = 30 V $\times$ 2.828 = 84.84 Vp-p

2. Solving for the peak voltage at the load:

   Vpeak = Vp-p$/2$ = 84.84 Vp-p$/2$ = 42.42 V

3. Solving for the DC voltage across capacitor C1:

   V DC(C1) = Vpeak = 42.42 V, or more precisely,

   V DC(C1) $\approx$ Vp $- (\Delta V_{C1}/2) = 42.42$ V $- 0.104/2 \approx 42.4$ V

4. Solving for the voltage drop across resistor R1:

   $V_{R1} \approx$ V DC(C1) $- V_{Z1} = 42.4$ V $- 15$ V $= 27.4$ V

5. Solving for the current through R1:

   $I_{R1} = V_{R1}/R1 = 27.4$ V$/2.2$ k$\Omega \approx 12.5$ mA

NOTE: For the unloaded condition, the full current of R1 flows through the zener diode.

6. Solving for the peak-to-peak ripple voltage across filter capacitor C1:

$$\Delta V_{C1} = T/C \times I_{RL} = \frac{T}{C} \times \frac{V_{R1}}{RL} = \frac{1/120}{1000\ \mu F} \times \frac{27.4\ V}{2.2\ k\Omega} = 8.33 \times 12.5\ mA = 104.2\ mVp\text{-}p$$

$$\Delta V_{C1} = I_{R1}/fC_1 = \frac{12.5\ mA}{120\ Hz \times 1000\ \mu F} = \frac{12.5 \times 10^{-3}}{120 \times 10^{-3}} = 104.2\ mVp\text{-}p$$

7. Therefore, a ripple voltage of 104.2 mVp-p exists across the filter capacitor. However, the ripple voltage across the zener will be considerably lower because it is stepped down by the voltage divider ratio of the dynamic resistance of Z1 and the R1 resistance.

For instance, the resistance of most zeners, found in the data sheets, is between 5 ohms and 30 ohms. However, a "ball park" estimate is that the dynamic resistance of a zener is approximately equal to the zener voltage: 20 V = 20 Ω, 5 V = 5 Ω, and so on. Therefore, if data is not available on the zener, this kind of estimate is valid until measurements can be made.

Hence, assuming that the 15 V zener is 15 ohms, then the ripple voltage across the zener will be:

$$\Delta V_{Z1} = \frac{\Delta V_C \times R_{Z1}}{R_{Z1} + R1} = \frac{104.2 \text{ mVp-p} \times 15 \text{ }\Omega}{15 \text{ }\Omega + 2200 \text{ }\Omega} = 0.7 \text{ mVp-p}$$

Therefore, the DC voltage across the zener, and to the load, is 15 V DC and the ripple voltage is less than 1 mVp-p or 0.7 mVp-p.

ZENER DIODE LOADED CONDITIONS

The zener will continue to provide a constant 15 volts of DC and less than 1 mVp-p of ripple voltage, for varying degrees of loading, as long as zener regulation is maintained. However, in order to make sure the zener does not come out of regulation, a minimum of 1 mA of DC current flow must be provided to the zener. Also, a resistor must be chosen to insure against too much current flow through the device when the load is removed. And the no load condition means that all the current flows through the zener so the zener must be able to dissipate the power safely.

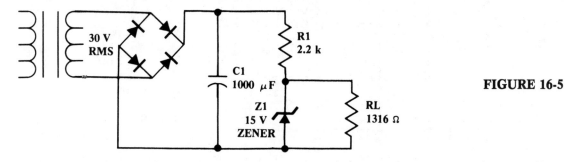

FIGURE 16-5

1. Solving for the maximum current flow into the load, maintaining a minimum 1 mA of zener regulation current:
   a. Solving for I_{R1}:

   $$I_{R1} = \frac{V \text{ DC}_{(C1)} - V_{Z1}}{R1} = \frac{42.4 \text{ V} - 15 \text{ V}}{2.2 \text{ k}\Omega} = 12.5 \text{ mA}$$

   b. Solving for I_{RL}, based on a minimum zener current of 1 mA:

   $$I_{RL} = I_{R1} - I_{Z1} = 12.5 \text{ mA} - 1 \text{ mA} = 11.5 \text{ mA}$$

2. Solving for the minimum load resistance possible for the 15 V zener, maintaining the 1 mA of regulation current:

   $$RL = V_{Z1}/I_{RL} = 15 \text{ V}/11.5 \text{ mA} \approx 1304 \text{ }\Omega$$

3. Solving for the power dissipated by the zener:
   a. For full load condition where only 1 mA of zener current flows:

   $$P_{Z1} = I_{Z1} \times V_{Z1} = 1 \text{ mA} \times 15 \text{ V} = 15 \text{ mW}$$

   b. No load condition where 12.5 mA of zener current flows:

   $$P_{Z1} = I_{Z1} \times V_{Z1} = 12.5 \text{ mA} \times 15 \text{ V} = 187.5 \text{ mW}$$

NOTE: The "rule of thumb" ratio for choosing the power capability of a device is a mimum of 2 : 1 over the actual power dissipation conditions. Therefore, the zener diode used in the circuit should have a standard dissipation parameter of 400 mW.

THE SERIES PASS REGULATOR

The main drawback to the zener diode regulator is that the maximum current into the load is limited by I_{R1}. For example, if a minimum zener current of 1 mA is maintained, then the maximum current into a load is 11.5 mA. For most applications, 11.5 mA is too low.

A solution to this problem is to connect a power transistor in an emitter follower configuration, as shown in Figure 16-6. Then, if the base current of the transistor is maintained at 11.5 mA and the beta of the device is 49, I_E will be approximately 575 mA from: $I_E = (\beta + 1) \times I_B = 50 \times 11.5$ mA $= 575$ mA. Also, because $V_Z = 15$ V, then $V_B = 15$ V and V_E is 14.4 V from: $V_E = V_B - V_{BE} = 15$ V $- 0.6$ V $=$ 14.4 V. Therefore, the RE resistor value can be as low as RE $= V_{RE}/I_E = 14.4$ V/575 mA ≈ 25 ohms without having the zener come out of regulation. The problem with this connection, however, is that the power transistor dissipation of about 14.6 watts is excessive and the ripple voltage at the output will be too high.

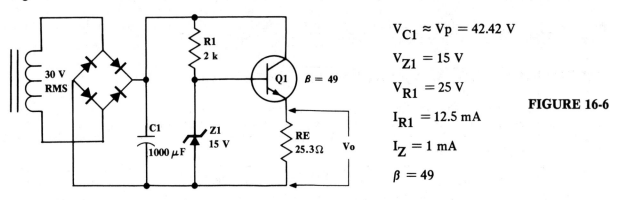

$V_{C1} \approx V_p = 42.42$ V

$V_{Z1} = 15$ V

$V_{R1} = 25$ V

$I_{R1} = 12.5$ mA

$I_Z = 1$ mA

$\beta = 49$

FIGURE 16-6

The step-by-step solution for the maximum IE conditions follow. Also, given in conjunction with the circuit of Figure 16-6 are the zener regulator circuit parameters and the transistor beta of 49, which has the effect of multiplying the base current by $(\beta + 1 = 49 + 1)$ 50.

1. Solving for the approximate base current from:

   $I_B = I_{R1} - I_Z = 12.5$ mA $- 1$ mA $= 11.5$ mA

2. Solving for the voltage at the emitter of Q1 from:

   $V_E = V_B - V_{BE} = 15$ V $- 0.6$ V $= 14.4$ V

3. Solving for the approximate emitter current of Q1, assuming a beta of 49.

   $I_E = (\beta + 1) \times I_B = (49 + 1) \times 11.5$ mA $= 50 \times 11.5$ mA $= 575$ mA

4. Solving for the minimum value for RE, where $V_E = V_{RE} = 14.4$ V and $I_E = 575$ mA

   $RE = V_{RE}/I_E = 14.4$ V/575 mA ≈ 25 Ω

5. Solving for the total current flow out of filter capacitor C1 from:

   $I_L(C1) = I_{R1} + I_{RL} = 11.5$ mA $+ 575$ mA $= 586.5$ mA

6. Solving for the ripple voltage across filter capacitor C1 from:

   $\Delta V_{C1} = T/C_1 \times I_L(C1) = \dfrac{1/120 \times 586.5 \text{ mA}}{1000 \text{ μF}} = \dfrac{4.887 \times 10^{-3}}{1 \times 10^{-3}} = 4.887$ Vp-p, or

   $\Delta V_{C1} = I_L(C1)/fC_1 = \dfrac{586.5 \text{ mA}}{120 \text{ Hz} \times 1000 \text{ μF}} = \dfrac{586.5 \times 10^{-3}}{120 \times 10^{-3}} = 4.887$ Vp-p

7. Solving for the more precise V DC(C1), knowing ΔV_{C1}, from:

$$V\ DC(C1) = V_p - \Delta V_{C1}/2 = 42.42\ V - 4.887\ V_{p\text{-}p}/2 \approx 42.42\ V - 2.44\ V \approx 40\ V$$

8. Solving for the V_{CE} of transistor Q1 from:

$$V_{CE} = V_C - V_E = 40\ V - 14.4\ V = 25.6\ V$$

9. Solving for the power dissipated by Q1 from:

$$P(Q1) = V_{CE} \times I_{RE} = 25.6\ V \times 575\ mA \approx 14.7\ watts$$

NOTE: The ripple at the output is approximately ΔV_{C1}, and if $R_{Z1} = 15\ \Omega$, then:

$$\Delta V_o \approx \Delta V_{Z1} = \frac{\Delta V_{C1} \times R_{Z1}}{R1 + R_{Z1}} = \frac{4.887\ V_{p\text{-}p} \times 15\ \Omega}{2.2\ k\Omega + 15\ \Omega} \approx 33.1\ mV_{p\text{-}p}$$

IMPROVED SERIES PASS REGULATOR CIRCUIT

In the previous circuit of Figure 16-6, both the power dissipated by Q1 and the ripple voltage developed across the zener diode were too high. Both of these problems can be minimized by lightening the load, which will draw less current. Another solution is to decrease the secondary voltage which will lower the V_{CE} voltage and lessen the power dissipated by Q1.

In Figure 16-7, the input voltage is halved by using a transformer secondary voltage of 15 V RMS instead of 30 V RMS. Secondly, a more realistic load resistor of 180 ohms, which only draws 80 mA, is used. However, establishing the load resistor first, does change the analysis procedure slightly, where the I_B is dependent on I_E and the beta of Q1. Hence, R1 must be selected to accommodate the total current of I_B and I_Z. Also, this approach uses $VDC(C1) \approx V_p$ in the initial calculations.

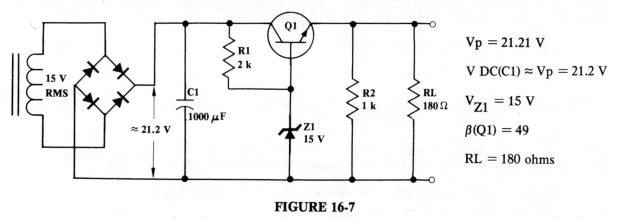

FIGURE 16-7

1. Solving for the peak and, hence, the approximate DC voltage developed across the C1 filter capacitor:

$$V_p = V\ DC(C1) = V\ RMS \sqrt{2} = 15\ V \times 1.414 \approx 21.2\ V$$

2. Solving for the voltage at the base of Q1 from:

$$V_{B1} = V_{Z1} = 15\ V$$

3. Solving for the approximate voltage drop across the R1 resistor from:

$$V_{R1} = V\ DC(C1) - V_{Z1} = 21.2\ V - 15\ V = 6.2\ V$$

4. Solving for the voltage at the emitter of Q1 with respect to ground (V_{E1} also equals $V_{R2} = V_{RL}$):

$$V_{E1} = V_{B1} - V_{BE1} = 15\ V - 0.6\ V = 14.4\ V$$

5. Solving for the voltage drop across the collector emitter of Q1 from:

$$V_{CE1} = V_{C1} - V_{E1} = 21.2\ V - 14.4\ V = 6.8\ V$$

6. Solving for the currents through R1, R2, and RL from:

$$I_{R1} = V_{R1}/R1 = 6.2\ V/2\ k\Omega = 3.1\ mA$$

$$I_{R2} = V_{R2}/R2 = 14.4\ V/1\ k\Omega = 14.4\ mA$$

$$I_{RL} = V_{RL}/RL = 14.4\ V/180\ \Omega = 80\ mA$$

7. Solving for the emitter and base current of Q1 (assuming a beta of 49):

$$I_{E1} = I_{R2} + I_{RL} = 14.4\ mA + 80\ mA = 94.4\ mA$$

$$I_{B1} = I_{E1}/(\beta + 1) = 94.4\ mA/50 \approx 1.89\ mA$$

8. Solving for the zener current, which is the current flow in R1 minus the base current of Q1:

$$I_{Z1} = I_{R1} - I_{B1} = 3.1\ mA - 1.89\ mA = 1.21\ mA$$

9. Solving for the total current capacitor C1 must deliver:

$$I_L(C1) = I_{R1} + I_{R2} + I_{RL} = 3.1\ mA + 14.4\ mA + 80\ mA = 97.5\ mA$$

10. Solving for the ripple voltage across C1, across Z1 (where Z1 is assumed to be 15 ohms), and at the output (where r_e is considered to be at 1 ohm):

$$\Delta V_{C1} = T/C_1 \times I_L(C1) = I_L(C1)/fC_1 = \frac{97.5\ mA}{120\ Hz \times 1000\ \mu F} = \frac{97.5 \times 10^{-3}}{120 \times 10^{-3}} \approx 812\ mVp\text{-}p$$

$$\Delta V_{Z1} = \frac{\Delta V_{C1} \times R_{Z1}}{R_{Z1} + R1} = \frac{812\ mVp\text{-}p \times 15\ \Omega}{2\ k\Omega + 15\ \Omega} \approx 6.05\ mVp\text{-}p$$

$$\Delta V_{RL} = \frac{\Delta V_{Z1} \times r_L}{r_e + r_L} = \frac{6.05\ mVp\text{-}p \times (1\ k\Omega\ /\!/\ 180\ \Omega)}{1\ \Omega + (1\ k\Omega\ /\!/\ 180\ \Omega)} \approx 6\ mVp\text{-}p$$

NOTE: The r_e of power transistors rarely goes below 1 ohm because of the base spreading resistance, which remains relatively constant regardless of changes in current. Therefore, $\Delta V_{Z1} \approx \Delta V_{RL}$.

11. Solving for the effective value of the AC ripple voltage at the load so that it can be compared to the DC voltage in terms of percent ripple.

$$V\ AC(Z1) \approx V\ AC(RL) = \Delta V_{Z1}p\text{-}p/2\sqrt{3} = 6.05\ mVp\text{-}p/2\sqrt{3} \approx 1.75\ mV$$

$$\%\ ripple = V\ AC/V\ DC \times 100 = 1.75\ mV/14.4\ V \times 100 \approx 0.012\%$$

12. Solving for the power dissipated by Q1 and the zener diode for this connection:

$$P_{Z1} = I_{Z1} \times V_{Z1} = 1.2\ mA \times 15\ V = 18\ mW$$

$$P_{Q1} = V_{CE} \times I_{E1} = 6.8\ V \times 94.4\ mA \approx 642\ mW$$

NOTE: Actually, the power dissipated by Q1 is slightly lower, because the more precise V DC(C1) is 20.8 volts from:

$$V\ DC(C1) = V_p - \Delta V_{C1}/2 = 21.2\ V = 812\ mVp\text{-}p/2 \approx 20.8\ V$$

Hence, the more precise V_{CE} is 6.4 V from:

$$V_{CE} = V\,DC(C1) - V_{RE} = 20.8\text{ V} - 14.4\text{ V} = 6.4\text{ V}$$

and the power dissipated by Q1 is 604 mW from:

$$P_{Q1} = V_{CE} \times I_{E1} = 6.4\text{ V} \times 94.4\text{ mA} = 604\text{ mW}$$

FURTHER STABILITY IMPROVEMENTS

Further improvements can be made to the series pass regulator to improve the stability and minimize the ripple voltage to the load. The first improvement is to add a diode in series with the zener diode to offset the DC voltage drift of the zener that can be caused by temperature. Since diodes and zener diodes have opposite temperature coefficients, one (and sometimes two) in-series diodes are used to achieve minimized DC voltage drift. The second improvement is the use of darlington pair transistors, which achieve a much larger beta, so that large variations in loads will have little effect on the base current of Q1. (Refer to Figure 16-8.) Therefore, I_Z will remain relatively constant with large load variations and the DC stability is improved.

The darlington pair series pass regulator of Figure 16-8 is shown drawn in two ways. In Figure 16-8(a), the circuit is drawn so it is easy to follow, and in Figure 16-8(b), the circuit is drawn in the standard way by which regulator circuits are normally illustrated in the literature.

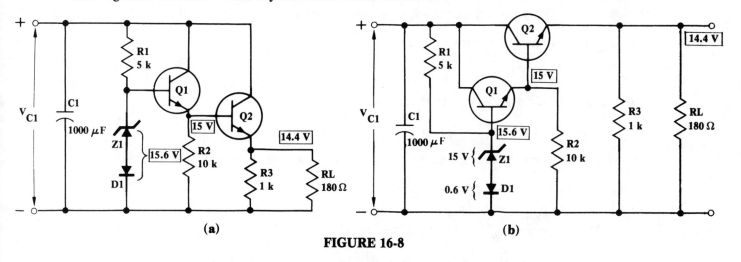

FIGURE 16-8

LABORATORY EXERCISE

OBJECTIVES
To investigate the zener diode series pass regulator circuit.

LIST OF MATERIALS AND EQUIPMENT
1. Transformer: 120 V : 24 V
2. Diodes: 1N4001 or equivalents (four)
3. Zener Diode: 15 V
4. Resistors (one each): 10 Ω   470 Ω   2.2 kΩ   3.3 kΩ   10 kΩ   10 kΩ pot
5. Capacitor: 1000 μF (one)
6. Voltmeter
7. Oscilloscope

PROCEDURE
Section I: Zener Diode Regulator
1. Connect the circuit as shown in Figure 16-9.

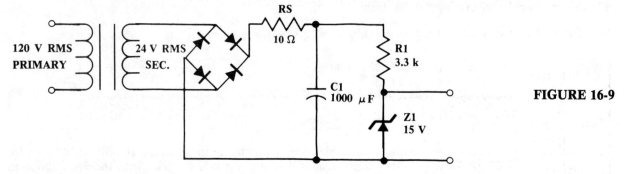

FIGURE 16-9

2. Across the secondary of the transformer:
   a. Measure the Vp-p with an oscilloscope.
   b. Measure the RMS voltage with a voltmeter.
3. Calculate the Vp across the filter capacitor from:

$$Vp = Vp\text{-}p/2 \quad \text{or} \quad Vp = \sqrt{2}\ RMS$$

4. Calculate the DC voltage across the filter capacitor from: $V\ DC(C1) \approx Vp$.

Measure the DC voltage drops across both the filter capacitor and the zener diode.

5. Calculate the voltage drop across the current limiting resistor R1 from:

$$V_{R1} = V\ DC(C1) - V_{Z1}$$

6. Calculate the AC ripple voltage across the filter capacitor from:

$$\Delta V_{C1} = T/C_1 \times I_L(C1) = I_L(C1)/fC_1 \quad \text{where:}\ I_{R1} = V_{R1}/R1\ \text{and}\ I_L(C1) = I_{R1}$$

Measure the AC ripple voltage across the filter capacitor and, then, convert to the effective V AC.

7. Calculate the AC ripple voltage across the zener diode from:

$$\Delta V_{Z1} = \frac{\Delta V_{C1} \times R_{Z1}}{R_{Z1} + R1}$$

NOTE: R_{Z1} can be obtained from the device's data sheet. However, if this information is not available the "rule of thumb" voltage-resistance equivalent approximation is acceptable until actual measurements can be made. For instance, V_Z at 15 V can be estimated to be 15 ohms. Also, when ΔV_{Z1} is very low, then the zener "noise" sometimes can be larger than the ΔV_{Z1}.

Measure the AC ripple voltage across the zener.

8. a. Monitor the sawtooth ΔV_{C1} discharge-charge ratio across the filter capacitor.
   b. Calculate the recurring DC charging current of the bridge from:

$$I\ DC(charging) = \frac{I_L(C1) \times \text{discharge time}}{\text{charge time}} \quad \text{where:}\ I_L(C1) = I_{R1}$$

   c. Monitor the peak recurring charging current across RS, the series 10 ohm resistor — bridge side of RS with respect to ground.
   d. Calculate the peak recurring charging current out of the bridge from:

$$Ip(charging) = Vp(recurring)/RS$$

9. Calculate the effective value of the AC ripple voltage across the zener from: $V\ AC(Z1) = \Delta V_{Z1}/2\sqrt{3}$

10. Calculate the percent ripple across the zener from: % ripple = $V_{AC(Z1)}/V_{Z1} \times 100$

11. Insert the calculated and measured values, as indicated, into Table 16-1.

TABLE 16-1		Vp-p SEC.	V RMS SEC.	Vp	V DC C1	V_Z	V_{R1}	I_{R1}
N.L.	CALC.	////	////		////	////	////	
	MEAS.			////				////
4.7 kΩ	MEAS.	////	////	////			////	////
470 Ω	MEAS.	////	////	////			////	////

		ΔV		I recurring		V AC(RMS)		% Ripple
		C1	Z1	DC	PEAK	C1	Z1	
N.L.	CALC.							
	MEAS.			////	////	////	////	////
4.7 kΩ	MEAS.			////	////	////	////	////
470 Ω	MEAS.			////	////	////	////	////

12. Connect the 4.7 k load resistor across the zener, as shown in Figure 16-10.
    a. Measure the output voltage across the zener.

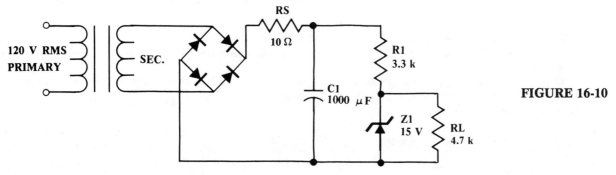

FIGURE 16-10

   b. Measure the ΔV across the zener.
   c. Measure the V DC across the filter capacitor.
   d. Measure the ΔV across the filter capacitor.
13. Insert the measured values into Table 16-1 and note any significant changes.
14. Connect the 470 Ω load resistor across the zener.
   a. Measure the output voltage across the zener.
   b. Measure the ΔV across the zener.
   c. Measure the V DC across the filter capactior.
   d. Measure the ΔV across the filter capacitor.
15. Insert the measured values into Table 16-1 and note any significant changes.
16. Connect the variable 10 kΩ potentiometer across the zener, as shown in Figure 16-11.
   a. Monitor the regulated output voltage (across the zener) with the oscilloscope while the potentiometer is set to approximately 10 kΩ.
   b. Vary the value of the pot until the output voltage changes. Just before it changes, or the zener comes out of regulation, note the rise in "noise".

NOTE: Zener diodes are extremely "noisy" devices and the noise increases dramatically just before it comes out of regulation. Step 16 demonstrates this fact.

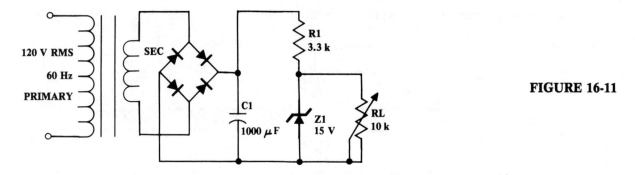

FIGURE 16-11

Section II: The Series Pass Regulator

1. Connect the circuit of Figure 16-12 to minimize the limitations of the zener diode. A 2N3055 power transistor does not require heat sinking. However, if a medium power device is used, heat sinking will probably be required. The calculation should indicate the power dissipation condition of Q1.

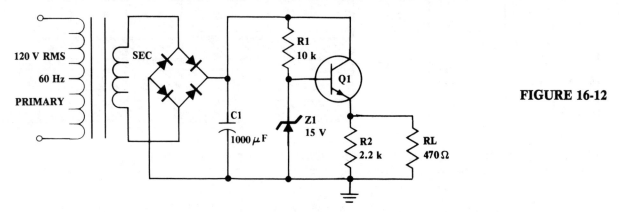

FIGURE 16-12

2. Monitor the DC voltage across the filter capacitor C1.
3. Monitor the DC voltage across the zener diode. The measurement is made at the base of Q1, with respect to ground.
4. Calculate and measure the voltage across the emitter resistor R2, which is also the voltage across the load resistor RL. The measurement is made at the emitter of Q1. Also, $V_{R2} = V_Z - V_{BE}$.
5. Calculate the V_{CE} of Q1 from:

$$V_{CE} = V_C - V_E \quad \text{where:} \quad V_C = V\ DC(C1)$$

6. POWER DISSIPATION:
   a. Calculate the power dissipated by Q1 from:

$$P_{Q1} = V_{CE} \times I_E \quad \text{where:} \quad I_E = V_E/R2 + V_E/RL$$

   b. Calculate the power dissipated by the zener diode Z1 from:

$$P_{Z1} = V_{Z1} \times I_Z \quad \text{where:} \quad I_Z = I_{R1} - I_B,\ I_B = I_E/\beta,\ \text{and}\ V_{R1} = V_{C1}(DC) - V_{Z1}$$

7. RIPPLE VOLTAGE:
   a. Calculate the ripple voltage across the filter capacitor C1 from:

$$\Delta V_{C1} = I_L(C1)/fC_1 \quad \text{where:} \quad I_L(C1) = V_{R2}/R2 + V_{RL}/RL + V_{R1}/R1$$

   b. Measure the ripple voltage across the filter capacitor C1.
   c. Measure the ripple voltage across the zener diode Z1.
   e. Measure the ripple voltage across the emitter resistor R2.
8. Insert the calulated and measured values into Table 16-2.

TABLE 16-2	V DC (C1)	V_{CE}	V_{Z1}	V_{R2}	C1	ΔV_{Z1}	RL	P_{Q1}	P_{Z1}
CALCULATED	/////		/////			/////			
MEASURED		/////						/////	/////

CHAPTER QUESTIONS

1. Name the two major advantages that the zener diode regulator has over the unregulated bulk power supply.
2. What is the major advantage of adding an emitter follower transistor to the zener diode regulator circuit?
3. When the emitter follower transistor is added, is the ripple voltage across the output load much different from across the zener diode? Explain why or why not.
4. What are the major improvements of the circuit of Figure 16-7 over the circuit of Figure 16-6?
5. Does the noise of zener diodes pose a problem when measuring for the developed ripple across the zener diode? Explain.
6. With reference to Figure 16-12, does disconnecting the 470 Ω load resistor cause the ripple voltage to increase or decrease? Explain why or why not.

CHAPTER PROBLEMS

1. Using the circuit and circuit component values of Figure 16-13, calculate:

   a. V DC(C1)
   b. I_{R1}
   c. I_{RL}
   d. I_{Z1}
   e. ΔV_{C1}
   f. ΔV_{Z1}

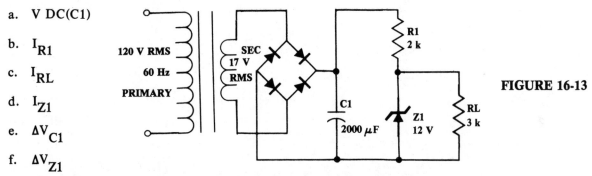

FIGURE 16-13

2. Using the circuit and circuit component values of Figure 16-14, calculate:

   a. V DC(C1)
   b. V_{R1}
   c. V_{R2}
   d. I_{R1}
   e. I_{R2}
   f. I_{RL}
   g. ΔV_{C1}
   h. ΔV_{Z1}
   i. ΔV_{RL}

FIGURE 16-14

3. Using the circuit and circuit component values of Figure 16-15, calculate:

   a. V DC(C1)
   b. V_{R1}
   c. V_{R2}
   d. I_{R1}
   e. I_{R2}
   f. I_{RL}
   g. ΔV_{C1}
   h. ΔV_{Z1}
   i. ΔV_{RL}

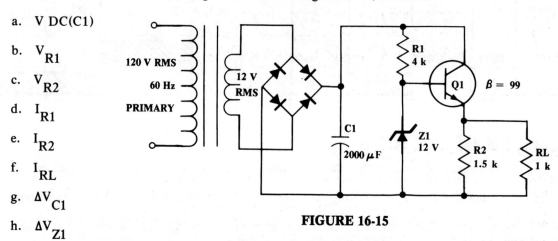

FIGURE 16-15

17 | SEMICONDUCTOR SWITCHES AND LATCHES

GENERAL DISCUSSION

Switches are basic controls used to make, break, and change connections in electrical circuits. The ideal switch has zero resistance in ohms across the contacts in a closed condition and infinite resistance in ohms across the contacts in an open condition. Mechanical switches come close to achieving ideal switch conditions. So, when a mechanical switch is used in series with a load resistor and voltage is applied, zero volts are dropped across the in-series load resistor in the open condition and zero volts are dropped across the switch contacts in the closed condition.

In other words, when the switch is closed, current flows in the circuit. Since the contact resistance is close to the ideal zero ohms, zero volts are dropped across the contacts and all of the applied voltage is dropped across the in-series load resistance. However, when the switch is in the open condition, no current flows and no voltage is dropped across the load resistor. Therefore, all of the applied voltage is dropped across the infinite resistance of the open contacts. Circuits with open and closed switch conditions for 12 volts of applied voltage are shown in Figure 17-1.

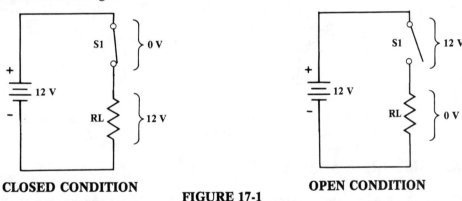

CLOSED CONDITION **FIGURE 17-1** **OPEN CONDITION**

The main difference between a switch and a latch is that a switch can be turned on and off, but once a latch is switched on, it remains on until it is released. A mechanical latch can be constructed by using a relay, a switch, and two voltage sources, as shown in Figure 17-2.

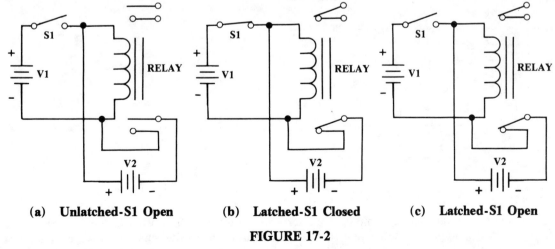

(a) Unlatched-S1 Open   (b) Latched-S1 Closed   (c) Latched-S1 Open

FIGURE 17-2

419

With reference to Figure 17-2, the contacts of the relay are normally open and they close only when voltage is applied across the relay coil to activate the relay. Therefore, when switch S1 is closed, the V1 voltage is applied across the coil, the contacts are closed, and V2 is also applied across the coil. However, when S1 is opened, the relay contacts remain closed because V2 continues to activate the relay, and the only way the relay can then be deactivated is to disconnect or lower the V2 voltage. The unlatched conditin is shown in Figure 17-2(a), the latched condition when S1 is closed is shown in Figure 17-2(b), and the continued latched condition when S1 is again opened is shown in Figure 17-2(c).

SEMICONDUCTOR SWITCHES

Bipolar and field effect transistors can both be used as semiconductor switches to replace mechanical switches. In fact, with regard to speed, contact bounce, and physical wear, semiconductor switches are superior to mechanical switches. However, with regard to ideal "open and closed" resistance conditions, transistors and FETS operate, but at less than ideal conditions. This is because the collector-emitter of transistors and the drain-source of FETS will always have some voltage drop (although small) in the "on" or "saturated" conditions, and there will always be some leakage current (although small) when the devices are in "cutoff" or in "open" condition.

In Figure 17-3(a), the bipolar transistors are shown in the not so ideal "on" and "off" conditions. In Figure 17-3(b), the JFETs are shown, where again the "on" and "off" conditions are not ideal.

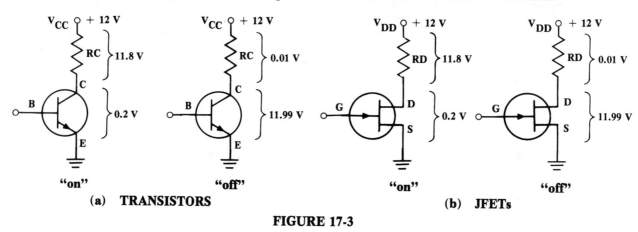

FIGURE 17-3

THE TRANSISTOR AS A SWITCH

Bipolar transistors can be biased into three distinct regions; the active region, when the device is used as an amplifier, and the saturation and cutoff regions, when the device is used as a switch. In the active region, the transistor is operating in the middle of the load line, as shown in Figure 17-4. In cutoff, the

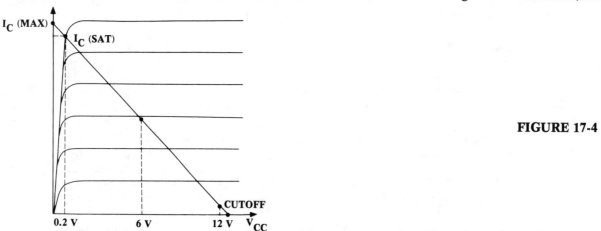

FIGURE 17-4

leakage current is so small that the voltage drop across V_{RC} is about 0.01 V and V_{CE} is 11.99 V. Hence, the V_{CE} in cutoff is approximately equal to V_{CC}. In saturation, as the characteristic curves illustrate clearly, the $V_{CE}(SAT)$ cannot be zero volts. Typical $V_{CE}(SAT)$ voltages are about 0.1 V to 0.2 V.

ACTIVE REGION

If the bipolar transistor is to be operated in the middle of the active region, with 6 volts dropped across the collector resistor and 6 volts dropped across the collector-emitter of the device, as shown in Figure 17-5 and Figure 17-6, the collector current needs to be 1.2 mA for an RC value of 5 kΩ. Assuming the beta of the transistor to be 100, then $I_B = 12\ \mu A$, from $I_B = I_C/\beta$, and RB is solved by obtaining V_{RB} from $V_{RB} = V_{CC} - V_{BE}$ and dividing V_{RB} by the base current I_B, as shown solved in conjunction with Figure 17-5.

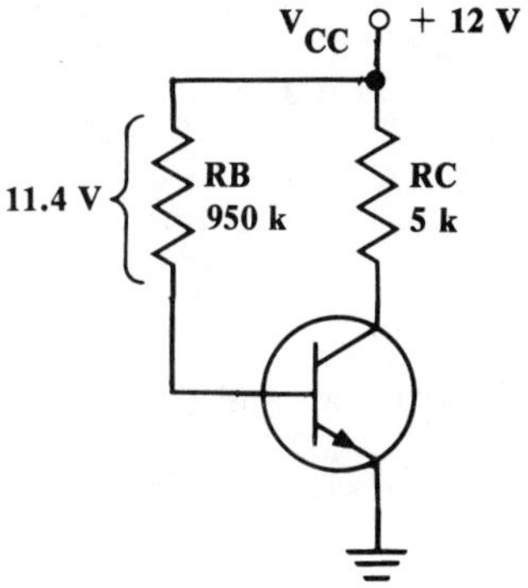

Given: $\beta = 100$, $V_{CE} = 6$ V, $V_{RC} = 6$ V, and RC = 5 kΩ

$I_C = V_{RC}/RC = 6\text{ V}/5\text{ k}\Omega = 1.2\text{ mA}$

$I_B = I_C/\beta = 1.2\text{ mA}/100 = 12\ \mu A$

$V_{RB} = V_{CC} - V_{BE} = 12\text{ V} - 0.6\text{ V} = 11.4\text{ V}$

$RB = V_{RB}/I_B = 11.4\text{ V}/12\ \mu A = 950\text{ k}\Omega$

FIGURE 17-5

Solving for the voltage drops of the circuit, once the values of RB, RC, and beta are known, as shown in conjunction with Figure 17-6.

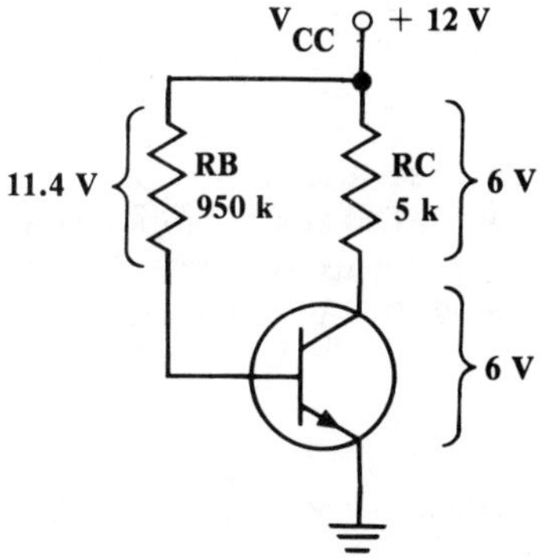

$V_{RB} = V_{CC} - V_{BE} = 12\text{ V} - 0.6\text{ V} = 11.4\text{ V}$

$I_B = V_{RB}/RB = 11.4\text{ V}/950\text{ k}\Omega = 12\ \mu A$

$I_C = \beta I_B = 100 \times 12\ \mu A = 1.2\text{ mA}$

$V_{RC} = I_C R_C = 1.2\text{ mA} \times 5\text{ k}\Omega = 6\text{ V}$

$V_{CE} = V_{CC} - V_{RC} = 12\text{ V} - 6\text{ V} = 6\text{ V}$

FIGURE 17-6

SATURATED REGIONS

Turning on, or saturating, the transistor is accomplished by increasing the collector current until the magnitude of V_{RC} is only a few tenths of a volt less than V_{CC}. Therefore, lowering the value of the base resistance RB increases the base current I_B, which in turn increases the collector current I_C. $I_C(SAT)$ is achieved when further increases in I_B do not increase the collector current I_C further. In the first of the two following examples, RB is selected to cause the device to just go into saturation. In the second example, RB is reduced further to force the device into further saturation.

In Figure 17-7, RB is lowered so that I_C is just at 2.36 mA, $V_{RC} = 11.8$ V, and $V_{CE}(SAT) = 0.2$ V. For a transistor with a beta of 100, the I_B is 23.6 μA and, in order to achieve the I_B of 23.6 μA, the base resistor RB should be at least 483 kΩ. These values are solved in the equations shown in conjunction with Figure 17-7.

In Figure 17-8, the value of RB is lowered further to 380 kΩ. This increases the base current I_B, but if $V_{CE}(SAT)$ remains at 0.2 V, then I_C also remains fixed. Therefore, the effect of increased base current I_B is to drive the transistor further into saturation, which will lower the $V_{CE}(SAT)$ only slightly. However, for explanatory purposes in the following example, $V_{CE}(SAT)$ is maintained at 0.2 V.

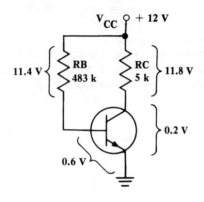

FIGURE 17-7

Given: $V_{CE}(SAT) = 0.2$ V, $\beta = 100$, and $RC = 5$ kΩ

$V_{RB} = V_{CC} - V_{BE} = 12$ V $- 0.6$ V $= 11.4$ V

$I_B = V_{RB}/RB = 11.4$ V$/483$ k$\Omega \approx 23.6$ μA

$I_C = \beta I_B = 100 \times 23.6$ μA $= 2.36$ mA

$V_{RC} = I_C R_C = 2.36$ mA $\times 5$ k$\Omega = 11.8$ V

$V_{CE} = V_{CC} - V_{RC} = 12$ V $- 11.8$ V $= 0.2$ V

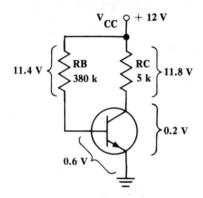

FIGURE 17-8

Given: $V_{CE}(SAT) = 0.2$ V, $\beta = 100$, and $RC = 5$ kΩ

$V_{RB} = V_{CC} - V_{BE} = 12$ V $- 0.6$ V $= 11.4$ V

$I_B = V_{RB}/RB = 11.4$ V$/380$ k$\Omega = 30$ μA

$V_{RC} = V_{CC} - V_{CE}(SAT) = 12$ V $- 0.2$ V $= 11.8$ V

$I_C = V_{RC}/RC = 11.8$ V$/5$ k$\Omega = 2.36$ mA

As shown in the graph of Figure 17-9, the increased I_B of 30 μA does little to effect the I_C at 2.36 mA if $V_{CE}(SAT)$ remains at 0.2 V. Notice, too, that the highest possible I_C is 2.4 mA and, if the beta of 100 for an I_B of 30 μA is used, I_C would have to be 3 mA, which is impossible. Hence, the effect of increasing I_B to 30 μA, with I_C remaining at 2.36 mA, is to lower $V_{CE}(SAT)$ slightly, from about 0.2 V to approximately 0.15 V.

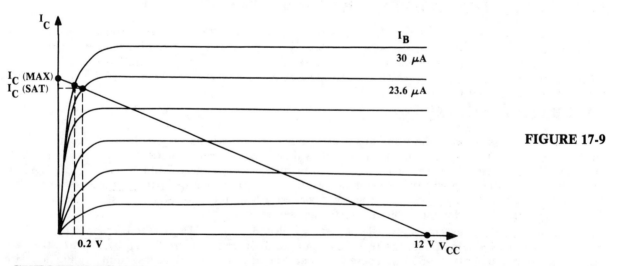

FIGURE 17-9

CUTOFF REGION

Transistor cutoff occurs when the base-emitter junction is reverse biased, the base-emitter junction is shorted together, or when the base is allowed to float by disconnecting the base resistor. Essentially, all three methods cut off the flow of base current and almost all of the collector current. Therefore, almost zero volts are dropped across the collector resistor and almost all of the V_{CC} is dropped across the collector-emitter of the device.

In Figure 17-10(a), a V_{BE} of -1 V is used and the collector current flow is the leakage current I_{CO} of the reverse-biased, base-collector junction. However, when the base is allowed to float, as shown in Figure 17-10(b), the leakage current can be as high as $(\beta + 1)I_{CO}$. I_{CO} is also given in literature as I_{cbo} and $(\beta + 1)I_{CO}$ is labeled I_{ceo}. In the solved example, $I_{CO} \approx I_{cbo}$ is given at 2 µA and I_{ceo} at 6 µA for purpose of illustration.

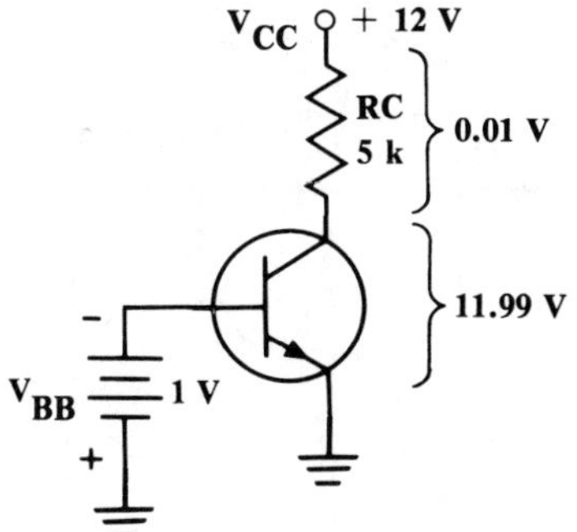

$V_{BE} = -1$ V

$I_B \approx 0$ mA

$I_C = I_{CO} \approx I_{cbo} = 2$ µA

$V_{RC} = I_C R_C = I_{CO} R_C = 2\,\mu A \times 5\,k\Omega = 0.01$ V

$V_{CE} = V_{CC} - V_{RC} = 12\,V - 0.01\,V = 11.99$ V

(a)

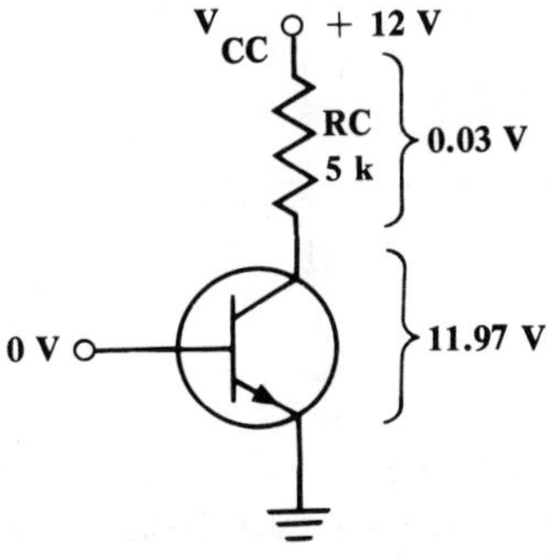

$V_{BE} = 0$ V

$I_B \approx 0$ mA

$I_C \approx I_{ceo} = 6$ µA

$V_{RC} = I_{CO} R_C = 6\,\mu A \times 5\,k\Omega = 0.03$ V

$V_{CE} = V_{CC} - V_{RC} = 12\,V - 0.03\,V = 11.97$ V

(b)

FIGURE 17-10

THE ALTERNATING CURRENT BIPOLAR SWITCH

If a transistor is biased into a saturated condition, V_{CE} is about 0.2 V and a voltmeter connected to the collector will read 0.2 V with respect to ground. However, if a negative voltage, with respect to ground, is then connected to the base (momentarily), the transistor will be forced into cutoff voltage. Then, the collector voltage, with respect to ground, should increase instantaneously to about 11.99 V. Both connections and conditions are shown in Figure 17-11.

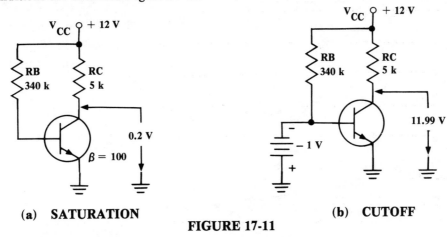

(a) SATURATION   (b) CUTOFF

FIGURE 17-11

In Figure 17-12, the diode processes only negative-going pulses to the input base of the transistor. Therefore, when the input signal is going positive, the transistor is in saturation, but, when the negative-going pulse is applied to the base, the transistor is in cutoff. If the input signal amplitude is large enough, the output signal should be a square wave alternating between the two extreme conditions of 11.99 V and 0.2 V, cutoff and saturation.

A representative circuit showing the input signal, the diode, the rectified negative-going pulse, the transistor, and the associated square wave output is shown in Figure 17-12(a). The output wave shape is 180 degrees out of phase with the input signal, as shown in Figure 17-12(b).

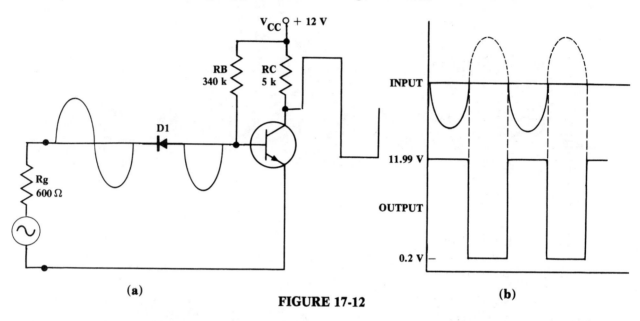

FIGURE 17-12

THE JFET AS A SWITCH

JFETs, like transistors, can be operated in the active region, saturated region, or cutoff region. In the active region, the device is operated between the two extreme conditions of cutoff and saturation. In the cutoff region, the channel of the JFET is considered pinched off and the drain current is, theoretically, zero milliamps. However, some leakage current will flow in the "pinchoff" channel condition and a slight amount of voltage will be dropped across the drain resistor. In saturation, the drain current is at a maximum and the V_{DS} is normally about 0.2 volts or less, if the RD resistor is large enough.

The JFET characteristic curves, showing the load line and the cutoff, linear, and saturated regions are shown in Figure 17-13. Notice that the larger the RD value, the lower the $V_{DS}(SAT)$.

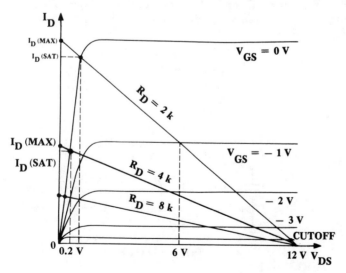

FIGURE 17-13

424

THE ACTIVE REGION

If the JFET is to be operated in the middle of the active region, so that $V_{RD} = 6$ V and $V_{DS} = 6$ V for a V_{DD} of 12 V, the circuit must have an RD value of 4 kΩ for an I_D of 1.5 mA. In the solved example shown in conjunction with Figure 17-14, the JFET is considered, theoretically, to follow square law principles, where $I_{DSS} = 6$ mA and the $V_p = V_{GS}(off)$ voltage is $|4\,V|$. The solved current and voltage drops are shown in Figure 17-14, where V_{GS} is selected at one-half $V_p = V_{GS}(off)$ so that I_D will be one-fourth I_{DSS}.

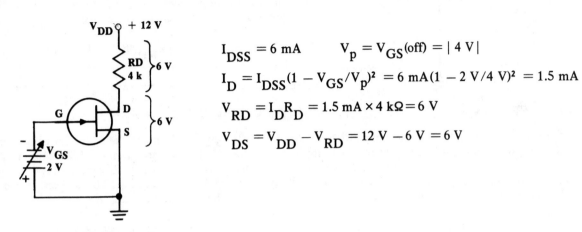

$I_{DSS} = 6$ mA $\qquad V_p = V_{GS}(off) = |4\,V|$

$I_D = I_{DSS}(1 - V_{GS}/V_p)^2 = 6\text{ mA}(1 - 2\,V/4\,V)^2 = 1.5$ mA

$V_{RD} = I_D R_D = 1.5\text{ mA} \times 4\text{ k}\Omega = 6$ V

$V_{DS} = V_{DD} - V_{RD} = 12\text{ V} - 6\text{ V} = 6$ V

FIGURE 17-14

THE CUTOFF REGION

At $V_{GS} = V_p = V_{GS}(off)$, the JFET is "pinched off", theoretically, so that zero milliamps flow in the channel. However, JFETs, like bipolar transistors, have some leakage current which, with regard to I_{DSS}, is extremely low. For example, as shown in Figure 17-15, if the low value of I_D at cutoff is 2.5 µA, then 10 millivolts will be dropped across the 4 kΩ resistor RD. Therefore, $V_{RD} = 11.99$ V, which is approximately equal to V_{DD}.

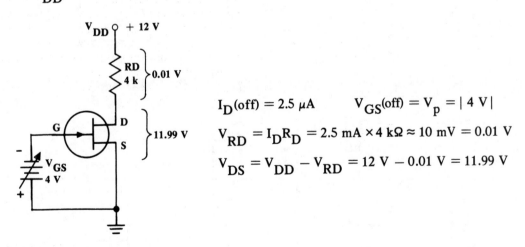

$I_D(off) = 2.5$ µA $\qquad V_{GS}(off) = V_p = |4\,V|$

$V_{RD} = I_D R_D = 2.5\text{ mA} \times 4\text{ k}\Omega \approx 10\text{ mV} = 0.01$ V

$V_{DS} = V_{DD} - V_{RD} = 12\text{ V} - 0.01\text{ V} = 11.99$ V

FIGURE 17-15

THE SATURATED REGION

Decreasing the VGS to about zero volts increases the I_D until the JFET enters saturation. Therefore, although the I_{DSS} of the device is 6 mA, as given in Figure 17-16, the maximum I_D current that can flow, as a result of the V_{DD} of 12 V and the RD of 4 kΩ, is 3 mA. Hence, since the $V_{DS}(SAT)$ is given as 0.2 V, then the I_D in saturation is 2.95 mA.

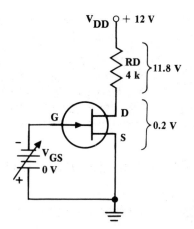

$I_{DSS} = 6$ mA $\quad V_p = V_{GS}(off) = |4\ V|$

$V_{RD} = V_{DD} - V_{DS}(SAT) = 12\ V - 0.2\ V = 11.8\ V$

$I_D(SAT) = V_{RD}/RD = 11.8\ V/4\ k\Omega = 2.95$ mA

NOTE: Although $I_{DSS} = 6$ mA, the maximum I_D current is 3 mA, because it is limited by V_{RD}/RD where $V_{DS} = 0\ V$ is assumed. Therefore:

$I_D(MAX) = V_{RD}/RD = 12\ V/4\ k\Omega = 3$ mA

FIGURE 17-16

THE JFET AS AN ALTERNATING CURRENT SWITCH

If the drain of the JFET is measured with a voltmeter, with respect to ground, and the device is repeatedly saturated and cut off, the voltmeter readings will alternate between approximately 0.2 V and 11.99 V. The connections for the JFET in cutoff and saturated conditions are shown in Figure 17-17. In Figure 17-17(a), the negative power supply ensures a cutoff condition while, in Figure 17-17(b), a short across the R2 resistor causes a zero volt V_{GS} condition and the device to be saturated.

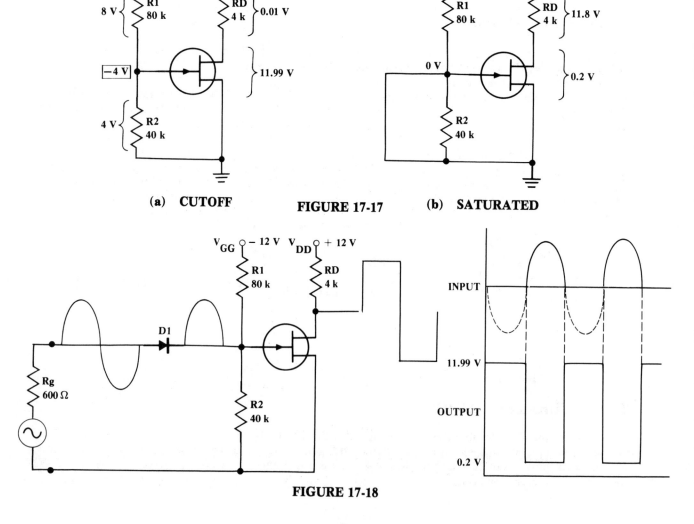

(a) CUTOFF   **FIGURE 17-17**   (b) SATURATED

FIGURE 17-18

Therefore, if an alternating current signal is applied to the input of the device, which causes the JFET to switch, alternately between cutoff and saturation, the ouput wave will be a square wave, as shown in Figure 17-18.

Essentially, the circuit of Figure 17-18 is DC biased so that the device is in cutoff with no input signal. However, when a sine wave is applied to the circuit, the diode processes only positive-going pulses which, for the period of the positive-going pulse, forces the JFET into saturation. During the absence of the positive-going pulse, or during the negative cycle of the pulse, the device is forced into cutoff. If the input signal is large enough, the output will be a square wave, alternating between the extreme cutoff and saturated conditions, as shown in Figure 17-18.

SEMICONDUCTOR LATCHES AND THE SILICON CONTROLLED RECTIFIER

There are two methods of obtaining a semiconductor latch. One method is by using bipolar transistors in a NPN-PNP connection, as shown in Figure 17-19(a), and the other is to use a silicon controlled rectifier (SCR), as shown in Figure 17-19(b).

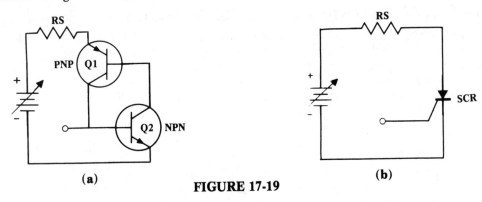

FIGURE 17-19

THE TRANSISTOR CONNECTED LATCH

In the transistor latch connection of Figure 17-20, the PNP and NPN transistors are connected so that the devices remain in cutoff until the applied DC voltage, which is slowly increased from an initial zero volt condition, is large enough to cause either of the collector-base junctions of Q1 and Q2 to break down. At breakdown, both devices will be forced into saturation. In other words, prior to junction breakdown, the current flow through Q1 and Q2 is the leakage current I_{ceo}, the voltage drop across V_{RS} is approximately zero volts, and the voltage drop across the PNP-NPN connection is approximately equal to the applied voltage. Then, at breakdown, the collector current increase of either of the transistors forces the other transistor into saturation because the collector current of one device is the base current of the other device. Hence, both transistors saturate and the voltage drop across the PNP-NPN connection falls to below 1 volt.

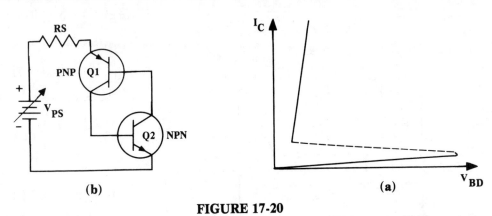

FIGURE 17-20

The latch circuit connection and the characteristic curve, depicting the increase voltage drop across the connection with little collector current and then the sudden voltage drop and the large increase in collector current as the device saturates, are shown in Figure 17-20. In this circuit, once the Q1,Q2 transistors are saturated, the devices will remain in saturation until the applied power supply voltage is momentarily disconnected, or lowered to a low voltage approaching zero volts. The disadvantage to the circuit connection of Figure 17-20 is that latching occurs only when the breakdown of either Q1 or Q2 is exceeded. However, if the base of Q2 is connected to the supply voltage by a variable resistor through a base resistor, as shown in Figure 17-21, the level of the base current can determine when latching will occur. For example, if the no base current connection of Figure 17-20 latches at 40 V, then applying base current will cause the latching to occur at a voltage lower than 40 volts. Essentially, the larger the base current, the lower the latching voltage. Therefore, if the applied voltage is set at 12 volts and the base voltage is slowly increased from an initial zero volt condition, the monitored voltage across the PNP-NPN connection would be approximately equal to the applied voltage of 12 volts until latching occurs. Then, the voltage across the connection would drop to below one volt. Again, once latching occurs, the circuit can be unlatched only by lowering or momentarily disconnecting the power supply voltage. For example, decreasing the base voltage will not unlatch the circuit. It can only cause latching and loses control once latching occurs.

UNLATCHED CONDITION

In the unlatched condition of the transistor latch circuit of Figure 17-21, the base voltage V_{BB} is at zero volts (initially), V_{PS} is at 12 V and, if a leakage current I_{ceo} of 10 μA is used, 10 mV or 0.01 V are dropped across RS and 11.99 V are dropped across the connection.

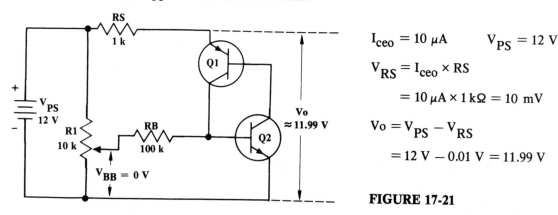

$$I_{ceo} = 10\ \mu A \qquad V_{PS} = 12\ V$$

$$V_{RS} = I_{ceo} \times RS$$

$$= 10\ \mu A \times 1\ k\Omega = 10\ mV$$

$$V_O = V_{PS} - V_{RS}$$

$$= 12\ V - 0.01\ V = 11.99\ V$$

FIGURE 17-21

LATCHED CONDITION

In the latched condition of the transistor latch circuit of Figure 17-22, the V_{BB} is increased until such time as latching occurs. The exact voltage level of V_{BB}, that causes the connection to latch, is unpredictable, because of the varying device characteristics. However, once V_{BE2} is forward biased, I_{C2} increases, I_{B1} increases, Q1 saturates (which causes Q2 to saturate), and latching is achieved.

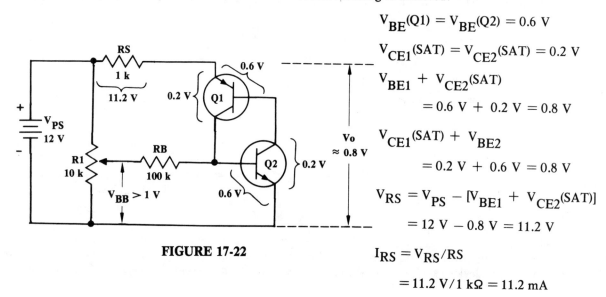

$$V_{BE}(Q1) = V_{BE}(Q2) = 0.6\ V$$

$$V_{CE1}(SAT) = V_{CE2}(SAT) = 0.2\ V$$

$$V_{BE1} + V_{CE2}(SAT)$$

$$= 0.6\ V + 0.2\ V = 0.8\ V$$

$$V_{CE1}(SAT) + V_{BE2}$$

$$= 0.2\ V + 0.6\ V = 0.8\ V$$

$$V_{RS} = V_{PS} - [V_{BE1} + V_{CE2}(SAT)]$$

$$= 12\ V - 0.8\ V = 11.2\ V$$

FIGURE 17-22

$$I_{RS} = V_{RS}/RS$$

$$= 11.2\ V/1\ k\Omega = 11.2\ mA$$

In the solved example of Figure 17-22, the $V_{CE}(SAT)$ conditions of Q1 and Q2 are shown at 0.2 V. Therefore, if the V_{BE} of both transistors is 0.6 V, either transistor path will provide a voltage path of 0.8 V and 11.2 V is dropped across the 1 kΩ resistor RS.

THE SCR AS A LATCH

The silicon controlled rectifier is a four layer diode with three leads that have characteristics similar to those of the PNP-NPN latch circuit shown in Figure 17-23. Therefore, in practice it is not necessary to construct a transistorized latch but, simply, to use an SCR. In fact, SCRs are constructed to operate at voltages in excess of 500 V and with currents in excess of 300 amps, which are much greater than transistor capabilities. Therefore, there are no advantages to using a transistor connection over the silicon controlled device.

For purpose of comparison, the schematic symbol and block diagram of an SCRs four-layer construction are shown with the PNP-NPN bipolar transistor simulated circuit construction, in both block and schematic diagram form, in Figure 17-23.

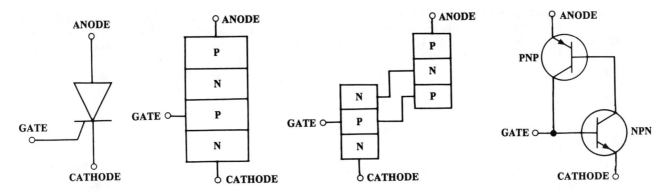

FIGURE 17-23

In Figure 17-24(a), the SCR is shown in a circuit connection similar to the bipolar transistor connection of Figure 17-20. The curves are a little different but the electrical results are similar. With enough gate current, the SCR latches at voltages as low as a few volts. And, like the transistorized version, once latching, or breakdown voltage, occurs, the gate no longer controls the circuit. Hence, the gate voltage or current only determine when the SCR latches, and unlatching must be accomplished by momentarily disconnecting the V_{PS} voltage or lowering the V_{PS} voltage or current through the SCR to a low level. The minimum current that must flow through an SCR to maintain a latched condition is called the holding current. The holding current of an SCR is nominally 1 mA.

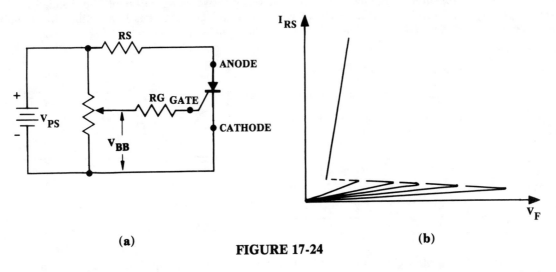

FIGURE 17-24

As shown in Figure 17-25, in the unlatched condition, almost all of the applied V_{PS} voltage is dropped across the SCR and only the device reverse leakage current or forward blocking current, as it is referred to, flows through the device and the series resistor RS. However, since the reverse forward blocking current is small, about 10 μA, the voltage drop across RS is about 10 mV.

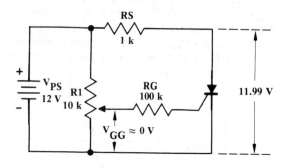

FIGURE 17-25

In the latched condition of the SCR latch circuit shown in Figure 17-26, the gate voltage (and hence current) has been increased, the forward voltage across the SCR is about 0.8 V, and the voltage drop across the series RS resistor is about 11.2 V. Again, after latching occurs, the gate loses control and only momentary interruption of the V_{PS} voltage or decreasing the current flow through the SCR below the holding current of about 1 mA will cause the latched condition to "unlatch".

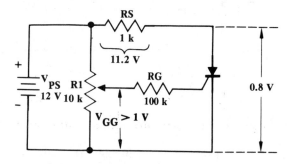

FIGURE 17-26

THE SCR IN THE ALTERNATING CURRENT MODE

In the DC mode, the SCR is primarily used as a latch or a switch and, in some instances, serves as a direct replacement for relays. In the alternating current mode, the SCR is widely used in phase control circuits such as light dimmers and motor speed controls. In fact, the name silicon controlled rectifier implies that the device is a controlled rectifier and, if the gate current is high enough to cause the SCR to latch at very low breakdown voltages, almost all of the positive-going pulse of the applied alternating current signal will reach the load. Hence, the SCR, at high gate current, acts like a diode in the half-wave rectifier circuit, and the VDC to the load is solved in the same way, as shown in Figure 17-27.

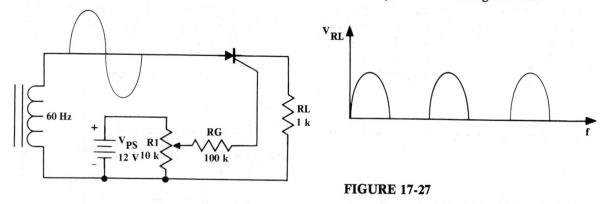

FIGURE 17-27

Operating the SCR in the AC mode is actually easier than operating it in the DC mode, because unlatching the SCR occurs 60 times per second with every negative-going wave, which occurs from about 180° to

360°, as shown in Figure 17-28.

Latching the SCR in the AC mode, as in the DC mode, is controlled by the gate current which, in turn, determines when breakover occurs. For example, if the gate current is too low, no signal is processed to the load. Once enough gate current flows, the output waveshape to the load can be varied between the theoretical triggering phase angle of zero degrees and 90 degrees. At zero degree triggering, the full half-wave pulse gets to the load. Then if the gate current is lowered, the triggering angle increases to a maximum of 90 degrees. In practice, the zero degree phase angle is closer to ten degrees, but the 90 degree phase angle is reasonably close to theory. The input wave and the output waveshapes across the load, for the zero degree and 90 degree triggered phase angles, are shown in Figure 17-28.

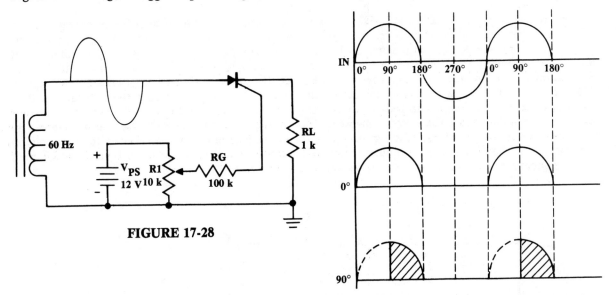

FIGURE 17-28

NOTE: The VDC at zero degrees is approximately equal to V_p/π and, for one-half pulse ninety degree triggering, the VDC is solved from $VDC \approx V_p/2\pi$.

CIRCUIT IMPROVEMENT AND SIMPLIFICATION

At maximum gate current, the breakover voltage required to cause latching is very low, and latching will occur, normally, within the first 10 degrees of the positive-going pulse. Once latched, the SCR will remain latched until about 180 degrees, when the input waveshape goes negative. Therefore, at the theoretical zero degree triggering phase, the full half-wave pulse translates to maximum power delivered to the load and, if the load is a motor, for full speed. However, as the gate current is lowered, the breakover voltage increases and latching occurs at a higher and higher voltage level. At 90 degrees, the minimum gate current flows, and one-half the DC voltage is developed across the load and, in the case of a motor, the speed is reduced. Lowering the gate current further unlatches the SCR, and no output signal is processed to the load.

Two improvements can be made to the circuit of Figure 17-28 that will simplify it and extend its firing angle beyond 90 degrees. In the circuit of Figure 17-29(a), the gate current is obtained from the applied signal voltage and, as shown in Figure 17-29(b), a capacitor is added to obtain a 45 degree phase shift to

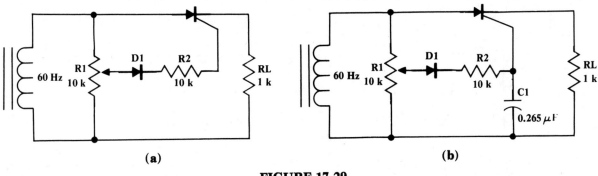

FIGURE 17-29

431

the gate. The 45 degree phase shift at the gate has the effect of delaying the triggering angle to 135 degrees and beyond.

With reference to Figure 17-29(b), capacitor C1 is selected so, at the line frequency of 60 Hz, the capacitive reactance is approximately equal to the 10 kΩ resistor R2, which will achieve the 45 degree phase shift. Therefore, since $X_C = 1/2\pi fc$, frequency is 60 Hz, and $X_C = R = 10$ kΩ, then C is:

$$C1 = \frac{1}{2\pi f X_C} = \frac{0.159}{60 \times 10 \text{ k}\Omega} = \frac{0.159}{60 \times 10^4} = \frac{0.159}{6 \times 10^5} = 0.265 \text{ }\mu F$$

In the circuits of Figure 17-29, diode D1 serves two purposes. It provides a DC gate current and it prevents the negative-going input pulse from exceeding the gate-to-source peak reverse voltage of the SCR. In Figure 17-29(b), the R_2C_1 components of 10 kΩ and 0.265 μF provide the 45 degree phase shift necessary for extending the phase triggering beyond 90 degrees, but the 45 degree phase shift can be monitored only if the gate loading is removed and the diode temporarily shorted to process the AC signal. Hence, the R_2C_1 time constant delays the triggering angle while C1 charges, but once the SCR latches, then C1 is discharged through the low impedance of the gate.

The input waveshape and the output waveshape across the load and the SCR for the 45 degree, 90 degree, and 135 degree phase angles are given in Figure 17-30. Notice that the SCR is on for that portion of the curve that the load is not on, and vice versa. Therefore, the total input wave is developed across both components for each of the phase shifts shown.

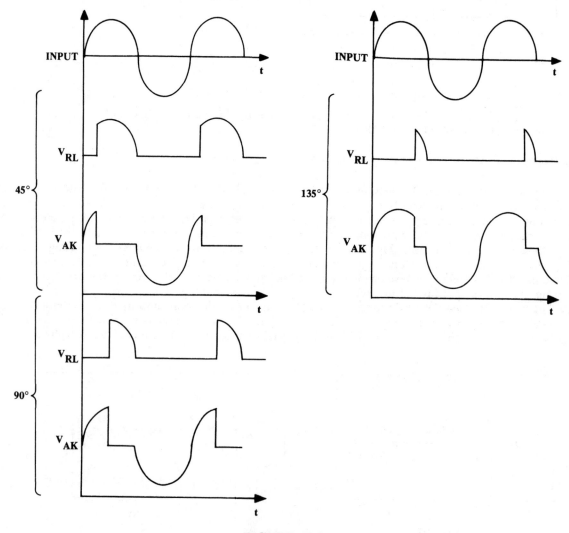

FIGURE 17-30

TRIACS AND DIACS

The SCR, like the half-wave rectifier circuit, processes only positive or negative waveshapes to the load. Therefore, if two SCRs are connected back-to-back, so that one SCR processes the positive-going wave and the other the negative-going wave, then the full positive and negative-going AC wave can be controlled. However, the triac and diac, shown in Figure 17-31, also respond to both positive and negative-going waves and can be used instead. The triac is similar to back-to-back SCRs and the diac, which protects the cathode-to-anode for peak reverse voltage, is similar to back-to-back diodes.

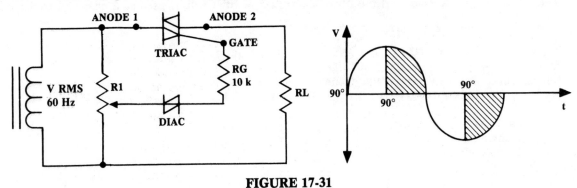

FIGURE 17-31

LIGHT EMITTING DIODE

Light emitting diodes are primarily constructed from gallium arsenide phosphide and emit a visible light when forward biased. LEDs, as they are referred to, emit different colors but the most widely used are red, green, and yellow. Doping techniques are primarily responsible for the color obtained. LEDs are similar to regular diodes, since they must be forward biased to emit light, but the voltage drop across LEDs are noticeably higher, from about 1.5 V to above 2 V. Also, the current required to obtain reasonable intensity is in the 20 mA range. Therefore, a V_F of 2 V at about 20 mA requires that a 500 ohm resistor be used in the circuit to obtain the necessary LED intensity, as shown in Figure 17-32.

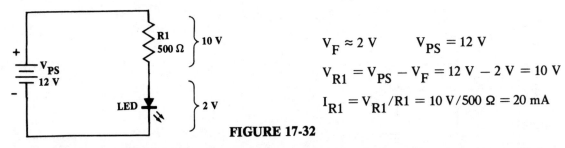

$V_F \approx 2\ V \qquad V_{PS} = 12\ V$

$V_{R1} = V_{PS} - V_F = 12\ V - 2\ V = 10\ V$

$I_{R1} = V_{R1}/R1 = 10\ V/500\ \Omega = 20\ mA$

FIGURE 17-32

The circuit of Figure 17-33 shows an SCR controlling the brightness of an LED. Therefore, beginning at full brightness, the current flow through the LED is 22 mA and it is reduced to about 8.5 mA at the 90 degree firing angle. The calculations are based on V_F remaining at a theoretical 2 volt condition as the firing angle is changed.

1. Full brightness — zero degrees

$VDC = V_p/\pi \approx 34\ V/3.1416 = 10.8\ V$

$V_{R3} = VDC - V_F = 10.8\ V - 2\ V = 8.8\ V$

$I_{R3} = V_{R3}/R3 = 8.8\ V/400\ \Omega = 22\ mA$

2. Ninety degrees

$VDC = V_p/2\pi = 34\ V/2\pi = 5.4\ V$

$V_{R3} = VDC - V_F = 5.4\ V - 2\ V = 3.4\ V$

$I_{R3} = V_{R3}/R3 = 3.4\ V/400\ \Omega = 8.5\ mA$

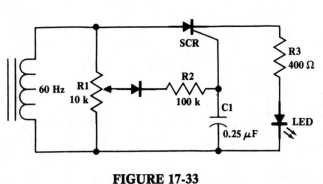

FIGURE 17-33

LABORATORY EXERCISE

OBJECTIVES

To investigate semiconductor switches and latches in direct current and alternating current applications. Bipolar and field effect transistors will be used as switches, bipolar transistors and the silicon controlled rectifier will be used as latches, and the silicon controlled rectifier will be used as a phase controller.

LIST OF MATERIALS AND EQUIPMENT

1. Transistors: 2N3904, 2N3906, and 2N5951 (JFET) or equivalents
2. Diodes: 1N4001 or equivalent and LED
3. Silicon Controlled Rectifier: C106 (General Electric) or equivalent
4. Resistors (one each): 470 Ω 1 kΩ 4.7 kΩ 10 kΩ 22 kΩ 27 kΩ 47 kΩ 100 kΩ
   10 kΩ Potentiometer
5. Capacitor: 0.22 µF (one)
6. Signal Generator
7. Power Supply: ± 12 volts
8. Oscilloscope

PROCEDURE
Section I: Switches
Part 1: The Transistor as a Switch

1. Connect the circuit of Figure 17-34, where the 100 kΩ base resistor RB and the 4.7 kΩ collector resistor RC are selected to ensure saturation.

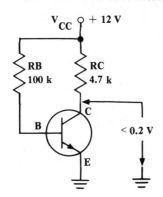

FIGURE 17-34

2. Measure the V_{CC}, the collector voltage V_C, and the base voltage V_B, all with respect to ground. Then, calculate:

$$V_{RC} = V_{CC} - V_{CE}(SAT) \qquad \text{where:} \quad V_{CE}(SAT) = V_C \qquad I_B = V_{RB}/RB$$

$$I_C(SAT) = V_{RC}/RC \qquad\qquad \beta = I_C/I_B$$

$$V_{RB} = V_{CC} - V_B \qquad \text{where:} \quad V_B = V_{BE}$$

3. Insert the calculated and measured values, as indicated, into Table 17-1.

TABLE 17-1	V_{CC}	V_C	V_B	V_{RC}	V_{CE}(sat)	I_C(sat)	V_{RB}	I_B	β
CALCULATED	//////	//////	//////						
MEASURED				//////	//////	//////	//////	//////	//////

434

4. Disconnect the base resistor, as shown in Figure 17-35(a), so that the device is in cutoff. To make sure that the device is in cutoff, the base-emitter junction can be reverse biased by the connection shown in Figure 17-35(b). Both methods are acceptable and should be tested to note any discernable V_{CE} voltage drop difference.

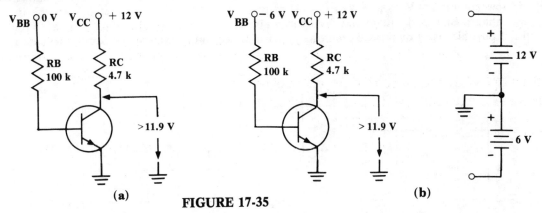

FIGURE 17-35

5. Measure V_{CC}, V_C, and V_B, all with respect to ground. Then, calculate:

$$V_{RC} = V_{CC} - V_C \quad \text{where:} \quad V_C = V_{CE}$$

$$I_C = V_{RC}/RC \quad \text{where:} \quad I_C \approx I_{ceo}$$

$$V_{CE} = V_C - V_E$$

NOTE: If the V_{BB} voltage of Figure 17-35(b) is used, measure V_B and calculate V_{RB} and I_B.

6. Insert the measured and calculated values, as indicated, into Table 17-2.

TABLE 17-2	V_{CC}	V_C	V_B	V_{RC}	I_C	V_{CE}	OPTIONAL		
							V_B	V_{RB}	I_B
CALCULATED	/////	/////	/////				/////		
MEASURED				/////	/////	/////		/////	/////

ALTERNATING CURRENT CONSIDERATIONS

1. Connect the transistor in a saturated mode as shown in Figure 17-36(a)., and test the voltage drops around the circuit to make sure the device is in saturation.

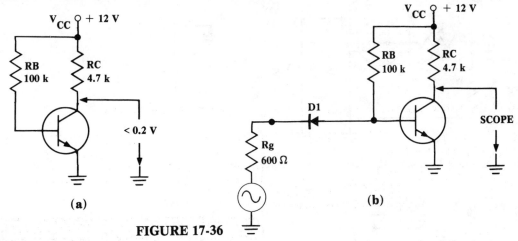

FIGURE 17-36

2. Connect a signal generator to the circuit, as shown in Figure 17-36(b). Use a 1 kHz sine wave input signal. Monitor the collector with an oscilloscope and increase the magnitude of the input signal until the output waveshape becomes a square wave. Measure the rise time and note. (See appendix.)

3. Monitor the magnitude of the negative-going input pulse at the base of the transistor and, then, the upper and lower limits ($\approx V_{CC}$ and $V_{CE}[SAT]$) of the output square wave. Sketch both the input and output waveshapes (sine and square wave) on the graph of Figure 17-37. Use 0.2 ms/DIV on the oscilloscope and, if possible, use a dual trace oscilloscope for a direct comparison of the input and output waves.

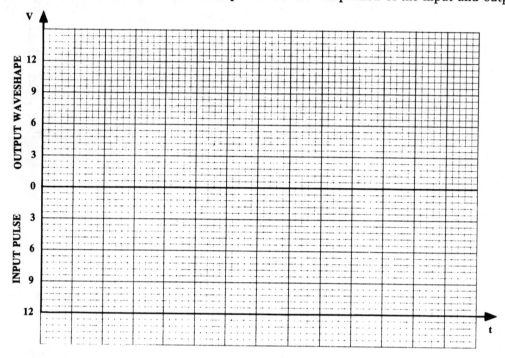

FIGURE 17-37

Part 2: The JFET as a Switch

1. Connect the circuit of Figure 17-38, where the R1,R2 voltage divider circuit ensures a cutoff condition by placing a larger than -4 V at the gate of the JFET.

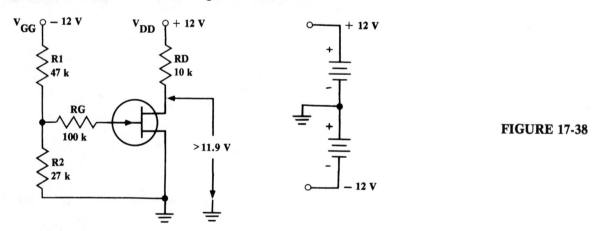

FIGURE 17-38

2. Calculate V_{R1} and V_{R2} from:

$$V_{R1} = \frac{V_{GG} \times R1}{R1 + R2} \qquad V_{R2} = \frac{V_{GG} \times R2}{R1 + R2} \qquad \text{where: } V_{R2} \approx V_G$$

3. Measure V_{DD}, the drain voltage V_D, and the gate voltage V_G, all with respect to ground. Calculate:

$$V_{RD} = V_{DD} - V_D$$
$$I_D = V_{RD}/RD$$
$$V_{DS} = V_D - V_S$$

4. Insert the calculated and measured values, as indicated, into Table 17-3.

TABLE 17-3	V_{DD}	V_D	V_G	V_{R1}	V_{R2}	V_{RD}	I_D	V_{DS}
CALCULATED	///	///	///					
MEASURED				///	///	///	///	///

5. Connect the circuit of Figure 17-39, which is similar to the circuit of Figure 17-38 except the R2 resistor has been shorted. The net effect of the circuits of Figure 17-39 is to make $V_{GS} = 0.0$ V, which will saturate the device. In saturation, the V_{DS} should be at least 0.3 V, or lower. However, RD must be at least 10 kΩ to accomplish a V_{DS} of less than 0.3 V for the 2N5951.

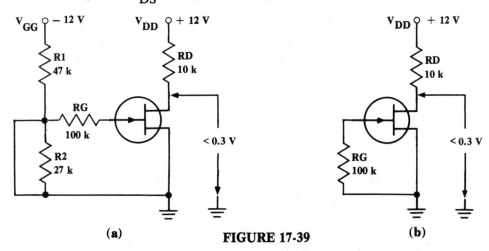

FIGURE 17-39

6. Measure V_{DD}, V_D, and V_G, with respect to ground. Then, calculate:

$$V_{RD} = V_{DD} - V_D \quad \text{where:} \quad V_D = V_{DS}(SAT)$$
$$I_D(SAT) = V_{RD}/RD$$
$$V_{DS}(SAT) = V_D - V_S$$

7. Insert the calculated and measured values, as indicated, into Table 17-4.

TABLE 17-4	V_{DD}	V_D	V_G	V_{RD}	I_D(sat)	V_{DS}(sat)
CALCULATED	///	///	///			
MEASURED				///	///	///

ALTERNATING CURRENT CONSIDERATIONS

1. Connect the JFET in a cutoff mode as shown in Figure 17-40(a), and test the voltage drops around the circuit to make sure the device is in cutoff.

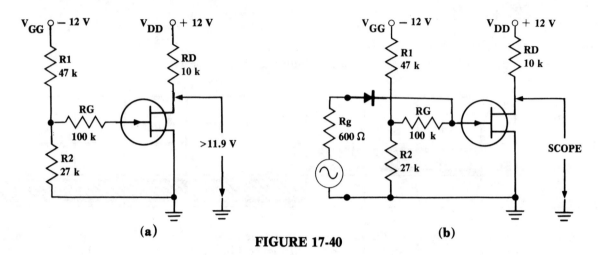

FIGURE 17-40

2. Connect a signal generator to the circuit, as shown in Figure 17-40(b), and use a 1 kHz sine wave input signal. Monitor the drain of the JFET with an oscilloscope and increase the magnitude of the input signal until the output waveshape becomes a square wave.

3. Monitor the magnitude of the positive-going input pulse at the gate and, then, the upper and lower limits ($\approx V_{DD}$ and $V_{DS}[SAT]$) of the output square wave. Sketch both the input and output waveshapes (sine and square wave) on the graph of Figure 17-41. Use 2 ms/DIV on the oscilloscope and, if possible, use a dual trace oscilloscope for a direct comparison of the input and output waveshapes. A nominal input to output ratio should be about 5:1.

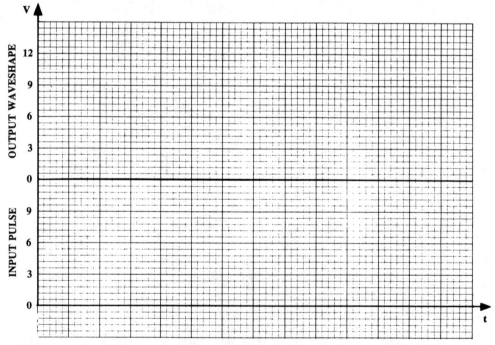

FIGURE 17-41

NOTE: JFETs, like the 2N5951, have large Vp voltages and do not saturate as easily as bipolar transistors unless the RD resistor is relatively large. Hence, an RD of 10 kΩ is used.

Section II: Latches

Part 1: Transistor Latches

A. UNLATCHED CONDITIONS

1. Connect the PNP-NPN latching circuit of Figure 17-42.

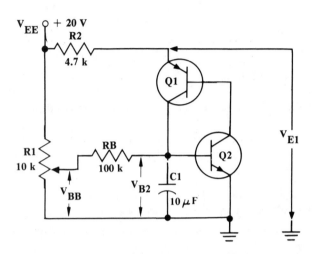

FIGURE 17-42

2. Initially, adjust the variable resistor R1 to bottom rotation, prior to connecting the 20 VDC VPS voltage to the circuit so that V_{BB} will equal zero volts. The zero volt V_{BB} condition ensures an initial unlatched circuit condition.
3. Monitor the voltage drop across the Q1,Q2 configuration, V_{E1} with respect to ground, and slowly increase the V_{BB} voltage until the voltage of V_{E1}, with respect to ground, drops below 1 volt, which indicates a latched circuit condition. Test latching, by varying the V_{BB} voltage from V_{EE} to zero volts, to show that the V_{BB} voltage level can only cause latching to occur and has no effect in unlatching the circuit.
4. Repeat Step 3 one more time and, this time, make measurements around the circuit before latching and, then, after latching occurs. To unlatch the circuit, momentarily disconnect the applied V_{PS} voltage.
5. At $V_{BB} = 0.0$ V and at $V_{PS} = 20$ V, measure the voltage drops across the Q1,Q2 configuration, or V_{E1} with respect to ground. Then, calculate:

$$V_{R2} = V_{EE} - V_{E1}$$

$$I_{R2} = V_{R2}/R2$$

$$V_{RB} = V_{BB} - V_{B2}$$

$$I_B = V_{RB}/RB$$

NOTE: Capacitor C1, at the base of Q2 to ground, minimizes the possibility of the circuit being prematurely latched by a transient pulse, when power is initially applied.

6. Insert the calculated and measured values, as indicated, into Table 17-5.

TABLE 17-5	UNLATCHED					LATCHED				
	V_{E1}	V_{R2}	I_{R2}	V_{RB}	I_{RB}	V_{E1}	V_{R2}	I_{R2}	V_{RB}	I_{RB}
CALCULATED	///					///				
MEASURED		///	///	///	///		///	///	///	///

B. LATCHED CONDITIONS

1. Increase the V_{BB} voltage until latching just occurs and measure V_{E1}, with respect to ground. Then, calculate:

$$V_{R1} = V_{PS} - V_{E1}$$
$$I_{R1} = V_{R1}/R1$$
$$V_{RB} = V_{BB} - V_{B2}$$
$$I_B = V_{RB}/RB$$

NOTE: Once the Q1,Q2 configuration is latched, $V_{B2} \approx 0.6$ V.

2. Insert the calculated and measured values, as indicated, into Table 17-5 for the latched conditions.

Part 2: The SCR as a Latch

A. UNLATCHED CONDITIONS

1. Connect the SCR, as shown in the circuit of Figure 17-43, and initially set the variable resistor R1 to bottom rotation so that $V_{GG} = 0.0$ V. The zero volt gate condition ensures an initial unlatched SCR condition when the V_{PS} of 20 volts is applied.

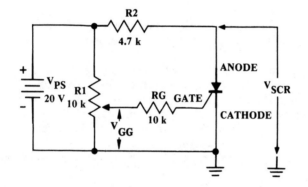

FIGURE 17-43

2. Measure the voltage drop across the SCR, at the anode with respect to ground, and the voltage at the gate with respect to ground. Calculate:

$$V_{R2} = V_{PS} - V_A \quad \text{where:} \quad V_A = V(\text{anode})$$
$$I_{R2} = V_{R2}/R2$$
$$V_{RG} = V_G - V_{GG} \quad \text{where:} \quad V_G = V(\text{gate})$$
$$I_{RG} = V_{RG}/RG$$

3. Insert the calculated and measured values, as indicated, into Table 17-6 for the unlatched condition.

TABLE 17-6	UNLATCHED						LATCHED					
	V_A	V_G	V_{R2}	I_{R2}	V_{RG}	I_{RG}	V_A	V_G	V_{R2}	I_{R2}	V_{RG}	I_{RG}
CALCULATED	▨	▨					▨	▨				
MEASURED			▨	▨	▨	▨			▨	▨	▨	▨

B. LATCHED CONDITIONS

1. Increase the V_{GG} voltage until latching just occurs, and then measure the voltage at the anode and the voltage at the gate, with respect to ground. Calculate:

$$V_{R2} = V_{PS} - V_A \quad \text{where:} \quad V_A = V(\text{anode})$$

$$I_{R2} = V_{R2}/R2$$

$$V_{RG} = V_{GG} - V_G \quad \text{where:} \quad V_G = V(\text{gate})$$

$$I_G = V_{RG}/RG$$

NOTE: Once the SCR latches, the gate voltage will be about 0.6 V and the V_{GG} should be slightly higher.

3. Insert the calculated and measured values, as indicated, into Table 17-6 for latched conditions.

C. HOLDING CURRENT MEASUREMENTS

1. Connect the SCR into the circuit shown in Figure 17-44 to measure the holding current.

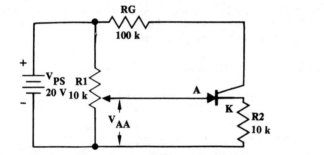

FIGURE 17-44

a. Initially, set V_{AA} to 20 volts to provide a latched condition.

b. Monitor the voltage across the 10 kΩ resistor R2 and disconnect the 100 kΩ gate resistor RG. Then, decrease V_{AA} until the SCR becomes unlatched. Repeat this procedure and obtain V_{R2} and V_{AA} just before the SCR unlatches. Use the V_{R2} and V_{AA} values to calculate:

$$V_{AK} = V_{AA} - V_K \quad \text{where:} \quad V_K = V(\text{cathode}) = V_{R2}$$

$$I_H = V_{R2}/R2$$

NOTE: Unlatching can occur at a holding current of lower than 0.3 mA. Therefore, V_{RE}, just before unlatching occurs, can be lower than 3 volts.

2. Insert the calculated and measured values, as indicated, into Table 17-7.

TABLE 17-7	V_{AA}	$V_{R2} = V_K$	V_{AK}	I_H
CALCULATED	/////////	/////////		
MEASURED			/////////	/////////

D. SCR ALTERNATING CURRENT CONSIDERATIONS

1. Connect the SCR into the circuit shown in 17-45, where a 6.3 V AC secondary voltage is used as an input. However, larger V RMS voltages can be used successfully.

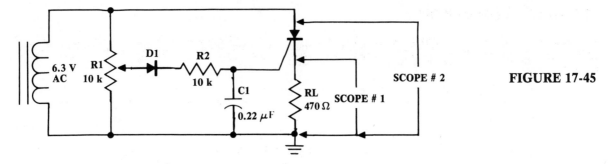

FIGURE 17-45

2. Monitor the output waveshape across the 470 ohms load resistor with an oscilloscope. If a dual trace oscilloscope is available, also monitor the peak-to-peak waveshape at the anode of the SCR, with respect to ground. Vary R1 and monitor the full half-wave pulse and, then, the 45°, 90°, 135°, and the eventual 180° no pulse conditions.

NOTE: C1 and R2 are selected to provide an initial 45° phase angle. Also, C1 can be removed to test the 90° trigger angle limit and to monitor the minimum (close to zero degree) phase triggering.

3. APPROXIMATELY ZERO DEGREE PHASE SHIFT: The complete positive-going, half-wave pulse appears at the load for the approximately zero degree phase shift. Therefore, just at full pulse conditions (prior to partial pulse conditions), monitor the Vp and Vp-p output waveshape on the oscilloscope. Vp should be slightly less than one-half Vp-p because of the voltage drop across the SCR.

    a. Calculate the DC voltage drop across the load resistor RL from: $VDC = Vp/\pi$

    b. Calculate the current flow in the load resistor RL from: $I_{RL} = V_{RL}/RL$

    c. Measure the DC voltage drop across the load resistor to verify the calculated voltage.

    d. Insert the calculated and measured values for the zero degree phase shift into Table 17-8.

4. NINETY DEGREE PHASE SHIFT: Continue to monitor the load with an oscilloscope and a DC voltmeter and adjust the R1 resistor until the phase shift is at 90 degrees and the VDC has dropped to one-half of the VDC at zero degrees full pulse phase shift.

    a. Calculate the VDC at the load from: $VDC = Vp/2\pi$.

    b. Calculate the curent flow through the load resistor RL from: $I_{RL} = V_{RL}/RL$

    c. Measure the DC voltage drop across the load to verify the calculated voltage.

    d. Insert the calculated and measured values for the 90° phase shift into Table 17-8.

5. Measure the VDC developed across the load resistor at 45° and 135° phase angles, and insert the measured 45° and 135° VDC conditions into Table 17-8.

6. Sketch the waveshapes at 0°, 45°, 90°, and 135°. Note the magnitude and frequency (time). Reference the output waveshape to the input waveshape.

TABLE 17-8	0° phase		45° phase	90° phase		135° phase
	V_{RL}	I_{RL}	V_{RL}	V_{RL}	I_{RL}	V_{RL}
CALCULATED			/////			/////
MEASURED		/////			/////	

Part 3: SCR Applications

1. Test the LED and 470 ohm series resistor of Figure 17-46(a) by applying a VPS of about 12 volts. The LED will not "turn on" unless it is in the forward biased direction. If it does not work in one direction, turn it around.

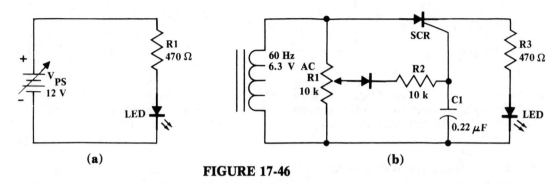

FIGURE 17-46

2. Connect the LED into the circuit shown in Figure 17-46(b). Vary the R1 resistor and visually monitor the increased and decreased intensity of the LED. A low voltage incandescent lamp can be used if an LED is not available.

CHAPTER QUESTIONS

1. Why are transistors and JFETs not considered perfect switches?
2. How does a latch work and how is it different from a switch?
3. What is the advantage of a triac over the SCR?
4. Why does 90° phase triggering occur at the lowest gate current in the theoretical zero to ninety degree triggering range?
5. What purpose does the diode serve in the gate circuit of the SCR during AC applications?
6. What is SCR holding current?
7. What is breakover voltage? Explain how breakover voltage can be varied?
8. In the laboratory, did the intensity of the LED (or low voltage incandescent lamp) increase or decrease with increased phase triggering? Explain why or why not.
9. In the laboratory, was the measured saturation voltage of bipolar transistors lower or higher than JFETs? Explain the difference, if any, with reference to the operating saturation currents.
10. Connecting the bipolar transistor in the inverse connection, where the collector and emitter terminals are reversed, is sometimes referred to as the perfect switch. Test the circuits of Figure 17-47 to see if this is a true statement.

FIGURE 17-47

CHAPTER PROBLEMS

1. Given: $V_{CE}(SAT) = 0.2$ V $RC = 3$ kΩ

   $\beta = 100$ $V_{CC} = 14$ V

   Solve for:

   a. $I_C(SAT)$

   b. $I_B(SAT)$ minimum

   c. RB at minimum $I_B(SAT)$ condition

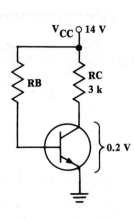

FIGURE 17-48

2. Given: V(secondary) = 30 V RMS

   RL = 100 Ω

   Solve for:

   a. V_{RL} at 0° phase triggering

   b. V_{RL} at 90° phase triggering

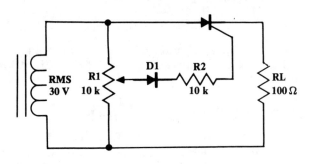

FIGURE 17-49

18 | INTRODUCTION TO MULTISTAGE CIRCUITS

GENERAL DISCUSSION

Multistage circuits are used to provide large voltage gains, current gains, and power gains that would not be possible with single stage circuits. Multistage cascaded circuits, where the output of one stage feeds the input of another, can be capacitor coupled, direct coupled, or transformer coupled, as shown in Figure 18-1. In this chapter, on the introduction to multistage circuits, only capacitor coupled two stage circuits will be used, because the direct current analysis of each circuit can be independently solved, and the alternating current circuit analysis is similar to the single stage AC analysis. Later, in more advanced course work, both direct coupled and transformer coupled circuits will be introduced.

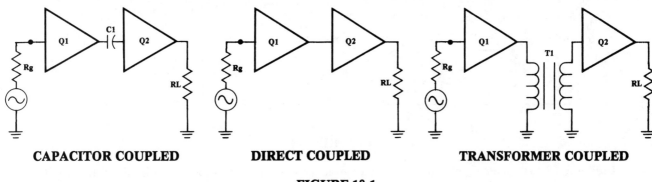

CAPACITOR COUPLED **DIRECT COUPLED** **TRANSFORMER COUPLED**

FIGURE 18-1

CAPACITOR COUPLED CIRCUITS

Two stage capacitor coupled circuits come in many forms. In Figure 18-2, the circuits shown are common emitter stages driving common emitter stages. Both of the circuits provide large voltage gains,

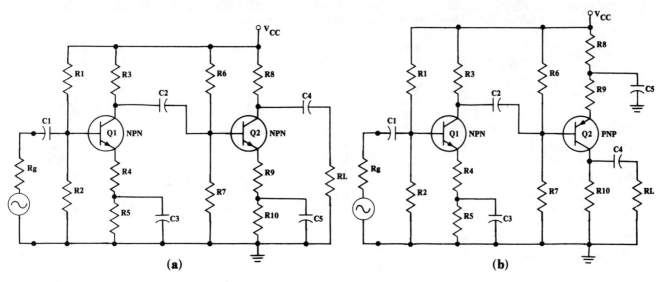

FIGURE 18-2

445

Essentially, the coupling capacitor for the circuits processes the alternating current signal between stages and, at the same time, it blocks the direct current flow between stages. Hence, the DC voltage distribution of each Q1 and Q2 stage is solved individually.

The capacitor coupled common emitter and common collector circuits of Figure 18-3 also use the characteristics of each stage to advantage. The common emitter circuit provides the voltage gain and the common collector circuit provides the impedance transfer. In Figure 18-3(a), the Q1 common emitter stage provides the voltage gain, while the Q2 common collector stage allows the signal to be applied to a relatively low output load resistance without large signal magnitude losses. In Figure 18-3(b), the Q1 common collector stage provides a high input impedance, while the Q2 common emitter stage provides a voltage gain to a moderate output impedance. Again, the circuit transistors can be either NPN or PNP, as long as proper polarities are observed.

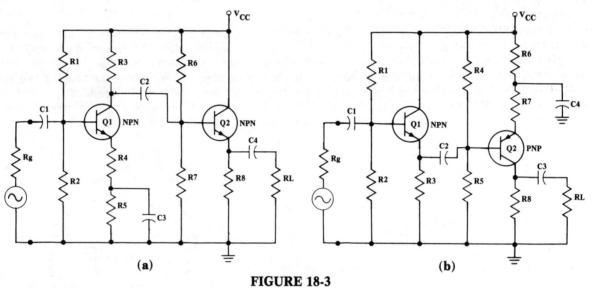

FIGURE 18-3

In Figure 18-4, two completely different capacitor coupled circuits are shown. In Figure 18-4(a), the two stage circuit is comprised of a common source stage driving a common emitter stage. The advantage to this circuit is that it is capable of extremely high input impedances and excellent voltage gains into moderate valued load resistances. In Figure 18-4(b), a common emitter stage is used to drive a complementary symmetry stage. The advantage of the complementary symmetry stage, which is essentially back-to-back emitter followers, is that it provides impedance transfer and, at the same time, shares the dissipated power that occurs when delivering power to the load resistor. The circuit is primarily used in power amplifier application. It is shown to illustrate one advantage of combining NPN and PNP transistors. In later course work, complementary symmetry circuits in high power amplifier circuits will be fully analyzed.

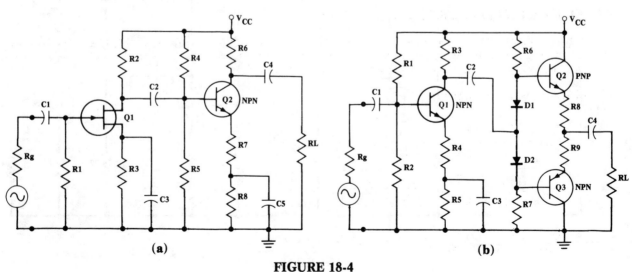

FIGURE 18-4

446

Three circuits will be analyzed in this chapter: the common emitter stage driving the common collector stage, the common source stage driving the common emitter stage, and the common emitter stage driving the complementary symmetry stage. Also, to facilitate the overall analysis, the circuits used in the examples, with the exception of the complementary symmetry circuit, are circuits used in the earlier chapters on single stage circuits.

SECTION I: THE COMMON-EMITTER, COMMON-COLLECTOR CIRCUIT

GENERAL DISCUSSION

The advantage of using a common emitter stage to drive a common collector stage is that the common emitter stage provides the voltage gain and the common collector stage is used to drive a relatively low output load resistance with minimum voltage gain loss. For example, if an unloaded common emitter Q1 stage has an unloaded voltage gain of 25, then connecting the output signal directly into a 4 kΩ load would cause the voltage gain to drop to less than 9. However, if the impedance transfer capability of the Q2 common collector stage is used to increase the 4 kΩ load to about 25.47 kΩ, the voltage gain of the entire circuit can be maintained at slightly less than 19, for a greater than 2 : 1 signal amplitude advantage.

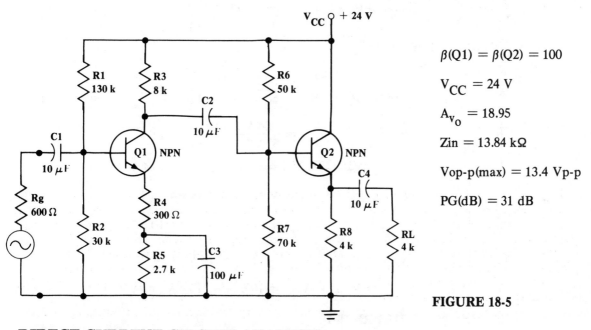

$\beta(Q1) = \beta(Q2) = 100$

$V_{CC} = 24\ V$

$A_{V_O} = 18.95$

$Z_{in} = 13.84\ k\Omega$

$V_{op\text{-}p}(max) = 13.4\ Vp\text{-}p$

$PG(dB) = 31\ dB$

FIGURE 18-5

DIRECT CURRENT CIRCUIT ANALYSIS

The direct current circuit analysis of the common emitter and common collector stages are solved the same as the individual single stage common emitter and common collector circuits of Chapter 8. Therefore, the DC analysis of each circuit is solved without further discussion.

A. NPN COMMON EMITTER STAGE

$$VR1 = \frac{V_{CC} \times R1}{R1 + R2} = \frac{24\ V \times 30\ k\Omega}{130\ k\Omega + 30\ k\Omega} = 19.5\ V$$

$$VR2 = \frac{V_{CC} \times R2}{R1 + R2} = \frac{24\ V \times 30\ k\Omega}{130\ k\Omega + 30\ k\Omega} = 4.5\ V$$

$$V_{RE} = V_{R2} - V_{BE_1} = 4.5\ V - 0.6\ V = 3.9\ V$$

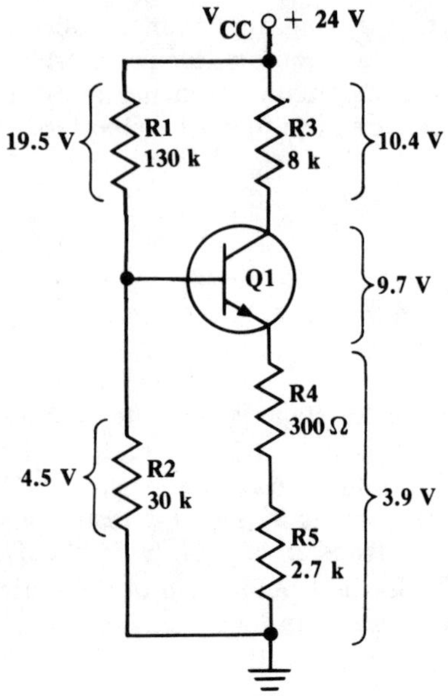

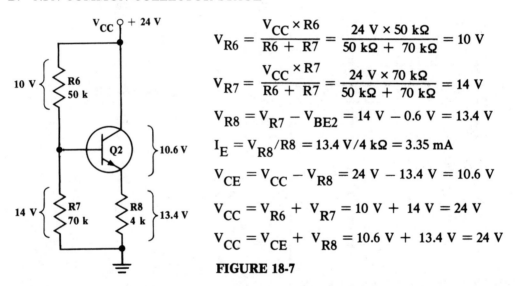

$$I_E = V_{RE}/R_E = \frac{V_{R4} + V_{R5}}{R4 + R5} = \frac{3.9 \text{ V}}{300 \text{ }\Omega + 2.7 \text{ k}\Omega} = 1.3 \text{ mA}$$

$$V_{R3} = I_C \times R_C \approx I_E \times R3 = 1.3 \text{ mA} \times 8 \text{ k}\Omega = 10.4 \text{ V}$$

$$V_{CE} = V_{CC} - (V_{R3} + V_{RE}) = 24 \text{ V} - (10.4 \text{ V} + 3.9 \text{ V}) = 9.7 \text{ V}$$

$$V_{CC} = V_{R1} + V_{R2} = 19.5 \text{ V} + 4.5 \text{ V} = 24 \text{ V}$$

$$V_{CC} = V_{R3} + V_{CE} + V_{RE} = 10.4 \text{ V} + 9.7 \text{ V} + 3.9 \text{ V} = 24 \text{ V}$$

FIGURE 18-6

B. NPN COMMON COLLECTOR STAGE

$$V_{R6} = \frac{V_{CC} \times R6}{R6 + R7} = \frac{24 \text{ V} \times 50 \text{ k}\Omega}{50 \text{ k}\Omega + 70 \text{ k}\Omega} = 10 \text{ V}$$

$$V_{R7} = \frac{V_{CC} \times R7}{R6 + R7} = \frac{24 \text{ V} \times 70 \text{ k}\Omega}{50 \text{ k}\Omega + 70 \text{ k}\Omega} = 14 \text{ V}$$

$$V_{R8} = V_{R7} - V_{BE2} = 14 \text{ V} - 0.6 \text{ V} = 13.4 \text{ V}$$

$$I_E = V_{R8}/R8 = 13.4 \text{ V}/4 \text{ k}\Omega = 3.35 \text{ mA}$$

$$V_{CE} = V_{CC} - V_{R8} = 24 \text{ V} - 13.4 \text{ V} = 10.6 \text{ V}$$

$$V_{CC} = V_{R6} + V_{R7} = 10 \text{ V} + 14 \text{ V} = 24 \text{ V}$$

$$V_{CC} = V_{CE} + V_{R8} = 10.6 \text{ V} + 13.4 \text{ V} = 24 \text{ V}$$

FIGURE 18-7

ALTERNATING CURRENT CIRCUIT ANALYSIS

The alternating current circuit analysis of the two stage circuit begins with the Q2 output stage, because the input impedance of the common collector Q2 stage is the effective load of the Q1 common emitter stage. Once the voltage gains, input impedance, and Vop-p(max) of each individual stage is known, then the combined A_v, Zin, Vop-p, and power gain of the two stage circuit are solved.

A. COMMON COLLECTOR STAGE — ALTERNATING CURRENT CIRCUIT ANALYSIS

$$A_v(Q2) = \frac{r_L}{r_{e_2} + r_L} = \frac{R8 // RL}{r_{e_2} + (R8 // RL)} = \frac{4 \text{ k}\Omega // 4 \text{ k}\Omega}{7.8 \text{ }\Omega + (4 \text{ k}\Omega // 4 \text{ k}\Omega)} = \frac{2 \text{ k}\Omega}{7.8 \text{ }\Omega + 2 \text{ k}\Omega} \approx 0.996$$

where: $r_{e_2} = 26 \text{ mV}/I_E = 26 \text{ mV}/3.35 \text{ mA} = 7.76 \text{ }\Omega \approx 7.8 \text{ }\Omega$

$$Zin = R6 // R7 // \beta(r_{e_2} + r_L) = 50 \text{ k}\Omega // 70 \text{ k}\Omega // 100(7.8 \text{ }\Omega + 2 \text{ k}\Omega)$$

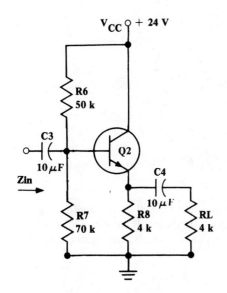

$Z_{in} = 50\text{ k}\Omega \mathbin{/\mkern-5mu/} 70\text{ k}\Omega \mathbin{/\mkern-5mu/} 200.78\text{ k}\Omega \approx 25.47\text{ k}\Omega$

$V_{op\text{-}p(max)} = 2V_{CE}$ or $2I_E r_L$ (whichever is smaller):

$2V_{CE} = 2 \times 10.6\text{ V} = 21.2\text{ Vp-p}$

$2I_E r_L = 2I_E(R8 \mathbin{/\mkern-5mu/} RL) = 2 \times 3.35\text{ mA} \times 2\text{ k}\Omega = 13.4\text{ Vp-p}$

Therefore, $V_{op\text{-}p(max)} = 13.4\text{ Vp-p}$

FIGURE 18-8

B. COMMON EMITTER STAGE — ALTERNATING CURRENT CIRCUIT ANALYSIS

$$A_V(Q1) = \frac{r_L}{r_{e1} + R4} = \frac{R3 \mathbin{/\mkern-5mu/} Z_{in}(Q2)}{r_{e1} + R4} = \frac{8\text{ k}\Omega \mathbin{/\mkern-5mu/} 25.47\text{ k}\Omega}{20\text{ }\Omega + 300\text{ }\Omega} = \frac{6087.8\text{ }\Omega}{320\text{ }\Omega} = 19.02$$

where: $r_{e1} = 26\text{ mV}/I_E = 26\text{ mV}/1.3\text{ mA} = 20\text{ }\Omega$

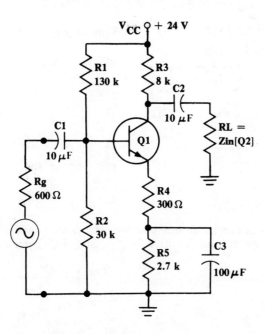

$Z_{in}(Q1) = R1 \mathbin{/\mkern-5mu/} R2 \mathbin{/\mkern-5mu/} \beta(r_{e_1} + R4) =$

$= 130\text{ k}\Omega \mathbin{/\mkern-5mu/} 30\text{ k}\Omega \mathbin{/\mkern-5mu/} 100(20\text{ }\Omega + 300\text{ }\Omega)$

$= 130\text{ k}\Omega \mathbin{/\mkern-5mu/} 30\text{ k}\Omega \mathbin{/\mkern-5mu/} 32\text{ k}\Omega = 13.84\text{ k}\Omega$

$V_{op\text{-}p(max)} = 2V_{CE}$ or $2I_C r_L$ (whichever is smaller):

$2V_{CE} = 2 \times 9.7\text{ V} = 19.4\text{ Vp-p}$

$2I_E r_L = 2I_E(R3 \mathbin{/\mkern-5mu/} RL) = 2I_E[R3 \mathbin{/\mkern-5mu/} Z_{in}(Q2)]$

$= 2 \times 1.3\text{ mA} \times (8\text{ k}\Omega \mathbin{/\mkern-5mu/} 25.47\text{ k}\Omega)$

$= 15.83\text{ Vp-p}$

Therefore, $V_{op\text{-}p(max)} = 15.83\text{ Vp-p}$

FIGURE 18-9

C. The overall voltage gain of the two stage circuit of Figure 18-5 is solved by multiplying the voltage gains of the Q1 and Q2 stages. The input impedance of the two stage circuit is simply the input impedance of Q1 alone. The $V_{op\text{-}p(max)}$ of the two stage circuit can be limited by either Q1 or Q2, because approximately the same peak-to-peak signal rides on both the collector output of the Q1 stage, and at the emitter of the Q2 emitter follower stage. Since the smallest $V_{op\text{-}p(max)}$ is the limiting factor, then $V_{op\text{-}p(max)}$ for the entire two stage circuit is determined by $2I_E r_L$, which equals 13.4 Vp-p. The power gain of the overall circuit is solved from the overall two stage voltage gain, the input impedance of Q1 and the 4 kΩ load resistor. Therefore:

$$A_{V_O} = A_V(Q1) \times A_V(Q2) = 19.02 \times 0.996 \approx 18.95$$

$$Zin = Zin(Q1) = 13.84 \text{ k}\Omega$$

$$Vop\text{-}p(max) = 2I_{E_2}r_L = 13.4 \text{ Vp-p}$$

$$PG = A_{v_o}^2 \times Zin(Q1)/RL = 18.95^2 \times 13.84 \text{ k}\Omega/4 \text{ k}\Omega = 1242.5$$

$$PG(dB) = 10 \log 1242.5 = 30.94 \text{ dB} \approx 31 \text{ dB}$$

SECTION II: THE COMMON SOURCE, COMMON-EMITTER CIRCUIT

GENERAL DISCUSSION

The main advantage of using the common source circuit as a driving stage is that it provides both a voltage gain and a very high input impedance of 100 kΩ, limited by the gate resistor R1 alone. And since the common emitter provides good voltage gains into moderate output resistance loads, the entire circuit of Figure 18-10 is capable of very high power gains.

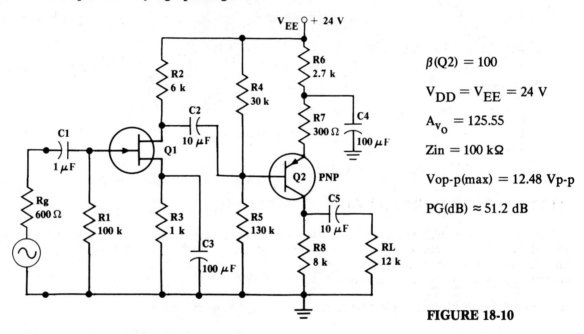

FIGURE 18-10

DIRECT CURRENT CIRCUIT ANALYSIS

The direct current circuit analysis of the common source and common emitter circuits are solved like the single stage common source circuit of Chapter 12 and the common emitter PNP circuit of Chapter 8. They are solved, therefore, without further explanation.

A. N CHANNEL COMMON SOURCE CIRCUIT

$$V_p = V_{GS}(\text{off}) = |4 \text{ V}|$$

$$I_{DSS} = 8 \text{ mA}$$

$$V_{GS} = -2 \text{ V}$$

$$I_D = I_{DSS}(1 - V_{GS}/V_p)^2 = 8 \text{ mA}(1 - 2 \text{ V}/4 \text{ V})^2$$
$$= 8 \text{ mA} \times 0.5^2 = 2 \text{ mA}$$

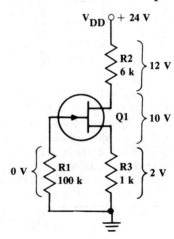

FIGURE 18-11

450

$$V_{R3} = I_D R3 = 2 \text{ mA} \times 1 \text{ k}\Omega = 2 \text{ V}$$

$$V_{R2} = I_D R2 = 2 \text{ mA} \times 6 \text{ k}\Omega = 12 \text{ V}$$

$$V_{DS} = V_{DD} - (V_{R2} + V_{R3}) = 24 \text{ V} - (12 \text{ V} + 2 \text{ V}) = 10 \text{ V}$$

B. PNP COMMON EMITTER CIRCUIT

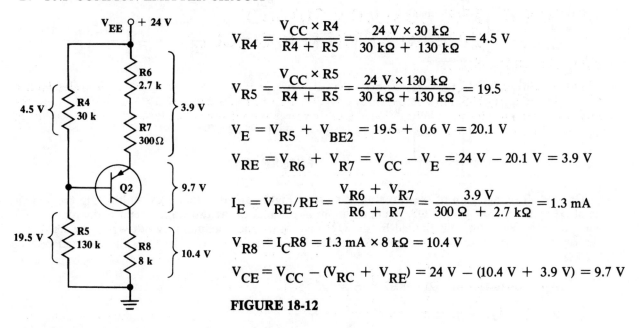

$$V_{R4} = \frac{V_{CC} \times R4}{R4 + R5} = \frac{24 \text{ V} \times 30 \text{ k}\Omega}{30 \text{ k}\Omega + 130 \text{ k}\Omega} = 4.5 \text{ V}$$

$$V_{R5} = \frac{V_{CC} \times R5}{R4 + R5} = \frac{24 \text{ V} \times 130 \text{ k}\Omega}{30 \text{ k}\Omega + 130 \text{ k}\Omega} = 19.5$$

$$V_E = V_{R5} + V_{BE2} = 19.5 + 0.6 \text{ V} = 20.1 \text{ V}$$

$$V_{RE} = V_{R6} + V_{R7} = V_{CC} - V_E = 24 \text{ V} - 20.1 \text{ V} = 3.9 \text{ V}$$

$$I_E = V_{RE}/RE = \frac{V_{R6} + V_{R7}}{R6 + R7} = \frac{3.9 \text{ V}}{300 \text{ }\Omega + 2.7 \text{ k}\Omega} = 1.3 \text{ mA}$$

$$V_{R8} = I_C R8 = 1.3 \text{ mA} \times 8 \text{ k}\Omega = 10.4 \text{ V}$$

$$V_{CE} = V_{CC} - (V_{RC} + V_{RE}) = 24 \text{ V} - (10.4 \text{ V} + 3.9 \text{ V}) = 9.7 \text{ V}$$

FIGURE 18-12

ALTERNATING CURRENT CIRCUIT ANALYSIS

The alternating current circuit analysis begins with the output stage Q1, since the input impedance for the Q2 stage is the effective load for the Q1 common source circuit. Next, the voltage gains and input impedances of each stage are solved, along with the Vop-p(max) of the output Q2 stage. Then, the combined A_v, Zin, Vop-p(max), and power gain of the overall circuit are solved.

A. PNP COMMON EMITTER CIRCUIT

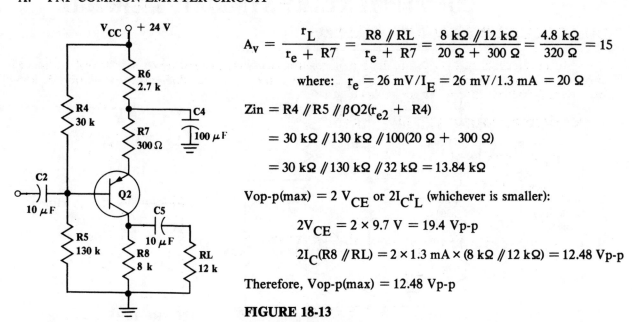

$$A_v = \frac{r_L}{r_e + R7} = \frac{R8 \text{ // } RL}{r_e + R7} = \frac{8 \text{ k}\Omega \text{ // } 12 \text{ k}\Omega}{20 \text{ }\Omega + 300 \text{ }\Omega} = \frac{4.8 \text{ k}\Omega}{320 \text{ }\Omega} = 15$$

where: $r_e = 26 \text{ mV}/I_E = 26 \text{ mV}/1.3 \text{ mA} = 20 \text{ }\Omega$

$$\text{Zin} = R4 \text{ // } R5 \text{ // } \beta Q2(r_{e2} + R4)$$

$$= 30 \text{ k}\Omega \text{ // } 130 \text{ k}\Omega \text{ // } 100(20 \text{ }\Omega + 300 \text{ }\Omega)$$

$$= 30 \text{ k}\Omega \text{ // } 130 \text{ k}\Omega \text{ // } 32 \text{ k}\Omega = 13.84 \text{ k}\Omega$$

Vop-p(max) = $2 V_{CE}$ or $2 I_C r_L$ (whichever is smaller):

$$2 V_{CE} = 2 \times 9.7 \text{ V} = 19.4 \text{ Vp-p}$$

$$2 I_C (R8 \text{ // } RL) = 2 \times 1.3 \text{ mA} \times (8 \text{ k}\Omega \text{ // } 12 \text{ k}\Omega) = 12.48 \text{ Vp-p}$$

Therefore, Vop-p(max) = 12.48 Vp-p

FIGURE 18-13

B. N CHANNEL COMMON SOURCE CIRCUIT

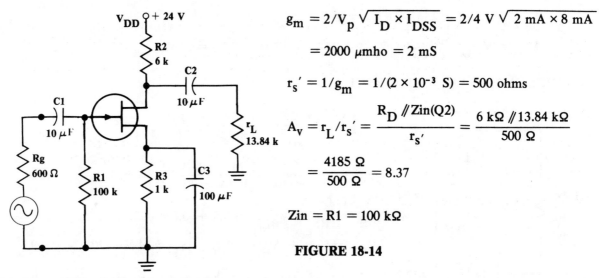

$g_m = 2/V_p \sqrt{I_D \times I_{DSS}} = 2/4 \text{ V} \sqrt{2 \text{ mA} \times 8 \text{ mA}}$

$= 2000 \text{ } \mu\text{mho} = 2 \text{ mS}$

$r_s' = 1/g_m = 1/(2 \times 10^{-3} \text{ S}) = 500 \text{ ohms}$

$A_v = r_L/r_s' = \dfrac{R_D \text{ // } Zin(Q2)}{r_s'} = \dfrac{6 \text{ k}\Omega \text{ // } 13.84 \text{ k}\Omega}{500 \text{ }\Omega}$

$= \dfrac{4185 \text{ }\Omega}{500 \text{ }\Omega} = 8.37$

$Zin = R1 = 100 \text{ k}\Omega$

FIGURE 18-14

C. The overall voltage gain of the two stage circuit of Figure 18-10 is found by multiplying the voltage gain of Q1 by the voltage gain of Q2. Then, the input impedance of the two stage circuit is the input to Q1 alone. Also, the Vop-p(max) is limited by the Q2 stage, since the maximum output signal occurs at the collector of Q2. The power gain of the two stage circuit is solved from A_{v_o}, Zin(Q1), and the load resistor of 12 kΩ.

$A_{v_o} = A_v(Q1) \times A_v(Q2) = 8.37 \times 15 = 125.55$

$Zin = Zin(Q1) = 100 \text{ k}\Omega$

$Vop\text{-}p(max) = 2I_{C2}r_L = 12.48 \text{ Vp-p}$

$PG = A_{v_o}^2 \times Zin(Q1)/RL = 125.55^2 \times 100 \text{ k}\Omega/12 \text{ k}\Omega = 131{,}375$

$PG(dB) = 10 \log 131{,}375 = 51.18 \text{ dB} \approx 51.2 \text{ dB}$

SECTION III: COMMON-EMITTER, COMPLEMENTARY-SYMMETRY CIRCUIT

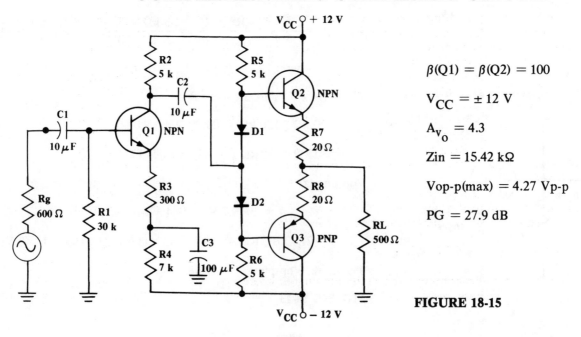

$\beta(Q1) = \beta(Q2) = 100$

$V_{CC} = \pm 12 \text{ V}$

$A_{v_o} = 4.3$

$Zin = 15.42 \text{ k}\Omega$

$Vop\text{-}p(max) = 4.27 \text{ Vp-p}$

$PG = 27.9 \text{ dB}$

FIGURE 18-15

GENERAL DISCUSSION

The main advantage of using a complementary symmetry circuit, instead of a single voltage follower circuit, is that each Q2 and Q3 output transistor processes only one-half of the input signal and, therefore, shares the power dissipated by the output transistors. Hence, the output circuit is used to deliver power to the load at a reasonably high efficiency and the common emitter, two power supply, NPN circuit is used to provide the voltage gain for the overall circuit.

DIRECT CURRENT CIRCUIT ANALYSIS

The direct current circuit analysis of the two power supply, common emitter circuit is like the single stage circuit analyzed in Chapter 9 and is given without further explanation. The complementary symmetry circuit, however, because it is newly introduced, is analyzed in considerable detail.

A. THE COMMON EMITTER, TWO POWER SUPPLY CIRCUIT

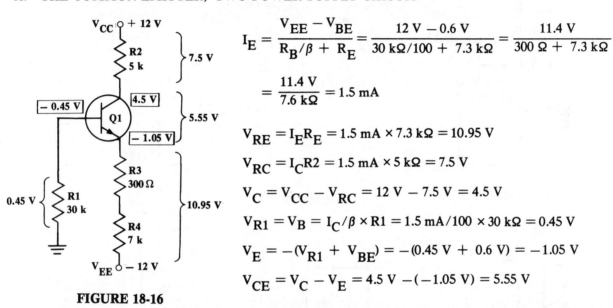

$$I_E = \frac{V_{EE} - V_{BE}}{R_B/\beta + R_E} = \frac{12\text{ V} - 0.6\text{ V}}{30\text{ k}\Omega/100 + 7.3\text{ k}\Omega} = \frac{11.4\text{ V}}{300\text{ }\Omega + 7.3\text{ k}\Omega}$$

$$= \frac{11.4\text{ V}}{7.6\text{ k}\Omega} = 1.5\text{ mA}$$

$$V_{RE} = I_E R_E = 1.5\text{ mA} \times 7.3\text{ k}\Omega = 10.95\text{ V}$$

$$V_{RC} = I_C R2 = 1.5\text{ mA} \times 5\text{ k}\Omega = 7.5\text{ V}$$

$$V_C = V_{CC} - V_{RC} = 12\text{ V} - 7.5\text{ V} = 4.5\text{ V}$$

$$V_{R1} = V_B = I_C/\beta \times R1 = 1.5\text{ mA}/100 \times 30\text{ k}\Omega = 0.45\text{ V}$$

$$V_E = -(V_{R1} + V_{BE}) = -(0.45\text{ V} + 0.6\text{ V}) = -1.05\text{ V}$$

$$V_{CE} = V_C - V_E = 4.5\text{ V} - (-1.05\text{ V}) = 5.55\text{ V}$$

FIGURE 18-16

B. COMPLEMENTARY SYMMETRY CIRCUIT

The complementary symmetry circuit can use either dual or single DC power supplies, but with a single power supply the load should be capacitor coupled so that no DC voltage appears across the load. The DC balance, especially important in the dual supply circuits, depends on making the NPN stage and the PNP stage mirror images of each other.

For instance, in the single power supply circuit of Figure 18-17(a), a balanced circuit has 12 V at the junction of R7, R8, and the coupling capacitor C2. Therefore, assuming VD1 = VD2 = 0.7 V and V_{BE2} = V_{BE3} = 0.6 V, then $V_{R7} = V_{R8} = 0.1$ V. Hence, $V_{E2} = 12.1$ V, $V_{B2} = 12.7$ V, $V_{E3} = 11.9$ V, and $V_{B3} = 11.3$ V. In the two power supply circuit of Figure 18-17(b), 0 V exist at the junction of the R7, R8, and RL resistors. Therefore, $V_{E2} = 0.1$ V, $V_{B2} = 0.7$ V, $V_{E3} = -0.1$ V, and $V_{B3} = -0.7$ V. The reason zero volts exist across RL is because, in a perfectly balanced circuit, the current flow through Q2, R3, and RL to ground is equal and exactly opposite to the current flow through Q3, R8, and RL. Hence, zero volts is dropped across RL.

In Figure 18-18(a), conventional current flow from + 12 V to ground is shown, while Figure 18-18(b) shows the conventional ground to −12 V current flow. Figure 18-18(c) shows the currents in the load cancel, because they are equal (theoretically) but opposite.

C. DIRECT CURRENT CIRCUIT ANALYSIS

Once the voltage drops around the circuit are known, where $V_{D1} = V_{D2} = 0.7$ V and $V_{BE2} = V_{BE3} = 0.6$ V, the emitter, base, resistor, and diode currents can be found. Theoretically, zero volts exist at both

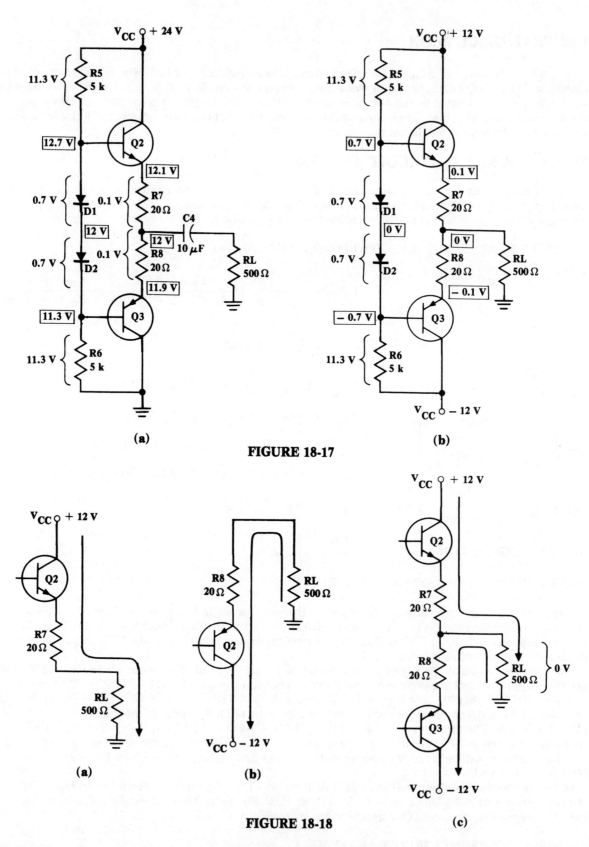

FIGURE 18-17

FIGURE 18-18

the junction of D1 and D2, and at the junction of R7, R8, and RL. Hence, $V_{R5} = V_{R6} = 11.3$ V, since $V_{D1} = V_{D2} = 0.7$ V, and $V_{R7} = V_{R8} = 0.1$ V, since $V_{B2} = V_{B3} = 0.6$ V. The solved currents and voltage drops are given in conjunction with Figure 18-19.

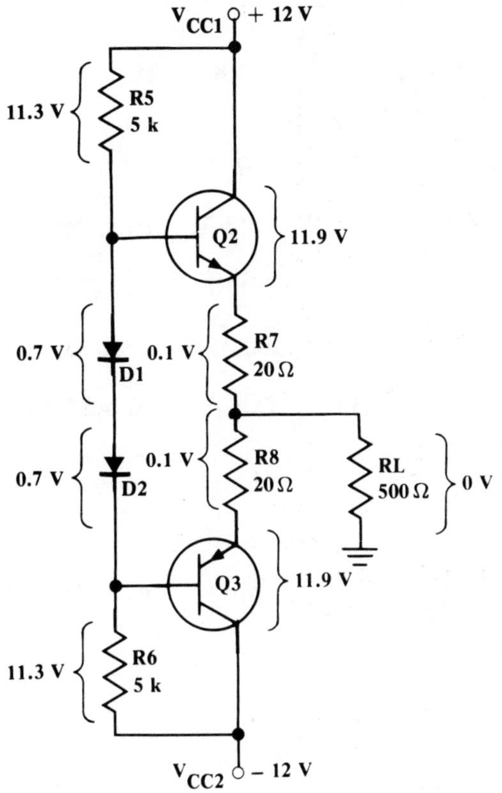

$$V_{D1} = V_{D2} = 0.7 \text{ V}$$

$$V_{BE2} = V_{BE3} = 0.6 \text{ V}$$

$$V_{R8} = V_{D1} - V_{BE2} = 0.7 \text{ V} - 0.6 \text{ V} = 0.1 \text{ V}$$

$$V_{R9} = V_{D2} - V_{BE3} = 0.7 \text{ V} - 0.6 \text{ V} = 0.1 \text{ V}$$

$$I_{E2} = I_{E3} = V_{R7}/R7 = V_{R8}/R8 = 0.1 \text{ V}/20 \text{ }\Omega = 5 \text{ mA}$$

$$I_{B2} = I_{B3} = I_{E2}/\beta(Q2) = I_{E3}/\beta(Q3) = 5 \text{ mA}/100 = 50 \text{ }\mu\text{A}$$

$$I_{R5} = I_{R6} = V_{R5}/R5 = V_{R6}/R6 = 11.3 \text{ V}/5 \text{ k}\Omega = 2.26 \text{ mA}$$

where: $I_{D1} = I_{D2} \approx I_{R5} = I_{R6}$

$$V_{CE2} = V_{C2} - V_{E2} = 12 \text{ V} - 0.1 \text{ V} = 11.9 \text{ V}$$

$$V_{CE3} = V_{EC3} = V_{E3} - V_{C3} = -0.1 \text{ V} - (-12 \text{ V})$$

$$= -0.1 \text{ V} + 12 \text{ V} = 11.9 \text{ V}$$

FIGURE 18-19

NOTE: $V_{R5} = V_{CC1} - V_{B2} = 12 \text{ V} - 0.7 \text{ V} = 11.3 \text{ V}$ where: $V_{B2} = V_{D1}$

$V_{R6} = V_{B3} - V_{CC2} = -0.7 \text{ V} - (-12 \text{ V}) = 11.3 \text{ V}$ where: $V_{B3} = V_{D2}$

ALTERNATING CURRENT CIRCUIT ANALYSIS

The alternating current circuit analysis for the circuit of Figure 18-15, like the coupled circuits previously analyzed, begins with the second stage, because the input impedance of the complementary circuit is the load resistance for the Q1 circuit. After the individual circuit voltage gains and input impedance are found, the overall A_{v_o}, Zin, Vop-p(max), power gain, power dissipated by Q1 and Q2, and power to the load are solved.

A. THE COMPLEMENTARY SYMMETRY CIRCUIT

The complementary symmetry circuit is, effectively, two back-to-back emitter followers, so the Q2 and Q3 stages are mirror images of each other. However, the Q2 stage processes only the positive-going signal and the Q3 stage processes only the negative going signal. Hence, when the positive-going signal is being processed, the Q2 stage is turned on, the Q3 stage is turned off. With the small resistance of the diodes D1 and D2 being safely ignored, the AC equivalent circuit for the positive-going signal is shown in Figure 18-21(a). When the negative-going signal is being processed, Q3 is on and Q2 is off. The equivalent circuit is shown in Figure 18-21(b).

The power dissipated by the Q2 and Q3 transistors is a function of both the no signal static DC and the rectified DC, that is caused when the AC signal is applied. Essentially, the initial (static) DC current of about 5 mA is established so that no portion of the applied AC signal will have to be used to turn on the base-emitter-diode juntions of Q2 and Q3. By properly DC biasing the base-emitter-diode junctions, the signal can be processed without distortion. However, applying an AC signal having a magnitude in excess of the V_{BE} DC voltages of about 0.6 V will cause the junction to be turned on further and the DC current in the load to increase.

Essentially, the DC current obtained when an AC signal is applied is the average DC current of the positive and negative-going waves, and it is solved like any rectified signal from: $I_{DC} = V_p/\pi R_L$. Recall from diode rectification that a positive-going pulse of 2.135 V is converted to VDC from: VDC =

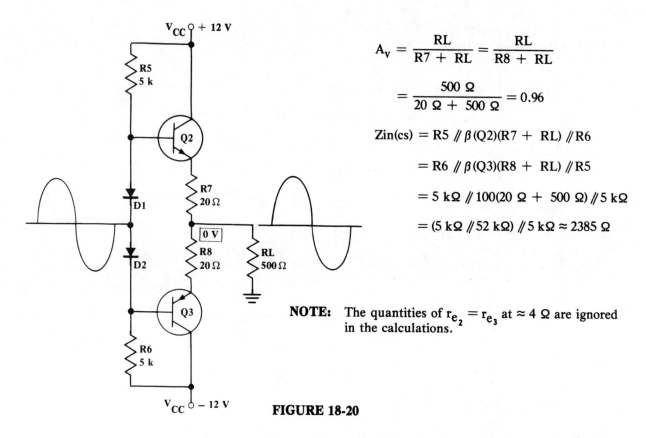

$$A_v = \frac{RL}{R7 + RL} = \frac{RL}{R8 + RL}$$

$$= \frac{500\ \Omega}{20\ \Omega + 500\ \Omega} = 0.96$$

$$Zin(cs) = R5 \mathbin{/\mkern-5mu/} \beta(Q2)(R7 + RL) \mathbin{/\mkern-5mu/} R6$$

$$= R6 \mathbin{/\mkern-5mu/} \beta(Q3)(R8 + RL) \mathbin{/\mkern-5mu/} R5$$

$$= 5\ k\Omega \mathbin{/\mkern-5mu/} 100(20\ \Omega + 500\ \Omega) \mathbin{/\mkern-5mu/} 5\ k\Omega$$

$$= (5\ k\Omega \mathbin{/\mkern-5mu/} 52\ k\Omega) \mathbin{/\mkern-5mu/} 5\ k\Omega \approx 2385\ \Omega$$

NOTE: The quantities of $r_{e_2} = r_{e_3}$ at $\approx 4\ \Omega$ are ignored in the calculations.

FIGURE 18-20

FIGURE 18-21

$Vp/\pi = 2.135/\pi \approx 680$ mV, and then to I DC from I DC = V DC/RL = 680 mV/500 Ω = 1.36 mA, as shown in Figure 18-22.

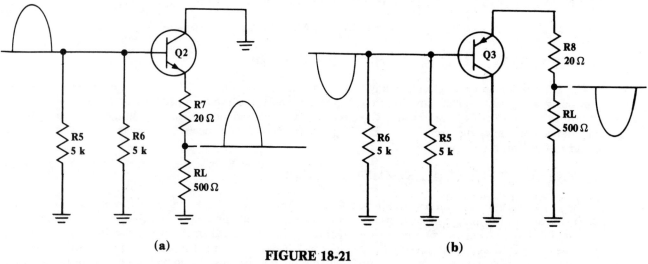

$Vp = Vp\text{-}p/2 = 4.27\ Vp\text{-}p/2 = 2.135$ V

$V\ DC = Vp/\pi \approx 2.135/3.14 \approx 680$ mV

$I\ DC = V\ DC/RL = 680\ mV/500\ \Omega = 1.36$ mA

FIGURE 18-22

Therefore, for a Vop-p(max) of 4.27 Vp-p, Q2 processes the positive-going 2.135 Vp signal and Q3 processes the negative-going 2.135 Vp signal. Also, since Q2 and Q3 alternately process the positive and negative-going waves, only the I DC for one pulse need be solved.

$$I\ DC = Vp/\pi R_L = 4.27\ Vp\text{-}p/(\pi \times 500\ \Omega) \approx 4.27\ Vp\text{-}p/(3.14 \times 500\ \Omega) = 1.36\ mA$$

Solving for the power dissipated by the combined ± 12 volt power supplies.

$$P_{PS} = 2V_{CC}(I_E + Vp/\pi R_L) = 24\ V(5\ mA + 1.36\ mA) = 152.64\ mW$$

Solving for the power dissipated by the load resistor:

$$P_{RL} = Vop\text{-}p^2/8RL = 4.27\ Vp\text{-}p^2/(8 \times 500\ \Omega) = 4.55\ mW$$

B. THE TWO POWER SUPPLY COMMON EMITTER CIRCUIT

The common emitter stage of Figure 18-15 was solved previously in Chapter 9. Therefore, the basic parameters are given again, in conjunction with Figure 18-23, without further explanation.

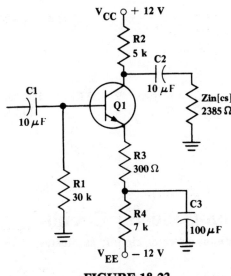

FIGURE 18-23

$$A_v = \frac{R2\ //\ Zin(cs)}{r_e + R3} = \frac{5\ k\Omega\ //\ 2385\ \Omega}{17.33\ \Omega + 300\ \Omega} = \frac{1422\ \Omega}{317.33\ \Omega} = 4.48$$

where: $r_e = 26\ mV/I_E = 26\ mV/1.5\ mA = 17.33\ \Omega$

$Zin(Q1) = R1\ //\ \beta(Q1) \times (r_e + R3) =$

$= 30\ k\Omega\ //\ 100 \times 317.33\ \Omega$

$\approx 30\ k\Omega\ //\ 31.73\ k\Omega = 15.42\ k\Omega$

Vop-p(max) $= 2\ V_{CE}$ or $2I_C r_L$ (whichever is smaller):

$2V_{CE} = 2 \times 5.55\ V = 11.1\ Vp\text{-}p$

$2I_C r_L = 2 \times 1.5\ mA \times 1422\ \Omega \approx 4.27\ Vp\text{-}p$

Therefore, Vop-p(max) $= 4.27\ Vp\text{-}p$

C. The overall gain of the two stage circuit of Figure 18-15 is solved by multiplying the gain of the Q1 stage by the gain of the complementary symmetry stage, the Zin is solved from the Zin of Q1 alone, the Vop-p(max) is limited by the Q1 stage, and the maximum power delivered to the load is a combination of Vop-p(max) and RL. Additionally, the power dissipated by Q2 and Q3 is solved for both no signal and maximum peak-to-peak signal conditions.

$A_{v_O} = A_v(Q1) \times A_v(Q2) = 4.48 \times 0.96 \approx 4.3$

$Zin = Zin(Q1) = 15.42\ k\Omega$

Vop-p(max) $= 2I_{C1} r_L = 4.27\ Vp\text{-}p$

$PG = A_{v_O}^2 \times Zin(Q1)/RL = 4.3^2 \times 15.42\ k\Omega/500\ \Omega \approx 570$

$PG(dB) = 10\ \log PG = 10\ \log 570 = 27.6\ dB$

Power dissipated by Q2 and Q3 and the combined power supplies for no signal condition:

$P(Q2) = P(Q3) = V_{CE} \times I_E = 11.9\ V \times 5\ mA = 59.5\ mW$

$P_{PS} = 2V_{CC} \times IDC = 24\ V \times 5\ mA = 120\ mW$

Power dissipated by Q2 and Q3 for maximum peak-to-peak voltage conditions of 4.27 Vp-p:

$$P(Q2) = P(Q3) = (P_{PS} - Po)/2 = (152.64 \text{ mW} - 4.55 \text{ mW})/2 \approx 74 \text{ mW}$$

where: $P_{PS} = 152.64$ mW and $Po = 4.55$ mW at 4.27 Vp-p.

NOTE: The complementary symmetry circuit is used primarily in power amplifier applications. Therefore, this small signal, low power circuit is not being used to its full capabilities. However, the principles of the complementary symmetry circuit remain the same regardless of the amount of power dissipated. Later, in power amplifier circuits, full analysis of the complementary symmetry circuit in many power amplifier circuit applications will be given.

LABORATORY EXERCISE

OBJECTIVES

To investigate capacitor coupled multistage circuits in three different circuit configurations, which include: the common emitter - common collector circuit, the common source - common emitter circuit, and the common emitter - complementary symmetry circuit.

LIST OF MATERIALS AND EQUIPMENT

1. Transistors (one each): 2N3904 (two), 2N3906, and 2N5951 or equivalents
2. Diodes: 1N4001 (two)
3. Resistors (one each except where indicated):

10 kΩ (two)	330 Ω	1 kΩ	3.3 kΩ	10 kΩ (two)	22 kΩ (two)	100 kΩ
220 Ω	470 Ω	2.2 kΩ	4.7 kΩ (three)	12 kΩ	47 kΩ	120 kΩ

4. Capacitors: 10 μF (three), 100 μF (two)
4. Power Supply
6. Voltmeter
7. Oscilloscope

PROCEDURE
Section I: The Common-Emitter, Common-Collector Circuit

1. Connect the NPN common emitter circuit driving a NPN common collector circuit, as illustrated in Figure 18-24.

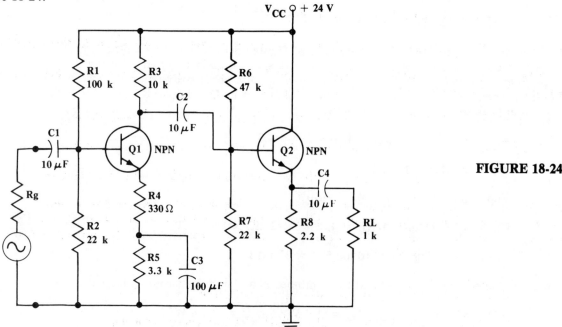

FIGURE 18-24

2. Calculate and measure the DC voltage drops of the entire circuit.
3. Insert the calculated and measured values, as indicated, into Table 18-1.

TABLE 18-1 Q1 STAGE	V_{R1}	V_{R2}	V_{R3}	V_{R4}	V_{R5}	V_{CE1}	I_{E1}	V_{CAP2}
CALCULATED								
MEASURED								
Q2 STAGE	V_{R6}	V_{R7}	V_{R8}	V_{RL}	V_{CE2}	I_{E2}	V_{CAP3}	V_{CAP4}
CALCULATED								
MEASURED								

4. Calculate the voltage gain of the Q2 stage from:

$$A_v(Q2) = \frac{r_L}{r_{e2} + r_L} = \frac{R8 // RL}{r_{e2} + (R8 // RL)}$$

5. Calculate the input impedance of the Q2 stage from:

$$Zin(Q2) = R6 // R7 // \beta(Q2)[r_{e2} + (R8 // RL)]$$

6. Calculate the voltage gain of the Q1 stage from:

$$A_v(Q1) = \frac{r_L}{r_{e1} + r_L} = \frac{R3 // Zin(Q2)}{r_{e1} + R4}$$

7. Calculate the input impedance of the Q1 stage from:

$$Zin(Q1) = R1 // R2 // \beta(Q1)(r_{e1} + R4)$$

8. Calculate the overall voltage gain of the two stage circuit from: $A_v(\text{overall}) = A_v(Q1) \times A_v(Q2)$

9. VOLTAGE GAIN: Apply an input signal of 100 mVp-p at 1 kHz to the circuit and monitor the voltage out at the collector of Q1 and at the emitter of Q2.

   a. Calculate the "measured" overall voltage gain from: $A_v(\text{overall}) = Vo(Q2)/Vin(Q1)$

   b. Calculate the "measured" voltage gain of the Q1 stage from: $A_v(Q1) = Vo(Q1)/Vin(Q1)$

   c. Calculate the "measured" voltage of the Q2 stage from: $A_v(Q2) = Vo(Q2)/Vin(Q2)$

10. Measure the input impedance of the Q1 stage using the series resistor RS technique. The same method can be applied to the Q2 stage, if further accuracy in calculating the voltage gain of Q1 and the overall voltage gain is required. This would be necessary if large discrepancies between measured and calculated values are being obtained.

11. Vop-p(max):
    a. Calculate the maximum peak-to-peak output voltage from:

    $$Vop\text{-}p(max) = 2V_{CE}(Q2) \quad \text{or} \quad 2I_E(Q2)(R8 // RL), \text{whichever is smaller.}$$

b. Measure the Vop-p(max).

12. POWER GAIN:

   a. Calculate the overall power gain of the circuit from; $PG = A_{v_o}^2 \times Zin(Q1)/RL$.

   b. Convert the power gain to dB.

13. Insert the calculated and measured values, as indicated, into Table 18-2.

TABLE 18-2	$A_v(Q1)$	$A_v(Q2)$	Zin(Q1)	Zin(Q2)	A_{v_o}	Vop-p(max)	PG	PG(dB)
CALCULATED								
MEASURED			/////	/////			/////	/////

Section II: The Common Source, Common-Emitter Circuit

1. Connect the self-biased, N channel FET circuit driving the PNP common emitter circuit shown in Figure 18-25.

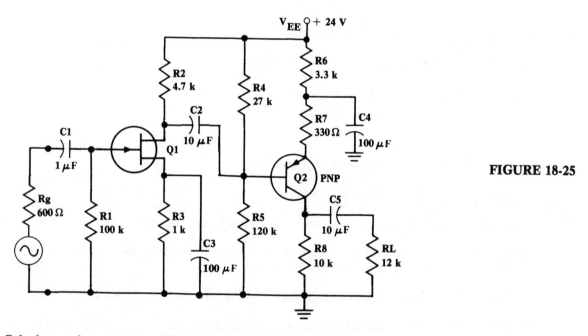

FIGURE 18-25

2. Calculate and measure the DC voltage drops of the entire circuit.
3. Insert the calculated and measured values, as indicated, into Table 18-3.

TABLE 18-3 Q1 STAGE	V_{R1}	V_{R2}	V_{R3}	V_{DS1}	I_{D1}	V_{CAP1}	V_{CAP2}	V_{CAP3}
CALCULATED								
MEASURED								

TABLE 18-3 Q2 STAGE	V_{R4}	V_{R5}	V_{R6}	V_{R7}	V_{R8}	V_{CE2}	I_{C2}	V_{CAP4}
CALCULATED								
MEASURED								

4. Calculate the voltage gain of the Q2 stage from:

$$A_v(Q2) = \frac{r_L}{r_{e2} + R7} = \frac{R8 \parallel R_L}{r_{e2} + R7}$$

5. Calculate the input impedance of the Q2 stage from:

$$Zin(Q2) = R4 \parallel R5 \parallel \beta(Q2)(r_{e2} + R7)$$

6. Calculate the voltage gain of the Q1 stage from:

$$A_v(Q1) = \frac{R2 \parallel Zin(Q2)}{r_s'}$$

7. Calculate the input impedance of the Q1 stage from: $Zin(Q1) = R1$

8. Calculate the overall voltage gain of the two stage circuit from: $A_v(\text{overall}) = A_v(Q1) \times A_v(Q2)$

9. VOLTAGE GAIN: Apply an input signal of 100 mVp-p at 1 kHz to the circuit and monitor the voltage out at the drain of Q1 and at the collector of Q2.

   a. Calculate the "measured" overall voltage gain from: $A_v(\text{overall}) = Vo(Q2)/Vin(Q1)$

   b. Calculate the "measured" voltage gain of the Q1 stage from: $A_v(Q1) = Vo(Q1)/Vin(Q1)$

   c. Calculate the "measured" voltage gain of the Q2 stage from: $A_v(Q2) = Vo(Q2)/Vin(Q2)$

10. Measure the input impedance to the Q1 stage using the series resistor RS technique. The same method can be applied to the Q2 stage, if further accuracy in calculating the voltage gain of Q1 and the overall voltage gain is required. This would be necessary if large discrepancies between measured and calculated values are being obtained.

11. Vop-p(max):
    a. Calculate the maximum peak-to-peak output voltage from:

    $$Vop\text{-}p(max) = 2V_{CE}(Q2) \quad \text{or} \quad 2I_C r_L = 2I_{C_2}(R8 \parallel R_L), \text{ whichever is smaller.}$$

    b. Measure the Vop-p(max).

13. Insert the calculated and measured values, as indicated, into Table 18-4.

TABLE 18-4	$A_v(Q1)$	$A_v(Q2)$	Zin(Q1)	Zin(Q2)	A_{v_o}	Vop-p(max)	PG	PG(dB)
CALCULATED								
MEASURED				/////			/////	

Section III: Common-Emitter, Complementary Symmetry Circuit

1. Connect the two power supply NPN common emitter circuit driving the Class AB complementary symmetry circuit shown in Figure 18-26.

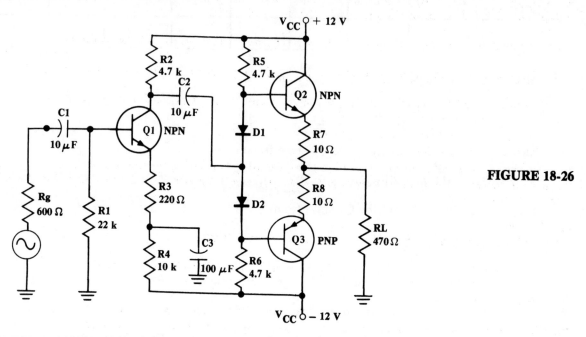

FIGURE 18-26

2. Calculate and measure the DC voltage drops of the entire circuit.
3. Insert the calculated and measured values, as indicated, into Table 8-5.

TABLE 18-5 Q1 STAGE	V_{R1}	V_{R2}	V_{R3}	V_{R4}	V_{CE1}	I_{C1}	V_{CAP1}	V_{CAP2}	V_{CAP3}
CALCULATED									
MEASURED									
Q2 STAGE	V_{R5}	V_{R6}	V_{D1}	V_{D2}	V_{CE2}	V_{CE3}	V_{R7}	V_{R8}	$I_{C2} = I_{C3}$
CALCULATED									
MEASURED									

4. Calculate the approximate voltage gain of the complementary symmetry circuit output from:

$$A_v(cs) \approx \frac{RL}{R7 + RL} = \frac{RL}{R8 + RL} \quad \text{where: } r_e \text{ can be safely ignored}$$

5. Calculate the approximate input impedance of the complementary symmetry stage from:

$$Zin(Q2) = [R5 \mathbin{/\mkern-5mu/} \beta(Q2)(R7 + RL)] \mathbin{/\mkern-5mu/} R6, \text{ or}$$
$$Zin(Q3) = [R6 \mathbin{/\mkern-5mu/} \beta(Q3)(R8 + RL)] \mathbin{/\mkern-5mu/} R5$$

NOTE: The r_e can be ignored, safely, in the calculations.

6. Calculate the voltage gain of the Q1 stage from: $A_v(Q1) = \dfrac{R2 \mathbin{/\mkern-5mu/} Zin(Q2)}{r_{e1} + R3}$

7. Calculate the input impedance of the Q1 stage from: $Zin(Q1) = R1 \mathbin{/\mkern-5mu/} \beta(Q1)(r_{e1} + R3)$

8. Calculate the overall voltage gain of the two stage circuit from: $A_v(\text{overall}) = A_v(Q1) \times A_v(cs)$

9. VOLTAGE GAIN: Apply an input signal of 100 mVp-p at 1 kHz to the circuit and monitor the voltage out at the collector of Q1 and across the load resistor RL.

   a. Calculate the "measured" overall voltage gain from: $A_v(\text{overall}) = Vo(RL)/Vin(Q1)$

   b. Calculate the "measured" voltage gain of the Q1 stage from: $A_v(Q1) = Vo(Q1)/Vin(Q1)$

   c. Calculate the "measured" voltage gain of the complementary symmetry stage from:

   $$A_v(cs) = Vo(RL)/Vin(cs) = Vo(RL)/Vo(Q1)$$

10. Measure the input impedance to the Q1 stage using the series resistor RS technique. The same method can be applied to the complementary symmetry stage if further accuracy in calculating the voltage gain of Q1 is required.

11. Vop-p(max)
    a. Calculate the maximum peak-to-peak output voltage from:

    $$Vop\text{-}p(max) = 2\, V_{CE}(Q1) \quad \text{or} \quad 2I_{C_1}(Q1)[R2 \mathbin{/\mkern-5mu/} Zin(cs)], \text{ whichever is smaller.}$$

    b. Measure the Vop-p(max)

NOTE: Because the complementary symmetry stage is essentially back-to-back emitter followers, the Vop-p(max) condition normally will be limited to the Q1 stage. That is, if the output of the complementary symmetry circuit is lightly loaded, as is the case for this circuit.

12. POWER GAIN:

    b. Calculate the overall power gain of the circuit from: $PG = A_v(\text{overall})^2 \times Zin(Q1)/RL$.

    b. Convert the power gain to dB.

13. Calculate the power dissipated by the Q2 and Q3 transistors, when no signal is applied, from:

    $$P(Q2) = P(Q3) = V_{CE2} \times I_{E2} \quad \text{where:} \quad V_{CE2} = V_{CE3} \text{ and } I_{E2} = I_{E3}$$

14. Calculate the power delivered by the power supplies for maximum peak-to-peak conditions from:

    $$P_{PS}(max) = 2V_{CC}(I_{E2} + Vp/\pi R_L) \quad \text{where:} \quad 2V_{CC} = 24\text{ V}$$

15. Calculate the power developed across the load resistor for maximum peak-to-peak conditions from:

    $$P_{RL} = Vop\text{-}p(max)^2 / 8RL$$

16. Insert the calculated and measured values, as indicated, into Table 18-6.

TABLE 18-6	$A_v(Q1)$	$A_v(cs)$	$Zin(Q1)$	$Zin(cs)$	A_{V_o}	Vop-p(max)	PG	PG(dB)	P_{Q2}(ns)	P_{PS} MAX	P_{RL}
CALCULATED											
MEASURED							/////	/////	/////	/////	/////

CHAPTER QUESTIONS

1. With reference to direct current, what is the advantage of capacitor coupled circuits over direct coupled circuits?
2. Give two reasons why multistage circuits are superior to single stage circuits.
3. Why is the common source circuit generally used as the first stage in a multistage circuit?
4. Common emitter and common collector circuits can be combined to provide superior two stage amplifier circuits. List two distinct advantages of the CC-CE or CE-CC circuit connections and explain why.
5. What advantage does the complementary symmetry circuit provide over a simple common collector circuit?
6. Which of the three circuits analyzed provides the largest power gain and why?

CHAPTER PROBLEMS

1. Given: $V_{CC} = 24$ V

    $\beta_{Q1} = \beta_{Q2} = 100$

    Solve for:

    a. V_{CE1} g. $Z_{in}(Q1)$
    b. V_{CE2} h. $Z_{in}(Q2)$
    c. I_{C1} i. A_{v_o}
    d. I_{C2} j. PG
    e. $A_V(Q2)$ k. $V_{op-p}(max)$
    f. $A_V(Q1)$

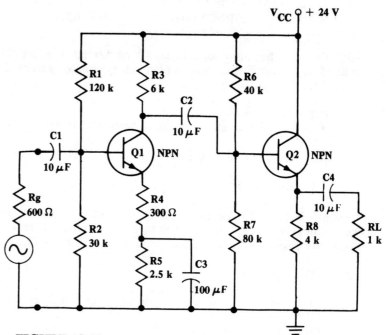

FIGURE 18-27

2. Given: $V_p = |3\ V|$

    $V_{GS} = -1$ V

    $I_{DSS} = 4.5$ mA

    $\beta_{Q2)} = 100$

    Solve for:

    a. I_{D1} h. A_{v_o}
    b. V_{DS} i. PG
    c. $A_V(Q1)$ j. $V_{op-p}(max)$
    d. V_{CE2}
    e. I_{C2}
    f. $A_V(Q2)$
    g. $Z_{in}(Q2)$

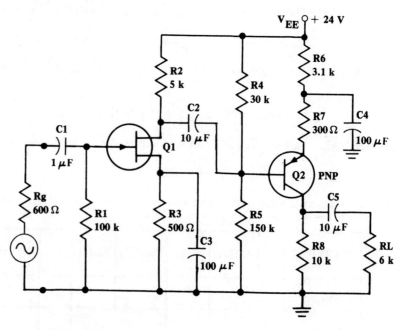

FIGURE 18-28

464

APPENDIX A: COMPONENTS REQUIRED TO WORK ALL THE EXPERIMENTS

ACTIVE DEVICES (one each except where indicated):

2N3904 or equialent NPN (two)
2N3906 or equivalent PNP
2N5951 or equivalent N channel JFET
2N3055 or equivalent NPN power transistor
1N4001 or equivalent diode (four)
1N4742 or equivalent zener diode
C106 (General Electric) or equivalent SCR

NOTE: All of the discrete devices listed above can be replaced with other devices having reasonably similar characteristics.

CAPACITORS (one each except where indicated):

0.01 μF   0.22 μF   1 μF   10 μF (two)   100 μF (two)

RESISTORS (one each except where indicated):

10 Ω	10 kΩ (two)
100 Ω	10 kΩ (two potentiometers)
220 Ω	12 kΩ
330 Ω	15 kΩ
470 Ω	22 kΩ
680 Ω	33 kΩ
1 kΩ (two)	47 kΩ
1.5 kΩ	100 kΩ (two)
2.2 kΩ	120 kΩ
2.7 kΩ	330 kΩ
3.3 kΩ	470 kΩ
4.7 kΩ	1 MΩ
6.8 kΩ	2.2 MΩ

APPENDIX B: POWER GAIN IN dB FORM

The power gain of a circuit is the ratio of power in to power out, and it is usually stated in decibel (dB) terms. Essentially, decibels are a logarithmic form of the additive process. They are convenient because they can convert a large cumbersome power gain to a simpler form. For example, a power gain of 79,450 converts to 49 dB.

Power gain can be converted to dB form through use of the formula: dB = 10 log PG. Since the dB form is a reference to power gain or loss, a power gain of 1 is 0.0 dB. For example, if the power input to an amplifier is 10 watts and the power output is also 10 watts, there is no power gain or loss — hence, 0.0 dB. However, if 100 watts were developed at the output for the same 10 watt input, a 10 dB gain occurs because: 10 log 100 W/10 W = 10 log 10 = 10 × 1 = 10 dB. Also, if the output power level were 20 watts for the 10 watt input, then a 3 dB gain occurs because: 10 log 20 W/10 W = 10 log 2 = 10 × 0.3 = 3 dB. Therefore, as shown in Table 1, a doubling of power gain causes a 3 dB gain.

TABLE 1

POWER GAIN (actual)	POWER GAIN (approximate)	dB
1	1	0
1.259	1.25	1
1.585	1.6	2
1.996	2	3
2.512	2.5	4
3.163	3.2	5
3.981	4	6
5.013	5	7
6.31	6.4	8
7.945	8	9
10	10	10

The table clearly shows that doubling the power gain from 1 to 2, 2 to 4, 3.2 to 6.4, etc. results in a power gain of 3 dB. Therefore, in every situation where power is doubled, a 3 dB power gain occurs.

Table 2 shows that a multiplication of power by 10 results in a 10 dB gain.

TABLE 2

$10 \log 1 = 10^0 = 0.0$ dB	for a power gain of	1
$10 \log 10 = 10^1 = 10$ dB	for a power gain of	10
$10 \log 100 = 10^2 = 20$ dB	for a power gain of	100
$10 \log 1,000 = 10^3 = 30$ dB	for a power gain of	1,000
$10 \log 10,000 = 10^4 = 40$ dB	for a power gain of	10,000
$10 \log 100,000 = 10^5 = 50$ dB	for a power gain of	100,000
$10 \log 1,000,000 = 10^6 = 60$ dB	for a power gain of	1,000,000

This table clearly shows that multiplication of power by 10 results in a 10 dB gain. Hence, a power gain of 10 results in a dB gain of 10, a power gain of 100 results in a dB gain of 20, etc.

Now, using a combinations of both methods:

TABLE 3: Converting Power Gain to dB

1. $3,200 = 3.2 \times 10^3 = 35$ dB
2. $50,000 = 5 \times 10^4 = 47$ dB
3. $16 = 1.6 \times 10^1 = 12$ dB
4. $8 = 8 \times 10^0 = 9$ db

TABLE 4: Converting from dB to Power Gain

1. $27 = 5 \times 10^2 = 500$
2. $48 = 6.4 \times 10^4 = 64,000$
3. $4 = 2.5 \times 10^0 = 2.5$
4. $33 = 2 \times 10^3 = 2,000$

While this method is not as accurate as results produced from tables, slide rules, or hand calculators, the results are reasonable "ball park" answers, and it gives insight into how power gain to dB and dB to power gain can be converted.

APPENDIX C: DEVICE DATA SHEETS

2N3903
2N3904

NPN SILICON ANNULAR♦ TRANSISTORS

... designed for general purpose switching and amplifier applications and for complementary circuitry with types 2N3905 and 2N3906.

- High Voltage Ratings — BV_{CEO} = 40 Volts (Min)
- Current Gain Specified from 100 μA to 100 mA
- Complete Switching and Amplifier Specifications
- Low Capacitance — C_{ob} = 4.0 pF (Max)

NPN SILICON SWITCHING & AMPLIFIER TRANSISTORS

MAXIMUM RATINGS

Rating	Symbol	Value	Unit
*Collector-Base Voltage	V_{CB}	60	Vdc
*Collector-Emitter Voltage	V_{CEO}	40	Vdc
*Emitter-Base Voltage	V_{EB}	6.0	Vdc
*Collector Current	I_C	200	mAdc
Total Power Dissipation @ T_A = 60°C	P_D	250	mW
**Total Power Dissipation @ T_A = 25°C Derate above 25°C	P_D	350 2.8	mW mW/°C
**Total Power Dissipation @ T_C = 25°C Derate above 25°C	P_D	1.0 8.0	Watts mW/°C
**Junction Operating Temperature	T_J	150	°C
**Storage Temperature Range	T_{stg}	-55 to +150	°C

THERMAL CHARACTERISTICS

Characteristic	Symbol	Max	Unit
Thermal Resistance, Junction to Ambient	$R_{\theta JA}$	357	°C/W
Thermal Resistance, Junction to Case	$R_{\theta JC}$	125	°C/W

*Indicates JEDEC Registered Data
**Motorola guarantees this data in addition to the JEDEC Registered Data.
♦Annular Semiconductors Patented by Motorola Inc.

Courtesy of Motorola Semiconductor Products Inc.

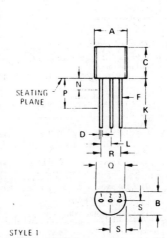

STYLE 1
PIN 1. EMITTER
2. BASE
3. COLLECTOR

DIM	MILLIMETERS		INCHES	
	MIN	MAX	MIN	MAX
A	4.450	5.200	0.175	0.205
B	3.180	4.190	0.125	0.165
C	4.320	5.330	0.170	0.210
D	0.407	0.533	0.016	0.021
F	0.407	0.482	0.016	0.019
K	12.700	-	0.500	-
L	1.150	1.390	0.045	0.055
N	-	1.270	-	0.050
P	6.350		0.250	
Q	3.430	-	0.135	-
R	2.410	2.670	0.095	0.105
S	2.030	2.670	0.080	0.105

CASE 29-02
TO-92

MOTOROLA INC. 1973   DS 5127 R2

*ELECTRICAL CHARACTERISTICS ($T_A = 25°C$ unless otherwise noted)

Characteristic		Fig. No.	Symbol	Min	Max	Unit
OFF CHARACTERISTICS						
Collector-Base Breakdown Voltage ($I_C = 10$ μAdc, $I_E = 0$)			BV_{CBO}	60	—	Vdc
Collector-Emitter Breakdown Voltage (1) ($I_C = 1.0$ mAdc, $I_B = 0$)			BV_{CEO}	40	—	Vdc
Emitter-Base Breakdown Voltage ($I_E = 10$ μAdc, $I_C = 0$)			BV_{EBO}	6.0	—	Vdc
Collector Cutoff Current ($V_{CE} = 30$ Vdc, $V_{EB(off)} = 3.0$ Vdc)			I_{CEX}	—	50	nAdc
Base Cutoff Current ($V_{CE} = 30$ Vdc, $V_{EB(off)} = 3.0$ Vdc)			I_{BL}	—	50	nAdc
ON CHARACTERISTICS						
DC Current Gain (1) ($I_C = 0.1$ mAdc, $V_{CE} = 1.0$ Vdc)	2N3903 2N3904	15	h_{FE}	20 40	— —	
($I_C = 1.0$ mAdc, $V_{CE} = 1.0$ Vdc)	2N3903 2N3904			35 70	— —	
($I_C = 10$ mAdc, $V_{CE} = 1.0$ Vdc)	2N3903 2N3904			50 100	150 300	
($I_C = 50$ mAdc, $V_{CE} = 1.0$ Vdc)	2N3903 2N3904			30 60	— —	
($I_C = 100$ mAdc, $V_{CE} = 1.0$ Vdc)	2N3903 2N3904			15 30	— —	
Collector-Emitter Saturation Voltage (1) ($I_C = 10$ mAdc, $I_B = 1.0$ mAdc)		16, 17	$V_{CE(sat)}$	—	0.2	Vdc
($I_C = 50$ mAdc, $I_B = 5.0$ mAdc)				—	0.3	
Base-Emitter Saturation Voltage (1) ($I_C = 10$ mAdc, $I_B = 1.0$ mAdc)		17	$V_{BE(sat)}$	0.65	0.85	Vdc
($I_C = 50$ mAdc, $I_B = 5.0$ mAdc)				—	0.95	
SMALL-SIGNAL CHARACTERISTICS						
Current-Gain—Bandwidth Product ($I_C = 10$ mAdc, $V_{CE} = 20$ Vdc, $f = 100$ MHz)	2N3903 2N3904		f_T	250 300	— —	MHz
Output Capacitance ($V_{CB} = 5.0$ Vdc, $I_E = 0$, $f = 100$ kHz)		3	C_{ob}	—	4.0	pF
Input Capacitance ($V_{BE} = 0.5$ Vdc, $I_C = 0$, $f = 100$ kHz)		3	C_{ib}	—	8.0	pF
Input Impedance ($I_C = 1.0$ mAdc, $V_{CE} = 10$ Vdc, $f = 1.0$ kHz)	2N3903 2N3904	13	h_{ie}	0.5 1.0	8.0 10	k ohms
Voltage Feedback Ratio ($I_C = 1.0$ mAdc, $V_{CE} = 10$ Vdc, $f = 1.0$ kHz)	2N3903 2N3904	14	h_{re}	0.1 0.5	5.0 8.0	$\times 10^{-4}$
Small-Signal Current Gain ($I_C = 1.0$ mAdc, $V_{CE} = 10$ Vdc, $f = 1.0$ kHz)	2N3903 2N3904	11	h_{fe}	50 100	200 400	—
Output Admittance ($I_C = 1.0$ mAdc, $V_{CE} = 10$ Vdc, $f = 1.0$ kHz)		12	h_{oe}	1.0	40	μmhos
Noise Figure ($I_C = 100$ μAdc, $V_{CE} = 5.0$ Vdc, $R_S = 1.0$ k ohms, $f = 10$ Hz to 15.7 kHz)	2N3903 2N3904	9, 10	NF	— —	6.0 5.0	dB
SWITCHING CHARACTERISTICS						
Delay Time	($V_{CC} = 3.0$ Vdc, $V_{BE(off)} = 0.5$ Vdc, $I_C = 10$ mAdc, $I_{B1} = 1.0$ mAdc)	1, 5	t_d	—	35	ns
Rise Time		1, 5, 6	t_r	—	35	ns
Storage Time	($V_{CC} = 3.0$ Vdc, $I_C = 10$ mAdc, $I_{B1} = I_{B2} = 1.0$ mAdc) 2N3903 2N3904	2, 7	t_s	—	175 200	ns
Fall Time		2, 8	t_f	—	50	ns

(1) Pulse Test: Pulse Width = 300 μs, Duty Cycle = 2.0%.
*Indicates JEDEC Registered Data

FIGURE 1 – DELAY AND RISE TIME EQUIVALENT TEST CIRCUIT

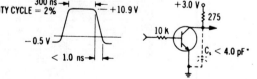

FIGURE 2 – STORAGE AND FALL TIME EQUIVALENT TEST CIRCUIT

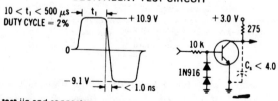

*Total shunt capacitance of test jig and connectors

MOTOROLA *Semiconductor Products Inc.*

Courtesy of Motorola Semiconductor Products Inc.

2N3905
2N3906

PNP SILICON SWITCHING & AMPLIFIER TRANSISTORS

PNP SILICON ANNULAR♦ TRANSISTORS

.... designed for general purpose switching and amplifier applications and for complementary circuitry with types 2N3903 and 2N3904.

- High Voltage Ratings — BV_{CEO} = 40 Volts (Min)
- Current Gain Specified from 100 μA to 100 mA
- Complete Switching and Amplifier Specifications
- Low Capacitance — C_{ob} = 4.5 pF (Max)

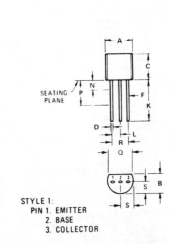

STYLE 1:
PIN 1. EMITTER
2. BASE
3. COLLECTOR

DIM	MILLIMETERS		INCHES	
	MIN	MAX	MIN	MAX
A	4.450	5.200	0.175	0.205
B	3.180	4.190	0.125	0.165
C	4.320	5.330	0.170	0.210
D	0.407	0.533	0.016	0.021
F	0.407	0.482	0.016	0.019
K	12.700	-	0.500	
L	1.150	1.390	0.045	0.055
N	-	1.270		0.050
P	6.350		0.250	-
Q	3.430	-	0.135	-
R	2.410	2.670	0.095	0.105
S	2.030	2.670	0.080	0.105

CASE 29-02
(TO-92)

*MAXIMUM RATINGS

Rating	Symbol	Value	Unit
Collector-Base Voltage	V_{CB}	40	Vdc
Collector-Emitter Voltage	V_{CEO}	40	Vdc
Emitter-Base Voltage	V_{EB}	5.0	Vdc
Collector Current	I_C	200	mAdc
Total Power Dissipation @ T_A = 60°C	P_D	250	mW
Total Power Dissipation @ T_A = 25°C Derate above 25°C	P_D	350 2.8	mW mW/°C
Total Power Dissipation @ T_C = 25°C Derate above 25°C	P_D	1.0 8.0	Watt mW/°C
Junction Operating Temperature	T_J	+150	°C
Storage Temperature Range	T_{stg}	-55 to +150	°C

THERMAL CHARACTERISTICS

Characteristic	Symbol	Max	Unit
Thermal Resistance, Junction to Ambient	$R_{\theta JA}$	357	°C/W
Thermal Resistance, Junction to Case	$R_{\theta JC}$	125	°C/W

Courtesy of Motorola Semiconductor Products Inc.

*Indicates JEDEC Registered Data.
♦Annular semiconductors patented by Motorola Inc.

©MOTOROLA INC., 1973   DS 5128 R2

ELECTRICAL CHARACTERISTICS ($T_A = 25°C$ unless otherwise noted.)

Characteristic		Fig. No.	Symbol	Min	Max	Unit
OFF CHARACTERISTICS						
Collector-Base Breakdown Voltage ($I_C = 10$ μAdc, $I_E = 0$)			BV_{CBO}	40	–	Vdc
Collector-Emitter Breakdown Voltage (1) ($I_C = 1.0$ mAdc, $I_B = 0$)			BV_{CEO}	40	–	Vdc
Emitter-Base Breakdown Voltage ($I_E = 10$ μAdc, $I_C = 0$)			BV_{EBO}	5.0	–	Vdc
Collector Cutoff Current ($V_{CE} = 30$ Vdc, $V_{BE(off)} = 3.0$ Vdc)			I_{CEX}	–	50	nAdc
Base Cutoff Current ($V_{CE} = 30$ Vdc, $V_{BE(off)} = 3.0$ Vdc)			I_{BL}	–	50	nAdc
ON CHARACTERISTICS (1)						
DC Current Gain		15	h_{FE}			
($I_C = 0.1$ mAdc, $V_{CE} = 1.0$ Vdc)	2N3905			30	–	
	2N3906			60	–	
($I_C = 1.0$ mAdc, $V_{CE} = 1.0$ Vdc)	2N3905			40	–	
	2N3906			80	–	
($I_C = 10$ mAdc, $V_{CE} = 1.0$ Vdc)	2N3905			50	150	
	2N3906			100	300	
($I_C = 50$ mAdc, $V_{CE} = 1.0$ Vdc)	2N3905			30	–	
	2N3906			60	–	
($I_C = 100$ mAdc, $V_{CE} = 1.0$ Vdc)	2N3905			15	–	
	2N3906			30	–	
Collector-Emitter Saturation Voltage		16, 17	$V_{CE(sat)}$			Vdc
($I_C = 10$ mAdc, $I_B = 1.0$ mAdc)				–	0.25	
($I_C = 50$ mAdc, $I_B = 5.0$ mAdc)				–	0.4	
Base-Emitter Saturation Voltage		17	$V_{BE(sat)}$			Vdc
($I_C = 10$ mAdc, $I_B = 1.0$ mAdc)				0.65	0.85	
($I_C = 50$ mAdc, $I_B = 5.0$ mAdc)				–	0.95	
SMALL-SIGNAL CHARACTERISTICS						
Current-Gain – Bandwidth Product			f_T			MHz
($I_C = 10$ mAdc, $V_{CE} = 20$ Vdc, $f = 100$ MHz)	2N3905			200	–	
	2N3906			250	–	
Output Capacitance ($V_{CB} = 5.0$ Vdc, $I_E = 0$, $f = 100$ kHz)		3	C_{ob}	–	4.5	pF
Input Capacitance ($V_{BE} = 0.5$ Vdc, $I_C = 0$, $f = 100$ kHz)		3	C_{ib}	–	1.0	pF
Input Impedance		13	h_{ie}			k ohms
($I_C = 1.0$ mAdc, $V_{CE} = 10$ Vdc, $f = 1.0$ kHz)	2N3905			0.5	8.0	
	2N3906			2.0	12	
Voltage Feedback Ratio		14	h_{re}			$\times 10^{-4}$
($I_C = 1.0$ mAdc, $V_{CE} = 10$ Vdc, $f = 1.0$ kHz)	2N3905			0.1	5.0	
	2N3906			1.0	10	
Small-Signal Current Gain		11	h_{fe}			–
($I_C = 1.0$ mAdc, $V_{CE} = 10$ Vdc, $f = 1.0$ kHz)	2N3905			50	200	
	2N3906			100	400	
Output Admittance		12	h_{oe}			μmhos
($I_C = 1.0$ mAdc, $V_{CE} = 10$ Vdc, $f = 1.0$ kHz)	2N3905			1.0	40	
	2N3906			3.0	60	
Noise Figure		9, 10	NF			dB
($I_C = 100$ μAdc, $V_{CE} = 5.0$ Vdc, $R_S = 1.0$ k ohm, $f = 10$ Hz to 15.7 kHz)	2N3905			–	5.0	
	2N3906			–	4.0	
SWITCHING CHARACTERISTICS						
Delay Time	($V_{CC} = 3.0$ Vdc, $V_{BE(off)} = 0.5$ Vdc, $I_C = 10$ mAdc, $I_{B1} = 1.0$ mAdc)	1, 5	t_d	–	35	ns
Rise Time		1, 5, 6	t_r	–	35	ns
Storage Time	($V_{CC} = 3.0$ Vdc, $I_C = 10$ mAdc, $I_{B1} = I_{B2} = 1.0$ mAdc) 2N3905	2, 7	t_s	–	200	ns
	2N3906			–	225	
Fall Time	2N3905	2, 8	t_f	–	60	ns
	2N3906			–	75	

* Indicates JEDEC Registered Data. (1) Pulse Width = 300 μs, Duty Cycle = 2.0 %.

FIGURE 1 – DELAY AND RISE TIME EQUIVALENT TEST CIRCUIT

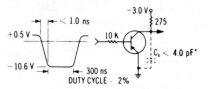

FIGURE 2 – STORAGE AND FALL TIME EQUIVALENT TEST CIRCUIT

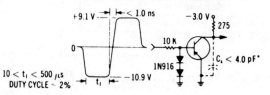

*Total shunt capacitance of test jig and connectors

Courtesy of Motorola Semiconductor Products Inc.

RCA POWER TRANSISTORS

2N3053 2N3054 2N3055
Including
40372 40389 40392

File No. 145

RCA-2N3053, 2N3054*, and 2N3055* are silicon n-p-n transistors intended for a wide variety of medium to high-power applications.

The 2N3053 is a triple-diffused planar type useful up to 20 MHz in small-signal, medium-power applications.

The 2N3054 and 2N3055 are **Hometaxial-base**** types useful for power-switching circuits, for series- and shunt-regulator driver and output stages, and for high-fidelity amplifiers.

* Formerly Dev. Type Nos. TA2402A and TA2403A, respectively.

**"Hometaxial" was coined by RCA from "homogeneous" and "axial" to describe a single-diffused transistor with a base region of homogeneous-resistivity silicon in the axial direction (emitter-to-collector).

ALSO AVAILABLE...

Type 40372 is a 2N3054 with a factory-attached heat radiator.

40372

Type 40389 is a 2N3053 with a factory-attached heat radiator.

Types 40372 and 40389 are intended for printed-circuit-board applications.

40389

Type 40392 is a 2N3053 with a factory-attached diamond-shaped mounting flange.

40392

SILICON N-P-N GENERAL-PURPOSE TYPES FOR INDUSTRIAL AND COMMERCIAL APPLICATIONS

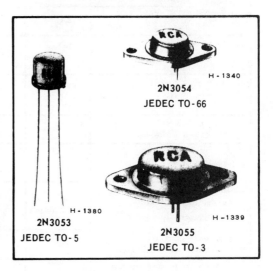

2N3054 JEDEC TO-66
2N3053 JEDEC TO-5
2N3055 JEDEC TO-3

- Maximum area-of-operation curves for DC and pulse operation
- Now possible to determine maximum operating conditions for operation free from second breakdown

2N3053
- Low leakage current (I_{CBO}) and wide beta (h_{FE}) range

2N3054
- $V_{CEV}(sus)$ = 90 V min.
- Low saturation voltage, $V_{CE}(sat)$ = 1.0 V (at I_C = 0.5 A)

2N3055
- High dissipation capability -- 115 W
- $V_{CEV}(sus)$ = 100 V min.
- Low saturation voltage, $V_{CE}(sat)$ = 1.1 V (at I_C = 4 A)

MAXIMUM RATINGS

Absolute-Maximum Values:		2N3053 40389, 40392	2N3054 40372	2N3055	
COLLECTOR-TO-BASE VOLTAGE	V_{CBO}	60	90	100	V
COLLECTOR-TO-EMITTER VOLTAGE:					
With -1.5 V (V_{BE}) of reverse bias	$V_{CEV}(sus)$	60	90	100	V
With external base-to-emitter resistance (R_{BE}) = 10 Ω	$V_{CER}(sus)$	50	—	—	V
= 100 Ω	$V_{CER}(sus)$	—	60	70	V
With base open	$V_{CEO}(sus)$	40	55	60	V
EMITTER-TO-BASE VOLTAGE	V_{EBO}	5	7	7	V
COLLECTOR CURRENT	I_C	0.7	4	15	A
BASE CURRENT	I_B	—	2	7	A
TRANSISTOR DISSIPATION:	P_T				
At case temperatures up to 25°C		5(2N3053) 7(40392)	29(2N3054) —	115 —	W W
At free-air temperatures up to 25°C		1(2N3053) 3.5(40389)	5.8(40372) —	— —	W W
At temperatures above 25°C, See Figs.		1, 2, & 5	2, 3, & 6	2, 4, & 6	
TEMPERATURE RANGE:					
Storage & Operating (Junction)		← -65 to 200 →			°C
LEAD OR PIN TEMPERATURE (During soldering): At distance ≥ 1/32" from seating plane for 10 s max.		255	235	235	°C

RADIO CORPORATION OF AMERICA
ELECTRONIC COMPONENTS AND DEVICES, HARRISON, N.J.

Trademark(s) ® Registered
Marca(s) Registrada(s)

Printed in U.S.A.
2N3053, 40389, 40392,
2N3054, 40372, 2N3055   8/66
Supersedes issue dated 10/64

Courtesy of Radio Corporation of America

2N3053 2N3054 2N3055
Including 40372 40389 40392

ELECTRICAL CHARACTERISTICS
Case Temperature (T_C) = 25°C, Unless Otherwise Specified

Characteristics	Symbol	TEST CONDITIONS						LIMITS						Units
		DC Collector Volts		DC Emitter or Base Volts		DC Current milliamperes		Types 2N3053 40389 40392		Types 2N3054 40372		Type 2N3055		
		V_{CB}	V_{CE}	V_{EB}	V_{BE}	I_C	I_E	I_B	Min.	Max.	Min.	Max.	Min.	Max.	
Collector-Cutoff Current	I_{CBO}	30					0		—	0.25	—	—	—	—	μA
	I_{CEV}		90		−1.5				—	—	—	1.0	—	—	mA
			100		−1.5				—	—	—	—	—	5.0	
At T_C = 150°C	I_{CEV}		30		−1.5				—	—	—	5.0	—	—	mA
			60		−1.5				—	—	—	—	—	10.0	
Emitter-Cutoff Current	I_{EBO}			4		0			—	0.25	—	—	—	—	μA
				7		0			—	—	—	1.0	—	—	mA
				7		0			—	—	—	—	—	5.0	mA
DC Forward-Current Transfer Ratio	h_{FE}		10			150[a]			50	250	—	—	—	—	
			4			500			—	—	25	100	—	—	
			4			4A[a]			—	—	—	—	20	70	
Collector-to-Base Breakdown Voltage	BV_{CBO}					0.1	0		60	—	—	—	—	—	V
Emitter-to-Base Breakdown Voltage	BV_{EBO}					0	0.1		5	—	—	—	—	—	V
						0	1		—	—	7	—	—	—	
						0	5		—	—	—	—	7	—	
Collector-to-Emitter Sustaining Voltage: With base open	$V_{CEO}(sus)$					100[a]		0	40	—	—	—	—	—	
						100		0	—	—	55	—	—	—	V
						200		0	—	—	—	—	60	—	
With base-emitter junction reverse biased	$V_{CEV}(sus)$				−1.5	100			—	—	—	—	100	—	V
With external base-to-emitter resistance (R_{BE}) = 10 Ω	$V_{CER}(sus)$					100[a]			50	—	—	—	—	—	
= 100 Ω						100			—	—	60	—	—	—	V
= 100 Ω						200			—	—	—	—	70	—	
Base-to-Emitter Voltage	V_{BE}		4			500			—	—	—	1.7	—	—	V
			4			4A[a]			—	—	—	—	—	1.8	
Base-to-Emitter Saturation Voltage						150		15	—	1.7	—	—	—	—	V
Collector-to-Emitter Saturation Voltage	$V_{CE}(sat)$					150		15	—	1.4	—	—	—	—	
						500		50	—	—	—	1.0	—	—	V
						4A[a]		400	—	—	—	—	—	1.1	
Small-Signal, Forward Current Transfer Ratio (At 20 MHz)	h_{fe}		10			50			5	—	—	—	—	—	
Gain-Bandwidth Product	f_T					200			—	—	800	—	—	—	kHz
						1 A			—	—	—	—	800	—	
Output Capacitance	C_{ob}	10					0		—	15	—	—	—	—	pF
Input Capacitance	C_{ib}			0.5			0		—	80	—	—	—	—	pF
Power Rating Test	PRT		39			3 A			—	—	—	—	—	1[b]	s
Thermal Resistance: Junction-to-Case	θ_{J-C}								35(max.) 2N3053		6(max.) 2N3054		—	1.5	°C/W
									25(max.) 40392		—				°C/W
Junction-to-Free Air	θ_{J-FA}								175(max.) 2N3053		30(max.) 40372		—	—	°C/W
									50(max.) 40389		—		—	—	°C/W

[a] Pulsed; pulse duration = 300 μs, duty factor = 1.8%. [b] At 115 W.

Courtesy of Radio Corporation of America

2N5949-53 N-Channel JFETs

Process 50

General Description

The 2N5949 thru 2N5953 series of N-channel JFETs is characterized for low frequency to VHF amplifiers requiring tightly specified I_{DSS} ranges.

Absolute Maximum Ratings (25°C)

Reverse Gate-Drain or Gate-Source Voltage	30V
Gate Current	10 mA
Total Device Dissipation at 25°C Case Temperature (Derate 2.88 mW/°C)	360 mW
Total Device Dissipation at 25°C Lead Temperature (Derate 4 mW/°C)	500 mW
Storage Temperature Range	−65°C to +150°C
Lead Temperature (1/16" from case for 10 seconds)	260°C

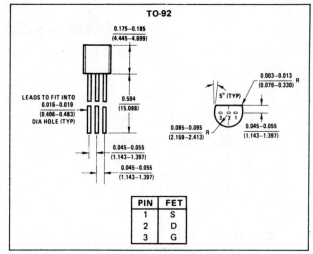

Electrical Characteristics (25°C unless otherwise noted)

PARAMETER		CONDITIONS		2N5949 MIN	2N5949 MAX	2N5950 MIN	2N5950 MAX	2N5951 MIN	2N5951 MAX	2N5952 MIN	2N5952 MAX	2N5953 MIN	2N5953 MAX	UNITS
I_{GSS}	Gate Reverse Current	$V_{GS} = -15V$, $V_{DS} = 0$			−1		−1		−1		−1		−1	nA
			$T_A = 100°C$		−200		−200		−200		−200		−200	
BV_{GSS}	Gate-Source Breakdown Voltage	$I_G = -1\mu A$, $V_{DS} = 0$		−30		−30		−30		−30		−30		V
$V_{GS(off)}$	Gate-Source Cutoff Voltage	$V_{DS} = 15V$, $I_D = 100$ nA		−3	−7	−2.5	−6	−2	−5	−1.3	−3.5	−0.8	−3	V
V_{GS}	Gate-Source Voltage	$V_{DS} = 15V$	$I_D = 1.2$ mA	−2.25	−6									V
			$I_D = 1$ mA			−1.8	−5							
			$I_D = 0.7$ mA					−1.3	−4.5					
			$I_D = 0.4$ mA							−0.75	−3			
			$I_D = 0.25$ mA									−0.5	−2.5	
I_{DSS}	Saturation Drain Current	$V_{DS} = 15V$, $V_{GS} = 0$, (Note 1)		12	18	10	15	7	13	4	8	2.5	5	mA
$r_{ds(on)}$	Drain-Source ON Resistance	$V_{GS} = 0$, $I_D = 0$	$f = 1$ kHz		200		210		250		300		375	Ω
g_{fs}	Common-Source Forward Transconductance	$V_{DS} = 15V$, $V_{GS} = 0$	$f = 1$ kHz	3.5	7.5	3.5	7.5	3.5	6.5	2	6.5	2	6.5	mmho
g_{os}	Common-Source Output Conductance				75		75		75		50		50	μmho
$Re(Y_{os})$	Common-Source Output Conductance				75		75		75		75		50	μmho
$Re(Y_{fs})$	Common-Source Transconductance	$V_{DS} = 15V$, $V_{GS} = 0$	$f = 100$ MHz	3.0	7.5	3.0	7.5	3.0	6.5	1.0	6.5	1.0	6.5	mmho
$Re(Y_{is})$	Common-Source Input Conductance				250		250		250		250		250	μmho
C_{iss}	Common-Source Input Capacitance	$V_{DS} = 15V$, $V_{GS} = 0$	$f = 1$ MHz		6		6		6		6		6	pF
C_{rss}	Common-Source Reverse Transfer Capacitance				2		2		2		2		2	pF
NF	Noise Figure	$V_{DS} = 15V$, $V_{GS} = 0$	$f = 100$ MHz, $R_G = 1$ kΩ		5		5		5		5		5	dB
			$f = 1$ kHz, $R_G = 1$ MΩ		2		2		2		2		2	
e_n	Equivalent Input Noise Voltage	$V_{DS} = 15V$, $V_{GS} = 0$	$f = 1$ kHz		100		100		100		100		100	$\frac{nV}{\sqrt{Hz}}$

Note 1: Pulse width 300 μs, duty cycle ≤ 3%.

Courtesy of National Semiconductor Corporation

Silicon Controlled Rectifier
Flat Pack Design
4 Amperes (RMS)   Up to 200 Volts

Model C106

150.9
Supersedes 150.9

PRODUCT FEATURES

The Type C106 Silicon Controlled Rectifier (SCR) has the following outstanding features:

LOW COST   Priced from 30-50 cents in volume

RELIABLE   Uses the proved planar passivated process (all junctions protected by a silicon dioxide layer)

SENSITIVE   Operates directly from low signal sensors such as thermistors, photo-conductive cells, etc.

VERSATILE   Designed for a variety of mount-down methods—printed circuit, plug-in socket, screws, or point-to-point soldering

RUGGED, COMPACT   Uses a solid plastic encapsulant in rectangular shape for high density packaging

C106 TYPE 1

C106 TYPE 2

C106 TYPE 3

C106 TYPE 4

(FULL SIZE)

MAXIMUM ALLOWABLE RATINGS

Type	Repetitive Peak Forward Blocking Voltage, V_{FXM} R_{GK} = 1000 Ohms T_J = −40°C to +110°C	Working and Repetitive Peak Reverse Voltage, V_{ROM}(wkg) and V_{ROM}(rep) T_J = −40°C to +110°C
C106Q1, C106Q2, C106Q3, C106Q4	15 Volts	15 Volts
C106Y1, C106Y2, C106Y3, C106Y4	30 Volts	30 Volts
C106F1, C106F2, C106F3, C106F4	50 Volts	50 Volts
C106A1, C106A2, C106A3, C106A4	100 Volts	100 Volts
C106B1, C106B2, C106B3, C106B4	200 Volts	200 Volts

RMS Forward Current, On-State _____ 4 Amperes
Rate of Rise of Forward Current (non-repetitive), di/dt (See Chart 9) _____ 50 Amperes/Microsecond
Peak Forward Current, On-State (repetitive) _____ 75 Amperes*
Peak One Cycle Surge Forward Current, Non-Repetitive, I_{FM} (surge) _____ 20 Amperes
I^2t (for fusing) _____ 0.5 Ampere^2 seconds (for times > 1.5 Milliseconds)
Peak Gate Power, P_{GM} _____ 0.5 Watt
Average Gate Power, $P_{G(AV)}$ _____ 0.1 Watt
Peak Gate Current, I_{GFM} _____ 0.2 Amperes
Peak Reverse Gate Voltage, V_{GRM} _____ 6 Volts
Storage Temperature, T_{stg} _____ −40°C to +150°C
Operating Temperature _____ −40°C to +110°C

*This rating applies for operation at 60 Hz, 75°C maximum tab (or anode) lead temperature, switching from 80 volts peak, sinusoidal current pulse width 10 μsec. minimum, 15 μsec. maximum.

Courtesy of General Electric

CHARACTERISTICS

Test	Symbol	Min.	Typ.	Max.	Units	Test Conditions
Reverse or Forward Blocking Current (All Types)	I_{RX} or I_{FX}	—	0.1	10	μA	$V_{RX} = V_{FX}$ Rated V_{ROM} (rep) Value $T_L = 25°C$, $R_{GK} = 1000$ Ohms
		—	10	100	μA	$V_{RX} = V_{FX}$ Rated V_{ROM} (rep) Value $T_L = 110°C$, $R_{GK} = 1000$ Ohms
Gate Trigger Current	I_{GT}	—	30	200	μAdc	$T_L = 25°C$, $V_{FX} = 6$ Vdc, $R_L = 100$ Ohms $R_{GK} = 1000$ Ohms
		—	75	500	μAdc	$T_L = -40°C$, $V_{FX} = 6$ Vdc, $R_L = 100$ Ohms $R_{GK} = 1000$ Ohms
Gate Trigger Voltage	V_{GT}	0.4	0.5	0.8	Volts DC	$T_L = 25°C$, $V_{FX} = 6$ Vdc, $R_L = 100$ Ohms $R_{GK} = 1000$ Ohms
		0.5	0.7	1.0	Volts DC	$T_L = -40°C$, $V_{FX} = 6$Vdc, $R_L = 100$ Ohms $R_{GK} = 1000$ Ohms
		0.2	—	—	Volts DC	$T_L = 110°C$, $V_{FX} =$ Rated V_{FXM} Value $R_L = 3000$ Ohms, $R_{GK} = 1000$ Ohms
Peak On-Voltage	V_{FM}	—	1.8	2.2	Volts	$T_L = 25°C$, $I_{FM} = 4$ Amperes Peak, Single Half Sine Wave Pulse, 2 Millisec. Wide
Holding Current	I_{HX}	0.3	1.0	3.0	mAdc	$T_L = 25°C$, $V_{FX} = 12$ Vdc, $R_{GK} = 1000$ Ohms
		0.4	2.0	6.0	mAdc	$T_L = -40°C$, $V_{FX} = 12$ Vdc, $R_{GK} = 1000$ Ohms
		0.14	0.6	2.0	mAdc	$T_L = 110°C$, $V_{FX} = 12$ Vdc, $R_{GK} = 1000$ Ohms
Latching Current	I_{LX}	0.3	1.5	4.0	mAdc	$T_L = 25°C$, $V_{FX} = 12$ Vdc, $R_{GK} = 1000$ Ohms
		0.4	3.0	8.0	mAdc	$T_L = -40°C$, $V_{FX} = 12$ Vdc, $R_{GK} = 1000$ Ohms
Critical Rate of Rise of Forward Blocking Voltage	dv/dt	—	8	—	Volts/Microsecond	$T_L = 110°C$, $V_{FX} =$ Rated V_{FXM} Value $R_{GK} = 1000$ Ohms
Turn On Time	$t_d + t_r$	—	1.2	—	Microseconds	$T_L = 25°C$, $V_{FX} =$ Rated V_{FXM} Value $I_{FM} = 1$ Ampere, Gate Pulse = 4 Volts, 300 Ohms, 5 Microseconds Wide.
Circuit Commutated Turn-Off Time	t_{off}	—	40	100	Microseconds	$T_L = 110°C$, rectangular current waveform. Rate of rise of current <10 amps/μsec. Rate of reversal of current <5 amps/μsec. $I_{FM} = 1$ Amp (50 μsec pulse). Repetition Rate = 60 pps. $V_{RXM} =$ Rated. $V_{RX} = 15$ Volts Minimum. $V_{FXM} =$ Rated. Rate of Rise Reapplied Forward Blocking Voltage = 5 Volts/μsec. Gate Bias = 0 Volts, 100 Ohms (during turn-off time interval).

The lead temperature (T_L) is measured in the center of the tab, 1/16 inch from the body on Type 1 and Type 3 devices, and in the center of the anode lead, 1/16 inch from the body on Type 2 and Type 4 devices.

OUTLINE DRAWINGS

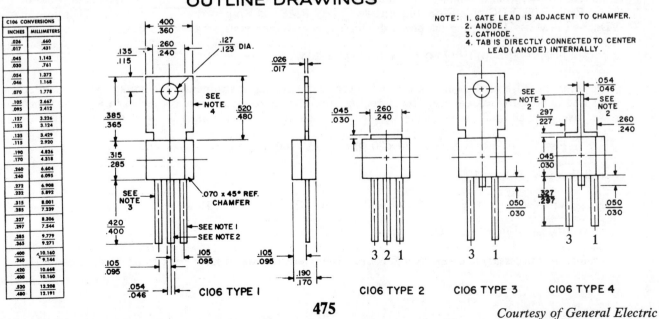

NOTE: 1. GATE LEAD IS ADJACENT TO CHAMFER.
2. ANODE.
3. CATHODE.
4. TAB IS DIRECTLY CONNECTED TO CENTER LEAD (ANODE) INTERNALLY.

Courtesy of General Electric

APPENDIX D: WAVEFORM RISE TIME

An ideal square wave has a flat top and 90 degree sides (slopes). However, as long as some capacitance exists in the circuit, rise time (and fall time) cannot be perfect. The rise time is described in literature as the time it takes for the wave to rise from 10% to 90% of maximum square wave amplitude. For example, if the ideal square wave amplitude is 1 volt, then the rise time is measured on the rise time slope between the voltage levels of 0.1 V and 0.9 V. Fall time, which is normally faster than rise time, is measured and calculated in a similar manner. The ideal square wave and the rise time t_r and fall time t_f are shown in Figure E-1.

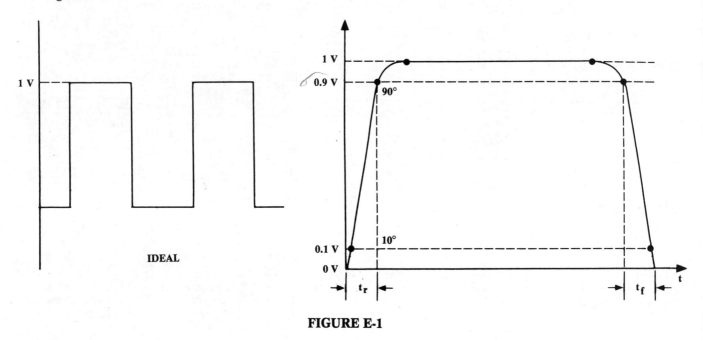

FIGURE E-1

APPENDIX E: POWER DISSIPATION CALCULATIONS

Power is solved from $P = V^2 R$ if the voltage is DC or RMS and $P = V^2/8R$ if the voltage is in peak-to-peak. Essentially, VDC = V RMS and $V_p = \sqrt{2} \times$ V RMS and $V_{p-p} = 2\sqrt{2} \times$ V RMS. Therefore, if VDC = V RMS = 10 V, then the power dissipated by a 20 ohm resistor is 5 watts from:

$$P = V^2/R = 10 \ V^2/20 \ \Omega = 100 \ V/20 \ \Omega = 5 \ W$$

However, because oscilloscopes are widely used as measuring instruments, the voltage value is often measured in peak-to-peak voltage. Therefore, the peak-to-peak voltage can be converted to RMS and then solved using the above formula, or the V_{p-p} can be used directly, but it must be divided by 8 as follows:

$$P = V_{p-p}^2/8R = 28.28 \ V^2/(8 \times 20 \ \Omega) = 800 \ V/160 \ \Omega = 5 \ W$$

Recall that if V RMS = 10 V, then:

$$V_p = \sqrt{2} \times 10 \ V = 14.14 \ V \quad \text{and} \quad V_{p-p} = 2\sqrt{2} \times 10 \ V. \text{ However,}$$

$$V_{p-p}^2 = 28.28^2 = 2.828^2 \times 10^2 = 8 \times 100.$$

Therefore, it is necessary to divide by 8 to convert the power developed from peak-to-peak voltage to RMS voltage.

APPENDIX F: PROBLEM ANSWERS (EVEN)

CHAPTER 1:

2. a. 8 V   b. 20 kΩ   c. 160 µA
4. a. 4 V   b. 8 V   c. −4 V   d. −12 V
6. a. 1.5 Vp-p   b. 0.5 Vp-p
8. a. 12 V, 6 V, 2 V, 0.0 V   b. 1.59 Ω   c. 3.33 Vp-p, 3.33 Vp-p, 0.0 Vp-p

CHAPTER 2:
2. a. 0.708 V   b. 9.2 V   c. 92 mA
4. a. 9.65 mA   b. ≈ 3.2 Ω   c. ≈ 3.19 Ω   d. ≈ 1.57 mVp-p

CHAPTER 3:
2. a. 10 V   b. 14 V   c. 7 mA
4. a. 280 Ω   b. 50 mA
6. a. 14 V   b. 14 mA   c. 9.9 Ω   d. ≈ 9 mVp-p

CHAPTER 4:
2. a. 3 V   b. 1.5 mA   c. 15 µA   d. 1.8 V   e. 2.4 V
4. a. 2.4 V   b. 1.5 mA   c. 6 V   d. 6 V   e. 3.6 V
6. a. 4.8 V   b. 1.6 mA   c. 3.2 V   d. 0.2 V   e. 20 µA   f. 80

CHAPTER 6:
Section I (page 135):
2. a. 3.9 mA   b. 19.5 V   c. 3.9 V   d. 0.6 V   e. 19.5 V
4. a. 1.2 mA   b. 12 V   c. 9.6 V   d. 21 V   e. 2.625 MΩ
6. a. 2 mA   b. 12 V   c. 4 V   d. −12 V   e. −4 V   f. −4.6 V   g. |19.4 V| h. |8 V|
Section II (page 145):
2. a. 1.2 mA   b. 6 V   c. 6 µA   d. 0.6 V   e. 10.8 V   f. 7.2 V
4. a. 0.95 mA   b. 9.5 V   c. 1.9 V   d. 5.7 V   e. −6.3 V   f. 2.5 V
Section III, Section IV (page 159):
2. a. 4.8 V   b. 19.2 V   c. 4.2 V   d. 12 V   e. 19.8 V   f. 12 V   g. 7.8 V
4. a. 9 V   b. 1.5 mA   c. 2.4 V   d. 3 V   e. 21 V   f. 140 kΩ
Chapter Problems (page 171):
2. $V_{RS} = 4.7$ V   4. $R3 = 6$ kΩ   6. $V_{R3} = 0.36$ V

CHAPTER 7:
Section I (page 207):
2. a. 21.67 Ω   b. 276.9   c. 2167 Ω   d. 27.7 k   e. 14.4 Vp-p
4. a. 21.67 Ω   b. 6.22   c. 31.4 kΩ   d. 405   e. 4.8 Vp-p
Section II (page 213):
2. a. 1.2 mA   b. 7.2 V   c. 2.64 V, 0.36 V   d. 13.8 V   e. 20.4 V   f. 7.2 V   g. 21 V
   h. 20.4 V   i. 21.67 Ω   j. 12.44   k. 31.57 kΩ   l. 407   m. 32.72   n. 9.6 Vp-p
Section III (page 221):
2. a. 1.8 mA   b. 5.4 V   c. 18.6 V   d. 18 V   e. 18 V   f. 0 V   g. 18.6 V
   h. 14.44 Ω   i. 0.993   J. 167.67 kΩ   k. 27.56   l. 27.75   m. 7.2 Vp-p
Section IV (page 230):
2. a. 1.04 mA   b. 10.4 V   c. 2.6 V   d. 11 V   e. 20.8 V   f. 10.4 V   g. 20.8 V
   h. 21.4 V   i. 25 Ω   j. 6.35   k. 415.8 Ω   l. ≈ 3.35   m. ≈ 6.93 Vp-p

CHAPTER 8:
Section I (page 262):
2(a). 1. 18.4   2. ≈ 11.49 kΩ   3. 12 Vp-p
2(b). 1. 231   2. 1.6 kΩ
4. 30

Section II (273):
2(a).

TABLE 8-9 METHOD	V_{R1}	V_{R2}	V_{R3}	V_{R4}	V_{CE}	I_E
Thevenin	16.15 V	7.85 V	0.558 V	6.693 V	≈ 17.3 V	1.116 mA
Reflected Resistance	16.05 V	7.95 V	0.565 V	6.785 V	≈ 17.2 V	1.113 mA

2(b).   48.1 kΩ
Section III (281):
2.   a.   429.2 Ω   b.   ≈ 14.75

CHAPTER 9:
Section I (page 291):
2(a). 1. 0.54 V   2. 9 V   3. 0.36 V   4. 13.5 V   5. 7.14 V   6. −6 V   7. 0.54
      8. 1.14 V
2(b). 1. 8.74   2. 12.5 kΩ   3. ≈ 25 dB   4. 6.75 Vp-p
Section II (page 296):
2.   5 Ω
Section III (page 301):
2(a). 1. 4.86 V   2. 0.54 V   3. 5.4 V   4. −6.6 V   5. 0.6 V   6. 7.2 V
2(b). 1. 5.724   2. ≈ 563 Ω   3. 3.12 dB   4. 6.48 Vp-p   5. 0.358

CHAPTER 10:
2. a. 0.5 mA   b. 4.5 V   c. 9 kΩ
4. a. 2 mA   b. 3 V   c. 1.5 kΩ
6. a. 500 Ω   c. 4 kΩ

CHAPTER 11:
2. a. 3.1 mA   b. 1.5 V   c. 12.4 V   d. 6.1 V   e. 0 V
4. a. 2.8 mA   b. 11.6 V c. 414.3 Ω   d. 7 V   e. 5.4 V   f. 0 V

CHAPTER 12:
2. 3
4. a. 0 V   b. 12 V   c. 8 V   d. 4 V   e. 4.8   f. 100 kΩ   g. 288 or 24.6 dB

CHAPTER 13:
2. a. 6 kΩ   b. 750 Ω   c. 2.545 Vp-p

CHAPTER 14:
a. 461.5 Ω   b. 4   c. 1.23 or ≈ 0.9 dB

CHAPTER 15:
2. a. 393 mV   b. 35.16 V   c. 11.72 mA   d. ≈ 70.7 V   e. 113.45 mV   f. 0.323%
4. a. 39.2 mV   b. 28.26 V   c. 9.42 mA   d. 56.57 V   e. 11.3 mV   f. 0.04%
6. a. 117.9 mV b. 35.3 V   c. 14.12 mA   d. 36.36 V   e. 34 mV   f. 0.096%

CHAPTER 16:
2. a. 24 V   b. 12 V   c. 11.4 V   d. 1.2 mA   e. 7.6 mA   f. 11.4 mA
   g. 84.2 mVp-p   h. 100.9 μVp-p   i. 100.67 μVp-p

CHAPTER 17:
2. a. 13.5 V   b. ≈ 6.75 V

CHAPTER 18:
a. 2 mA   b. 13 V   c. 7.39   d. 10.6 V   e. 1 mA   f. 11.5 V   g. 14.15 kΩ
h. 84.97   i. 50.8 dB   j. 7.5 Vp-p

INDEX

AC beta, 117
AC equivalent circuits
  Bipolars, 231-235, 247, 267, 277, 285, 293, 298
  JFETS, 347
  Passive circuits, 18-19
AC ground, 18, 21, 173, 176-177
Active region
  JFET, 305
  Transistor (see linear region)
Alpha (a), 72-73
Alpha crowding, 180
Amplification (definition), 173
Amplifiers
  Capacitor coupled, 445-446
  Common base, 222-226, 275-277, 297-298
  Common collector, 214-217, 264-270, 252-294
  Common drain, 355-363
  Common emitter, 174-213, 240-258, 282-284
  Common gate, 369-374
  Common source, 335-353
  Complementary symmetry, 446, 452-459
Amplitude, maximum voltage swing, 191-193
Atoms, germanium and silicon, 25

Bandwidth, 248-249
Barrier (junction) potential, 28
Base biasing, 128-132
Base emitter diode resistance (r_e), 175
Beta DC, 78, 117
Bode plot, 248-249
Breakdown voltage
  Diodes, 30-32
  JFETS, 304
  SCR (breakover), 425-432
  Transistors, 118-119
  Zener, 54-56
Bridge rectifier circuit, 395-400
Buffer amplifier
  See emitter and source follower circuits
Bulk resistance, 34

Capacitors
  Bypass, 17, 173, 184-185
  Compensating, 264
  Coupling, 17, 445-446
Characteristic curves
  Bipolars, 81-87, 111, 115-118
  Diodes, 30-32
  JFETs, 304-311
  SCRs, 429
  Zeners, 54-57
Collector feedback biasing, 161-162
Comparison bipolar-JFETs, 342-344
Corner frequency, 248
Covalent bonding, 26
Current divider equation (Norton's Theorem), 12
Current gain, 99, 181

Current limiting resistor, 57
Curve tracer measurements
  Bipolar, 115-122
  Diode, 49-50
  JFET, 312-314, 320
  Zener, 67

DC load lines
  Diode, 36
  JFET, 341
  Transistor, 88-92, 118
Decibel (dB), Appendix B
Design, simplified circuit, Appendix D
Depletion region, 305
Doping, 26-27
Drain output curves, 304
Dynamic resistance
  Diode, 38-39
  Zener, 60-61

Emitter follower (see common collector)
Equivalent circuits (simplified)
  CB hybrid Pi (π), 237, 277, 298
  CC hybrid Pi (π), 239, 267, 293
  CE hybrid Pi (π), 236, 247, 285
  CS hybrid Pi (π), 347

Forward bias PN junction, 28
Four layer diode, 429
Free electron, 26
Frequency response, 248-249.
Full wave bridge rectifier, 395-400

Ground loops, 20, 379
Ground reference, 14

Hybrid parameters, 231-232
Half wave rectifier, 378-387
Hole flow, 27

Input current, 176, 236
Input impedance, 176-177
Input voltage, 175-176
Inverse connection of transistors, 91-92

Junction PN, 27-28
JFET (introduction) 302

Knee of curve
  Diode, 30
  Zener diode, 54

Latch
  Relay, 419-420
  SCR, 429-430
  Transistor, 427-429
Light emitting diode (LED), 433

Linear analysis, 30-31
Load lines,
  Alternating current, 193
  Direct current, 36

Maximum peak-to-peak voltage swing, 180, 187-193
Majority-minority carriers, 29-30

Norton's equivalent theorem, 12, 245

Output impedance,
  Common collector circuit, 269-270
  Common drain circuit, 364
  Common emitter circuit, 258
  Common source circuits, 346

Phase inversion, 180
Pinch off voltage, 305-307
Piecewise linear analysis, 34
Peak inverse voltage, 382, 385
Peak voltage (Vp), 379, 380, 383
PNP universal biasing, 154-156
PN junction, 27-28
Power dissipation, 272
Power gain, 254, Appendix B
Power supply filtering, 382-385

Quiescent point, 177-180

Rectifier diode, 379
Regulated power supply, 407-411
Reverse bias,
  Diode, 27-32
  Zener diode, 54-57
Reverse saturation current,
  Diode, 29
  Transistor, 77

Ripple voltage, 383
RMS-Vp-p-Vp conversion, 379, Appendix F

Saturation region, transistor, 90-91, 421-422
Silicon controlled rectifier (SCR), 429-432
Solid state switches, 420-432
Source follower, 359-368
Square law formula, 307
Stability,
  Alternating current, 182
  Direct current, 166
Stray capacitance, 248
Surge current, 385
Swamping resistor, 182-185

Thevenin's equivalent theorem, 11, 149-151, 169-170, 245
Transfer curve, 310-311
Transformers, 378
Transistor, mode of operation,
  Cutoff, 90-91
  Inverse, 91-92
  Linear, 90
  Saturated, 91

Voltage divider equations
  Alternating current, 16
  Direct current, 7, 16
Voltage doubler, Appendix G
Voltage tripler, Appendix G
Voltage reference diode (see zener diode),

Zener diode, 654, 405
Zener diode regulator circuit, 407-409
Zo (output impedance) derivation, 246